Computer-Assisted Microscopy
The Measurement and Analysis of Images

Computer-Assisted Microscopy

The Measurement and Analysis of Images

John C. Russ
North Carolina State University
Raleigh, North Carolina

Plenum Press • New York and London

Library of Congress Cataloging-in-Publication Data

Russ, John C.
    Computer-assisted microscopy : the measurement and analysis of
  images / John C. Russ.
        p.   cm.
    Includes bibliographical references.
    ISBN 0-306-43410-5
    1. Image processing. 2. Microscope and microscopy--Data
  processing. 3. Optical pattern recognition. I. Title.
  TA1632.R87 1990
  502'.8'20285--dc20   89-70945
                                                                 CIP

First Printing—April 1990
Second Printing—May 1991
Third Printing—March 1992

© 1990 Plenum Press, New York
A Division of Plenum Publishing Corporation
233 Spring Street, New York, N.Y. 10013

All rights reserved

No part of this book may be reproduced, stored in a retrieval system, or transmitted
in any form or by any means, electronic, mechanical, photocopying, microfilming,
recording, or otherwise, without written permission from the Publisher

Printed in the United States of America

Preface

The use of computer-based image analysis systems for all kinds of images, but especially for microscope images, has become increasingly widespread in recent years, as computer power has increased and costs have dropped. Software to perform each of the various tasks described in this book exists now, and without doubt additional algorithms to accomplish these same things more efficiently, and to perform new kinds of image processing, feature discrimination and measurement, will continue to be developed. This is likely to be true particularly in the field of three-dimensional imaging, since new microscopy methods are beginning to be used which can produce such data.

It is not the intent of this book to train programmers who will assemble their own computer systems and write their own programs. Most users require only the barest of knowledge about how to use the computer, but the greater their understanding of the various image analysis operations which are possible, their advantages and limitations, the greater the likelihood of success in their application.

Likewise, the book assumes little in the way of a mathematical background, but the researcher with a secure knowledge of appropriate statistical tests will find it easier to put some of these methods into real use, and have confidence in the results, than one who has less background and experience. Supplementary texts and courses in statistics, microscopy, and specimen preparation are recommended as necessary.

This text was originally created for use in teaching both a regular semester course and a one-week summer short course in image analysis. Although aimed initially at students in materials science and engineering, the courses have consistently attracted people from the life sciences, veterinary and medical schools, food sciences, forest products, geology, and archaeology, and so more examples and terminology from those fields have been incorporated. Many of the same methods, and indeed the same computer systems, can be used for macroscopic applications ranging up to astronomy and remote sensing, but the terminology used here is primarily that of the microscopist.

Of course, there are many kinds of microscopy. These include not only the familiar light and electron microscopes, but also ion, acoustic, X-ray, magnetic resonance, and other devices, and even analytical instruments not usually thought of as microscopes that nevertheless produce two- or three-dimensional arrays of data that can be treated and understood as images. Not all of the techniques covered here are appropriate to all of these kinds of images, but most of the useful methods are covered.

There is no substitute for actually using these methods, and no incentive better than the need to perform a real task. The reader or student with access to a source of images of specimens that are of real interest, and some computer-based image analysis system, should "try out" as many of the various operations as possible to better understand their consequences, as each subject is considered.

Finally, the user of these systems and methods should be alert to an important side-effect of studying this material - it should also make you a better observer. As you learn what the computer "sees" in images, you will learn to see some of it yourself. This will assist in selecting the proper algorithms for processing, discrimination and measurement, as well as forcing you to be a more careful microscopist, producing the best possible images for analysis.

This text, with all its figures and tables, was prepared on a Macintosh computer and printed directly on a Laserwriter, so any errors are solely my own responsibility. Special thanks are due to Chris Russ (Analytical Vision, Inc., Raleigh, NC) who has helped to develop methods and write many of the programs that execute these image analysis algorithms (also on the Macintosh), to John Matzka (Plenum Publishing Corp.) who has patiently tried to educate me in the preparation of book manuscripts, and to Helen Adams, who has long understood and tolerated my compulsion to undertake writing projects like this one.

John C. Russ
Raleigh, NC
February, 1990

Contents

Chapter 1
Introduction 1

The importance of images 1
    Man's primary source of information; space probes and computer icons
Why measure images? 1
    To describe structure, not to study human vision or for robotic control
Computer methods: an overview 3
    Mimic human vision if possible; comparison vs. measurement; accuracy vs. precision
Implementation 5
    Languages; parallel vs. serial architectures; importance of algorithms
Acquisition and processing of images 8
    Image sources; digitization; processing to emphasize or suppress information
Measurements within images 9
    Global and feature measurements; projection and section images; data interpretation
More than two dimensions 11
    Surfaces (roughness, stereoscopy) and volumes (serial sections, tomography)

Chapter 2
Acquiring Images 13

Image sources 13
    Control of illumination and viewpoint; surfaces vs. volumes; magnification scale
Multi-spectral images 16
    Visible light; other wavelengths; color coordinates; other modalities (electrons, X-rays)
Image sensors 20
    Raster format; slow and fast scans; video; standard video cameras
Digitization 23
    Signal voltages; RS-170; analog to digital conversion
Specifications 25
    Number of bits (grey levels); number of pixels (resolution)
References 31

Chapter 3
Image Processing 33

Point operations 33
    Transfer functions and look-up tables; enhanced visibility; rescaling; selective emphasis
Time sequences 38
    Subtraction; comparison and tracking; discrete differences; motion flow
Correcting image defects - averaging to reduce noise 42
    Weighted and true averaging; effect of number of frames; Kalman averaging
Reducing noise in a single image 45
    Neighbor smoothing (kernel operations); blurring and distortion; median and rank filter

Frequency space 48
Transform methods; periodic noise; high and low pass filters; periodic structures
Color images 56
RGB, YIQ and HIS encoding; processing of luminance information
Shading correction 57
Uneven illumination; variable specimen thickness or orientation; nonuniform detectors
Fitting backgrounds 59
Polynomial fitting; large kernel smoothing; rank filters to eliminate features
Rubber sheeting 61
Aligning images to each other or known geometry; bicubic model; pixel interpolation
Image sharpening 64
Laplacian and similar kernel operations; Fourier space filtering
Focussing images 67
Range finders; high pass filtering (quasi-real-time maximization)
References 68

Chapter 4
Segmentation of Edges and Lines 71

Defining a feature and its boundary 71
Contiguous pixels; region inside a boundary; derivatives as edge locators
Roberts' cross edge operator 75
Local pixel comparison; edge detection by magnitude and direction
The Sobel and Kirsch operators 77
Neighborhood operators; choice of kernels for directional derivatives; edge enhancement
Other edge-finding methods 80
Difference or Laplacian of Gaussian; simulation of human vision system
Other segmentation methods 89
Contour lines; edge following; region growing; split and merge
The Hough transform 92
Mapping of lines to points; linear and circular arrangements of points and features
Touching features 95
Fitting of lines or arcs; convex or watershed segmentation
Manual outlining 96
Pointing devices, bias and errors in drawing and filling
References 97

Chapter 5
Discrimination and Thresholding 99

Brightness thresholds 99
Image brightness histograms; Adjustment of upper and lower limits; boundary pixels
Thresholding after processing 101
Alteration of contrast; gradient and rank operations; texture in images
Selecting threshold settings 105
Using the brightness histogram; peaks and valleys
The need for automatic thresholding 107
Reproducibility; variation in overall illumination; changes in image contents
Automatic methods 108
Survey and criteria; fixed starting points; special rules at black and white limits
Histogram minimum method 109
Lowest point in histogram vs. fitting to peaks, smoothing; overlapping peaks

Contents

Minimum area sensitivity method 111
   Least change in phase areas with thresholds; problem of false local minima
Minimum perimeter sensitivity method 112
   Least change in phase perimeter with thresholds; assumes smooth boundaries
Reproducibility testing 115
   Changes in measured dimensions with repetitive digitization and thresholding
Fixed percentage setting 116
   Rule-of-thumb for gradient or edge images
Color images 117
   Change in color (hue or saturation) as well as intensity
Encoding binary images 119
   Chord or run-length encoding, boundary representation or chain code
Contiguity 122
   Connecting feature parts; 4- or 8 neighbor rules for features and background
References 127

Chapter 6
Binary Image Editing 129

Manual editing 130
   Addition or erasure of pixels, filling of holes, selection of features or regions
Combining images 131
   Boolean logic rules (AND, OR, ExOR, NOT); measurement templates; X-ray maps
Neighbor operations 134
   Morphological operations (erosion, dilation, opening, closing); neighbor coefficients
Skeletonization 141
   Neighbor rules; iterative methods; background skiz; nodes and branches
Measurement using binary image editing 148
   Size measured by erosion; gradients and clustering; masking; template matching
Covariance 151
   Binary or grey scale images; preferred orientation; autocorrelation; frequency space
Watershed segmentation 153
   Ultimate eroded points; Euclidean distance map; separation of touching features
Mosaic amalgamation and fractal dimensions 161
   Fractal dimensions of boundaries; Richardson plot; measurement methods
Contiguity and filling interior holes 169
   Feature boundaries and neighbors
References 173

Chapter 7
Image Measurements 175

Reference areas 175
   Area fraction; number density; hierarchies; units; intercept length
Boundary curvature 179
   Tangent count; inflection points; pixel orientations; gradient images
Feature measurements 181
   Size measures: area, diameter; filled and convex area
Perimeter points 183
   Convex polygon; Pythagorean and chain code perimeter; Feret's diameter
Length and breadth 185
   Max. and min. Feret; width and fiber length; ellipsoid volume and surface

Radius of curvature 192
    Derived parameters; iterative solution
Image processing approaches 195
    Circle fitting; the Hough transform
Counting neighbor patterns 199
    Pixel patterns; total vs. net curvature; skeletons and boundaries
Shape 200
    Dimensionless ratios of size parameters; formfactor and roundness; holes
Corners as a measure of shape 205
    Skeletons and boundaries; not a local operation
Harmonic analysis 206
    Unrolling; Fourier expansion; coefficients used for classification
Position 210
    Centroid; density weighting; moments and orientation angle
Neighbor relationships 212
    Inside; outside; touching; overlapping; alignments
Edge effects 216
    Effective count as a function of size; guard area
Brightness 217
    Optical density; uniformity, contrast and texture
References 218

Chapter 8
Stereological Interpretation of Measurement Data 221

Global measurements 222
    Phases; volume fraction by area, lineal and point counts; notation
Global parameters 224
    Surface area per unit volume; grain size; random sampling; precision
Mean free path 229
    Two and three dimensions; microscopic and astronomic application
Problems in 3-D interpretation 231
    Topology; the disector for direct 3-D sampling; number per unit volume
Feature specific measurements 233
    Spheres and circular intercepts; model-based unfolding; ellipsoids; limitations
Distribution histograms of size 239
    Mean based on number of volume; anisotropic structures
Interpreting distributions 241
    Plotting axes; normal distributions; inadequacy of descriptive statistics
Nonparametric tests 243
    Rank order and cumulative sum methods; preferred for image data
Cumulative plots 247
    Undersize and oversize plots; log and probability scales; two-way plots
Plotting shape and position data 253
    Nonlinear scales; gradients; angle (rose) plots
Other plots 257
    Shape vs. size; correlation and significance; neighbor distances; clustering
References 264

Chapter 9
Object Recognition 267

Locating features 268
    Template matching; cross correlation
Parametric object description 269
    Size, shape, brightness, etc.; comparison to human recognition
Distinguishing populations 272
    Multidimensional parameter space; regression; finding important directions
Decision points 275
    Bayesian statistics; distribution histograms
Other identification methods 279
    Cluster analysis; nearest neighbor search; predefined classes; fuzzy logic
An example 283
    SEM images of phytoliths from corn plants and species identification
Comparing multiple populations 285
    Production rules for simple character recognition
An example of contextual learning 291
    Automatic recognition of chaotic shapes (mixed nuts)
Other applications 299
    Industrial quality control; forensics; pathology; surveillance
Artificial intelligence 301
    Classic and fuzzy expert systems; cluster analysis; kNN search; neural nets
References 304

Chapter 10
Surface Image Measurements 309

Depth cues 309
    Stereo vision; object precedence; surface shading; relative position
Image contrast 310
    Lambertian light scattering; shape-from-shading reconstruction
Shape from texture 313
    Local surface gradients; reconstruction of surfaces; occluding boundaries
The scanning electron microscope 317
    Secondary and backscattered electrons; shape-from-shading reconstruction
Line width measurement 324
    Metrology for micrometer-sized structures; profile interpretation; modelling
Roughness and fractal dimensions 331
    Surface roughness; 2.D fractal dimension; relation to 1.D boundaries; texture
Other surface measurement methods 343
    Photometric stereo; structured light; confocal light microscopy; range images
References 346

Chapter 11
Stereoscopy 351

Principles from human vision 351
    Vergence and parallax
Measurement of elevation from parallax 352
    Shift method (aerial photos); tilt method (SEM images)
Presentation of the data 357
    Elevation profiles; contour maps; isometric displays; range maps
Automatic fusion 363
    Cross correlation; lessons from human vision; edge and point matching
Stereoscopy in transparent volumes 372
    3-D coordinates for analysis; locating surfaces; depth resolution
References 375

Chapter 12
Serial Sections ... 377

Obtaining serial section images ... 377
Mechanical sectioning; grinding; ablation; distortion and nonuniformity
Optical sectioning ... 381
Confocal light microscope; acoustic methods; tomography
Presentation of 3-D image information ... 383
Voxels (volumetric data) vs. surfaces
Aligning slices ... 387
Translation and rotation; stretching; fiducial marks; automatic methods
Displays of outline images ... 389
Wire frame; hidden line removal; depth coding; perspective; stereo views
Surface modelling ... 394
Tesselation and shading; surface rendering
Measurements on surface-modelled objects ... 399
Surface area and object volume; extensions of 2-D parameters
Voxel displays ... 402
All data preserved; transparency; arbitrary section planes; internal surfaces
Measurements on voxel images ... 405
Volume and surface area; density; internal voids; defining surfaces
Network analysis ... 408
Transitivity matrices; number of neighbors and paths
Connectivity ... 414
Topological properties in 3-D; flow in networks
References ... 416

Chapter 13
Tomography ... 419

Reconstruction ... 419
Frequency space vs. backprojections; algebraic methods; optimization
Instrumentation ... 425
Generations of medical equipment; resolution; beam hardening; electrons
3-D Imaging ... 431
Direct reconstruction from area projections; voxel and surface displays
References ... 436

Chapter 14
Lessons from Human Vision ... 439

The language of structure ... 441
Exterior surfaces vs. volumetric and internal displays
Illusion ... 443
Reveal mechanisms for shading correction, line completion, image processing
Conclusion ... 449
Toward "better seeing" by man and machine
References ... 450
For further reading ... 450

Index ... 451

Chapter 1

Introduction

The importance of images

Mankind's principal means of interacting with his environment is visual. In teaching students, I sometimes encounter those who express themselves by saying "I hear what you're telling me" or "I grasp that idea," but most of the time the expression is "I see what you mean." In fact, students who are principally auditory or tactile learners sometimes have real problems in dealing with engineering or scientific material that is presented in textbooks heavy with diagrams and graphs. Most of us learn visually. As age diminishes the acuity of our senses, we use eyeglasses commonly, hearing aids occasionally, and practically never any prosthetic aids for any of the remaining senses. The Chinese proverb that a picture is worth 1000 words probably underestimates.

This affects the kinds of scientific research we do, as well. For instance, the recent space probes to Comet Halley carried a number of sophisticated and important instruments, ranging from magnetometers to mass spectrometers. But it was by the pictures they returned that we judged their success (and this has been true of most of our other space projects as well). Many scientific instruments directly produce pictorial images (such as electron microscopes); others that do not usually have some type of graphics display (for instance to show a spectrum), expecting the operator to be able to extract meaning more readily from this than from a list of numbers.

This influence is even being felt in unexpected places. The "hottest" recent development in computers, popularized by the Macintosh, is the use of "icons," little pictures representing programs or data, which the user can recognize and point to instead of having to read words from the screen (Figure 1-1).

The evolution of man's visual apparatus has made it our most important and relied-upon sense by tailoring it to extract meaning from images. Approximately 60 percent of the sensory inputs to the brain come from the visual system. Not all animals have that reliance: bats use sound echolocation, fish have a pressure sensing organ we don't even possess, some snakes sense heat, birds and some bacteria respond to magnetic fields, and many animals have a sense of smell that communicates important information about the world around them. We find it hard to imagine what the world "looks like" to such an animal; in fact the word "imagine" itself carries with it an implicit visual metaphor.

Why measure images?

The fact that humans can easily interpret images does not mean we should not - or do not need to - use computerized methods for image measurement. In fact, it increases our desire to do so. One purpose can be to better understand the visual process itself, by duplicating or emulating its responses. That is not our goal here.

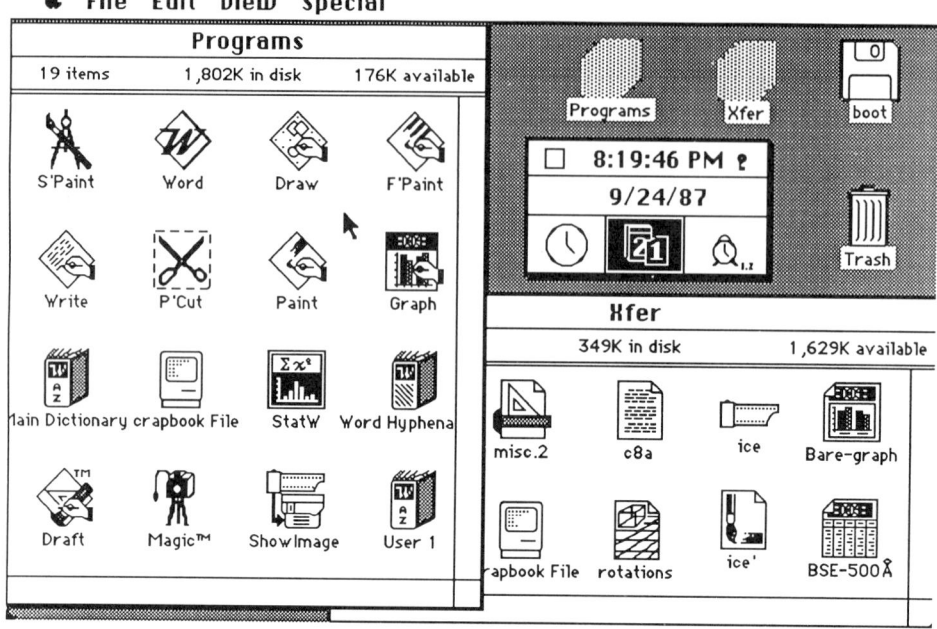

Figure 1-1: Example of a Macintosh screen with icons.

Computerized "understanding" of images, for instance the kind of real-time interpretation of a changing scene that allows us to guide a car down a road, requires a massive investment of computer resources and even so can deal with only very simplified situations. That is not our goal, either.

But recognizing, counting and measuring the size, shape, position, density, and other similar properties of particular objects in an image is something that is well within the power of mini- and microcomputers, and can be done by the computer relatively quickly with excellent reproducibility. Images in a form suitable for acquisition and analysis are produced by a variety of instruments, and computerized measurement can be used to extract specific information from the images much more accurately and reproducibly than a human can without such aid. In fact, human observers tend not to do this very well, with results that vary from observer to observer and from time to time. This perhaps reflects the fact that in normal situations, humans rarely need to exactly measure an object in an image. Instead, they can interact with their environment, to bring a comparison object or ruler into play, for example.

Computer image measurement is less easily distracted from what is important by trivia in the image, and is also better than a human observer at paying attention to all of the details present. It doesn't get bored, and it makes no (or at least very few and usually explicit) assumptions.

On the other hand, humans are very good at recognizing objects, often based on very incomplete or unconventional images, and this capability is much harder to program into the computer. There is considerable evidence that humans literally "turn things over" in the mind, to obtain the best viewpoint for examination or comparison, as shown in Figure 1-2. This is beyond the capability of most computer methods now.

Introduction

The process of image measurement involves an enormous reduction in the amount of data, by selecting from the original image those objects and parameters that are important. An original image may represent a million separate points stored in the computer (in the human eye, there are more than 150 million individual receptors on the retina). But the desired information may be as simple as (e.g.) the number of white blood cells on a slide, the size (width, etc.) of a device in an integrated circuit, the variation in the amount of a phase near the surface of a metal, or even just the presence of a tumor in an X-ray image. This selection and reduction is at the heart of image analysis and measurement. It is achieved by ignoring irrelevant information.

Computer methods: an overview

Most of the images that we will deal with here are single, two-dimensional ones, much like a single-eye look at of some real world view. In many cases we will further limit ourselves to a monochrome (black, shades of grey, and white) image rather than full color. The later chapters will deal with the additional information that can be obtained from multiple images, either used in a stereo (two-eyed) sense, or a series of parallel sections through an object, or projections in many directions as used in tomography. But even when several images are involved, each single one is usually dealt with to some extent separately so we are justified in first considering how to work with an individual image.

Many of the computer methods use algorithms that either consciously or accidentally mimic many aspects of human vision. For convenience, it is usual to separate the human

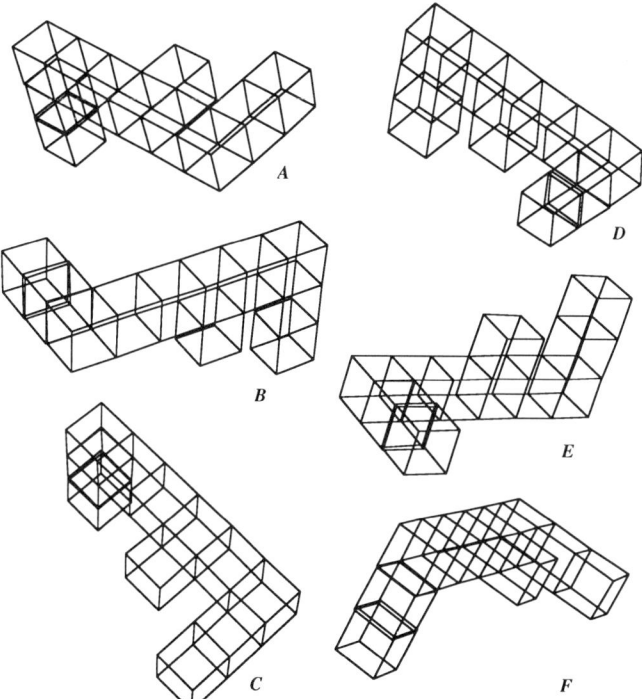

Figure 1-2: Which of these objects are the same? The length of time required to decide is proportional to the angular difference in the object orientation.

visual process into "early" and "late" vision. The former roughly corresponds to the processing of information in the retina and the neural networks close to it, before the higher level data is transmitted to the brain, while the latter refers to the further unravelling of the information in the brain, where more "learned" facts about the world can be brought into play. The distinction also has a rough correspondence to the distinction between extracting low level information from the image, such as the presence, location and orientation of edges, boundaries, and perhaps objects ("early" vision), as compared to the use of this information to "understand" a scene and the relationships between the objects present in it.

Most of our computer methods for image analysis and measurement use algorithms related to "early" vision. There is another large and active field, for instance connected with research in robotics and artificial intelligence, that seeks to understand and describe scenes, but we will not be dealing with it here. Fortunately, most of the images that are important for analysis and measurement are not "general" in nature, but are obtained in highly controlled situations where much is known about the specimen and the viewing conditions.

The examples are generally taken from the field of microscopy, but we will see that the same techniques apply (with a few additions and restrictions) to astronomy, remote sensing (satellite photos), and so forth. In all of these cases, we generally know beforehand that the subjects are (for example) flat surfaces cut through a material, or projected images through a thin section of the sample, and that the image contrast is primarily produced by some particular interaction such as light scattering or absorption, secondary electron or X-ray production, and so forth. This greatly simplifies the interpretation of the image. It also permits some computer modelling to predict the images that should be obtained from particular structures and objects.

The reproducibility of computerized image analysis and measurement methods can be far better than that of a human observer because the algorithms overtly ignore much of the content of the image, and the sensors and discriminators respond only to the image itself. Humans are influenced by many other things (hunger, emotional response to stress, etc.) that clog up the neural pathways and "take our minds off" the job at hand. This causes us to miss things in images that would otherwise be obvious (just after an argument, I might drive through a stoplight because I didn't "see" it). Likewise, we respond to information in the image other than that which we need to see (I might also drive through the stoplight because I was busy watching the bikini-clad blonde on the corner).

{As another example, I might see the stoplight, but knowing that it was 2 a.m., that there were no headlights visible in either direction, and that I was in a hurry, I might go through the light anyway. That isn't computer vision, it's Artificial Intelligence (AI).}

The often observed result is that reproducibility tests on the same images show large variations between different persons, or the results from a single person at different times of the day or week. Computer-based measurement does not show this pattern, and the variations are more or less directly tied to simple statistical patterns of fluctuation, so that the errors can be predicted and in many cases controlled.

On the other hand, this reproducibility does not imply accuracy. The speedometer in my car always reads 55 at a particular speed (reproducibility), but that is of little value in arguing with the trooper who pulls me over and says his radar clocked me at 63 (accuracy, presumably). Some of the techniques we employ require calibration against

Introduction

some other external source, such as known standard specimens or mathematical simulation, to produce accurate results.

Implementation

It is easy to be distracted from the real purpose of image analysis and measurement by the hardware and software used for its implementation. There are computer-based systems in existence, some of them commercially available, that employ many widely different approaches to similar problems. Some of the more obvious differences are sequential vs. parallel processing, hardware vs. software calculations, and various computer languages.

These differences are all unimportant compared to the choice of appropriate algorithms to carry out the desired method. In principle, any result obtained by a massively parallel computer employing hardware array processors and programmed in LISP can also be obtained using a conventional sequential computer with only software calculations and using Basic. There may be a noticeable difference in the ease with which the programmer initially implemented the method or the convenience with which it can be modified, and there may be a significant difference in the speed with which it can be applied, but the result should be indistinguishable.

Serial computers (the classic "von Neumann" architecture) perform one operation at a time, albeit at the rate of a few million per second. Data values are fetched from memory to a central processor, combined with other values, and written back to memory. This produces a bottleneck that limits the number of calculations that can be performed per unit time. Particularly for images, in which we very often want to perform the same operation or series of operations for every point in the picture, or perhaps to perform the same measurement for every object, it is particularly attractive to find some way to bypass this bottleneck.

Parallel computer architectures are actively being developed which allow this. Generally, they employ a fairly large number of identical processors (often single chips, in this era of large scale integrated circuits), each with its own program memory and perhaps its own data memory, and some means of communicating with the other processors. Two particular arrangements have been especially used: the processor array and the cosmic cube (Figure 1-3). In both cases, each processor might, for example, be

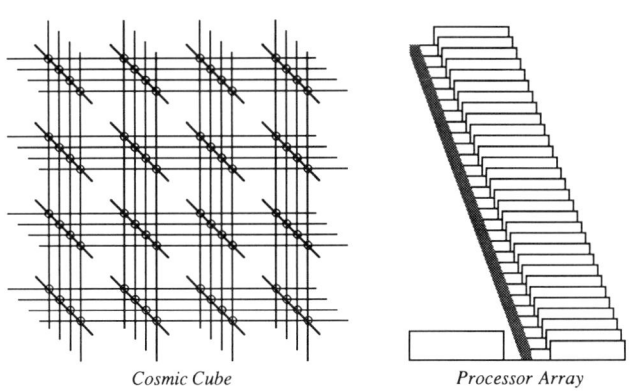

Cosmic Cube *Processor Array*

Figure 1-3: Two different parallel computer architectures.

given a portion of the image to work on. The image itself might either be written into the dedicated data memory of the processors, or each processor might also be able to access the main memory holding the image.

As the processors do their work simultaneously, they sometimes need results from the work of other processors (for instance to get information on neighboring points that lie in a different segment of the image belonging to another processor). This information is provided by communication between processors. In the cosmic cube architecture, each processor has a direct communication link with its neighbors. In the processor array case, there is another computer, nominally the boss who assigns the tasks and sends the data, but in actual practice more the mailman who carries messages back and forth. In either case, the total computation time is reduced nearly in proportion to the number of processors assigned to the task.

Assigning each processor a different portion of the image is not the only way that parallel computation can be organized, of course. In some cases it is more attractive to assign each object to a different processor, to classify it and measure its parameters and to determine inter-object comparisons, spacings, etc. from its neighbors. This method is especially attractive when object recognition is required, or when periodicities or other relationships between objects are to be found.

It is difficult to apply these parallel methods to general purpose computing, where tasks are generally quite varied and non-repetitive. But for image analysis, as for a few other problems such as some simulations, the application of parallelism is more direct and the results easier to achieve. For the particular case of image processing (to be more fully defined shortly), there are two other "parallel" cases that can be mentioned.

First is the array processor (distinct from the processor array). This is a special purpose arithmetic unit that works under the direction of the central processing unit (usually a conventional sequential computer) to carry out particular repetitive tasks at very high speed. It is often used for images, because the data can be sent to the array processor as a sequence of values (for instance, the brightnesses along lines in the image) to perform simple operations like addition, multiplication, subtraction, etc. and the results written back to memory. The specialized nature of the array processor and the low level of its operations allows it to be much faster than the main general purpose processor for this purpose. However, the array processor is rather inflexible and hard to program for any but the simplest operations. While it can speed up some image operations, it is inapplicable to many others.

Another rather new development is the use of so-called neural nets within the image sensor itself. This corresponds in a very simplified way to the operation of the human visual system (which we will encounter in the final chapter), in which there are many additional cells connected to the outputs of each few retinal receptors to compare and combine their responses to light in ways that extract specific information (such as the location, orientation and motion of edges), so that only the reduced data is sent on to the brain. This allows the human to respond very quickly to particular types of stimuli, and using electrical devices to form the same types of connections in the image sensor can also produce simple computerized devices that recognize and respond to certain features.

Both of these types of devices are rather specialized for our needs in general image analysis and measurement, although it is quite possible that applications research using a flexible general purpose system on a particular problem will result in a design that successfully applies these tools to individual situations.

Introduction 7

Similarly, the use of any parallel rather than sequential computer system is handicapped by the greater difficulty in programming it for a wide variety of methods and images. Any computation or algorithm that can be performed with a parallel system can also be done sequentially (the converse is not true), so it is often best (and certainly cost effective) to start with a conventional sequential system, and only consider a more specialized parallel setup when the method has been demonstrated and tested, and the increase in speed justifies the considerable additional expense for equipment and programming.

The same considerations apply, perhaps even more strongly, to the choice of a computer language (Figure 1-4). This selection often borders on religion, at least in the strength of individual beliefs. The "best" choice is most often dictated more by the computer to be used than by the application or genuine need. For small machines, languages such as C, Pascal, Fortran or Basic are well supported by libraries of utility functions (to perform low level operations like writing to the display, operating peripherals such as printers and disk drives, and so forth, and performing mathematical computations). Larger machines may add LISP or other languages that allow more complex statements to be easily written in a richer vocabulary and syntax.

But the old adage "Real programmers can write Fortran in any language" holds. The differences between the languages are minor if not trivial, and any of them are capable of expressing the algorithms that will be described. These will be given in a general notation, and anyone intending to implement them on a particular machine can use any language that is available and appropriate. Most of the more repetitive operations that must deal with the large number of points in a picture will generally need to be made as fast as possible, and this usually calls for the use of machine language for the lowest level functions. Since most images are stored within the computer in the form of one byte per picture point, even the differences between 8-bit, 16-bit or 32-bit (or more) computers is often unimportant in processing and calculations for this application.

All of the examples shown in this book have been obtained using a system employing a dedicated microcomputer (Apple or Macintosh with various accessories and peripherals). Most of the processing times are a few seconds or less, which is usually short compared to the time needed to acquire or record the image, but obviously falls short of "real time" response in which the user can manipulate the imaging device while observing the final result. The same algorithms can also be used in much faster dedicated systems for real time recognition and response, if it is required (for instance in some military applications).

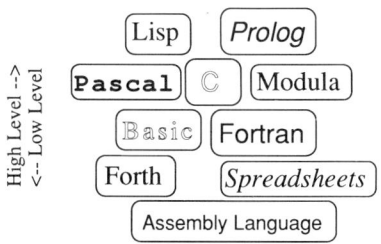

Figure 1-4: Some common high and low level languages. High level languages offer a richer vocabulary, more control structures and built-in functions. Low level languages deal with the hardware directly and can give greater execution speed.

Acquisition and processing of images

The process of image analysis and measurement has many parts, and for convenience and (hopefully) clarity these will be dealt with one at a time, in more or less sequential order. Not all situations require all of these steps, but all are potentially available to deal with particular problems. They are summarized in Figure 1-5.

Image acquisition is the process by which an analog image (in which brightness and perhaps color vary continuously with position) is digitized and placed in storage in the computer (where our algorithms can get at it). The image stored in memory is referred to as a "grey scale" image, although it may sometimes have color or other information within it, to emphasize its distinction from a black and white, or binary image. It is dealt with in Chapter 2.

Image processing (covered in Chapters 3 and 4) encompasses a broad range of operations that are often dealt with as an end in themselves, but which we will treat only as they may make it easier to subsequently perform analysis or measurement. This subject includes all processes that start with an image and finish with an image. The brightness (or color) of individual picture points may be changed depending upon the original value, or on the brightnesses of neighboring points, or other points in the image, or points in another image, or the image as a whole may be shifted, rotated or stretched. The reason for these operations is usually to increase the visibility of some type of feature or information, almost invariably at the expense of other information in the picture which is not presently of interest. It is also possible to use these operations to correct for known

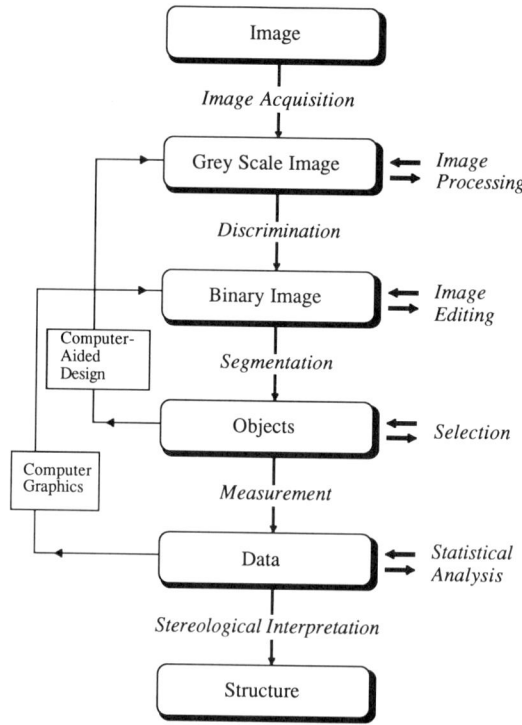

Figure 1-5: An overview of the steps in image analysis dealt with in this text.

defects in the way the image was obtained, such as nonuniform illumination or an oblique angle of view, or correcting an out-of-focus condition or blur due to motion.

A binary image is distinct from a grey-scale image because each point is either white or black. Normally this is a guide to which points lie within the features we are interested in measuring and which do not, in other words, to separate figure from ground. Binary images take up much less storage space, just one bit per picture point instead of a byte or even more, and are most often more suitable for measurement than grey scale images. However, they have lost all of the original brightness information.

The most common way of converting a grey scale image to a binary one is by discrimination or thresholding (discussed in Chapter 5), selecting a range of brightness values that correspond to the features of interest, which are then made white while the remainder becomes black. Of course, this need not be done on the original grey scale image as it was first digitized. All of the image processing operations can be used to alter the original brightness values in ways that make it easier to select threshold values for the discrimination that will successfully separate figure from ground. There are other methods as well, which locate and utilize the boundaries between figure and ground, and there are some measurements that can be made directly on the original grey image. A distinction can be made between discriminating figure from ground based on brightness and the more general problem of segmenting an image into specific features and objects represented either by the picture points that lie within the features or by the feature boundaries.

Binary images may themselves need some further work before they can be measured or used to represent the individual objects of interest. Chapter 6 discusses the various editing operations available. There is a broad class of "morphological" operators that can be applied to binary images to smooth the outlines of features, separate touching regions or connect broken lines, isolate particular shape information, and so forth. Another class of operations involves logical combinations of several binary images (obtained using different initial images of the same scene, or using different processing and discrimination of the same original grey image) to assist in selecting the features of interest, or to define particular relationships such as one type of feature lying within or adjacent to another. In some cases, manual editing of the binary image to erase, connect or modify features is also used when no practical automatic method can deal with the subtleties of the original image. This allows a human operator to apply independent knowledge about the structure(s) present.

Once the binary image shows the separate objects present, then there may still be one additional step needed before measurement data are saved. This is the selection of the particular objects that are of interest. In many images there are several kinds of objects present, and only one or a few types are to be measured. The distinction can sometimes be made automatically based on some measurement parameters or combinations of parameters, and we will see that the computer may be able to learn for itself which ones to use. In other cases, the operator will choose the interesting objects by pointing at the image using some appropriate device such as a mouse, light pen, drawing tablet, etc.

Measurements within images

Finally, the actual measurements can be performed (Chapter 7). These fall initially into two classes: measurements that are global and apply to the entire field of view, and those that apply individually to each object within the image. The former "field specific" parameters include such things as the number of objects, their total area fraction in the image, the amount of boundary separating them from their surroundings, and its curvature. The latter "feature specific" parameters can be divided broadly into four categories: measures of object size, shape, position, and density (or brightness). For each of these, there are many individual measures. Some are very general in application, such as area or length, while others find their principal use in highly specific situations, such as radius of curvature.

When measurements are made of several objects, all of a single nominal type, we usually want to perform some kind of statistical analysis on the data. This includes finding mean and standard deviation, relationships between different parameters (for instance, whether there is a significant change in shape with size), and distributions of number of objects as a function of a parameter (for instance, size, shape or orientation). Many of these results are also presented graphically, as an aid to the user in understanding them.

Finally, the measurement data (either global or feature specific) are generally interpreted to obtain a meaningful description of some aspect of the structure they represent. The principles for this interpretation comprise the field of stereology, the study of three dimensional structure and its relationships to two-dimensional images. This is dealt with in Chapter 8. Somewhat different models are used for different image types, for instance planes cut through a solid and projected or shadow images of a collection of objects (Figure 1-6).

As an example of each, consider looking at objects such as spheres or tori or even more complicated shapes, embedded in a matrix. If the matrix is transparent, we can see

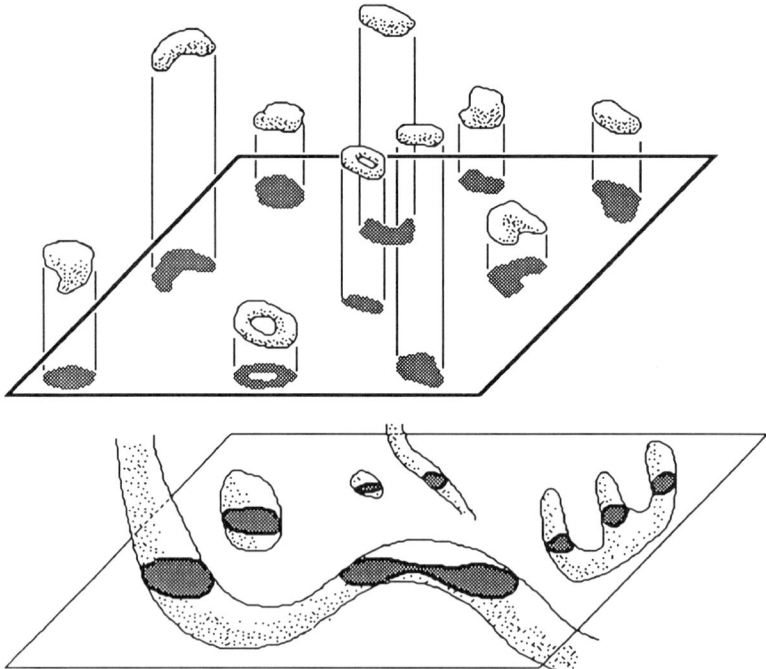

Figure 1-6: Examples of images produced by projection and sectioning. Both require interpretation to estimate the statistical properties of the three-dimensional structure that is represented.

through a reasonable thickness of the material so that the full extent of each object is evident. On the other hand, there may be overlaps, and concavities may be obscured due to individual object orientation. This is a projected image, encountered in transmission light and electron microscopy. It is also quite similar in treatment to that produced by particles dispersed on a substrate (the matrix medium is air).

A very different image would be produced by passing a plane through an opaque matrix containing the same types of objects. On the surface revealed by this cut, we would see sections through the objects that revealed indentations in their shapes, but would not usually see the greatest diameter of each object. Most sections would be much smaller than the maximum size, and might disguise the objects' shapes as well. Similar objects might give rise to sections of quite different size and shape, depending on their orientation with respect to the plane.

It is still possible to estimate the size of the objects considered as a population, even though they cannot be measured individually. In either of these two cases, the basic principles are derived by mathematical techniques such as geometric probability.

These mathematical modelling methods work in the opposite way to the interpretive path we have just followed. They start by assuming some structure or objects, and then simulate what the measurements or image appearance would be. This allows models and data tables to be established that can subsequently be applied to real structures. Modelling has an important role in this field that we will touch on only lightly here.

For the sake of completeness, it is also worth mentioning that unlike the procedures described here, which reduce images to a few measurement values that describe some aspects of structure, there are other computer techniques which often use rather similar hardware, that work in the opposite way. That is, they start with rather compact lists of measurement or dimension data and use computer graphics to generate images. The images may in turn be either binary or grey scale (or colored), and may represent either two- or three-dimensional views. This includes, for instance, CAD/CAM programs. Later chapters will touch lightly on some of the aspects of this computer image generation as a way to present data obtained from image analysis, such as surfaces within three-dimensional volumes.

Chapter 9 considers how the measurement data can be used by the computer to distinguish different classes of things. Automatic feature recognition and identification provides a type of expert system in which knowledge based on measurements from a small training set of objects is applied to (and refined by) a larger population. It has applications in research as well as quality control and surveillance.

More than two dimensions

Not all images are of planar sections. Irregular surfaces can also be measured using image analysis techniques. Chapter 10 shows ways to characterize and compare the roughness of surfaces, for instance in the form of a fractal dimension, or to measure relief and model the effect of topography on brightness. This is being applied to metrology of integrated circuits. More complete measurement of topography requires two images, in much the same way that humans use stereoscopic vision to judge distances. Chapter 11 shows the basic principles, which can be used either for manual measurement and computer-assisted reconstruction, or in some cases for completely automated measurement and conversion to elevation maps.

Another way to use multiple images is in the form of a series or parallel sections through the structure. Serial sections are commonly obtained in the biological sciences, but are equally applicable to materials, geology, and other fields. Most work to date has simply found useful ways to reconstruct images and for presentation, but some unique measurements can also be obtained from this data as discussed in Chapter 12.

One of the foremost new techniques for acquiring serial section images is tomography (the "CAT" scan). Chapter 13 describes the techniques for obtaining these images of internal structure from multiple projections, and deals with some of the more novel extensions of this technology to non-medical applications, including electron microscopy.

The final chapter puts the subject of computerized image measurement into the context of human vision, by introducing optical illusions. Many illusions that fool the human eye do not fool computerized image analyzers. But the "faults" of the human system may point to new possibilities for processing and measurement algorithms that allow very efficient and robust ways to extract meaning from complex and subtle images. If nothing else, optical illusions and a consideration of how people see provides an important lesson in educating our intuition to look at images in a more objective and analytical way, and will allow us to make better use of computerized tools for image measurement.

Chapter 2

Acquiring Images

Image sources

Computerized image analysis and measurement can be applied to almost any type of image, but the most successful applications are ones in which the conditions of illumination and viewpoint are well known and controlled, and there is some prior knowledge about the classes of objects that will be encountered. In particular, images are usually of surfaces, although they may not necessarily be restricted to flat surfaces, or of objects within a transparent volume of some known thickness. The later situation includes observing objects such as particles sitting on a substrate.

The scale of the image is less important. The same basic techniques apply to images obtained with microscopes or telescopes, at the extremes. A telescope image might be used to study the distributions of stellar types, by color type and position, for instance. Images from space probes and satellites are routinely analyzed to classify terrain, crops, etc. (Sabins, 1987), or to recognize and track targets of military potential. At a more familiar scale, macroscopic objects are often examined using these methods. Industrial quality control often requires examination of components or materials for recognition and measurement, for instance (Russ et al., 1988). Medical diagnosis also relies on images (X-rays, "CAT" scans, and so forth).

Another important class of measurements uses a light table and sensor to examine macroscopic transparent samples. These can include medical X–ray films, as well as gels as used for chromatography, electrophoresis and other one- or two-dimensional chemical separation techniques (Yakin et al., 1982; Schneider & Klose, 1983; Hahn, 1983; Giddings, 1984; Jansson et al., 1983; Russ et al., 1985; Zeineh & Kyriakidis, 1987; Hruschka et al., 1983). The technique is often referred to as scanning densitometry. These tables most often use a monochromatic light source such as a low power laser with a photodiode or photomultiplier detector, and physically translate the sample to measure the entire image. Either a linear detector array and a one-dimensional scan (Figure 2-1), or a single point detector and a full two-dimensional raster scan can be used. The former is faster, the latter gives the most precise measurement because the light source and detector are consistent. Other scanner configurations, including placing the image on a cylinder and scanning it in a lathe, have been used (Moore, 1968).

At smaller scales, the light microscope functions at magnifications from a few times to about 1000x, using either incident (reflected) or transmitted light, possibly filtered to be of one wavelength (color) or polarized. Transmitted light is often used for biological specimens in particular, and selective chemical stains are used to darken or color features or structures of interest. The petrographic microscope used by geologists also uses transmitted light, usually polarized, to reveal mineral phases in thin polished sections of rock.

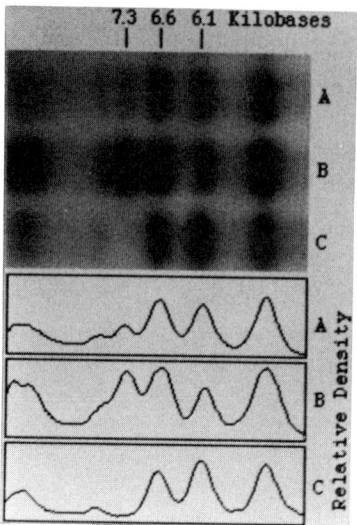

Figure 2-1: Scans of density along several columns in a protein separation gel.

The incident light microscope is principally used for materials examination, because most metals, ceramics and electronic materials are opaque. However, with appropriate surface polishing and chemical (or other means such as thermal or ion beam) etching, the internal structures are revealed. The incident light microscope is also used, at much lower magnifications because of the limited depth of field, to examine rough surfaces such as fractures, particulates, fabrics, coatings, and so forth.

The light microscope has been perhaps the most important scientific instrument of all time in both biological and materials research, again emphasizing the reliance that we place upon images to learn about the things that interest us. Modern light microscopes are available with a variety of types of optics and illumination, which routinely approach theoretical limits of performance (Rochow & Rochow, 1978). There has been little effort, by and large, to measure the images these instruments produce. This is primarily due to the time consuming nature of performing such measurements by hand, and to a lesser extent to the fear of dealing with the necessary mathematical, stereological and statistical relationships that require consideration. Fortunately, the advent of practical computerized image analysis and measurement systems has eased much of that pain (Inoue, 1986). Examples used throughout this text include many light microscope images.

When higher magnification or greater depth of field (the ability to keep sharp focus over a rough surface or through a thick section) is required than the light microscope can deliver, the next logical step is the electron microscope. There are two principal types, which form their images in entirely different ways. The transmission electron microscope (TEM) is analogous to the light microscope, except that electrons are used instead of light photons (they have a much shorter wavelength, and are capable of forming higher resolution images). This means that the sample and the beam must be in vacuum, and that "lenses" used to deflect and focus the electrons work with magnetic fields rather than pieces of glass with varying indices of refraction. The image is produced by the electrons striking a phosphor surface (or a piece of photographic film). It can then be viewed or acquired in the computer using a video camera, as will be discussed below.

The scanning electron microscope works quite differently. A fine beam of electrons strikes the surface of the sample (or in the case of the scanning transmission microscope, passes through a very thin specimen), producing signals such as secondary or backscattered electrons, X-rays, and so forth. The beam is scanned in a raster pattern (much like a common TV system) over the specimen, while the signal from a selected detector is used to control the brightness of the display cathode ray tube. This produces a time varying voltage that can be sent directly to the computer, similar to the scanning densitometry table mentioned above. Either by using a fairly rapid scan or a long persistence phosphor, this image can be viewed by the user. For image analysis, the signal is instead fed directly to the computer for digitization as discussed below (Stewart et al., 1986).

The scanning electron microscope commonly has several detectors, capable of producing images with the various signals produced when the incident electron beam strikes the specimen surface. Backscattered high energy electrons give a high-contrast, deeply shadowed representation of a rough surface, and show elemental contrast (proportional to the average atomic number) on chemically heterogeneous materials. The secondary electron signal, consisting of electrons initially bound to atoms in the sample and knocked loose by the incident high energy electrons, produce a higher resolution image of rough surfaces. The appearance is similar to viewing a surface in diffuse light, except that edges and protrusions are strongly highlighted.

Conventional X-ray images are formed by selecting an energy (or wavelength, depending upon the type of detector used) corresponding to a particular element, and then placing a single dot on the display screen at the location of the scanning beam whenever an X-ray of that energy is recorded. These images are sometimes difficult to use for measurement, and will be discussed in a subsequent chapter. They require rather different processing steps than more common images, and are often used in combination with other images which are less noisy and have better resolution and edge definition. Less common, but more readily interpreted, is an image in which the brightness of each picture point is proportional to the X-ray intensity (and more-or-less proportional to concentration) for the element.

Other types of images are less commonly encountered, such as cathodoluminescence (the emission of visible light from molecules excited by the electron beam), and electron beam induced conductivity (used for instance to locate junctions in semiconductor devices). These generally have high contrast and are straightforwardly analyzed and measured.

There are other analytical instruments that also produce images by one or the other of the two methods used for electron microscopes. For instance, the Ion Microscope uses an incident beam of high energy ions to knock atoms loose from the surface of the specimen and ionize them. These are then accelerated and passed through a mass spectrometer to select ones corresponding to a particular element or isotope and focussed on a fluorescent screen. The image can be picked up with a video camera.

Secondary Ion Mass Spectrometry (SIMS) is carried out using the scanning principle, by forming a finely focussed beam of ions that is rastered over the sample, and the intensity of secondary ions sputtered off the surface and selected by element is collected with a Faraday cage or similar detector. As with the SEM, this image is directly available for computer acquisition. A very similar technique, although with vastly different types of detectors and energies of particles, is used for the Scanning Auger Microscope, the Scanning Acoustic Microscope, and so forth. There is even a scanning

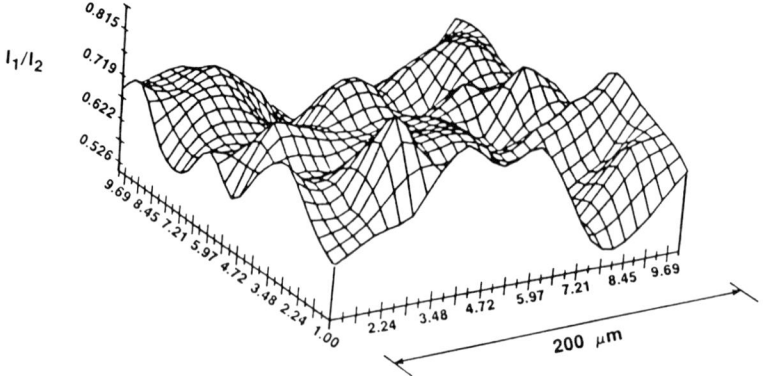

Figure 2-2: X-ray diffraction data presented as a two-dimensional array (Engler et al., 1988).

light microscope that uses laser light and performs the sequential scanning operation mechanically (it has much greater depth of field than the conventional light microscope). It is also possible to use a light microscope with a laser to ionize a small portion of the specimen for analysis in a mass spectrometer. With a step-scanning stage, this instrument can also be used to produce images showing the distribution of various elements (Wilk & Hercules, 1987).

Indirect creation of images from various kinds of analytical data can be performed. For instance, infrared reflectance and Raman spectroscopy can be used to identify materials by their intermolecular bonding. The translation of a multiphase specimen in a microdiffractometer can be used to measure X-ray diffraction patterns, from which the relative intensity of diffraction lines corresponding to particular phases can be used to produce an image brightness related to relative concentration (Figure 2-2). These images often have many fewer pixels, with limited counting precision producing "noise" in the brightness values due to counting statistics, but many different images (for instance corresponding to different element settings) of the same area. They are used more for global measurements than feature specific ones (for instance, the area fraction of a phase rather than the dimensions of individual objects).

Another way to record diffraction information is the rather familiar selected area electron diffraction (SAD) pattern (Figure 2-3) from the transmission electron microscope (TEM). Although it is recorded using film and the same camera equipment normally used for recording more conventional images, it is easy to dismiss this type of data as being something different from a normal image since we do not recognize it from common experience. However, the two-dimensional recording of brightness data is indeed an image, and image analysis methods are appropriate for extracting the desired measurement data from these unusual types of patterns as well.

Multi-spectral images

All of these image sources produce at least one kind of image, and many are capable of producing several. It could be something as simple as the many different color sensors on board a satellite (infrared detectors, for instance, are particularly sensitive to the absorption of light by the chlorophyll in plants, so that areas with plant cover appear dark). Or it could be the images of different elements or isotopes produced by the scanning electron microscope or ion microscope, showing the spatial distribution of each species over the same region of the specimen surface.

In principle, all of these images are available to the computer system. By using the information in several such images, much more elaborate techniques can be employed for identifying features and selecting ones to be measured. Many systems cannot simultaneously acquire all of the possible images, and many imaging systems cannot simultaneously produce them (different optics or detectors may be needed, for instance). But in most cases it is perfectly equivalent to acquire them sequentially.

There are exceptions of course. A satellite will have moved between acquiring one image and the next. The ion microscope has eroded away the surface in forming one image, and the next image is at a slightly greater depth (perhaps only a few atomic layers) in the sample. Many of these defects in the correspondence of the sequential images can be corrected for in the image analysis system.

For instance, one of the image processing steps we will consider performs "rubber sheeting" on images to correct for geometric distortion. It can be used to align the common portions of two different satellite images. Later, in discussing serial section images, we will see that a sequence of many images (as from the ion microscope) can be used to interpolate in the depth direction. This makes it possible to record several images in sequence, and then by interpolation find the intensities (proportional to concentrations) for several elements at some intermediate depth.

Remote sensing images are somewhat different in their purpose than microscope images. Classification of terrain based on spectral information is usually more important than feature measurement, and object recognition (for instance military surveillance) as discussed in Chapter 9 may require extensive pattern matching.

One particular type of "multi-spectral" image that is often of interest is the "true-color" image that corresponds to our normal vision. The human eye has sensors with peak sensitivities to red, green and blue light (Figure 2-4). Combinations of those colors

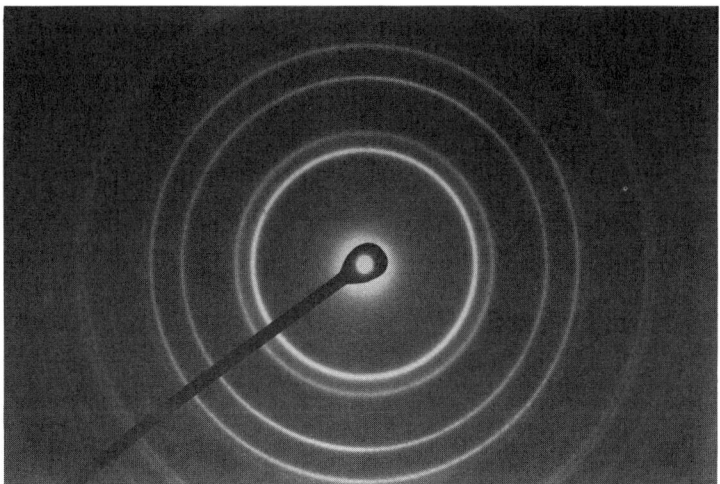

Figure 2-3: Selected area electron diffraction pattern from gold foil.

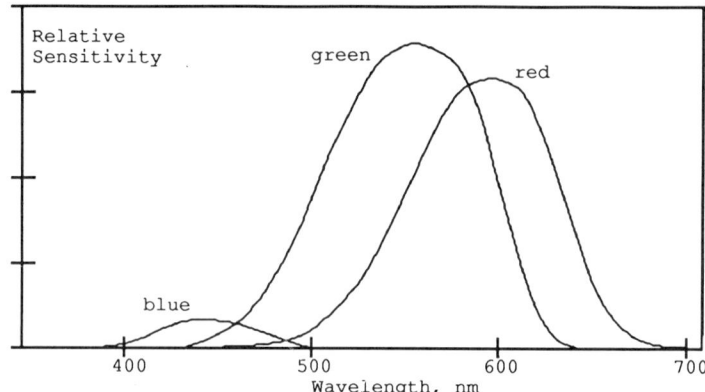

Figure 2-4: Diagram of the spectral response of cones in the human eye.

are interpreted as a continuous range of color. It is possible to acquire images in which the color information is preserved. This is most simply done by using a sensor such as a color video camera, which also produces red, green and blue information. There are two basic types of color cameras. The less expensive ones use a single tube with three types of phosphors. Of course, this limits how closely the sensors can be spaced. The more expensive (e.g., broadcast quality) cameras use beam splitters and three monochrome tubes. These offer better spatial resolution but require convergence adjustments to keep the images in registration. If the three signals are acquired separately (either simultaneously in a single image scan, or sequentially using different filters and a monochrome camera), they can be stored as three corresponding images and suitable software can use the information for identification and demarcation of features.

It should be noted that this is less common than acquiring monochrome images. For one thing, it takes three times as much storage space for the three images. For another, the software to analyze the three images in combination is difficult to design because there is almost too much information and it is hard to design algorithms that always use it

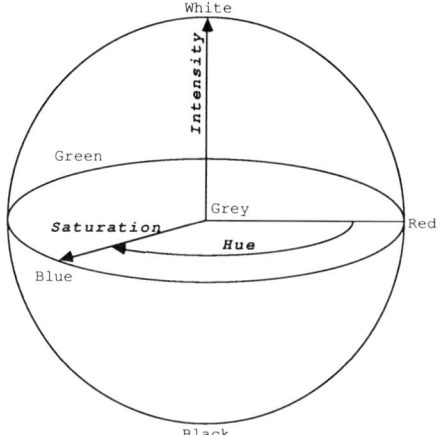

Figure 2-5: HIS color coordinate space.

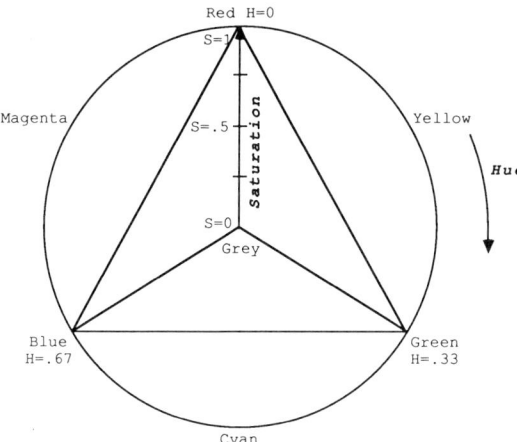

Figure 2-6: Saturation and hue in a constant intensity plane in HIS space. Notice that mixtures of red, green and blue cannot produce fully saturated yellow, cyan or magenta. The hue value varies from 0 to 1 as it goes from red through green (1/3) and blue (2/3), and back to red.

correctly. We will discuss the use of color images for segmentation and discrimination, but for the most part will assume that any processing and measurement are performed on monochrome images. This may mean that similar processing is performed on each of the three (RGB) images in a color set, or that different steps are used for each of them, as appropriate.

When color itself is to be measured, it is necessary to provide some means of calibration for the sensor and acquisition process. It is generally impractical to use more than a small number of color standards, and to perform intensity interpolation between them to measure others. The most common arrangement of sensors is RGB (Red Green Blue). The placement of the red, green and blue points on the color triangle shows that most of the perceivable colors can be produced by a linear combination of these three. It is also possible to equivalently characterize colors using CMY (Cyan Magenta Yellow, subtractive colors used for printing, whereas RGB is appropriate for cathode ray tube displays). YIQ is another set of three values employed in commercial TV broadcasting. The advantages of the YIQ method are that the brightness (Y) value can be used alone by monochrome receivers, and that the bandwidth for the color components can be less than for the brightness (Foley & Van Dam, 1984; Ballard & Brown, 1982).

Another particularly useful set of color coordinates is known as HSV (hue, saturation and brightness value), HIS (Hue, Intensity and Saturation) or HLS (hue, lightness and saturation). Hue describes the color (red, orange, yellow, green, blue, purple), while saturation is a measure of how much of the color is present (the difference between red and pink), and brightness value distinguishes between light and dark (Figures 2-5 and 2-6). These coordinates provide some particular advantages for distinguishing features, particularly when there may be shading may be present. All of these various scales can be interconverted by simple multiplication and addition programs, as shown in Tables 2-1 and 2-2.

These alternate color definition schemes may be quite important for the display of color graphics information (depending on the type of display hardware or hardcopy

Table 2-1: Interconversion of RGB and YIQ color scales

$$Y = 0.299\ R + 0.587\ G + 0.114\ B$$
$$I = 0.596\ R - 0.274\ G - 0.322\ B$$
$$Q = 0.211\ R + 0.523\ G + 0.312\ B$$

$$R = 1.000\ Y + 0.956\ I + 0.621\ Q$$
$$G = 1.000\ Y - 0.272\ I - 0.647\ Q$$
$$B = 1.000\ Y - 1.106\ I + 1.703\ Q$$

Table 2-2: Conversion from RGB color coordinates to HIS coordinates

$$H = [\pi/2 - \arctan\{(2 \cdot R - G - B)/\sqrt{3} \cdot (G-B)\} + \pi\ ; G<B] / 2\pi$$
$$I = (R + G + B) / 3$$
$$S = 1 - [\min(R,G,B) / I]$$

printer used), but they are much less often used for image acquisition. The RGB sensor is by far the most common, and color images are usually stored in a form that is effectively three monochrome (grey scale) images, one corresponding to each color. Some processing operations (for instance, smoothing to reduce noise) are simpler in HIS space, but these values can be calculated when needed and the results transformed back to RGB for storage. It would be preferable for HIS coordinates to be used directly in image analysis computers, instead of the less useful RGB, but this is not commonly done at the present time except in the field of remote sensing (satellite and aerial photography).

Regardless of whether there is a single monochrome (grey scale or black and white) image, or many multi-spectral images, we expect them to correspond to the same physical view, which is called the "field." A sequence of images that vary, for instance in depth or time, showing some differences in the placement or characteristics of objects, presents different problems for analysis and will be dealt with separately in the chapter on serial sections.

Image sensors

The scanning electron microscope, and a few similar instruments that function in the same way, directly produce a time sequence of intensity values along a raster pattern covering the image field. But most images are continuous, and acquired as a time-varying signal using a video camera. The U. S. standard video format provides a 525 line image, at the rate of 30 frames per second (actually 60 interlaced half frames per second, each of 262 lines). The format for the voltage pattern sent by these cameras is defined as RS-170 (see below). Color information can be superimposed on the monochrome signal, or not, and can be measured by the computer, or not, as required.

There are special cameras available that scan a higher number of lines, as many as 1000 or 2000. These are almost exclusively monochrome, although the use of color filters and acquiring three sequential scans can be used to provide color information. These cameras are considerably more expensive than standard video cameras.

Most video cameras have a standard threaded mount ("c-mount") for lens attachment, and many microscopes (especially light microscopes) provide for the direct attachment of these cameras without any lens. In effect, the optics of the microscope becomes the lens, and the image is focussed directly onto the sensing element of the camera.

Video cameras are of two basic types: those using a picture tube, and ones with a solid state sensor. There are almost countless varieties of each. It will suffice to consider the most common of each type here.

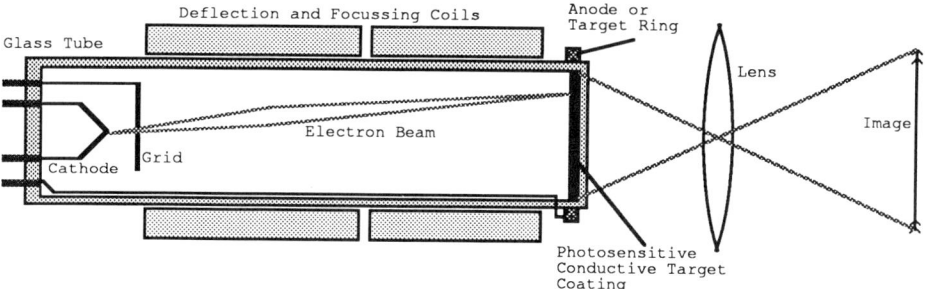

Figure 2-7: Functional diagram of a vidicon tube.

The vidicon picture tube (Figure 2-7) is used in many inexpensive industrial and surveillance cameras. Light focussed on the front surface of the tube strikes a photosensitive coating, usually consisting of a large number of resistive globules of antimony sulfide whose resistance decreases with the intensity of the light striking them. This coating is on the inside of the tube, and a low voltage electron beam is scanned over the coating, from an electron gun with an electrostatic or electromagnetic deflection system at the rear of the tube. The electrons in the beam strike the layer, charging the inside surface. The layer functions as a capacitor, inducing a current in the cathode that is the video signal, varying with the illumination on each point along the scan.

Other similar devices such as the nuvicon or plumbicon work in similar ways, but use a different coating material, interpose an additional photosensitive layer or an intermediate step in which the light produces electrons from a photoemitter which charge up the photocathode. The charge limits the current absorbed from the electron beam, producing a variation proportional to the illumination intensity. These tubes generally offer greater sensitivity to low light levels than the vidicon, and a gamma value equal to one. This value is the slope of the curve relating output voltage to input light intensity, on a log-log scale. Vidicon tubes have gamma values less than one, which can complicate the calibration of density measurements. On the other hand, vidicon tubes have a spectral sensitivity similar to the human eye while most of the other tube types are more sensitive in the red end of the spectrum. The eye also has a logarithmic response to brightness.

Most image tubes will dissipate the charge over time, even without the incident scanning beam, and so cannot truly function as integrators for very dim images. The resolution of these tubes is limited by the spreading of charge in the photocathode layer. This tendency for bright regions or spots to "bloom" or grow in size does have some advantages. For instance, images of dust particles on semiconductor wafers can not only reveal particles smaller than the resolution limit of the optics, it can even measure them because the scattered light intensity is proportional to size, and causes a proportional blooming of the spot on the image tube which allows measuring a size value that can be calibrated to actual particle size (Cooper & Rottmann, 1988).

Expensive color cameras use multiple tubes with colored filters and perhaps different photocathode materials. Inexpensive cameras use a single tube, in which the photocathode material consists of spots or stripes of different phosphors or other materials that respond to different colors of light. As the beam scans, it measures the intensity of each color light sequentially. Of course, this further reduces the resolution of the images obtained with the tube.

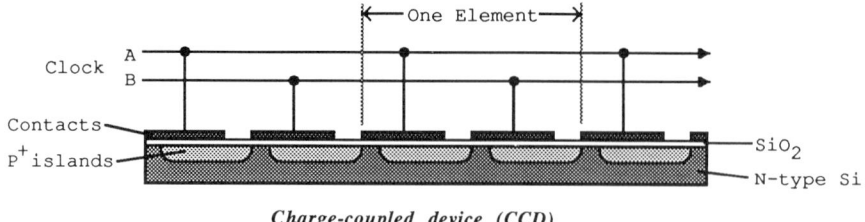

Figure 2-8: Schematic diagram of a solid-state camera.

The solid state sensing devices in common use are the CCD (charge coupled device, Figure 2-8) and CID (charge injection device). Both use what amount to separate devices for each picture point that is to be measured. It is convenient to imagine each point as a small field effect transistor (FET). Incident light photons provide enough energy to raise valence electrons to the conduction band, and these then penetrate from the source to the drain region where they are held until collected. The collection is accomplished by circuitry that sequentially scans the individual cells (sometimes directly, sometimes by shifting all of the information within one scan line from cell to cell, as in a shift register, and thus reading the line).

These devices are much more rugged, and smaller and lighter than tube cameras. They are also immune to stray magnetic fields (common to many electron and ion microscopes) which disrupt the scanning beam in the tube. Finally, they can collect charge without significant loss (especially when cooled), and are thus especially useful for dim images that may need to be integrated for quite a long time (for instance in astronomical work). On the other hand, the resolution is limited by the spacing of the individual cells, and is not yet so good as a tube camera of equal cost. This is especially true for color cameras, in which one-third of the detectors along each row are designed with sensitivity to red, green or blue light. Finally, the individual sensors may not all be absolutely identical in sensitivity. This can be corrected by programs that record a uniform image from the tube and use it to adjust all subsequent images, but it adds one additional processing step.

In addition to standard video technology, there are the high resolution cameras mentioned above and slow scan devices such as densitometry tables or scanning electron microscopes that use a higher number of scan lines, and perhaps a much slower scanning speed. Acquiring a high quality X-ray image from an SEM may take hundreds of seconds, for instance. This poses no special problem for the acquisition circuitry in the computer, but it may make it more difficult to utilize standard components, such as video recorders that are otherwise quite useful for collecting images from instruments (perhaps in a remote location) for later analysis, or for transmitting images from remote sensors to the computer.

Another important but often overlooked problem that can plague video cameras, or any raster scanning input device is distortion of the scan. The most common types of distortion are called pincushion and barrel distortion, as indicated in Figure 2-9. These can usually be corrected in the optical or tube design itself more economically than by using image processing to stretch or squeeze the image after acquisition. Another problem that sometimes arises with standard video cameras is a systematic misalignment of alternate scan lines, resulting from the interlace that will be discussed below.

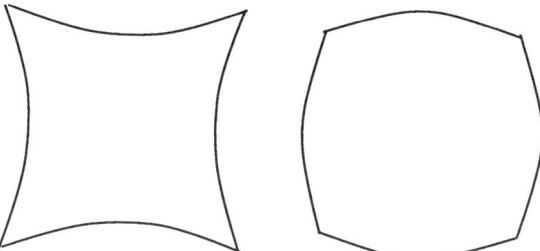

Figure 2-9: Pincushion and barrel distortion of a square image region.

Digitization

The signal produced by the camera or other sensor is a voltage (in the case of an RS170 signal, there is additional information which we will ignore that defines the timing of the raster). For storage in the computer, this must be digitized, and this requires an analog to digital converter. Standard A/D converters are of three basic types: the rundown type in which the voltage is allowed to charge up a capacitor, and the time to discharge provides a measure of the voltage; the successive approximation type in which a voltage is generated by the device and compared to the voltage to be measured, and then adjusted upwards or downwards in decreasing steps (which are counted) until the desired precision is reached; and the flash ADC in which a series of separate voltages are generated and added together to match the value to be measured.

The first of these is practically never used for image digitization because it is too slow, and the time required for measurement varies with the magnitude of the voltage. The second method is often used for slow scan devices, but is not fast enough to keep up with video rate images. The flash ADC can measure hundreds of points along each scan line, producing a series of values to be stored in memory. This is also called a "frame grab" because an entire image can be digitized during one frame or raster scan (actually because of the interlacing of two scans into a single frame, it may apply to either a single or a double scan depending on the particular device).

Frame grab or flash ADC's are inherently noisier and less precise devices than the slower types, but they are still usually adequate for image measurement purposes. Somewhat better results can be obtained by acquiring several successive frames and averaging them, which reduces the random variation in values.

For a standard video signal (RS170), digitization is accomplished by measuring the voltage at closely spaced intervals along each scan line (Figure 2-10). The video signal

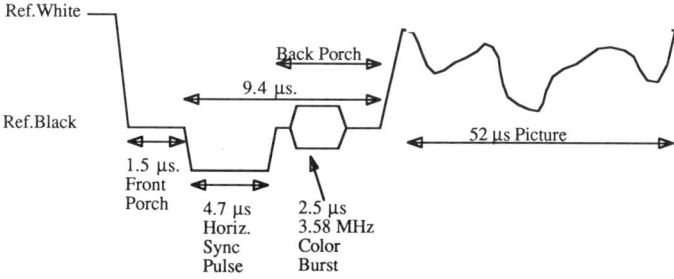

Figure 2-10: Schematic diagram of RS170 video signal waveform.

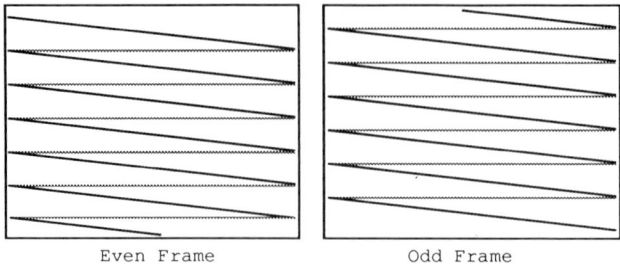

Figure 2-11: Successive half-frame video scans.

consists of 2 "half frames" (Figure 2-11), which alternately cover the even and odd lines of a 525 line frame each 60th of a second (so that the complete image is repeated 30 times per second). Not all of the lines are picture, due to the vertical retrace to get back to the top. It is possible to measure as many as 240 lines in each half frame, but because most images are "overscanned" (cover a larger area than that expected to be displayed on the receiver), only about 200 lines are really useful in most cases.

Each scan line contains a horizontal sync pulse that identifies where everything is located. A typical device to digitize a video signal is a single monolithic chip, containing a hybrid of analog and digital electronics. It locks onto the RS-170 signal using the horizontal sync pulse to control its timing. (A horizontal sync pulse that lasts longer than 6.6 μs signals the vertical sync, from which line addresses are counted.) In RS-170A, a color burst is added after the horizontal sync pulse. This provides a phase reference for the 3.58 MHz color signal that is superimposed on top of the brightness values.

9.4 μs after the horizontal sync pulse, the active picture data begin. Normally, the full range from black to white covers a 700 mV amplitude. A high frequency clock is used to count the time from the sync pulse to the start of picture data, and then to space the measurements along the scan line, which lasts only 52 μs. For instance, a 12 MHz clock can acquire 512 samples along the line. Of course, these may not all represent independent pixels in the original image, depending on the camera and transmission quality. Commercial video with a 3.2 MHz bandwidth can only have a maximum of about 330 pixels in the full width of the image (which is still enough for quite high quality images with recognizable fine detail).

To obtain a 512x512 image, two successive frames are needed (Figure 2-11). If the line address (number of horizontal sync pulses since the vertical retrace) is doubled and a value that switches from 0 to 1 on each frame is used as the least significant bit, then the line address will write each line from one frame into either the even or odd lines of memory, and two frames will produce a complete image. This would actually have about 512 (wide) x 400 (high) pixels, but if the 512 measurements were spaced across the full width of the scan line, the resulting pixels would not be square, but would have an aspect ratio (height to width ratio) of 4:3, making them unsuitable for image measurement.

When the image acquisition cards are intended for real image analysis applications, proper clock frequencies are used to sample at rates that produce square pixels. Because the video image is not square, this produces images that have pixel dimensions such as 640x480, 512x384, or 256x192, instead of 512x512. Except for complicating the use of Fast Fourier transforms to process these images, and perhaps the internal routines to address pixels within the storage array, this has no effect on subsequent image processing or analysis.

There are three principal approaches to averaging many scans over the image. One is to simply add all of the brightness values for each point together, and then finally divide by the number of scans to get an average value. This is an optimum method since the precision of measurement improves as the square root of the number of scans, but it requires more memory to hold the summed intensities than any single image requires and so it may not be possible depending upon the design of the system. A second method is continuous averaging, in which each new image is added to the stored image in declining proportions so that the total remains fixed, and can be continuously viewed during acquisition.

The third method is called exponential averaging, and limits the magnitude of the sum. As more values are added together than the computer can hold, they are continuously divided down. This means that the last values count more than the first ones, and a limit is reached beyond which no additional improvement in noise reduction can be achieved. In spite of this theoretical limitation, the practical advantage of any averaging method is usually achieved with a small number of successive images (perhaps 4 to 16) so either method may be satisfactory, and offer a significant improvement over a single frame. These methods are mentioned again in Chapter 3 in conjunction with noise reduction in images.

Specifications

The idea of picture points has been raised several times, and in the case of the CCD camera it is particularly clear that there is a smallest unit of resolution. When the image is stored in the computer, the degree of quantization becomes even more apparent. Each memory location in the computer (usually a byte, 8 bits) stores the brightness information from a picture point. The number of bytes then corresponds to the number of points and effectively defines the image resolution (Figure 2-12).

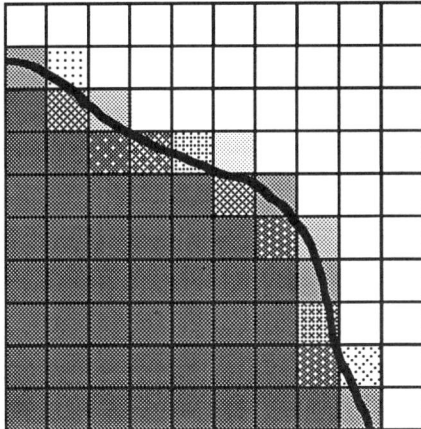

Figure 2-12: Pixels average over regions in the image, so that pixels straddling a feature edge take on brightness values intermediate between those of the feature and surroundings.

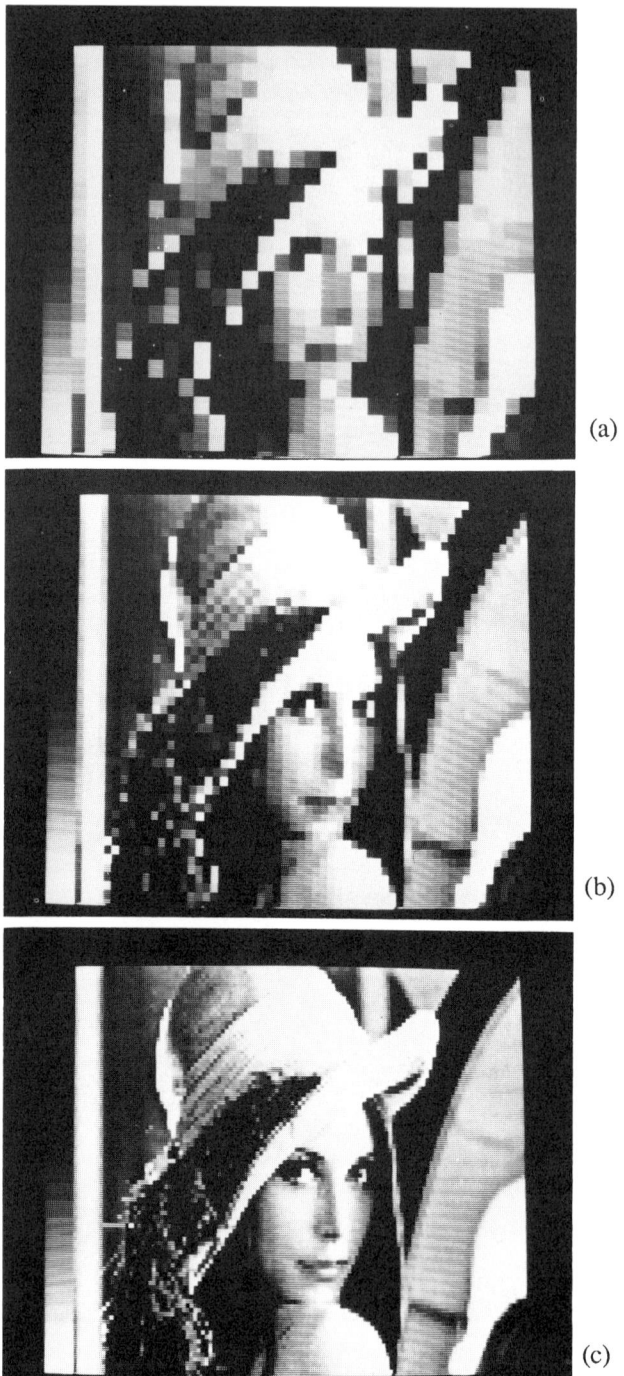

(d)

Figure 2-13: Examples of the effects of number of pixels on a grey scale image. Successive images are a) 32, b) 64, c) 128 and d) 256 pixels wide. Beyond 256 the changes are hardly distinguishable.

Of course, this assumes that the imaging device and the ADC do not further limit the resolution, but in most cases this is a safe assumption. Using a lot of bytes to store the image increases the costs of image analysis. Not only does the computer memory itself cost money (images occupy a *lot* of memory, as we will see), but more important, all of the processes that operate on the image take time, roughly proportional to the number of points stored. This goes up as the square of the number of points along each scan line.

Each picture point, or picture element, is referred to as a "pixel." Pixels are usually square, although a hexagonal arrangement is occasionally used. Each square pixel has a numerical value that represents the average brightness of the image over the area that the pixel subtends on the original continuous image. The number of pixels then defines the image resolution for all subsequent activities (Figure 2-13).

The number of bits used to store the brightness information defines the "depth" of the image (Figure 2-14). Since computer storage is often organized in bytes (8 bits), this produces a capacity for 256 brightness or grey levels. This is far more than the human eye can distinguish in a monochrome image (20-30 at best), and more than can be contained in a photographic print (negatives are much better than prints). It is also reasonable in terms of the performance of high speed ADC's, which typically produce 64 or 256 distinct values (6 or 8 bits). When full color images are acquired, they are usually stored with three bytes per pixel, giving 8 bits each for red, green and blue.

The resolution for most present day imaging systems is in the range of 256 to 1024 pixels wide. A 256x256 pixel image, using one byte to store each pixel brightness value, requires 65,536 memory locations (64K), or about as much memory as personal computers had in total a few years ago. A 1024x1024 image in full color would occupy 3 Megabytes, beyond the addressing capability of most personal computers now. Within the computer this data is stored as pixels, but for transmission other more compact formats are often used. Digital techniques have largely replaced analog ones for transmission of images. Instead of individual pixel brightnesses, some of the simple techniques are to send just the differences between successive pixels instead of entire

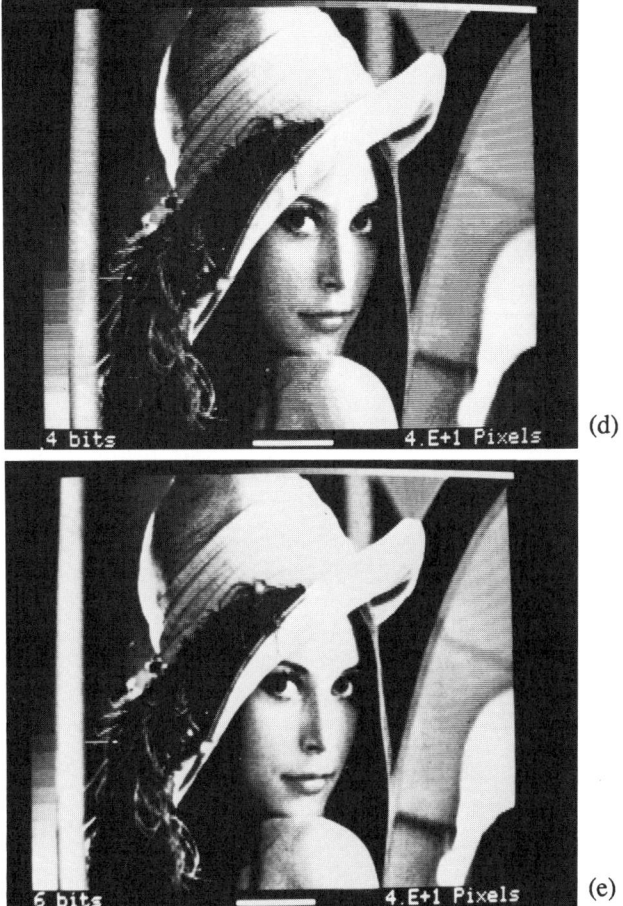

Figure 2-14: Effect of the number of bits on the image. Successive images have a) 2, b) 4, c) 8, d) 16 and e) 64 grey levels. Beyond 64 levels the changes are hardly distinguishable.

values, or run length encoding in which the number of identical pixels is sent instead of each individual value. For many typical images in which successive pixels are likely to be the same or similar, these methods can produce a significant reduction in the storage or transmission costs of images. These are discussed in later chapters, along with some of the more complex methods based on frequency transforms.

It is convenient, but incorrect, to refer to the number of pixels as the resolution of the images. From an information-content point of view, more than two pixels are required to define the location of any feature or boundary. Shannon's criterion states that the highest frequency (smallest spacing) that can be determined from a digitally sampled signal is half the sampling rate (the pixel spacing). In later chapters discussing the measurement of feature size and location, we must remember that dimensions approaching the size or spacing of the individual pixels are suspect. But the word "resolution" has become associated with pixel dimensions in this field, and will be used here for consistency.

For a variety of practical reasons, the most common image resolution in current use is 512 pixels wide, 8 bits (1 byte) deep. The hardware needed to acquire and store such an image, and even perform some of the processing steps to be discussed below can be placed on a single circuit board and plugged into many computers, and the processing and measurement time is reasonable.

Many of these systems use separate memory to hold the image. Usually this memory is also used to generate a display of the stored image data, and allow the values to be accessed by the host computer. This can be accomplished with multi-ported memory, or (more cheaply) by restricting each type of access to a specific time slot. Even lower cost systems use the systems main memory to hold the image data, and the digitized values from the ADC must be handled by the central processing unit as they are acquired. These distinctions produce no fundamental difference in the kinds of operations that are possible with the systems, however.

It is important, though, to realize what the consequences of this finite image resolution are. The primary limitation is on the size of objects that can be studied. Depending on what information is required about each object (such as its size, shape, density, etc.), a certain minimum number of pixels must be covered by it. For density measurements, for instance, the pixels along the edge cannot be used (they represent an average of the object's brightness and the surroundings), so there must be enough

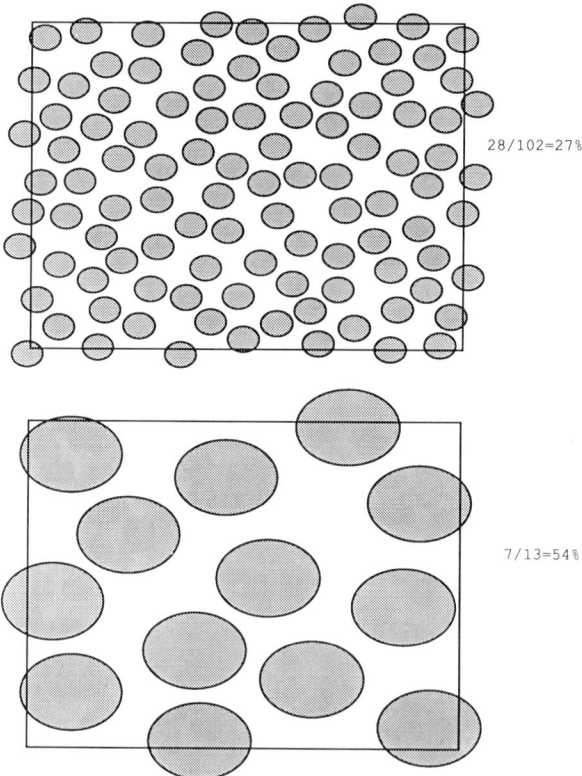

Figure 2-15: A greater percentage of large features than small ones in a field touch the edge and cannot be measured.

internal pixels to give a good average and perhaps a good estimate of the variation and even the pattern of variation. For shape measurements, the irregularities along the edges of the feature must be well enough defined to be accurately measured. In general, this may mean that tens or even hundreds of pixels are needed to adequately define the smallest features of interest.

But then the largest object expected to be present in the image must be considered (Figure 2-15). If it is too large, or covers too many pixels, it has a high probability of intersecting the edge of the field. In that case, it cannot be measured (because its full extent is not known). Features larger than about 10% of the total area of the image are at some risk from this limitation. Consider then the case of a 512x512 image (262,144 pixels). If the largest "safe" object for measurement covers 25,000 pixels and the smallest covers 250 pixels, the areas of features can vary by a factor of about 100. This means that the linear sizes can only range over about a factor of 10.

For smaller images the limitation is tighter, and for larger ones it relaxes slightly, generally in proportion to the square root of the number of pixels. But it is often necessary with computerized image analysis to select an image magnification carefully and deal with features only within a fairly narrow size range. If much larger or smaller objects are also to be studied, separate images at different magnifications should be used.

It is very instructive to compare this performance to that of the human eye. Instead of the 250,000 pixels described above, there are more than 150 million individual receptors in the human eye. No computerized image analysis system is likely to approach that capacity for some time. Furthermore, the total depth of the human sensor (the number of "bits") is much larger; the eye can respond over some 9 orders of magnitude change in brightness (although in the darker portion of this range it loses color detection capability).

This agrees with our everyday experience. Our built-in imaging capability can easily deal with objects that range from the size of millimeters (at a distance of a meter or so) to ones at least 1000 times larger (a meter in size). But we employ additional devices that alter the image magnification in order to view things that are kilometers or micrometers in extent.

References

D. H. Ballard, C. M. Brown (1982) Computer Vision, Prentice Hall, Englewood Cliffs, NJ

D. W. Cooper, H. R. Rottmann (1988) *Particle Sizing from Disk Images by Counting Contiguous Grid Squares or Vidicon Pixels* Journal of Colloid and Interface Science 126, 251-259

P. Engler, R. L. Barbour, J. H. Gobson, M. S. Hazle, D. G. Cameron, R. H. Duff *Imaging with Spectroscopic Data* Advances in X-ray Analysis vol. 32, Plenum Press, New York (in press)

J. D. Foley, A. Van Dam (1984) Fundamentals of Interactive Computer Graphics, Addison Wesley, Reading MA

J. C. Giddings (1984) *Two Dimensional Separations: Concept and Promise* Anal. Chem. 56 1259A-1270A

E. J. Hahn (1983) *Autoradiography: A review of the basic principles* American Laboratory (July 1983) 64-71

W. R. Hruschka, D. Massie, J. D. Anderson (1983) *Computerized Analysis of Two-Dimensional Electrophoretograms* Anal. Chem. 55 2345-2348

S. Inoue (1986) <u>Video Microscopy</u> Plenum Press, New York NY

P. A. Jansson, L. B. Grim, J. G. Elias, E. A. Bagley, K. K. Lonberg-Holm (1983) *Implementation and application of a method to quantitate 2-D gel electrophoresis patterns* Electrophoresis <u>4</u> 82-91

J. G. Moik (1980) *Digital Processing of Remotely Sensed Images*, NASA SP-341

G. A. Moore (1968) *Automatic Scanning and Computer Processes for the Quantitative Analysis of Micrographs and Equivalent Subjects* in <u>Pictorial Pattern Recognition</u>, Thompson Book Co, Washington DC

T. G. Rochow, R. G. Rochow (1978) <u>An Introduction to Microscopy by means of Light, Electrons, X-rays, or Ultrasound</u> Plenum Press, New York NY

J. C. Russ, W. D. Stewart, J. C. Russ (1985) *Densitometric image measurements*, Amer. Lab, <u>Apr 85</u>, 41

J. C. Russ, W. D. Stewart, J. C. Russ (1988) *The measurement of macroscopic images*, Journal of Food Technology <u>42</u>(2), 94-102

F. F. Sabins, Jr. (1987) *Remote Sensing: Principles and Interpretation* (2nd edition) W. H. Freeman, New York

W. Schneider, J. Klose (1983) *Analysis of two-dimensional electrophoretic protein patterns using a video camera and a computer: I. The resolution power of the video camera* Electrophoresis <u>4</u> 284-291

W. D. Stewart, J. C. Russ, J. C. Russ (1986) *Passive SEM-microcomputer interface for acquisition of electron images and X-ray maps*, <u>Microbeam Analysis 1986</u> (A. D. Romig, ed.), San Francisco Press, 141

Z. A. Wilk, D. M. Hercules (1987) *Organic and Elemental Ion Mapping Using Laser Mass Spectrometry* Anal. Chem. <u>59</u> 1819-1825

H. M. Yakin, H. Kronberg, H-G. Zimmer, V. Neuhoff (1982) *Photometric evaluation of slab gels II: Delimitation and integration of peaks in one-dimensional electropherograms* Electrophoresis <u>3</u> 244-254

R. A. Zeineh, G. Kyriakidis (1987) *Computer-aided soft laser scanning densitometry*, American Laboratory News <u>Jan 87</u>, 16

Chapter 3

Image Processing

In this chapter we deal with the techniques of image processing. These are operations that start with a grey scale (or color) image and return another grey scale image. The next chapter will deal with some additional techniques that operate on grey-scale images for purposes of locating feature edges in the context of isolating features for measurement.

By definition, image processing includes those methods that start with an image (an array of pixels, each with a brightness or "grey scale" value or perhaps with color information) and end with an image. Usually, the resulting image is of similar size (number of pixels and number of grey levels). The brightness of the resulting pixels will, in most cases, have been modified using rules that take into account the original value of the pixel and its neighbors, or perhaps its position or spatial relationship to many other pixels. Sometimes, several images are combined to produce a new image. The various principal methods of processing are often described as working in the spatial, time or frequency domains. We need not be too concerned with these terms for now. They will be defined in due course as examples of the various methods are discussed.

The various kinds of rules that will be described include ones based on theoretical principles as well as rather ad-hoc empirical techniques. They are all intended to produce a modified or processed image that emphasizes some aspect of the original image at the expense of others. This is, in effect, a filtering technique that selects certain kinds of image data. Most often, the features that are desired are objects, edges, boundaries, or some other defined structure. Depending on what is considered to be of interest, it is often possible to construct a suitable processing operation to extract it from the background (i.e., uninteresting) information in the image, so that it can be measured. It is sometimes necessary to operate several times on the same original image, using different processing operations, to isolate different types of information, such as structures of different size or orientation.

Point operations

The simplest technique that needs to be mentioned only briefly replaces the value of each pixel by a new value that depends only on the original value. This is called a "point" operation to distinguish it from "neighborhood" and "global" operations as will be discussed below. For instance, if the original brightness values covered only a small fraction of the total range available, then new values can be substituted which increase the image contrast. This is a one-to-one substitution, or a "point transformation" in which the resulting pixel brightness depends only on the original point and not on any other information in the image. The relationship between original and replacement brightness values is often called a transfer function, and can alter the appearance of an image as shown in Figure 3-1.

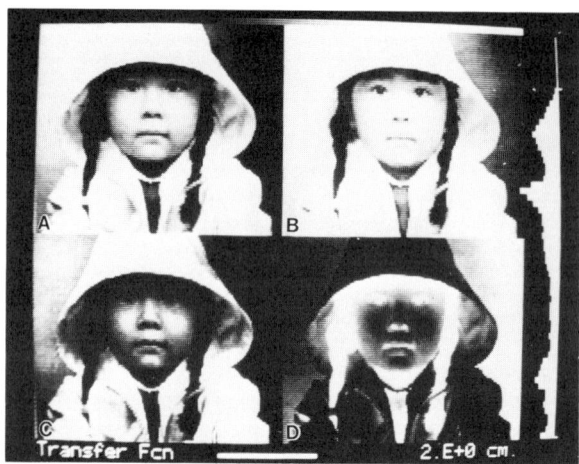

Figure 3-1: Examples of different transfer functions applied to an image:
a) original, b) logarithmic, c) histogram equalization, d) negative.

The function need not be linear, of course. In fact, nonlinear transfer functions are often used. A logarithmic function mimics the way that most photographic film works, and can be described by a "gamma" value that gives the slope of the semi-log plot. Positive values of gamma compress the range of displayed brightness in the bright end of the range while expanding the darker portion. Of course, this can be applied in addition to contrast expansion by selecting only a narrow range of values. It is also sometimes advantageous to use negative transfer functions in which contrast is reversed.

Almost any non-linear transfer function may be useful for particular images, but one of special interest is the so-called histogram equalization method. With this method, an equal number of pixels in the final image have each available shade of grey. The shape of the transfer function can be straightforwardly determined from the brightness histogram of the original image. The method expands the contrast in regions of high gradients, so that details are easier to see. Figure 3-2 shows several types of transfer functions. Notice that a transfer function can permit a different number of grey levels for its output, and this is sometimes used with display devices.

Special contrast transfer functions are sometimes used which are not monotonic. For instance, a linear (or log) function may be used between values that are not at the minimum and maximum value for the original, with reversal of the values beyond that range. This enables detail in extreme shadow areas to be displayed. Another technique that assists the viewer in seeing gradients of brightness is contouring, in which selected brightness pixels are set to extreme bright or dark brightness values (or to different colors). The result is to superimpose brightness contour lines on the image (Figure 3-3).

None of these techniques really changes anything in the original image that affects subsequent measurements. Indeed, with many modern systems it is not even necessary to operate on the image itself to produce the transformation. The use of "Look Up Tables" or LUT's to implement a transfer function is much quicker and completely nondestructive. With this method, the transfer function itself (a list of the display brightnesses for every original brightness value) is written to the hardware that produces the display, and is used to control the video output. This makes it very easy to change the display appearance.

Image Processing

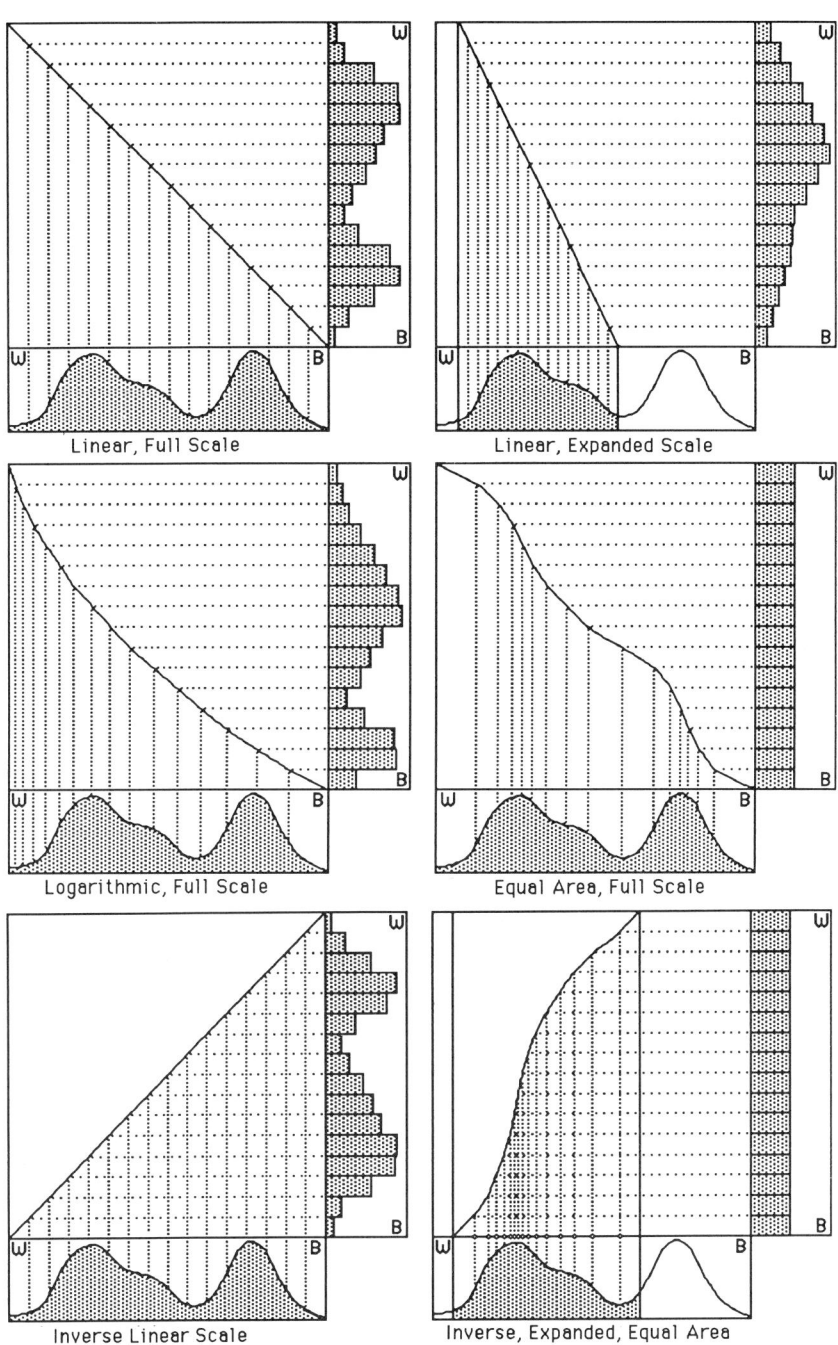

Figure 3-2: Transfer functions to alter the grey scale of an image.

Figure 3-3: Contour lines superimposed on an image.

LUT's are especially useful with color displays. The human eye is only capable of distinguishing a small number of grey levels, perhaps 20-30 under good conditions. But it can easily distinguish hundreds or thousands of colors. False color or pseudo-color displays are therefore a powerful way to accentuate small brightness variations. Normally, assigning color values to each brightness value is accomplished with an appropriate LUT.

For instance, if the original image has been digitized and stored with 256 brightness levels, and if the display is produced using a Red-Green-Blue (RGB) monitor, then it is common to use 8 bit DAC's for each color. This provides 256 possible levels for each of the primary colors, or a total of 256 cubed = 16 million possible color combinations, any of which can be assigned to each grey value in the image.

False or pseudo color displays are widely used, especially for astronomical images (such as the recent Halley's comet photos). They can emphasize gradients and communicate a great deal of information. When mis-used, however, false color can seriously confuse the eye, hiding the original information present in the image.

It is generally wise to use a smooth variation of color assignments versus brightness, to avoid or minimize this problem. This can be done by selecting a series of colors that follow some smooth line on the color chart, which is technically known as the chromaticity diagram. Some often-used color sequences are: a) a rainbow, proceeding from deep red through orange, yellow, green, cyan, blue, and violet, all with the same brightness value; b) a spiral starting at dark violet and proceeding around the diagram to red, while simultaneously increasing the white component of the color; or c) a color-temperature variation from dark red through orange and yellow to white, and beyond to blue-white.

The use of LUT's for display should be distinguished from the use of LUT's for input. Some ADC cards used for image digitization allow a transfer function to be specified which assigns a brightness to be stored to each possible measured brightness value. If the ADC range is greater than the available memory, this can be used to expand contrast either globally or selectively. This situation arises in few cases, however. For instance, with the scanning electron microscope (SEM), the slow speed of pixel

Image Processing

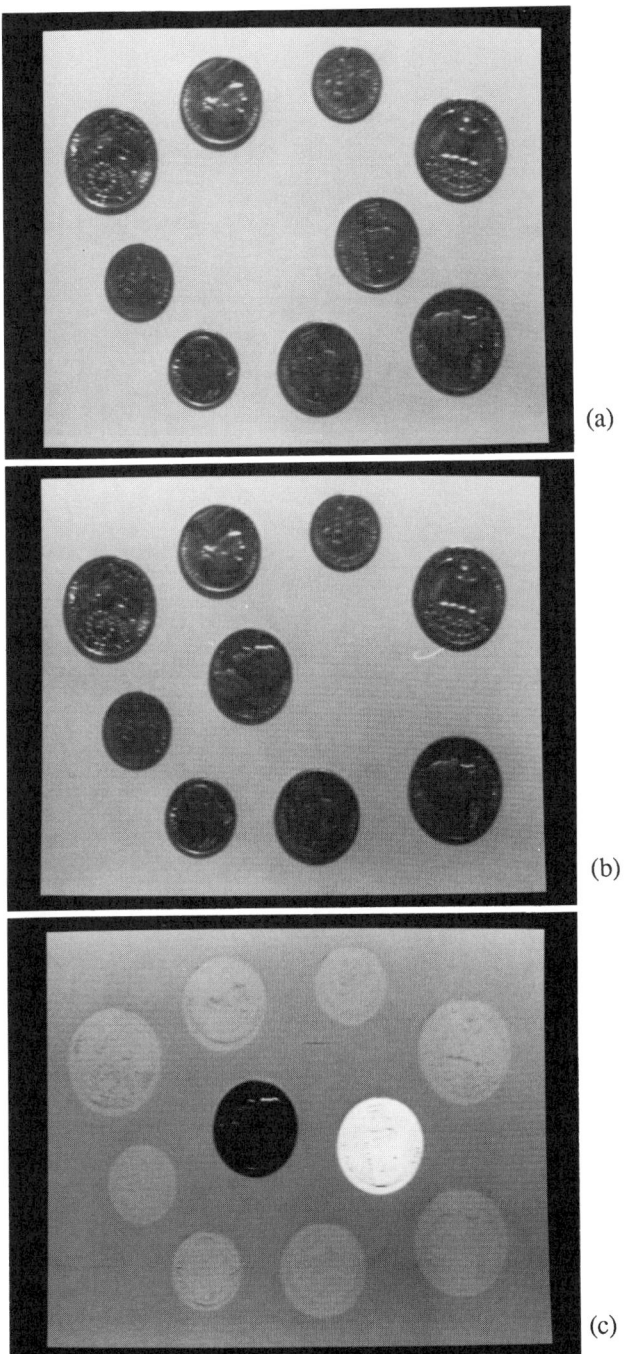

Figure 3-4: Subtraction of coin images showing difference: a) original, b) one coin moved, c) difference.

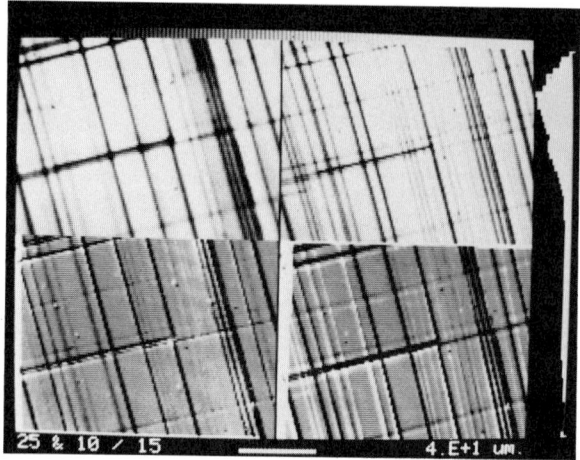

Figure 3-5: SEM Ebic images at 10 and 25 keV (top) and their ratios to a 15 keV image (below).

acquisition allows higher precision ADC's to be used, with 10 or 12 bit precision. From a quick sampling of the image, the approximate maximum and minimum values can be used to establish a LUT that produces 8 bit (256 level) stored brightness values that cover the actual image brightness range. However, with video rate image acquisition the ADC depth is typically 8 bit, and there is nothing to gain from an input LUT.

Another application for input LUT's arises in scanning densitometry. Density values are a logarithmic function of the light absorption,. With a linear detector to read the light intensity, a large number of bits (12 or 14) are required in the ADC to achieve adequate resolution in the high density regions. This produces much more information in the brighter areas than is needed. Using a logarithmic input LUT allows the brightness values to be directly converted to density, producing values that can easily be stored in 6 or 8 bits for subsequent manipulation, display and measurement.

Time sequences

One rather simple example of processing to extract a particular kind of information from images involves a time sequence. If two or more images of the same region are acquired, they can be compared to find differences that reveal motion of objects, or new or missing objects, or other similar changes that have occurred between the two images (Figure 3-4). Images representing information at different depths can be subtracted to reveal changes, producing higher depth resolution than any single image. Figure 3-5 shows an example of electron beam induced conductivity (EBIC) images in the SEM acquired at different accelerating voltages, which causes the electrons to penetrate to different depths, and the difference image showing lattice defects.

One of the major uses of this technique is for quality control of manufactured parts such as printed circuit boards. If an image of a "good" board is stored, then it can be compared to an image of each subsequent board to check for missing components (Figure 3-6). This comparison presumes that the two images are in registration, so that components, and the edges of the board and other marks, line up. If this cannot be done with mechanical alignment of the boards, then the "rubber sheeting" method described below may be used beforehand to prepare the images for comparison.

Image Processing

The comparison itself is usually performed by subtracting one image from the other. At each pixel, the brightness value of one image is subtracted from that of the other. It is necessary to deal with the possibility that the result of this subtraction may produce a negative value, which would lie outside the original range (e.g., 0...255) of brightnesses. The simplest solution to this problem is to divide the difference value by two and add 128, which will guarantee resulting values in the same range as the original images. This kind of problem often arises in dealing with the result of image processing, and is called rescaling. The alternatives to arbitrary rescaling are to clip the values that fall outside the legal range, or to use absolute values to deal with negatives, or to find the actual minimum and maximum values produced by the image processing (in this case, by subtraction), and then rescale the results with a transfer function to just fit the range of stored values as shown in Figure 3-7.

The drawback to this is that either the operation must be performed twice (once to find the limits, the second time to produce the values which are immediately scaled and

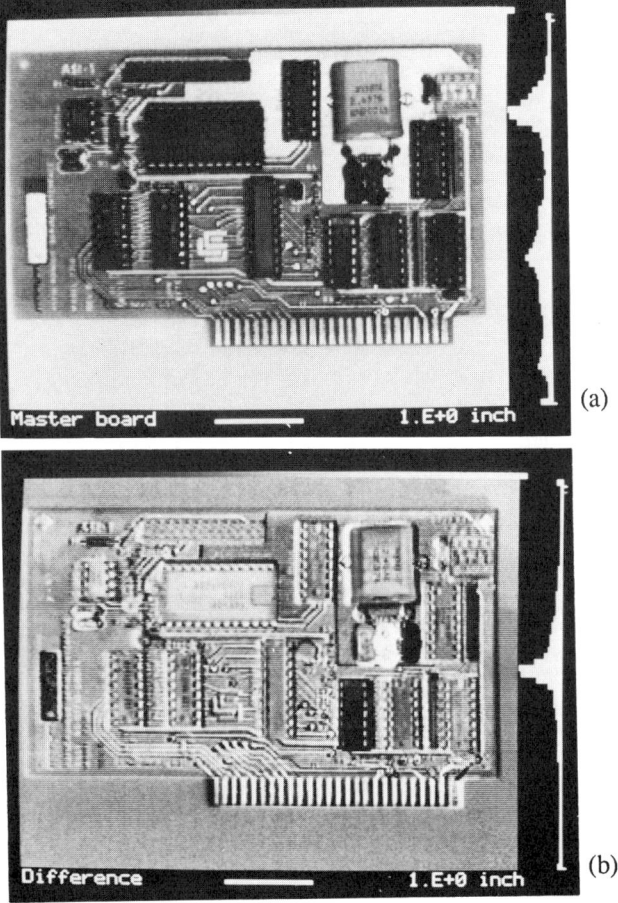

Figure 3-6: Subtraction to compare two printed circuit cards: a) master image, b) difference between master and another card with missing components.

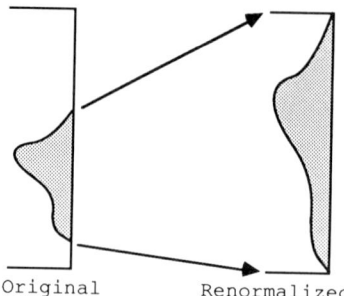
Figure 3-7: Rescaling the brightness range based maximum and minimum values.

stored) or to provide an intermediate storage memory that has much greater depth (more bits per pixel) to hold the larger values, so that they can be rescaled and stored in a separate operation. Both methods are used, depending on the details of hardware organization in individual systems. The second method seems initially to be more efficient, but it fact both methods require going through the data array twice. The second method is most often used when frequency domain processing (discussed below) is used (because this requires intermediate storage with considerably expanded bit depth) and/or an array processor is used (because this does rescaling very quickly).

Returning to the application to image comparison by looking at the difference, after two images have been subtracted, the difference is near zero (which is probably shown as an intermediate grey value) in regions that match, and close to white or black where objects have appeared or disappeared. Depending on which image was subtracted from which, black or white objects will either represent the new or missing features, or vice versa. The measurement of the objects, or perhaps their recognition, etc., can use simple discrimination methods (discussed in Chapter 5) to isolate the objects, since they are much brighter or darker than their surroundings. The result of the image processing is then the desired isolation of the objects of interest.

Differences between sequences of images are useful at all scales, and for more than determining motion. Satellite images using the Thematic Mapper (which records images at several wavelengths) can be taken of the same area at different times of the day to measure the change in the energy distribution from rocks as they heat and cool during the diurnal cycle. The results can be used to identify different rock types, and to map their distributions.

Time sequences of images are very useful for studying motion, as well. If objects can be located in each of two or more images, then their locations can be plotted as a function of time. This method is used, for example, to monitor the motion of animals in cages, or fish in tanks. However, this requires distinguishing the objects in the image, locating the object (for instance by its centroid), identifying which object is which if there are several in the image, and dealing with ones that may enter or leave the field of view.

These steps are all possible (we will discuss later ways to identify objects, for instance based on shape), but may be time consuming and indeed may not be necessary for some purposes. For some purposes, human digitization of points is a useful way to perform time sequences. The best known example is the use of manual marking of the location of elbows, hands, etc. from a sequence of video frames of an athlete to study patterns of motion, and perhaps suggest improved ways to swing a golf club, jump a hurdle, etc.

Image Processing

Another approach to time sequencing is known as "motion flow." In motion flow analysis, pixels are compared from one image to another (perhaps in a lengthy series of images) to see where each pixel has moved, without regard to what object the pixel belongs to. Usually, only those pixels which represent boundaries are compared, for efficiency. The boundary points are located using the same methods as used for edge location (discussed in the next chapter), or for stereo pair fusion (discussed in Chapter 11). In either case, the criterion for "interesting" pixels is that they are markedly different from some or all of their neighbors, and represent discontinuities in the image.

Given a moderately short list of points (at least as compared to the number of points in the entire image), it is possible to compare the points in the two images to match them up. This is usually based on the similarity of the brightness patterns around the points (the details are the same as those discussed for stereo fusion).

Once this has been done, each chosen pixel has a displacement vector associated with it, giving the direction and magnitude of its motion between the two images. This is in effect a new image, in which the pixel value encodes direction and/or magnitude (and both may be stored in two derived images). These images may be subject to further processing to smooth noise as discussed below. In any case, the resulting image is one in which the motion of boundaries, and thus presumably of objects, can be observed directly. Applications of this technique to study the motion of clouds or sea ice in weather satellite images have been especially successful (Figure 3-8).

Motion flow is also used to study the orientation of surfaces, by looking at the direction of variation of brightness. This is done by producing a derived image containing the direction and perhaps magnitude of the neighbor pixel that is most different from the original pixel. These vectors are quite different for different surfaces of objects, and thus are a powerful tool to distinguish these surfaces in an image. Like many other image processing operations we will consider, motion flow has some rather interesting and direct parallels to the functioning of the human visual system.

This should not be surprising, as the visual system is highly evolved to detect interesting things in images, and in the case of motion flow or image differencing, a little

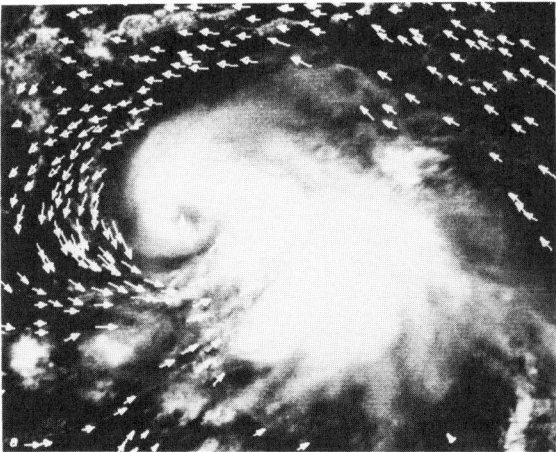

Figure 3-8: Motion flow applied to a satellite image of clouds in a tropical storm (Ballard & Brown, 1982).

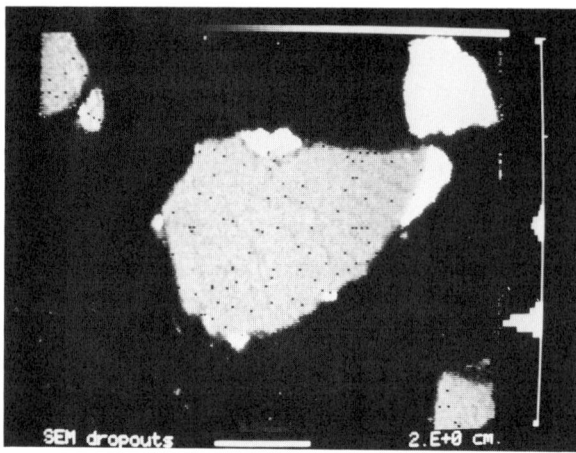

Figure 3-9: SEM image with "dropout" pixels due to noise in scan generator.

thought about the purposes of vision will confirm that in many situations it is things that move that are "interesting." The same criterion applies to many other applications, including surveillance and tracking.

Correcting image defects - averaging to reduce noise

The major use of image processing in conjunction with image measurement, is to correct specific defects in the as-acquired image. The most common problems are associated with noise in the original image, or nonuniform illumination or other systematic variation in brightness. There are several methods for dealing with each of these.

Noise may be either random ("stochastic") or periodic. Random noise is most familiar as "snow" in a TV image. Pixels are either randomly increased or decreased in brightness compared to their proper value, or may be "dropped out" as shown in Figure 3-9. Dropout pixels are missing from the acquired image, and are usually set by the hardware to either the darkest or brightest value available.

When it is practical to do so, most random noise problems can be solved by averaging together many image acquisitions. This kind of signal averaging depends upon the randomness of the noise itself, so that the problem pixels occur in different locations on each subsequent image frame. If this is the case, then by acquiring many images and adding them together, the effects of the noise will be averaged out and disappear (Figure 3-10). A few systems allow averaging together many images, for instance sequential SEM frames, in which each pixel total is separately divided by the number of values that were acquired. This effectively solves the dropout problem.

Of course, the use of image averaging is only appropriate when the image is stationary. Fortunately, this is often the case with microscope images, and even with remote sensing in some special cases. If sequential images are displaced in some known way, as by the motion of the spacecraft, then it is sometimes possible to acquire a series of images separately, use "rubber sheeting" to align them, and then perform the averaging. But the most common type of image averaging is carried out by directly adding the pixel brightness values as the images are acquired.

Image Processing

This requires having a memory depth (number of bits) large enough to hold the sum. If a 256 level (8 bit) value is produced by the ADC, then the use of two bytes (16 bits) per pixel (a practical choice dictated by the hardware organization of computer memory) will allow up to 256 images to be added together. It is unusual to require so many, as most averaging is limited to a much smaller number of frames.

This is partly because of the time required, but is primarily a recognition of the fact that averaging reduces noise only in proportion to the square root of the number of frames averaged. This fact comes from simple statistical considerations. What we are doing is trying to determine the mean brightness value for each pixel. The residual noise is measured by the standard deviation of that mean. The mean is determined as the total of all readings divided by the number of readings. The standard deviation is given by

$$\sigma = \sqrt{\frac{\sum B_i^2 - n \cdot B_{mean}^2}{(n-1)}}$$

where n is the number of readings averaged. The square root means that the precision, or standard deviation of the measurement increases only slowly as the number of readings is increased. Increasing the number of readings from 4 to 16 will halve the residual noise. To again halve it would take 64 readings. For this reason, a practical choice to limit the number of frames to be averaged is usually about 16. With a typical video signal (30 frames per second) and a flash ADC, this takes from 1/2 to several seconds (sometimes it

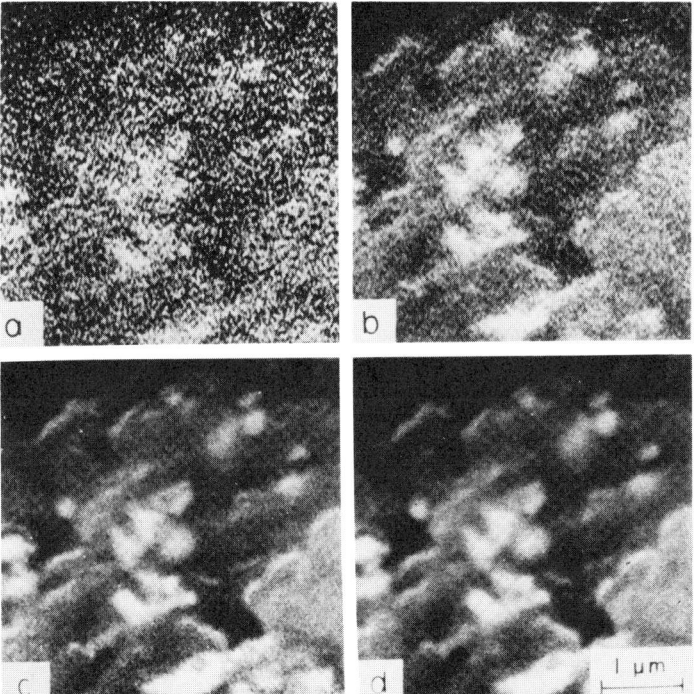

Figure 3-10: Effect of averaging a)1; b)10; c)100; d)1000 frames on signal-to-noise ratio in an SEM image.

is not possible to acquire every frame, depending on the hardware design and the need to store things in memory or synchronize on the video signal).

After the desired number of frames has been acquired, then the total brightness value is either divided by the number of acquisitions, or simply rescaled so that the minimum and maximum brightness values fit within the storage range for the image (usually this is a value of 0...255 for brightness, requiring 8 bits). If simple division by the number of frames is performed, there is a further advantage to using a number that is a power of 2 (2, 4, 8, 16, ...). This allows the division to be performed just by shifting bits in the computer memory, which is much faster than any real-number division operation.

The rescaling operation, mentioned before and often referred to in these image processing methods, simply finds the maximum and minimum values (MAX and MIN respectively) in the array, and then for each pixel calculates the new brightness as

$$\text{RANGE} \cdot (\text{VAL} - \text{MIN}) / (\text{MAX} - \text{MIN})$$

where RANGE is the desired full scale brightness range and VAL is the original brightness. It is very easy to accomplish this division and storage after the data have been acquired. With more difficulty, the averaging (at least to the extent of dividing by the number of frames acquired) may be done after each frame, with the result copied to a display device or second memory, so that a live display of the averaged image can be viewed, but usually the additive averaging method only produces a viewable image when the acquisition process is complete.

If a live image must be displayed during acquisition, or if less memory is available per pixel, then another choice is available. This is usually referred to as exponential averaging. One method for performing this continuous averaging uses exponential weighting, so that each frame is multiplied by a constant equal to $1 - k$ and added to k times the prior total. The result is that the most recently acquired image is more important that the older ones, and in fact the weighting is such that earlier images decline exponentially in importance. Ultimately, as limited by the number of bits available in the sum, they are completely forgotten. Because the maximum total remains fixed, it is usually possible to present a continuous display of the averaged image as it is acquired. Depending on the value of k, this results in improvement in signal to noise ratio until a saturation level is reached, with a minimal storage requirement. However, it is markedly inferior to the performance of a system in which the weighting constant is not a constant but varies with each frame. In that case, the result is identical to the averaging method. Figure 3-11 shows a comparison of averaging and the fixed constant exponential weighting method (Erasmus, 1982).

The hardware for accomplishing this "Kalman" averaging works by counting the frames and adding in decreasing proportions of the new image. For the nth frame, $1/n$ times the new image and $(n-1)/n$ times the stored image are added together. This can either be done digitally with a fast processor, or an analog signal produced by a D/A converter can be combined with the incoming signal using an op-amp circuit.

When real time averaging is done, the appearance of the image is very interesting. Initially noisy images improve continuously (although with decreasing rate of improvement because of the square root phenomenon mentioned above). It is usually not wise to rely on the visual appearance of the image to judge when it is suitable for measurement however, as the human eye is very tolerant of image noise itself, and images that "look" acceptable may still present problems for locating edges or measuring objects.

Image Processing

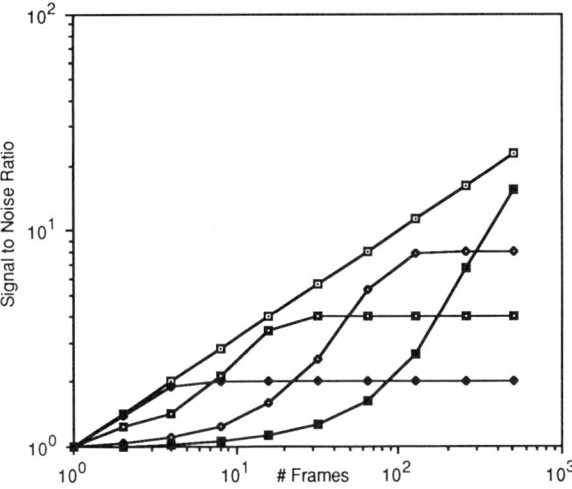

Figure 3-11: Improvement in Signal to Noise ratio with number of frames for averaging (straight line) and exponential weighting with several constants.

Reducing noise in a single image

In many cases, averaging of many successive frames from a live image source is not possible because the image is not constant or stationary, or it may already have been acquired and stored. If random noise (as defined above) is still present, then the image may be directly processed to improve it (Castleman, 1979; Pratt, 1978; Rosenfeld, 1982; Rosenfeld & Kak, 1982). By analogy to other signal processing methods, we might expect that "smoothing" would be a useful tool in this effort, and although it is not actually the preferred method, we can profitably begin by describing it. Within the image, each pixel has a brightness value. It also has 8 neighbors in the most common case of a rectilinear raster scan (for the less common hexagonal pixel array, there would be six neighbors, and in some cases with the rectangular array it is useful to think only of the four neighboring pixels above, below, left and right of the central one, which are somewhat closer than the diagonally oriented ones).

In any case, the information in the image is usually of significantly greater size than the individual pixels, so it may be expected that neighboring pixels would usually be a part of the same surface or object. If this is the case, then averaging can also be performed using the neighbors. Since many of the operations that will be described in this and the next few chapters make use of neighbor pixels, it will help to have a handy way to identify them. We will call the central pixel P_0, and the neighbors $P_1...P_8$ as shown in the diagram.

```
1  2  3
8  0  4
7  6  5
```

With this notation, the simplest way to average the information in neighboring pixels to smooth noise in the image would be to replace the value of each pixel by the average of it and its neighbors, or

$$P_0' = \frac{1}{9} \cdot \sum_{n=0}^{8} P_i$$

Note that this must be done simultaneously for all of the pixels in the image, and that the "old" brightness values of all pixels are used to compute the sums. This is most often accomplished by using two image memories, one for the original and one for the derived image, although is it possible to use less memory by copying only a few image lines in auxiliary memory space and writing the derived image back into the same space.

The simple averaging just described is usually not used. Center-weighting can be used to produce a Gaussian smoothing, for instance by calculating

$$P_0' = (1/16) \cdot \{ 1 \cdot (P_1+P_3+P_5+P_7) + 2 \cdot (P_2+P_4+P_6+P_8) + 4 \cdot P_0 \}$$

Other weighting factors can also be devised. In effect, they allow making the original pixel value more important, and neighboring pixels less important in proportion to their distance from the central pixel. In the same way, averaging over greater distances is also accomplished. We refer to the operation described above as using a 3x3 kernel of weighting factors, which would usually be written as

$$\begin{vmatrix} 1 & 2 & 1 \\ 2 & 4 & 2 \\ 1 & 2 & 1 \end{vmatrix} \cdot 1/16$$

Other sizes or weights are written similarly (Edwards, 1982), so that for example 5x5 or 7x7 smoothing kernels might be written as shown below. Notice that the total is generally a power of two, which speeds up the calculation just as it did for averaging several images.

$$\begin{vmatrix} 1 & 2 & 3 & 2 & 1 \\ 2 & 3 & 4 & 3 & 2 \\ 3 & 4 & 4 & 4 & 3 \\ 2 & 3 & 4 & 3 & 2 \\ 1 & 2 & 3 & 2 & 1 \end{vmatrix} \cdot 1/64 \quad \begin{vmatrix} 1 & 1 & 2 & 2 & 2 & 1 & 1 \\ 1 & 2 & 2 & 3 & 2 & 2 & 1 \\ 2 & 2 & 4 & 7 & 4 & 2 & 2 \\ 2 & 3 & 7 & 12 & 7 & 3 & 2 \\ 2 & 2 & 4 & 7 & 4 & 2 & 2 \\ 1 & 2 & 2 & 3 & 2 & 2 & 1 \\ 1 & 1 & 2 & 2 & 2 & 1 & 1 \end{vmatrix} \cdot 1/128$$

The fact is that smoothing with these weighted functions is rarely used, because it degrades the sharpness of real edges or discontinuities in the image. In addition to averaging noisy values, smoothing will average together values that really are different, where they are adjacent. This means that the boundaries of surfaces and objects will be "smeared out" and the brightness step across them will be reduced in magnitude. Both of these things are exactly contradictory to our usual purpose, which is to find those boundaries for measurement or identification purposes. Therefore, some other approach to smoothing is usually preferred (Nagao & Matsuyama, 1979; Chin & Yeh, 1983).

The best method to remove random noise from an image is to apply a "median filter" (Huang, 1979; Russ & Russ, 1986). This also works with each pixel and its neighbors, usually in a 3x3 pattern. The brightness values of the 9 pixels are ranked in order, and the median value (the fifth brightest in this example) is used to replace the original brightness for the central pixel. Again, the process is carried out simultaneously for all pixels in the image, so that the values that are ranked are the original ones.

Median filters may be applied repeatedly to an image until there are no further changes. They will smooth out noisy but grossly uniform areas, because isolated random noise pixels will rarely have a median brightness value and will be replaced by one of the

Image Processing 47

neighbor values. Median filters will even correct images for dropouts - entirely missing pixels that sometimes result from uneven or noisy scan generators (Bovik, 1987). At the same time, true discontinuities such as surface edges or object boundaries will not be broadened, shifted or reduced in contrast (Figure 3-12).

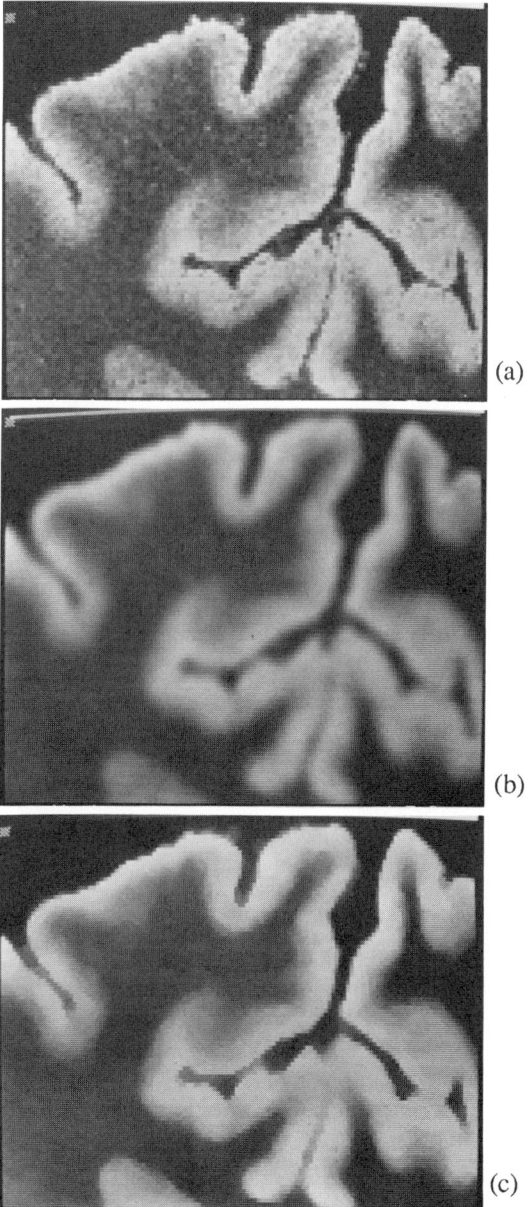

Figure 3-12: Comparison of median filtering and smoothing applied to an autoradiogram of a cat brain. a) original image; b) smoothed (note loss of sharpness and decrease of contrast at edges); c) application of a median filter.

While it is an ideal processing tool in terms of its results, the median filter does present one difficulty for image processing - it is very time consuming. The ranking procedure is not amenable to using an array processor, and is computationally demanding. Also, the operation may iterate an unknown and unpredictable number of times before the pixel values all remain constant. The time needed to apply a median filter is often used as a measure of the speed of an image processing system, and times ranging from a few seconds to more than a minute are not uncommon.

Frequency space

The operations just described, application of a smoothing kernel and median filtering, are "spatial domain" operations because they deal with the pixels and their neighbors in their physical or spatial arrangement within the image. The other principle class of image processing operations take place in the "frequency domain". A complete derivation of the Fourier transform, or other frequency transforms such as the Hadamard, Cosine, and other methods, is beyond the intended scope of this text. The interested reader is referred to the references for this background discussion (Pratt, 1982; Erasmus et al., 1980; Hashimoto, 1986; Saxton et al., 1979).

Processing of images in the frequency domain is most often used to remove periodic noise, blurring or other artefacts that can be traced to the image source. The most familiar example of this involves the use of the Fourier transform. For an image represented by a two-dimensional array of brightness values $f(x,y)$, it is possible to compute the transform $F(u,v)$ as

$$F(u,v) = \int_{-\infty}^{+\infty}\int_{-\infty}^{+\infty} f(x,y)\, e^{-2\pi i (ux + vy)}\, dx\, dy$$

The transform contains all of the information in the original image, which can in fact be recovered by performing the same operation with f and F interchanged. The u and v variables represent frequencies, and the transform codes the information in the image in terms of the frequencies in the original. The values of the F function are complex, and may either be expressed as a real and an imaginary part or more often as a magnitude and a phase. The image shown as a representation of F is usually the magnitude value.

The chief advantage of processing the transform image rather than the original is that it simplifies convolutions (the point-by-point application of a neighbor operator to combine every pixel in the original image with its neighbors). This means that it is only necessary to multiply the transform of the image by the transform of the operator, pixel by pixel, which takes much less computation (neglecting the original cost of transforming the image to frequency space, of course).

Mathematically, a convolution (the application of one of the 3x3 or larger "kernels" to each pixel in an image) can be written as

$$g(x,y) = \sum_{i=1}^{m}\sum_{j=1}^{n} h(i,j) \cdot f(x-i, y-j)$$

where $f(x,y)$ is the original image, $h(m,n)$ is the kernel, and $g(x,y)$ is the resulting image. In frequency space, the same operation is simply

$$G(u,v) = H(u,v) \cdot F(u,v)$$

where F, H and G are the frequency transforms of the image, the kernel and the result. The derived image $g(x,y)$ is then recovered by an inverse transformation. The H function

Image Processing

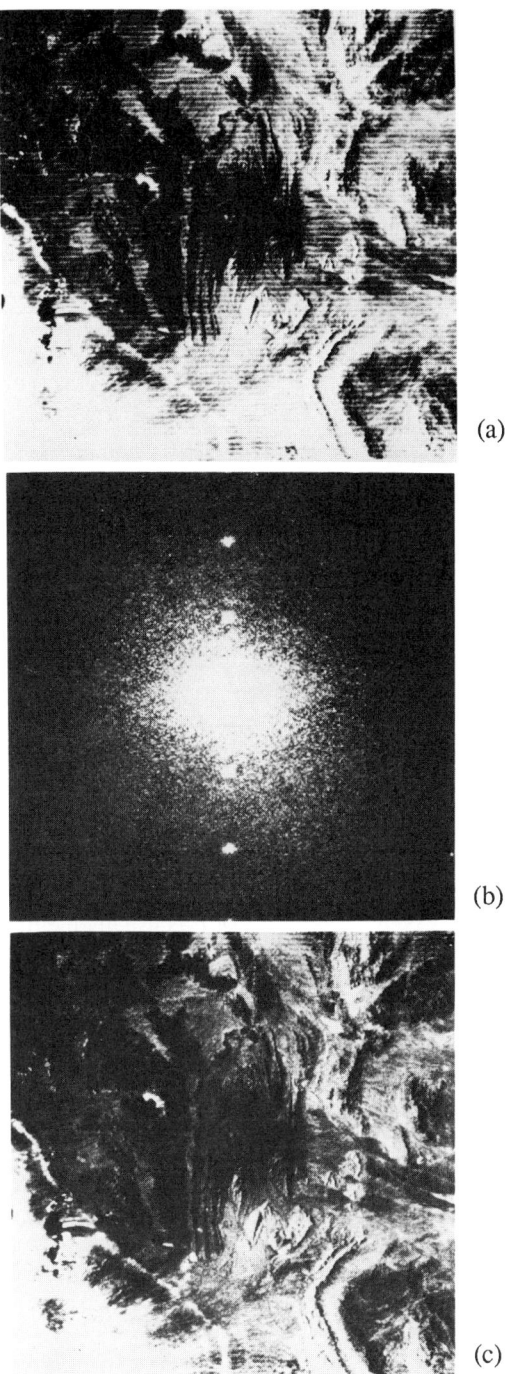

Figure 3-13: Application of a Fourier transform to remove periodic noise: a) original image, b) 2-D transform, c) filtered and re-transformed image.

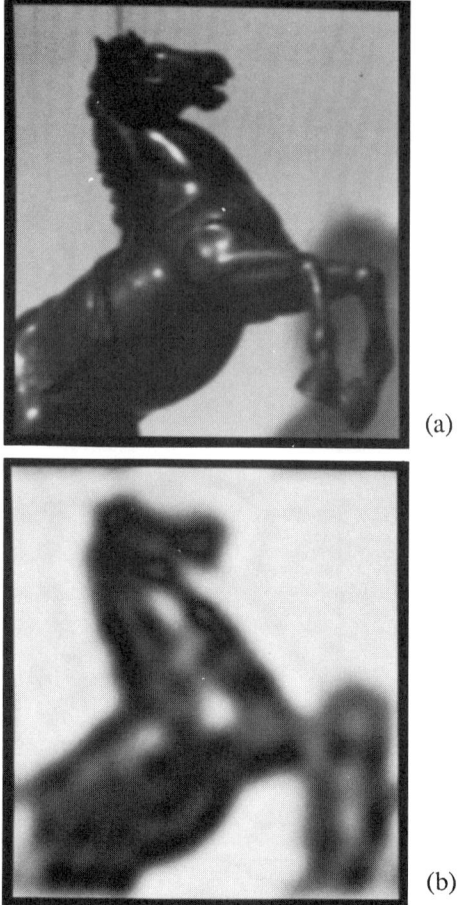

Figure 3-14: Image of a statue and the result of frequency filtering: a) original; b) low-pass filtered; c) high-pass filtered; d) 2-D transform (shown as an isometric view).

may be obtained either by calculation or measurement. When the source of image blurring or other defects is due to the image acquisition hardware, it is often possible to measure it directly with a point source image. This produces a blurred image containing the image defects, whose frequency transform gives the H function. In other cases, the H function can be calculated from the image itself, or various standard functions are used for particular purposes.

For instance, we saw before that it is possible to smooth an image by adding together the brightness value of each pixel and its neighbors (times appropriate weighting constants). In frequency space the same result is achieved by multiplying the magnitude of the $F(u,v)$ function by a filter H. The filter consists of a circularly symmetrical array of values that decrease with radius from the $u = 0, v = 0$ point. The result is to dampen the higher frequency information in the transform. When the product is re-transformed to the spatial domain, the recovered image has been smoothed.

The filter shapes correspond to different sets of weighting values in the neighborhood operators used in the spatial domain, but because they are expressed directly in terms of

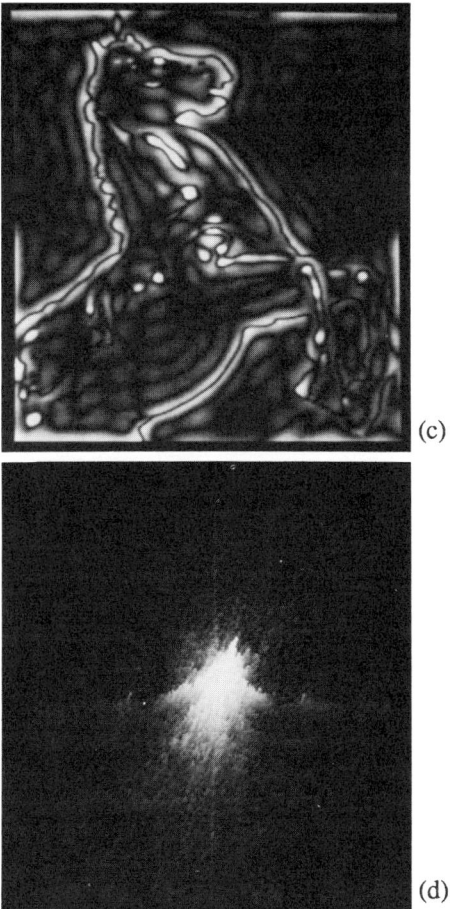

(c)

(d)

frequency, the distance between points in the spatial domain, it is often easier to understand them and predict what they will do (Mastin, 1985). It is also possible to keep high frequencies and remove or suppress low frequencies to "sharpen" images. We will see how this is done in the spatial domain in Chapter 4. In the frequency domain, if the artefacts and blurring produced by the imaging system are known, then from a transform of just the artefacts (produced by imaging a point), an optimal filter to remove those effects can be designed. When this information is not available, it is possible to estimate the optimal filter from the image itself. Various assumptions about the random nature of noise in the original image lead to the Wiener or inverse filters, which are widely used.

The design of special filters to remove certain frequencies allows noise superimposed on an image to be removed selectively, and this has been of great importance in processing images transmitted from satellites and space probes. Unique optimized filters may be customized for each image, in these cases. Many filters are symmetrical, but if extraneous noise in the image runs only in a particular direction then a filter can be designed to specifically eliminate it. The example in Figure 3-13 shows an image and its Fourier transform, with the retransformed result after applying a filter to remove periodic noise.

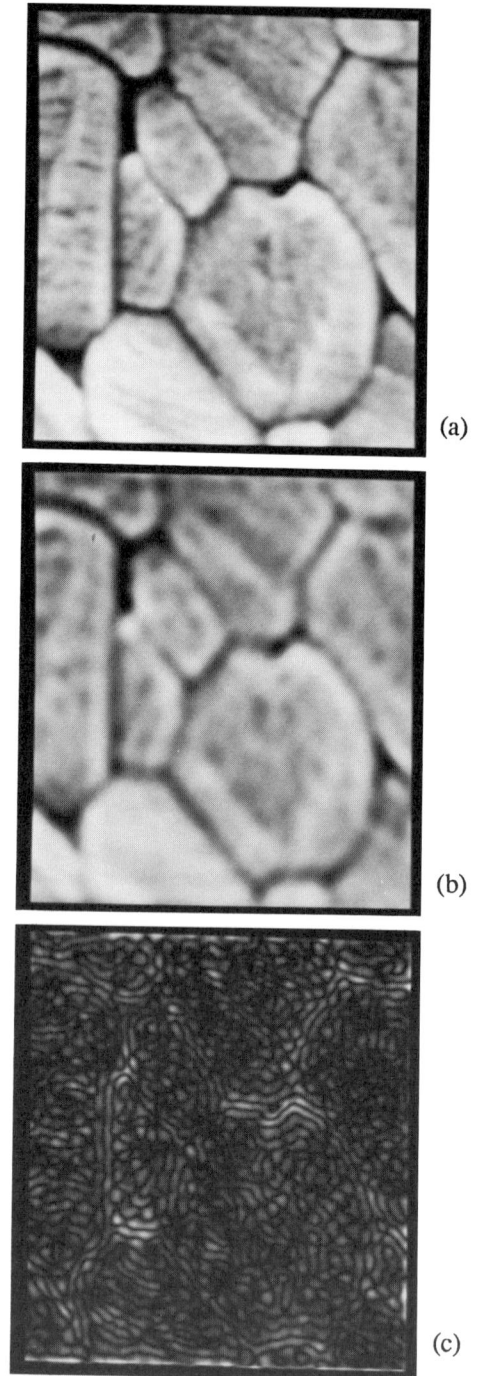

Figure 3-15: SEM image of thermally etched alumina the the result for frequency filtering: a) original; b) low-pass filtered; c) high-pass filtered.

By selectively removing or diminishing values in particular regions of the frequency space, high or low frequency information can be suppressed in the re-transformed image. Usually this is referred to as a high-pass or low-pass filter, depending on which frequencies are unchanged. In its simplest form, this method uses an annular aperture to cut off the low frequencies or a circular one to cut off high frequencies. Better results are obtained by using more gradual transitions in value. Figures 3-14 and 3-15 show the results of applying these filters to images, first a picture of a real-world object (a cast statue of a horse), and then a microscope image (the thermally etched grain structure of a ceramic viewed in the SEM.

Fourier transform methods are especially useful for enhancing periodic structures in images. Figure 3-16a shows a high-resolution TEM image of the atomic lattice in a mineral (potassium feldspar). Applying a filter to the frequency-space transform that passes only intermediate frequencies (an annular filter that blocks both the central portion of the transform and the extremities) produces the result in Figure 3-16b, in which the atom positions are more clearly defined. Using a filter that selects only a symmetrical pair of spots in the frequency transform is equivalent to selecting lines with a particular spacing and orientation, and produces the image of Figure 3-16c in which the edge dislocation is clearly evident (Zheng & Gandais, 1989). This technique must be used with care, since the selection and positioning of the filter defines the resulting retransformed image, and can eliminate real information or create artefacts.

The key to the practical application of frequency domain processing is the availability of fast and efficient algorithms to perform the transformation (Johnson & Jain, 1981). For the Fourier transform, the FFT algorithm is well known. It can be straightforwardly applied to images because the integrations (or summations in the discrete case) in x and y are separable. In other words, the FFT of each line in the image is obtained first, producing an intermediate result. Then the FFT of each column in this intermediate image is obtained, to yield the final transformed image.

The Fourier transform is not the only way to transform an image, nor even necessarily an optimum one. Other sets of basis functions can be used instead of the sine/cosine terms of the Fourier method. A few of the more popular alternate methods are the discrete cosine transform (in which only cosine functions are used), the Hadamard transform (in which the set of Walsh functions shown in Figure 3-17 are used), the Hankel transform (in which Bessel functions are used), and the Karhunen-Loéve

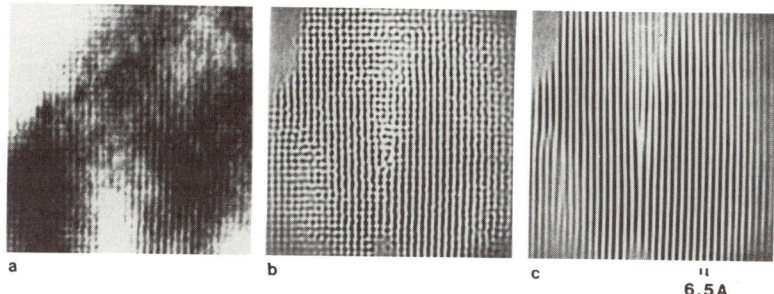

Figure 3-16: Locating regularly spaced structures with a frequency transform: a) TEM Image of (001)[110]/2 dislocation in potassium feldspar observed in the [100] direction; b) result of application of annular filter in frequency domain; c) retransformation of two symmetrical spots selected from frequency domain.

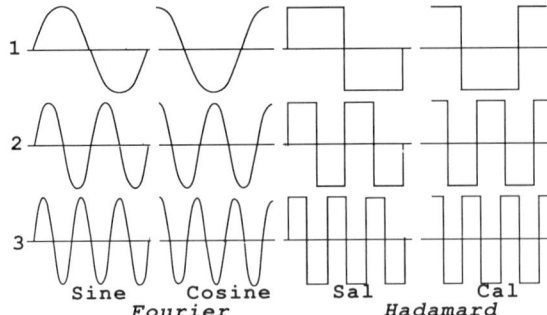

Figure 3-17: The trigonometric functions *sin* and *cos* used in the Fourier transform compared to the Walsh functions *sal* and *cal* used in the Hadamard transform. The advantage of the latter is that they do not involve any floating point arithmetic, and so speed up the calculation process.

transform (in which eigenfunctions are computed for each image, based on its statistical properties).

The Karhunen-Loéve transform can be shown to be optimum in a particular sense: the maximum amount of information in the original image is contained in the fewest number of terms of the expansion of the transform. This makes it especially suitable for image compression, although all of the methods are used for this purpose (the others are nearly as good if the images are random, and can be computed with great efficiency using well understood algorithms). Since the transform image is as large as the original, it would not seem to offer any advantages for storage or transmission. However, if the higher order terms can be discarded with little loss in image quality, then it becomes possible to reduce the amount of information that must be sent to allow reconstruction of the image. Some of the deep space probes return images using less than 2 bits per pixel using these image encoding techniques, which are then restored to full high-quality images.

These transform methods are also used in reconstruction of 2- and 3-D tomographic images from projection. This is discussed in Chapter 13, but the methods for obtaining the Fourier transform to move from the spatial domain to the frequency domain, and the reverse, are identical.

What the transform accomplishes is to map all repetitive arrangements of information in the image into a new image, so that the brightness of locations in the new image reflects the spacing and orientation of the original repetitive points. This allows the repetitive information to be separated from the non-repetitive, by applying a filter as discussed above. There are two ways that this can then be used, of course. One is to recover repetitive information from a noisy image (for instance, images of atoms viewed in the electron microscope are regularly arranged in a crystal structure, but the images are often very noisy because only a few electrons contribute to the image; frequency domain processing can produce smoothed images of the atomic arrangement as shown in Figure 3-18). The second is to remove any periodic or repetitive noise in the image (an example is interference bands that often appear in transmitted images).

One of the more unusual applications of Fourier analysis is to determine the resolution of electron micrographs. The power spectrum of the transformed image exhibits a drop at a frequency whose inverse is the smallest feature spacing that can be

Image Processing 55

discerned. Measured on the entire image, this value is more meaningful than finding individual points on the image and measuring their spacing (McCarthy, 1987).

Frequency transforms are costly in both time and memory. For a typical 8 bit deep original image, the transform (which consists of both a real and imaginary part) will require at least 32 and preferably more bits. Unless special arithmetic processors are used, the large number of trigonometric function values requires take a long time to compute or look up in a table. Various optimized forms of the two-dimensional fast Fourier transform have been published, and can be used in large or small computers, but are not fast. Adding special hardware is required for most serious use of these techniques.

For some particular types of images, the repetitive nature of the information makes it possible to achieve results equivalent to frequency space transforms using superposition. For example, in the example of the atomic lattice image viewed in the TEM, it seems reasonable to expect the neighborhood around each atom to be the same. Instead of selecting the repetitive information from the frequency domain, it may be easier to regard the image as a large number of individual images, and to superimpose them to obtain an average image with less noise. In some cases, this averaging includes not only the many copies of the various sites, but also all of the rotationally equivalent positions of the images. This approach can produce quite noise-free images (Figure 3-18). The major

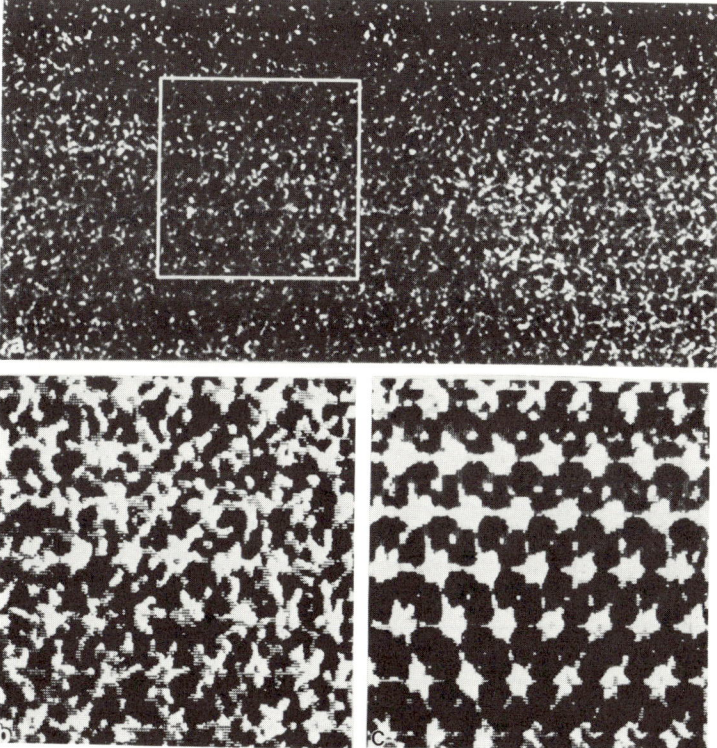

Figure 3-18: example of TEM high resolution image of a biological macromolecule (a) processed using Fourier transform filtering (b) and periodic image smoothing by correlation (c).

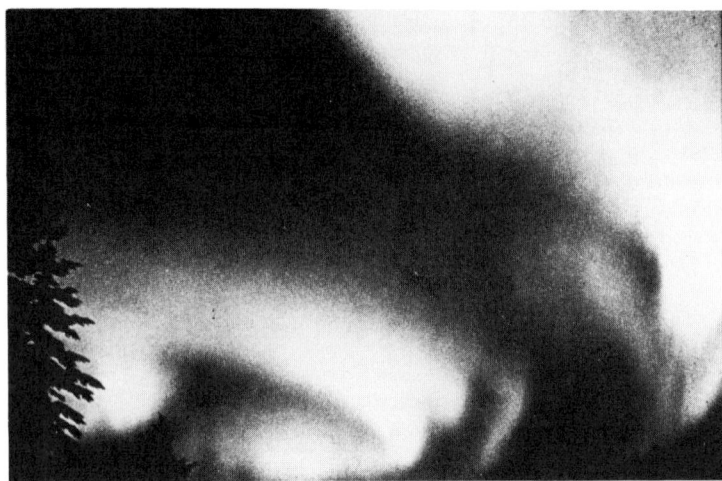

Figure 3-19: The aurora borealis, an example of an image in which color has direct meaning.

danger is that if a repeat distance or direction or a rotational symmetry is assumed to be present when it is in fact not, the averaging process will still yield a convincing, noise free image of artefacts. Of course, this also applies to filtering in frequency space.

Color images

So far, the discussion of noise reduction in images has considered monochrome images, but this restriction is not essential. For color images, most of the same techniques can be applied to the individual red, green and blue image planes separately, with good results. It should be noted, however, that noise in a color image is much more distracting than in a black and white one. This can be verified by trying to watch a "snowy" TV program caused by poor reception. In color it may be virtually impossible to watch, but by turning off the color, the black and white image may be much more acceptable. This is because the human eye does a pretty good job of ignoring noise in images when the noise causes only a brightness variation (we also cheerfully ignore rather extreme shading variations on surfaces, including shadows cast by other objects). But if there is independent noise present in the separate R, G and B image planes, this results not in a brightness variation in the final image, but rather a change in the color.

The easiest way to visualize this noise is as a change in the direction of the color vector at noisy locations in the image. While individual filtering of the separate image planes can reduce this effect, it is often better to carry out the filtering in a different space, in which the direction of the color vector itself is subjected to a median filter. This requires first converting the image (normally stored in RGB format) to another form, such as hue/intensity/saturation (HIS). The result, however, is a markedly superior filtering of the noise.

One example of the use of this technique is with extremely sensitive color cameras, in which three SIT (silicon-intensified-target) tubes are used for the RGB signals. These very high gain tubes are inherently very noisy, and the noise is uncorrelated. They are used for such high sensitivity work as studying the aurora borealis (Figure 3-19). The original video images are significantly improved by computerized noise reduction, and this can be accomplished without changing the colors (which are studied because they are indicative of the particular ions involved in the electron showers).

Shading correction

In an ideal image, features which represent the same thing (a surface at a particular orientation, or a particular type of object) should have the same brightness no matter where in the image they lie. This ideal situation can be realized in some cases, most notably in the light or electron microscope if it has been carefully aligned. In many cases, however, the illumination is nonuniform or strikes the object at an angle. In many more cases the video camera or other device used to record the image is itself not uniform in response. For many inexpensive vidicons, this takes the form of a brighter signal at the center, and darker response from the corners of the field of view. This exacerbates the common illumination defect in images called vignetting, in which light from the periphery passes obliquely through lenses and is more strongly absorbed, also making the edges and corners darker. It has already been mentioned that many solid-state CCD cameras have pixel-to-pixel variations in sensitivity due to the construction of the individual diodes, so that the response to a perfectly uniform grey image is not uniform.

Another class of problems may arise from the sample itself. Many specimens viewed in the microscope by transmitting light (or electrons) through them are not uniform in density or thickness. This produces an overall shading of the background brightness. When specimen surfaces are viewed, they may not be perfectly flat. The most common example is spacecraft images of planets, which are roughly spheroidal and therefore have a different average brightness in different locations depending on the direction of the sun. All of the sources of variation shown in Table 3-1 can frustrate attempts to recognize, select or measure edges or objects for measurement.

When these effects are present, it is not possible to directly discriminate features to be measured based on their absolute brightness (by setting lower and upper brightness limits to define the features) as shown in Figure 3-20. It is also impossible to relate the brightness (or color) information to the properties of objects in the images, such as density. There are three basic approaches to deal with this problem. We will consider two of them here, and the third in the next chapter under the general topic of using derivative images to find local edges or discontinuities, regardless of such gradual background variations.

The first approach, which is by far the fastest and easiest when it can be applied, is to measure a control or background image. In some cases, this can be done by acquiring an

Table 3-1. Sources of nonuniform image brightness

Nonuniform illumination
    Microscope misalignment
    Angled illumination
    Variation in distance to source
    Vignetting
Nonuniform detector
    Edge-Center variation
    Pixel-to-Pixel variation
Specimen variations
    Density or thickness
    Non-planar surfaces

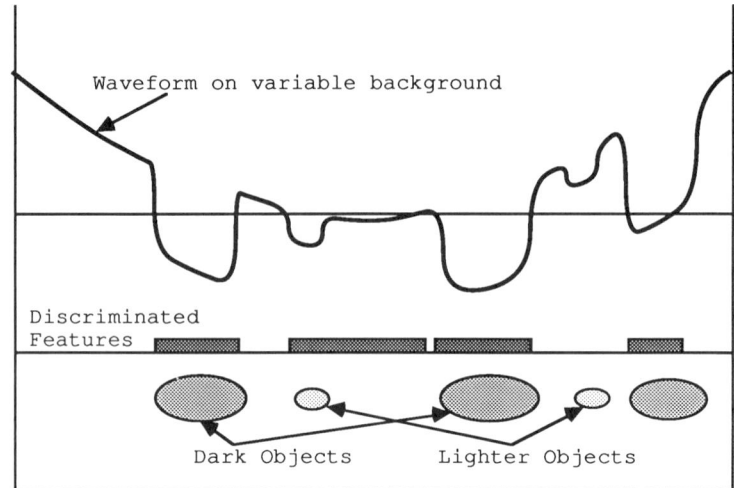

Figure 3-20: Two different kinds of feature with different intrinsic brightnesses, superimposed on a non-uniform background level, and the erroneous results obtained when fixed brightness threshold values are set for discrimination.

image of a reference grey card (for surface illumination) or of the light source itself (for the case of transmission microscopy, for example). In either case this image is stored and available to level subsequent images as they are acquired.

The levelling can be accomplished in one of two ways. Sometimes it is appropriate to subtract the background image, pixel by pixel, from the new one, and sometimes the ratio of the two values is required. For the case of subtraction, the mechanics are just the same as the differencing of two sequential images, discussed earlier. The difference image requires rescaling, either using the treatment discussed before, or sometimes simply by adding to each pixel brightness value the difference between the mean brightness of the acquired and background images, which also serves to bring the difference into the range of values that can be stored.

Ratioing, or dividing one image by another, produces values with a range of magnitudes that are real numbers. Special attention is needed to avoid division by zero, since zero brightness, or black is a legal brightness value. This can be handled either by adding one to the divisor, or replacing zeros by ones before dividing. The reason to choose subtraction or division depends on the how the image is obtained. Most film and some video cameras have a logarithmic response to changes in brightness, while some cameras are linear. If the device is linear, then ratioing or division is the proper model to remove the background signal, while in the logarithmic case, the division of the actual brightnesses is accomplished by subtraction of the recorded logarithmic signal.

Storing a control background image is practically the only way to deal with such problems as cell-to-cell variation in CCD cameras, and some very complex nonuniformities in illumination as can be encountered in the electron microscope (for instance if the filament is not well centered or fully saturated). In some cases (again, the CCD camera is a good example) the background image can be acquired once, very carefully, stored, and re-used for the life of the camera. In other cases (such as the microscope illumination problem) the background image is of only brief utility. But in either case it represents the most direct and exact way to cancel out the variations present.

Fitting backgrounds

Sometimes it is not practical or even possible to measure the background image. The most common example is that of curved surfaces or transmission samples of nonuniform thickness. No "dummy" specimen can be used to obtain an identical image without any features, which could be used as a background. In some other cases, such as vidicon response, the degree of nonlinearity is dependent on the overall scene illumination and it is not practical to duplicate the setup used for measurement. In these cases, a second approach to background levelling may be used.

This relies on the fact that most images consist of a great deal of background surrounding the features of interest, and that the variation in the background brightness usually varies smoothly as a function of position. Furthermore, depending on the particulars of the situation, the form of the variation is likely to be known. For the nonuniform sample thickness, for instance, the brightness should vary linearly or logarithmically in one direction (corresponding to a wedge-shaped sample). A good general model for background variation is a polynomial expressing brightness as a function of position in the general form

$$B = a_0 + a_1 \cdot x + a_2 \cdot x^2 + ... + b_1 \cdot y + b_2 \cdot y^2 + ... + c_1 \cdot x \cdot y + c_2 \cdot x^2 \cdot y + ...$$ and so forth.

In practice, the quadratic form of this expression (all terms in x, y or their cross products with power less than or equal to 2) usually is adequate to fit the background in many real situations. Obtaining the best values for the coefficients is a fairly straightforward task. If background points are selected on the image (more points than the number of coefficients, well dispersed and sampling all parts of the image), then a set of coefficients can be obtained by linear least squares fitting, solving the simultaneous equations by matrix inversion. Figure 3-21 shows the application of this levelling to a TEM image.

The only time consuming part of the job is to locate suitable background points to perform the fit. Sometimes this must be done manually, with the user selecting the points (using some pointing device such as a light pen, mouse, graphics tablet, etc.). In other cases, an automatic selection of the points can be made by using a derivative test. Regions of the image can be selected using a Laplacian or other nondirectional derivative (discussed below in detail as an image sharpening tool). All of these regions which satisfy some arbitrary criteria (such as approximate brightness, or size) are then considered to be background regions. Their centroid locations, maximum gradient value, and/or mean brightnesses are then used to perform the fit. It is typical to use 20-40 points for background fitting, in order to get robust values for the coefficients.

Another possible way to produce a background image should be mentioned for the sake of completeness. In some rather specialized cases, it is possible to apply extreme smoothing to an image to produce a background. This generally only works when the image is predominantly background with small objects present, and the smoothing kernel is quite large (e.g. 21x21).

A closely related method, applicable in similar situations, uses a rank filter (of which the median filter is an example we have already encountered). The rank filter is used to search the neighborhood of each pixel (again in a region that is larger than any object of subsequent interest) to find the brightest or darkest (depending on whether the objects are darker or lighter than the background, respectively) pixel. The value of this extreme brightness is saved in place of the original. The result is a smooth background which can be used to level the original image (Bright & Steel, 1986).

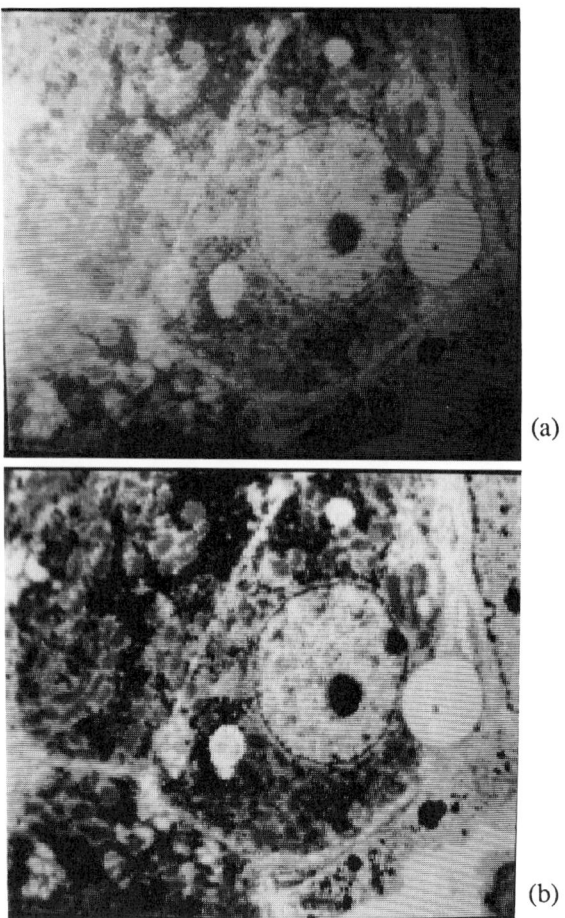

Figure 3-21: Original TEM image (a) of a biological section of nonuniform thickness, and the levelled image (b) with uniform brightness for similar structures.

There are other approaches to shading correction. One, which was originally developed when computer power was relatively scarce, slow and expensive, takes place in the analog domain (that is, before the image is digitized). If the signal from the detector (camera, etc.) is passed through an appropriate RC filter circuit, a derivative signal is formed. Combining this with DC restoration produces an image in which only large brightness discontinuities are permitted. The method works only along the direction of scan lines, does not deal well with images in which the size of features varies, and is little used now.

A more modern approach uses a stored background image, either measured or fitted, that may be of much lower resolution that the actual acquired images. These stored values are sent to a DAC (digital to analog converter) whose output is subtracted from the incoming signal voltage using an operational amplifier. The reduced number of "pixels" used in the stored background image is less a function of memory requirements (memory is now pretty inexpensive) and more due to the behavior and response of the DAC and its

Image Processing

associated electronics. This method is much less accurate and less flexible than the digital approach, and cannot deal with variations on a smaller scale than the stored background image, but has the advantage of working in "real time" as the image is acquired.

Rubber sheeting

Mention has been made several times of the occasional need to align multiple images to get them in registration before they can be combined in various ways or used for measurement. In addition, images are obtained of curved surfaces (particularly ones such as planets viewed from spacecraft) or inclined ones. An example of the latter is that many locally flat surfaces such as integrated circuits may be inclined to produce good quality scanning electron microscope images, which are to be measured to determine the width of exposed lines and so forth. In all of these cases, a method is desired to stretch the image so that it becomes the ideal view, which is a flat map surface viewed normally from above, as shown in Figure 3-22.

The method used to perform this is called rubber sheeting or image warping. To accomplish an accurate rectification of an acquired image, it is necessary to know what the shape of the actual surface is, or what the viewpoint of the camera is with respect to the surface. The two ways of obtaining this information are either to know it from independent sources (the tilt of the specimen stage, the trajectory of the spacecraft, an assumed spherical shape for the planet, etc.) or to determine it from the image itself. The means for accomplishing the latter will be deferred until the next chapter on locating edges, especially using the Hough transform. For the present, we will assume that the equation describing the modification is known. For many practical purposes, such as tilted or rotated surfaces or sequential images that are slightly offset, the relationship is very simple. We will also ignore such related problems as how to map a spherical surface onto a plane, and assume that some practical mapmaking compromise has been decided upon that suits the immediate purpose at hand.

One case in which the distortion introduced by the imaging system is known, and rubber-sheeting is required to justify the images, is commonly encountered in aerial

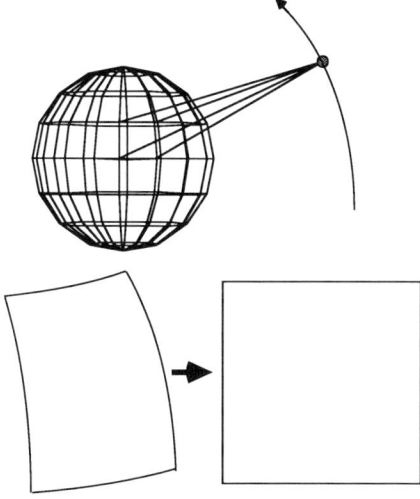

Figure 3-22: Viewing a curved or inclined surface and transformation to a flat map.

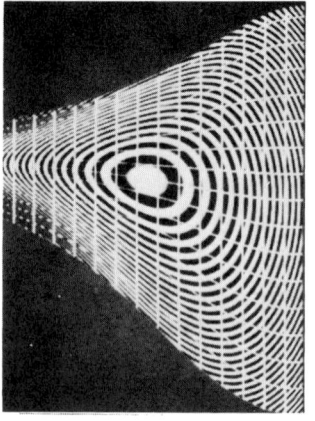 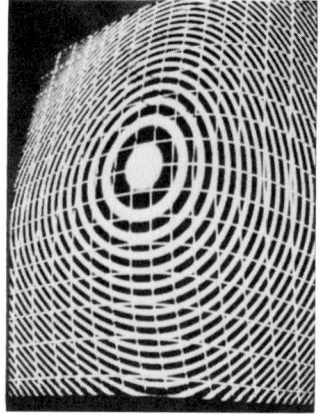

Figure 3-23: Warping of an image possible with a bicubic equation.

photography. Slit cameras produce a distorted image due to the motion of the airplane, which is corrected afterwards.

Frequently the warping of the image is sufficiently smooth that it can be expressed as a comparatively simple set of equations that relate the address in the target array $T(x,y)$ as a function of the locations in the source array $S(x,y)$ which can be written as two functions U and V

$$T(x,y) = \{U[S(x,y)], V[S(x,y)]\}$$

in which the functions are well approximated by polynomials, and can in fact be applied in real time by dedicated "warping" hardware (this has now become common in television advertising, for example). A very common and quite flexible method used is the bicubic polynomial warp (Figure 3-23), in which all possible cubic terms involving x, y and their products appear. This requires 16 coefficients for each function, a total of 32 for the full warping.

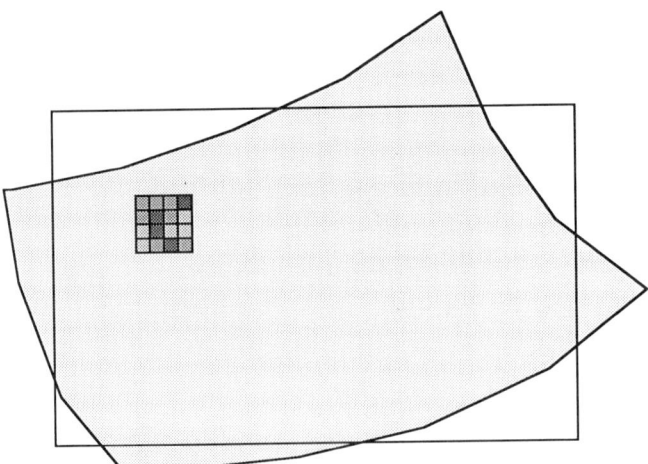

Figure 3-24: Fitting transformed coordinates onto a distorted image.

Image Processing

$$U(x,y) = \begin{vmatrix} x^3 & x^2 & x & 1 \end{vmatrix} \cdot \begin{vmatrix} a_0 & a_1 & a_2 & a_3 \\ a_4 & a_5 & a_6 & a_7 \\ a_8 & a_9 & a_{10} & a_{11} \\ a_{12} & a_{13} & a_{14} & a_{15} \end{vmatrix} \cdot \begin{vmatrix} y^3 \\ y^2 \\ y \\ 1 \end{vmatrix}$$

If this is the case, then in effect we have two images (the original one that was acquired, and another one which we are going to create), and a known addressing relationship between the pixels, as shown in Figure 3-24. Specifically, each pixel in the target image corresponds to some coordinates in the original image. These coordinates will not, in general, be integers. In other words, they will lie "between" pixels in the original image. There are several methods to obtain a brightness value for the pixel in the rectified image being created. The simplest (and least used) is to take the brightness value of the pixel in the original image that is nearest to the converted coordinates. The concept of nearest in this case must be understood in a Cartesian sense, where diagonal distances are calculated using the Pythagorean relationship Distance = $\sqrt{(dx^2 + dy^2)}$. This simple method is fast, and does the least smoothing of the resulting image (which may be important to preserve edges), but often produces "jaggies" in which lines appear broken up or moiré effects appear.

A better method for many purposes is to interpolate a new brightness value from the 4 (or more) neighbor points in the original. The most common technique is bilinear interpolation, in which the brightness is assumed to vary linearly between the points above and below, and left and right of the coordinates of the pixel. Linear interpolation is easy to perform (as shown in Figure 3-25) and produces a brightness value that can be mapped into the rectified image. The result appears quite smooth, with lines that do not show aliasing effects, but edges are somewhat smoothed by the procedure.

More complex weighted fits to 4 or 8 nearest neighbor points or polynomial fits can also be used, with the usual cost in computation. These are particularly useful when expanding the resolution ("zooming") to magnify a portion of the original image. In this case, several pixels in the new image may be interpolated from the same set of neighbors in the original, and linear interpolation will blur the sharpness of edges. This can be overcome by the use of higher order fits.

These techniques have been used extensively to process images from planetary probes, so that the various images can be justified and registered to create large mosaics of entire planet hemispheres. The images are also usually matched in brightness and contrast, using background levelling, so that the joins in the mosaic are nearly invisible. The amount of computation in these tasks is significant, but not in proportion to the effort required to obtain the images. For more routine measurement operations, simpler

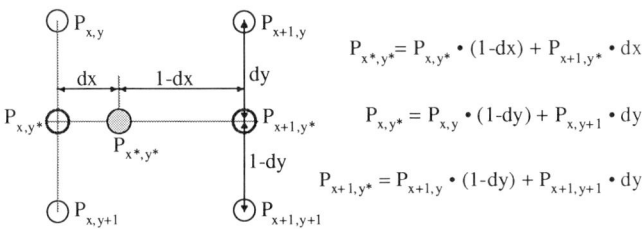

Figure 3-25: Bilinear interpolation.

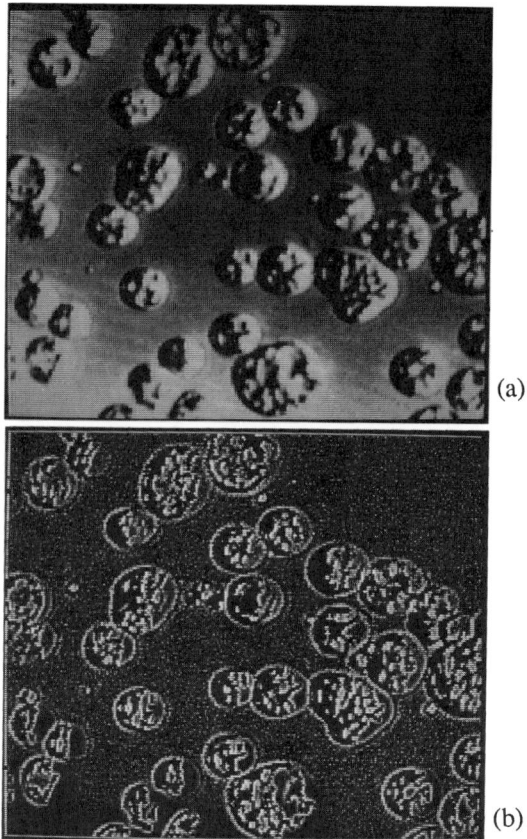

Figure 3-26: Examples of Laplacian used as a high pass filter to highlight boundaries between regions of different brightness around the edges and within bubbles in a nonuniformly illuminated image: a) original; b) processed.

procedures are usually adequate. However, spurred by the use of image warping as an attention-getting advertising tool in television, hardware is now becoming available that can perform image warping or rubber sheeting of video images in real time.

Image sharpening

The discussion above of smoothing an image introduced the idea of using a 3x3 (or larger) kernel of multiplying factors. Actually, as we will see, these kernels or operators have a great many other applications in image processing as well. One that fits nicely into the scope of this chapter is image sharpening. Consider an operator of the form

$$\begin{vmatrix} -1 & -1 & -1 \\ -1 & +8 & -1 \\ -1 & -1 & -1 \end{vmatrix}$$

This is a Laplacian operator, a nondirectional second derivative that gives a zero result on any uniform or smoothly varying region of the image, but gives a large response (either positive or negative) at edges, lines or points. As we will see later, it mimics rather closely the inhibition methods used in the human visual system. For the moment, our interest in this operator (and others like it) centers on the fact that is responds strongly to discontinuities, irrespective of their orientation.

When a Laplacian operator is applied to an image, the result is a derived image in which edges, lines, and point discontinuities appear and uniform areas are suppressed (Figure 3-26). This is the reason that such an operator can be used to locate uniform regions for the automatic selection of background points for levelling as discussed above.

If the Laplacian of the image is added back to the original image, it has the effect of enhancing edges. The overall contrast range of the original image is reduced (by half, if the two images are added in equal proportions). The remaining available contrast range is used to highlight the edges and discontinuities (Figure 3-27). This process is usually called edge sharpening or crispening of images. It does not improve the images from the standpoint of measurement, but does make them look better to the human observer. The same operation can be carried out in frequency space using appropriate filters (Figure 3-28).

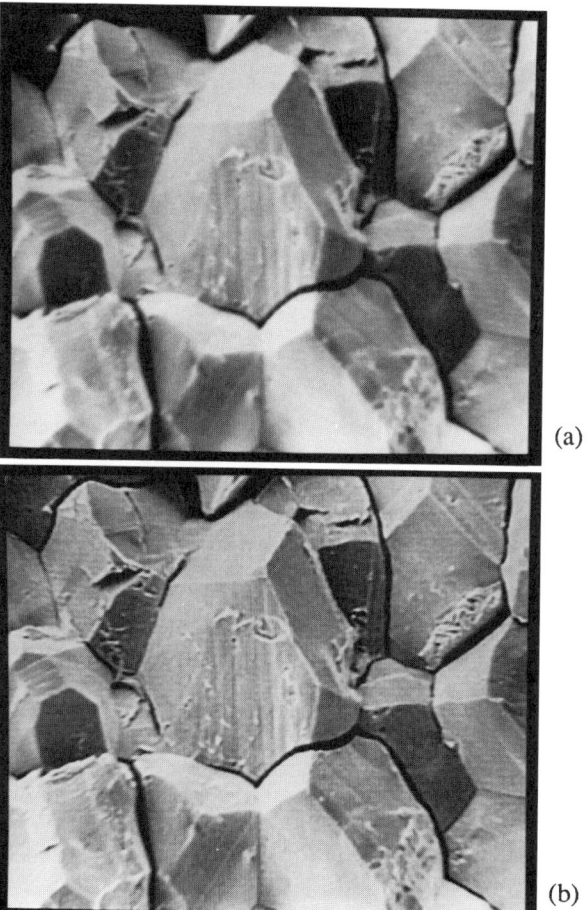

Figure 3-27: Laplacian operator used to enhance images in an SEM image of an alumina fracture surface: a) original; b) processed.

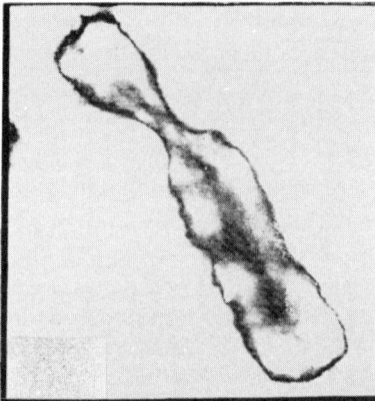

Figure 3-28: Outlining a chromosome similar to that produced by a Laplacian, but performed by a convolution in frequency space: a) original; b) Fourier filtered.

This improvement is present because the human visual system is highly concerned with edges, and the "early vision" operators in our retinas select such discontinuities to transmit to the brain. This is why cartoons and line drawings work as representations of objects which do not, after all, really have black lines around them. But by increasing the contrast at the edges and suppressing the remainder of the image contrast, the addition of a Laplacian of an image to the original produces a result that appears to be sharper than the original (Figure 3-29). Laplacian operators are also used to control contrast in images, by suppressing gradual changes and enhance detail in both bright and dark areas.

The application of the Laplacian kernel operator can accomplish the addition as well, by adding 1 to the central multiplier value. Some typical sharpening operators are shown below. Notice that they do not need to be limited to 3x3 size, and that the values in the filter may change sign at the extremes. The "shape" of these operators will be discussed more in the next chapter, and more types of derivative kernels shown. For the moment it is adequate to understand what the Laplacian operator does, and that it is a direct two-dimensional analogue of the one-dimensional derivative methods of Savitsky and Golay (1964), which are routinely applied to linear spectra.

$$\begin{vmatrix} -1 & -1 & -1 \\ -1 & 9 & -1 \\ -1 & -1 & -1 \end{vmatrix} \quad \begin{vmatrix} +1 & -2 & +1 \\ -2 & 5 & -2 \\ +1 & -2 & +1 \end{vmatrix} \quad \begin{vmatrix} -8 & -5 & -2 & -5 & -8 \\ -5 & +3 & +9 & +3 & -5 \\ -2 & +9 & +33 & +9 & -2 \\ -5 & +3 & +9 & +3 & -5 \\ -8 & -5 & -2 & -5 & -8 \end{vmatrix}$$

The use of sharpening operators is intended only to improve the visual appearance of images, not as a precursor to further processing. Furthermore, since the Laplacian is in effect a "high pass filter", it may accentuate the amount of noise in an image. The operator is more sensitive to lines than to edges, and more sensitive to points than to lines. This is an important defect for applications in which our interest is in the edges present, and we would prefer not to increase the noise due to isolated points, but as mentioned before the human eye is rather tolerant of this kind of noise, and the sharpening operators have therefore enjoyed some areas of active use.

Focussing images

Related to sharpening of images is the subject of optimizing image focus. Of course, this requires that there be some control mechanism from the computer, such as a motor control on the specimen stage of the microscope or the lens of the camera, or an electronic control for the current flowing in the lens coils of an electron microscope. The details of how this control is exerted need not concern us here. But it is appropriate to consider how well the computer-based system can detect the point of optimum focus, and whether this can be done in a short enough time for practical use.

There are presently a variety of quasi-real-time analog focussing devices being used, even in consumer goods such as 35 mm cameras. In a few cases these utilize a beam of light (sometimes not in the visible range) which is projected from a source, reflected from the object, and then detected at a corresponding point within the camera. Optimizing focus then consists simply of scanning the focus device until the spot of light falls on the

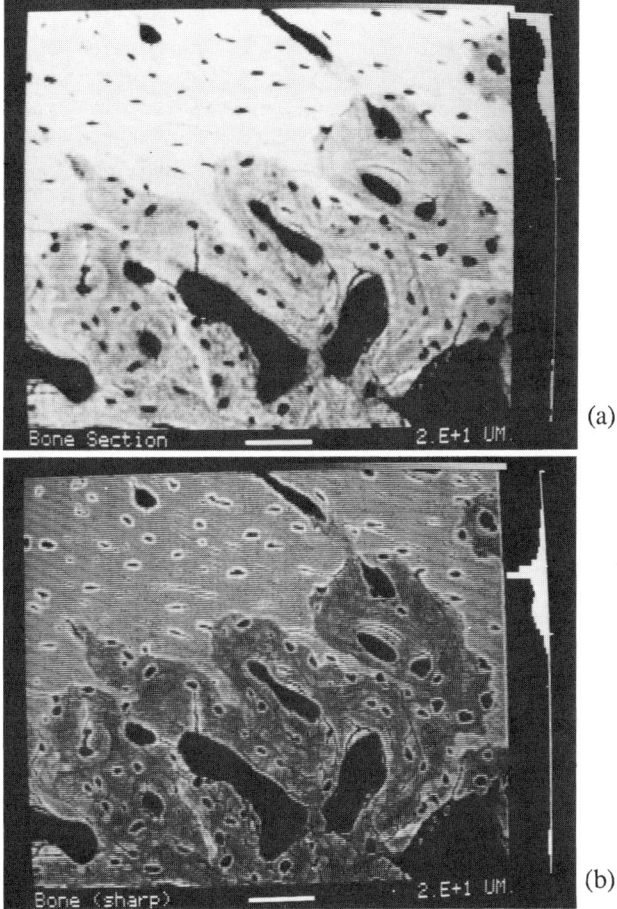

Figure 3-29: Laplacian operator used to compress overall brightness range in LM image of bone section while retaining edge contrast: a) original; b) processed.

detector. Another class of methods use light that passes through the edges of the lenses, falling outside the image area. For scanned images, a third approach passes the video signal through a high-pass filter. Maximizing the signal strength at high frequencies is taken to indicate the point of maximum sharpness in the image, and hence the point of best focus.

$$F_1 = \sum_i \sum_j (g_{i,j} - g_{i+1,j})^2$$
$$F_2 = \sum_i \sum_j g_{i,j}^2 - \sum_i \sum_j g_{i,j} \cdot g_{i+1,j}$$
$$F_3 = \sum_i \sum_j (g_{i,j} - \mu)^2$$
$$F_4 = \sum_i \sum_j g_{i,j} \cdot g_{i+1,j} - \sum_i \sum_j g_{i,j} \cdot g_{i+2,j}$$
$$F_5 = \sum_i \sum_j g_{i,j} \cdot g_{i+1,j} - N \cdot M \cdot \mu^2$$

Since our emphasis here is on digitized and stored images, it is possible to consider methods for determining optimum focus based on digital computation using the brightness values in the images. Vollath (1988) has summarized several algorithms that all attempt to define a quantity that is maximized based on the high-spatial-frequency variations of brightness in the image. Using g as the grey scale brightness of each pixel, sums are formed over the image (either the entire image or some rectangular target region of dimension $M \cdot N$ within it). Maximizing the value of F using any of the equations shown should produce an optimum focus point (μ is the mean brightness value in the sampled region).

Each of these can be carried out either along or perpendicular to the scan lines. The bandwidth of the video signal may limit high frequency variation along the lines, whereas overlapping line spacing may limit it in the perpendicular direction. The sensitivity of these different algorithms to noise, the ease of computation, uniformity of response to image regions with high or low contrast or brightness, and the ability of the algorithm to sharply define an optimum, all vary. No single optimum solution exists. Even more computationally demanding methods can be performed using the frequency transform of the image.

References

D. H. Ballard, C. M. Brown (1982) Computer Vision Prentice Hall, Englewood Cliffs, NJ

A. C. Bovik (1987) *Streaking in Median Filtered Images* IEEE Trans. Acoust. Speech and Signal Proc. ASSP-35 #4, 493-502

D. S. Bright, E. B. Steel (1986) *Bright-field image correction with various image-processing tools* Microbeam Analysis 1986 (A.D.Romig, W.F.Chambers, ed.) San Francisco Press, 517-520

K. R. Castleman (1979) *Digital Image Processing*, Prentice Hall, Englewood Cliffs, NJ

R. T. Chin, C-L. Yeh (1983) *Quantitative Evaluation of some Edge-Preserving Noise Smoothing Techniques* Computer Vision Graphics and Image Processing 23 67-91

T. R. Edwards (1982) *Two-Dimensional Convolute Integers for Analytical Instrumentation* Anal. Chem. 54 1519-1524

S. J. Erasmus (1982) *Reduction of noise in a TV rate electron microscope image by digital filtering* J. of Microscopy 127 29-37

S. J. Erasmus, D. M. Holburn, K. C. A. Smith (1980) Scanning 3/4 273
H. Hashimoto (1986) Electron Microscopy Technique 2 1
T. S. Huang (1979) *A fast two-dimensional median filtering algorithm* IEEE Trans ASSP-27 13-18
L. R. Johnson, A. K. Jain (1981) *An Efficient Two Dimensional FFT Algorithm* IEEE Trans. Patt. Anal. Mach. Intell. PAMI-3 #6 698-701
G. A. Mastin (1985) *Adaptive Filters for Digital Image Noise Smoothing: An Evaluation* Computer Vision Graphics and Image Processing 31 102-121
J. J. McCarthy (1987) *Techniques for computer aided imaging in SEM and STEM* Microbeam Analysis 1987, San Francisco Press
M. Nagao, T. Matsuyama (1979) *Edge Preserving Smoothing* Computer Graphics and Image Processing 9 394-407
W. K. Pratt (1978) *Digital Image Processing*, John Wiley, New York
A. Rosenfeld (1982) *Survey: Picture Processing* Comp. Graphics and Image Processing 19 35-75
A. Rosenfeld, A. C. Kak (1982) *Digital Picture Processing*, Academic Press, London
J. C. Russ, J. C. Russ (1986) *Image processing for the location and isolation of features*, Microbeam Analysis 1986, San Francisco Press, 501
A. Savitsky, M. J. E. Golay (1964) *Smoothing and Differentiation of Data by Simplified Least Squares Procedures* Anal. Chem. 36, 1627-1639
W. O. Saxton, T. J. Pitt, M. Horner (1979) *Digital Image Processing: The Semper System* Ultramicroscopy 4 343-354
W. O. Saxton, T. L. Koch (1982) *Interactive Image Processing: Organization*, J. Microscopy 128, 69
D. Vollath (1988) *The influence of the scene parameters and of noise on the behaviour of automatic focusing algorithms* J. Microscopy 151, 133
Y. Zheng, M. Gandais (1989) *An application of high-resolution transmission electron microscopy to the study of fine structure of dislocations in potassium feldspars*, J. Electr. Micr. Tech. 11, 234-237

Chapter 4

Segmentation of Edges and Lines

In order to recognize or measure objects in images, it is necessary to distinguish them from their surroundings. This is the familiar problem of separating figure from ground, which we can also generalize to include separating features from other, touching ones.

Defining a feature and its boundary

What is a feature? Two definitions that each seem reasonable are:

1) a region of contiguous pixels that share some property, or

2) a region inside some boundary (or adjacent to some other feature).

These two definitions are not identical, as shown in Figure 4-1. In some cases it will be appropriate to describe the regions on the basis of the pixels which comprise them, and when this can be done it is generally the easiest method. A typical and almost trivial example is an image of dark objects well dispersed and resting on a bright background, or vice versa. The darkness of the pixels within the objects easily discriminates them from the background. Many examples will be shown in the next chapter.

In other cases, the pixels comprising the objects are not different from their surroundings, and it is only the discontinuities at the edges that can be used to distinguish them. A common example is a polished metal surface, on which the individual metal grains are the same in brightness, but the boundaries between them are dark because they

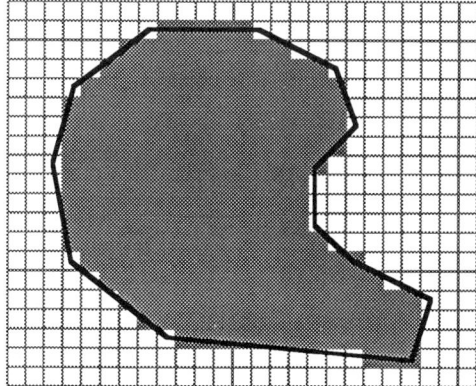

Figure 4-1: Feature composed of pixels vs. one defined by its outline.

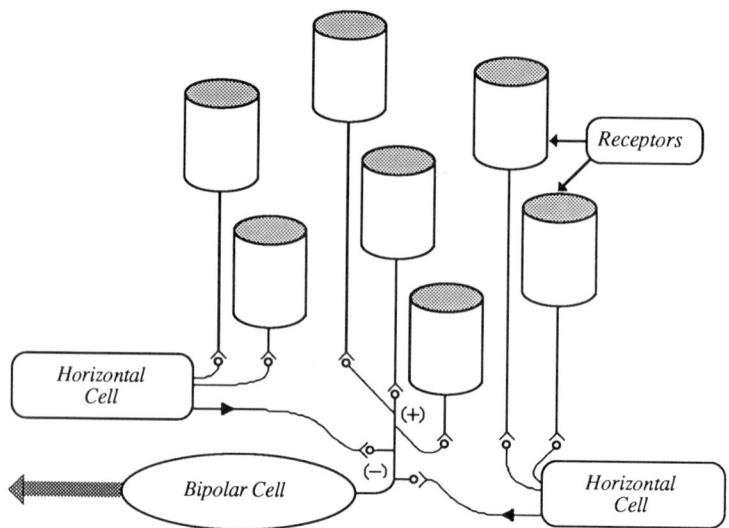

Figure 4-2: A simple line detector formed by lateral inhibition between neighboring receptors.

are softer (due to imperfections in the atomic crystal structure) and so have been etched out either chemically or by polishing so that they reflect less light.

The first of these definitions is most often used in automatic video image analyzers, in comparison to computer-assisted measurement systems in which the human operator outlines features using some pointing or tracing device connected to the computer. We will reserve the pixel discrimination method, in which the selection and delineation of features can be performed based on pixel brightness, color or some similar property to the next chapter. It is generally the easier technique to apply, and also more readily understood. However, it is not the way that the human visual system works (Marr & Hildreth, 1980; Hildreth, 1983).

The human eye has, as was mentioned before, an enormous number of individual light receptors in the retina. If each of these simply transmitted the brightness of light striking it to the brain, it would require a very large number of connections and would simply defer the problem of interpreting the data. Nature has instead developed a more distributed processing scheme in which only a smaller number of higher level pieces of data must be passed on. In fact, within the retina there are additional cells, known to physiologists by specific names such as amacrine cells and horizontal cells, which connect to several retinal cells (and to each other) to extract this higher level information. For instance, it is possible to arrange a simple neural circuit to detect an edge oriented in a particular direction. Figure 4-2 shows an example, in which the response of several cells are combined, some positively and some negatively. When a uniform region is viewed, there is no output from this detector. When an edge oriented in the "wrong" direction is viewed, there is also very little output. Only when the edge lies in the proper alignment with the detector does it produce the desired signal.

Other lateral connections between receptors are used to detect edges running in other directions (probably as many as 36 distinct directions have their own fairly low-level

Segmentation of Edges and Lines

detectors in the human eye). Other patterns of connections look for corners, and so forth (Figure 4-3). The human visual system has enough cells and lateral connections to allot more than 25,000 various kinds of feature detectors for each of the 150 million receptors, so some of these can be quite specific. A later chapter discusses in more detail how we can train a detector to look for a particular feature.

The heart of this method is inhibition. The positive and negative connections on the "logic" cells allow neighboring "pixels" or receptors to be considered. This is, in fact, just the same technique as was used in the Laplacian kernel operator mentioned in the preceding chapter. In that case the weighting factors which are negative represent inhibition.

It is now clear that the Laplacian operator as defined before uses lateral inhibition in all directions to locate isolated unique points with great efficiency (which is why it increases the noise in an image), and only secondarily serves as an edge enhancer. But using the idea of a local kernel of 3x3 or larger size, with appropriate weights, we can duplicate some of the most basic functions possible with lateral neural connections in the retina.

$$\begin{vmatrix} -1 & -1 & -1 \\ -1 & 9 & -1 \\ -1 & -1 & -1 \end{vmatrix} \quad \begin{vmatrix} +1 & -2 & +1 \\ -2 & 5 & -2 \\ +1 & -2 & +1 \end{vmatrix} \quad \begin{vmatrix} -8 & -5 & -2 & -5 & -8 \\ -5 & +3 & +9 & +3 & -5 \\ -2 & +9 & +33 & +9 & -2 \\ -5 & +3 & +9 & +3 & -5 \\ -8 & -5 & -2 & -5 & -8 \end{vmatrix}$$

Laplacian edge-finding operators

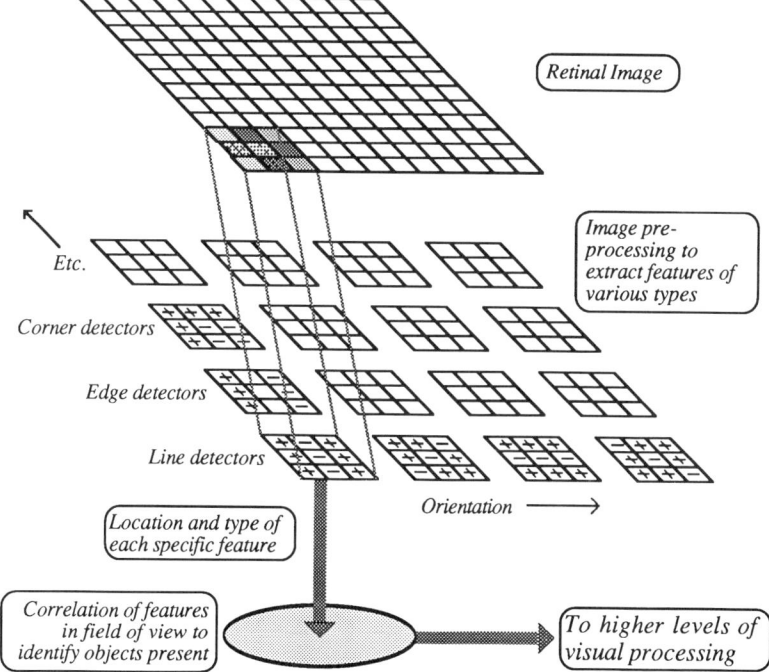

Figure 4-3: Finding many features from a pixel array.

If the Laplacian operator is not a good edge detector, how can be make a better one? It will help to consider what kinds of edges we may encounter, and what their properties are. Some of the kinds of boundaries we might consider are easiest to think about in one dimension (for instance, along a scan line in the image that crosses a feature edge):

1) a sudden change from one brightness value to another.

2) a narrow dark (or light) band separating two uniform regions of the same (or different) brightness.

3) adjacent light and dark bands resulting from shading or side illumination

Various derivatives of brightness across these edges are shown in Figure 4-4. For these cases, the first derivative is not a good choice, because the result is different depending on whether the boundary or the regions are dark or light, and the order in which they are encountered. The second derivative, however, is not sensitive to absolute brightnesses or their order. Nor is the square of the first derivative. In general, we can use any even derivative, or any even power of an odd derivative, to detect an edge (Konomopoulos, 1982). Examples of the use of derivatives to reveal detail are shown in Figures 4-5 and 4-6.

An important caution in the use of any derivative operator is that prior smoothing, as discussed in the previous chapter, may displace the edges and produce bias in subsequent measurement. The median filter is superior to smoothing operators in this regard (Berzins, 1984; Yang & Huang, 1981).

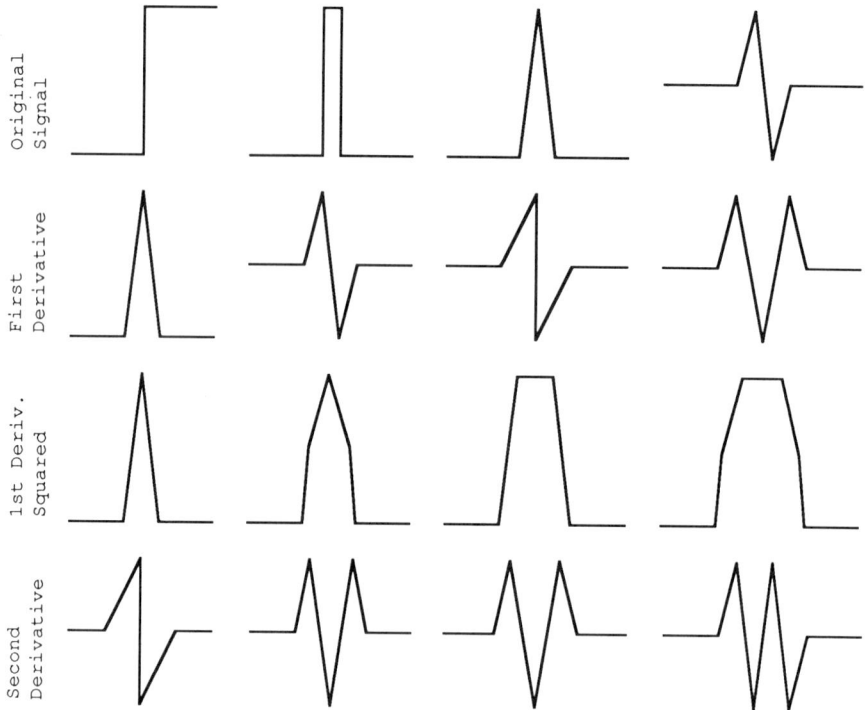

Figure 4-4: Brightness traces with derivatives.

Segmentation of Edges and Lines

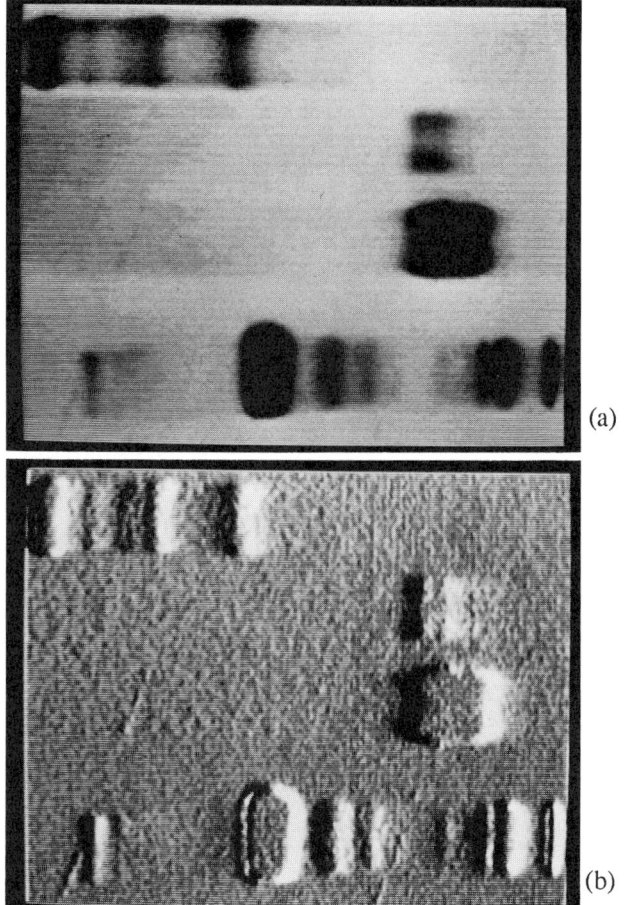

Figure 4-5: Derivative applied to electrophoresis gel (a) showing enhancement of detail along the column direction (b).

Roberts' cross edge operator

How can a derivative be obtained from a two-dimensional image? One way is to take several derivatives in different directions and then combine them (Castleman, 1979; Pratt, 1978; Rosenfeld & Kak, 1982). The first effort in this direction was the Roberts' cross operator. It consists of two derivatives at right angles, estimated digitally as the different between pairs of pixels.

As shown in Figure 4-7, these two differences are the first derivative of brightness in two perpendicular directions, each oriented at 45 degrees to the principal grid of points. Being first derivatives, they are not directly suitable for use as an edge detector. But if we combine them as $\sqrt{(D_1^2 + D_2^2)}$ then we have an even power, and also a single value that combines both directions so as to give a uniform response to an edge in any direction.

Furthermore, it is also possible to determine the orientation of the local edge. The direction of maximum gradient in brightness is given by

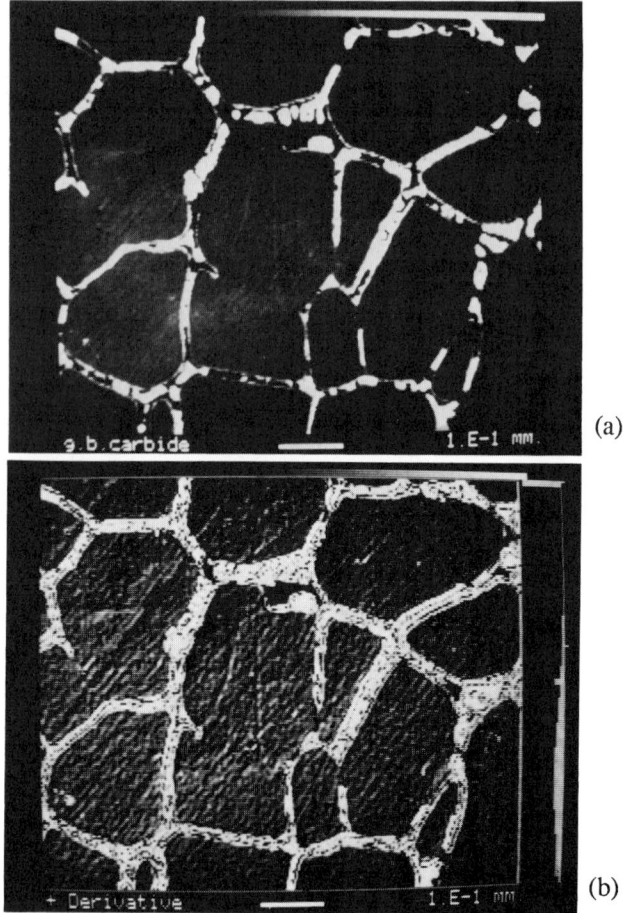

Figure 4-6: Derivative applied to a polished metal alloy (a), showing surface scratches (b).

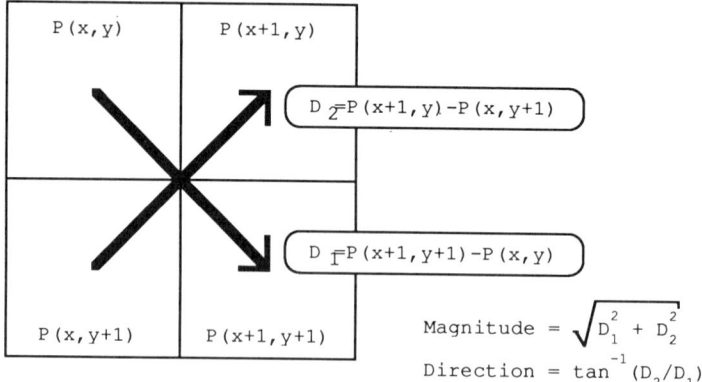

Figure 4-7: Roberts' cross operator.

Segmentation of Edges and Lines

$$\tan^{-1}(D_2/D_1) - \pi/4$$

and the edge is at right angles to this direction.

This method suffers from several practical drawbacks. First, the square root function is a demanding one for the computer (recall that this operation must be performed for every pixel in the image). This is sometimes overcome by using various approximations or shortcuts. For instance, it is not really essential to use the square root to determine the function's maxima, since the sum of squares will do as well. An even more drastic simplification is to use the maximum value of the two derivatives, which avoids all of the arithmetic. However, this makes the detector significantly more sensitive to edges in the 45 degree directions than to those aligned with the pixel grid.

The major difficulties with the Roberts' cross are that it calculates the edge parameter between pixels rather than aligned with the pixel grid, and that by using the difference between pairs of pixels, it is rather sensitive to any noise present in the image. For these reasons, other operators such as the Sobel and Kirsch have come into greater use.

The Sobel and Kirsch operators

The Sobel operator also obtains derivatives, using a family of 3x3 operators as shown below:

$$\begin{vmatrix} -1 & 0 & 1 \\ -2 & 0 & 2 \\ -1 & 0 & 1 \end{vmatrix} \quad \begin{vmatrix} -2 & -1 & 0 \\ -1 & 0 & 1 \\ 0 & 1 & 2 \end{vmatrix} \quad \begin{vmatrix} -1 & -2 & -1 \\ 0 & 0 & 0 \\ 1 & 2 & 1 \end{vmatrix} \quad \text{etc.}$$

Derivative Operators used in the Sobel Edge Finding Method

With these operators, all nine pixels (the central pixel and its immediate neighbors) are taken into account, which provides some smoothing of any noise in the image. Also, the derivative is calculated at the location of the central pixel rather than at an offset position. By rotations of the operator, it is possible to perform the operation in each of 8 directions. However, 4 of these are redundant except for a sign change. Also, the remaining orientations which are at 45 degrees are not quite equivalent to the 90 degree ones because the pixel spacing is not the same in this direction. We will encounter this problem repeatedly with square pixel grids, and it is one of the principal reasons why hexagonal grids are sometimes used (the six neighbor pixels are all at the same distance from the central pixel).

For our present purposes, it is only necessary to use two Sobel derivatives, either the 90 degree or 45 degree pair. Their magnitudes can be combined just as the Roberts' cross derivatives were, to obtain a magnitude and direction for the edge. Usually, this magnitude value (rescaled as necessary) is used to create a new, derived image in which the brightness is a measure of the "edgeness" of each pixel. The brightest pixels will mark the edges, and can be discriminated as described in the next chapter. Figure 4-8 illustrates the application of a Sobel operator to a fragment of an image.

It is also possible to establish a threshold, perhaps to include the 15% or so of the pixels with the greatest magnitude values, and acquire a histogram of edge orientations using the arc tangent relationship mentioned before. This information is very useful to describe the possible anisotropy of edges in the image. The information is usually presented as a graph, showing the frequency of each orientation in some practical number of steps (perhaps in 5 or 10 degree increments). This can be plotted as a compass rose

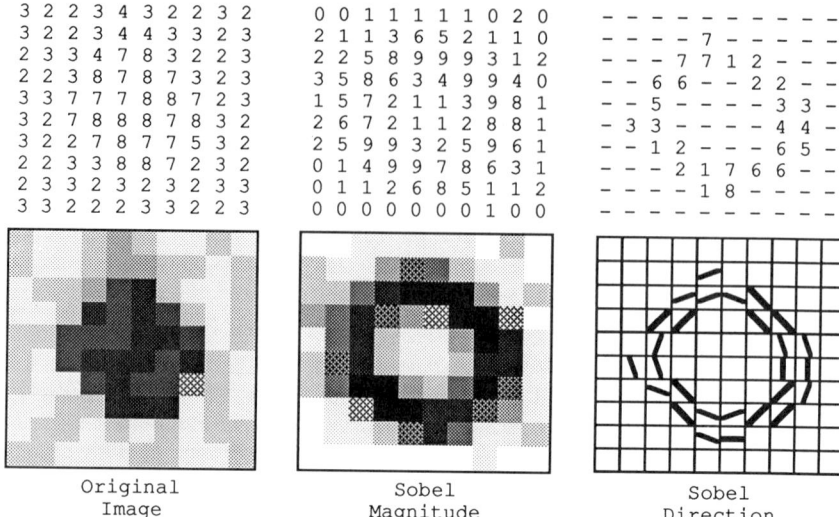

Figure 4-8: Numerical example of application of a Sobel operator to an image fragment. Values from 0 to 9 indicate brightness of each pixel.

(as in Figure 4-9), much the same way as wind directions or geological fault orientations are plotted on a map. Alternatively, mathematical calculations can be performed to describe statistically the degree of preferred orientation of the edges in the image. For either case, the measurement has yielded a few numerical values that describe one aspect of the image (preferred orientation of boundaries) directly from the grey scale image. It is common to sum together data from many image fields to better represent the entire specimen being examined.

The restriction of the measurement to those points with the greatest magnitude value for the edge is reasonable, but the use of any particular value as a threshold, or of any particular percentage of the points in the image, is arbitrary and may produce bias. A somewhat safer method is to use the magnitude value as a weight to sum the histogram of orientations. The non-edge pixels with low magnitudes generally produce a background in the histogram that averages out (measurements of preferred orientation are based on

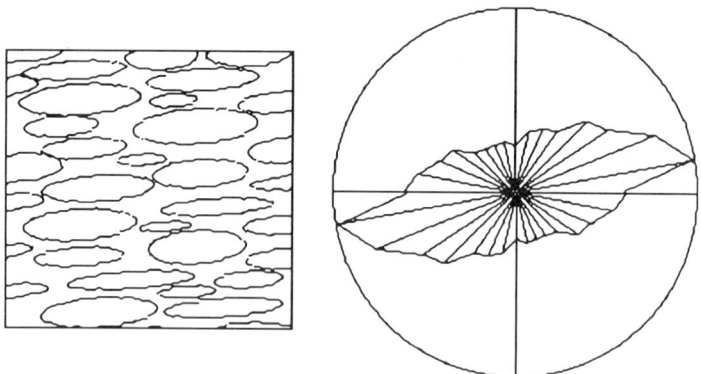

Figure 4-9: Rose of orientation for edge image.

Segmentation of Edges and Lines

the difference in histogram peak height in the maximum and minimum directions). If it does not, it usually indicates a texture in the image (for instance a set of oriented scratches on the surface of a polished specimen) that is itself of interest.

The Sobel operator is widely used to process images (Sobel, 1970). As an edge enhancement tool it is superior to the Laplacian, since it is more sensitive to edges than to points (Smith et al., 1977). By adding the Sobel edge image (the magnitude formed as the square root of the sum of squares of the original derivatives) to the original grey image in various proportions, images with a crisp visual appearance can be obtained. The edges themselves, discriminated as discussed in the next chapter, can be used for object recognition and counting.

Of course, while we call the Sobel operator (and the Roberts' cross, and the Kirsch, to be discussed next) edge detectors, they are really more accurately described as gradient operators. The magnitude of the response is to the steepest gradient in brightness of the image. That means they can also be used to locate flat or uniform grey regions, which is how they are employed in the automatic location of possible background regions for levelling of nonuniformly illuminated images, as discussed in the previous chapter.

Other sets of weighting factors are sometimes used in the Sobel method. A few examples are shown. Notice that some of these produce first derivatives, and some approximate second derivatives. In the latter case, there are effectively only four directions in which the operator can be used instead of 8, but as before only two are really needed.

$$\begin{vmatrix} -1 & 0 & 1 \\ -1 & 0 & 1 \\ -1 & 0 & 1 \end{vmatrix} \quad \begin{vmatrix} -1 & -1 & 0 \\ -1 & 0 & 1 \\ 0 & 1 & 1 \end{vmatrix} \quad \begin{vmatrix} -1 & -1 & -1 \\ 0 & 0 & 0 \\ 1 & 1 & 1 \end{vmatrix} \quad \text{etc.}$$

$$\begin{vmatrix} -1 & -1 & 1 \\ -1 & 1 & 2 \\ -1 & -1 & 1 \end{vmatrix} \quad \begin{vmatrix} -1 & -1 & -1 \\ -1 & 1 & 1 \\ -1 & 1 & 2 \end{vmatrix} \quad \begin{vmatrix} -1 & -1 & -1 \\ -1 & 1 & -1 \\ 1 & 2 & 1 \end{vmatrix} \quad \text{etc.}$$

$$\begin{vmatrix} -1 & 2 & -1 \\ -1 & 2 & -1 \\ -1 & 2 & -1 \end{vmatrix} \quad \begin{vmatrix} -1 & -1 & 2 \\ -1 & 2 & -1 \\ 2 & -1 & -1 \end{vmatrix} \quad \begin{vmatrix} -1 & -1 & -1 \\ 2 & 2 & 2 \\ -1 & -1 & -1 \end{vmatrix} \quad \text{etc.}$$

$$\begin{vmatrix} -1 & 2 & -1 \\ -2 & 4 & -2 \\ -1 & 2 & -1 \end{vmatrix} \quad \begin{vmatrix} -2 & -1 & 2 \\ -1 & 4 & -1 \\ 2 & -1 & -2 \end{vmatrix} \quad \begin{vmatrix} -1 & -2 & -1 \\ 2 & 4 & 2 \\ -1 & -2 & -1 \end{vmatrix} \quad \text{etc.}$$

Alternate Derivative Operators used in the Sobel Method

The importance and practical utility of the Sobel operator is demonstrated by the development and recent availability of chips which can perform a real-time Sobel operation on video images. It is necessary to retain the two most recent scan lines, to combine with the current scan line as it as acquired. The output is always one scan line delayed with respect to the incoming signal. These devices are used in some military and surveillance image analysis systems.

Even better than the Sobel in its sensitivity to weak edges, but more time consuming and computer intensive to apply, is the Kirsch operator. This performs the derivative operation in each of 8 directions and retains the maximum value. The Kirsch derivative operators are shown below.

$$\begin{vmatrix} -3 & -3 & 5 \\ -3 & 0 & 5 \\ -3 & -3 & 5 \end{vmatrix} \quad \begin{vmatrix} -3 & 5 & 5 \\ -3 & 0 & 5 \\ -3 & -3 & -3 \end{vmatrix} \quad \begin{vmatrix} 5 & 5 & 5 \\ -3 & 0 & -3 \\ -3 & -3 & -3 \end{vmatrix} \quad \text{etc.}$$

Derivative Operators used in the Kirsch Method

Notice that the central pixel is not used at all, but only the 8 neighbors. The weights produce a zero value on uniform or uniformly varying regions, and a strong response at edges. The direction with the greatest magnitude (in absolute value) is taken as the direction of maximum gradient, and the value is saved as the strength of the edge. As with the Sobel operator, the resulting derived image can be discriminated using the methods of the next chapter, to obtain an image of just the edges.

Figures 4-10 through 4-12 illustrate the use of edge-finding gradient operators to delineate boundaries in several types of images.

Other edge-finding methods

There are other edge-detecting operators besides the Sobel and Kirsch. One was specifically designed to mimic the way in which the human visual system works, based on physiological studies. Marr (1982) and his co-workers found that the eye uses smoothing techniques (similar to those discussed in the previous chapter) in which the smoothing distance (equivalent to the kernel size) varies. This distance is roughly equivalent to the maximum size of features which are to be ignored. By forming several different smoothed images using different scales, edges of different size structures can be extracted from the image. This is done by taking the difference of two successive smoothed images. The difference image (using the subtraction technique with rescaling discussed previously) shows the edges and boundaries present at the scale of the smoothing used (Figure 4-13). In fact, the human eye apparently does this at several different sets of scales over more than an order of magnitude range of sizes (Figure 4-14) and passes the edge information thus obtained up to higher processing levels, where the edges are used to form a "primal sketch" that will be discussed more in a later chapter.

While it is far from obvious, there is an alternative way to get virtually the same result as using the difference of two Gaussian smoothing operations (called a DOG) This is the LOG or Laplacian of a Gaussian. If the smoothed image is subjected to the Laplacian operator, the edges are also selected (because the smoothing operation has suppressed points and lines). Since the weights for the two operators can be combined into a single operation, it is possible to obtain the LOG image at a given resolution scale in a single pass. The zero-crossings of the resulting image mark the edges of features.

Still another family of edge-locating techniques uses the neighborhood patterns of brightness in ways not easily set forth as kernels. These include min-max and local entropy methods which are discussed more fully in Chapters 5 and 11 respectively. With most of the derivative edge-finding methods, one problem that must be dealt with is the continuity of the edges (or more particularly, the potential lack of continuity). The DOG and LOG methods tend to minimize this problem, as compared to the Sobel and similar operators. However, with any of these techniques it is not uncommon to find line segments that do not connect to form a continuous outline around a feature or object.

Segmentation of Edges and Lines 81

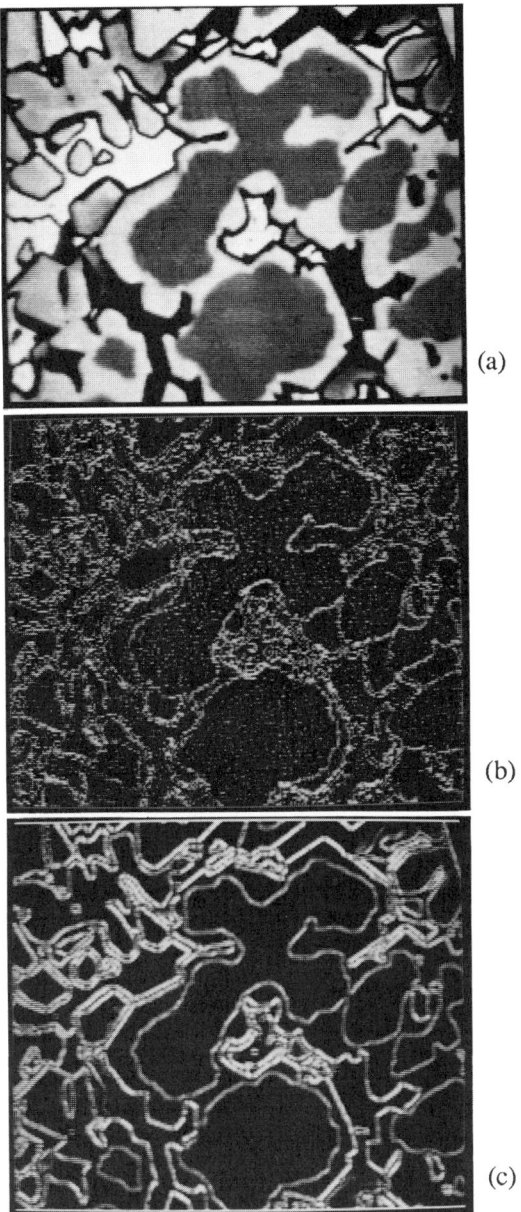

Figure 4-10: A metal alloy (a) showing contrast between phases, and the application of a Laplacian (b) which produces an increase in noise but poor edge definition; and a Sobel operator (c) which gives superior edge definition.

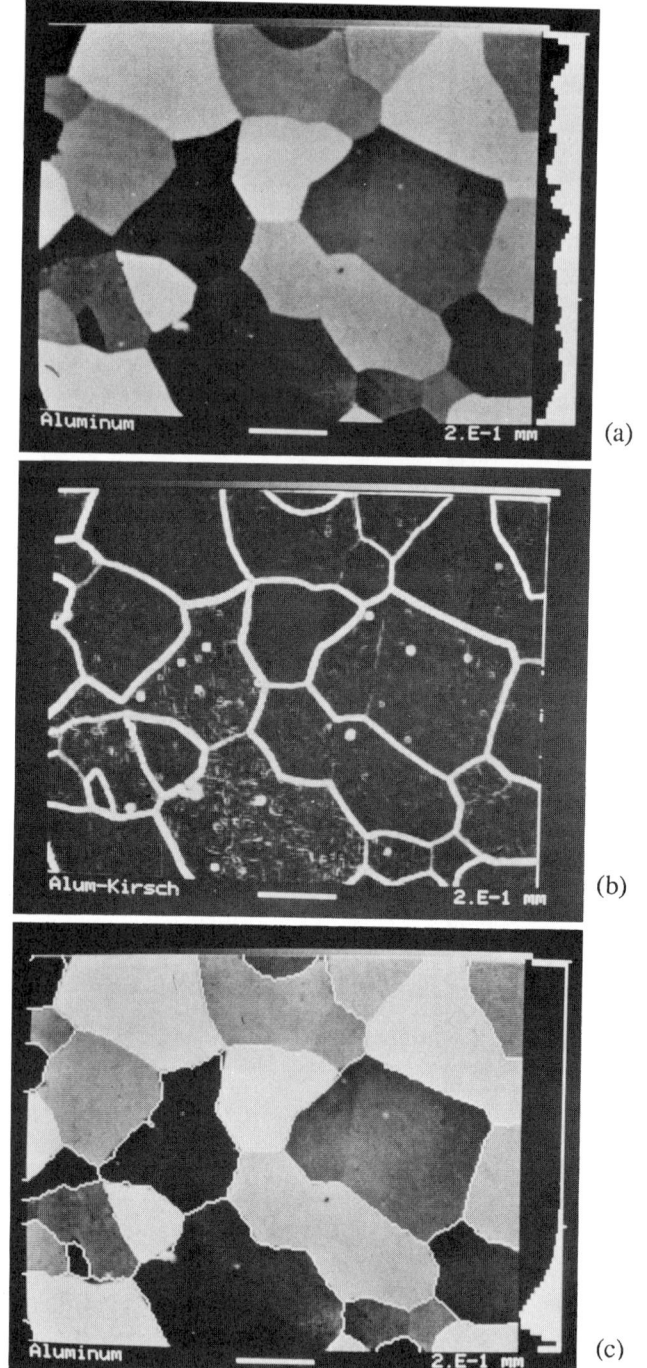

Figure 4-11: Application of Kirsch operator to extract grain boundaries in aluminum alloy (light microscope image). The original grains (a) show contrast because of crystallographic orientation. The Kirsch operator gives a large magnitude at the edges (b), which can be thresholded and thinned as discussed in later chapters. The edges are shown in (c) superimposed on the original image.

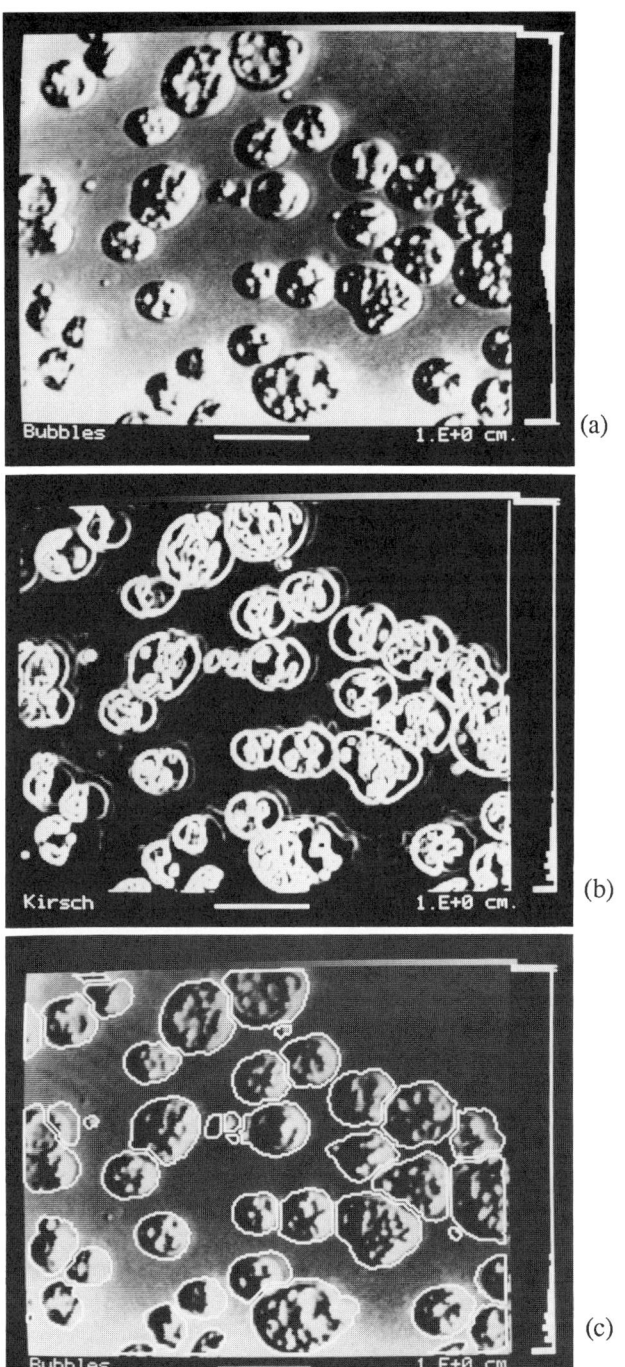

Figure 4-12: Image of bubbles rising in a liquid. The original boundaries (a) have varying contrast because of the side lighting, and the background is nonuniform. A Kirsch operator (b) produces the outlines shown superimposed on the original image (c).

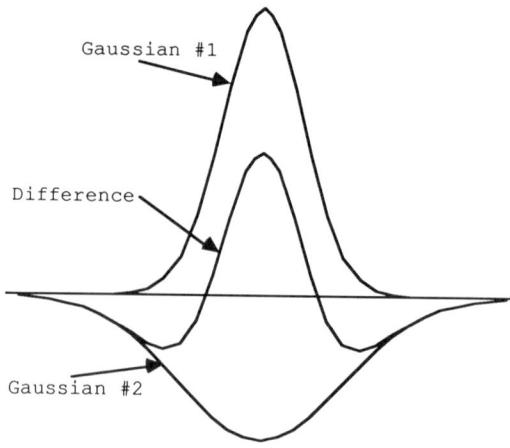

Figure 4-13: The difference between two Gaussian smoothing filters has positive central weights surrounded by negative weights.

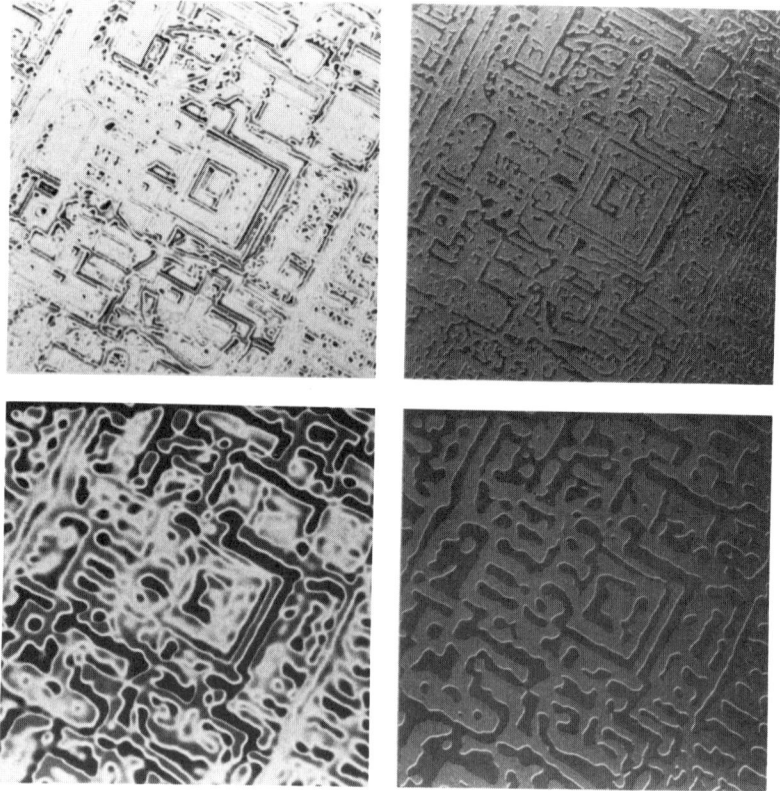

Figure 4-14: Example of LOG zero crossings at different resolution scales (SEM image of an IC).

Segmentation of Edges and Lines 85

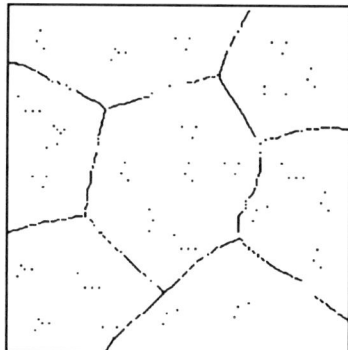

Figure 4-15: Example of incomplete or broken edge lines.

This is due to the fact that the methods discussed detect edge points using purely local comparisons between pixels, and there is no guarantee that the points will merge into connected lines (Figure 4-15). Individual objects may have brightness patterns that produce a sharp discontinuity as compared to background along some parts of their periphery, but not along others. This is particularly the case when there is side lighting or other nonuniform illumination, or the image is very noisy (Figures 4-16).

There are several approaches to connecting incomplete line segments into continuous outlines. These include methods that belong to scene understanding, or "late vision" which we will not deal with here because they are computationally very demanding, and are generally oriented toward breaking a scene up hierarchically into families of regions which are then grouped into facets of surfaces of objects, and finally into objects based on precedence (i.e. some objects can be placed in front of others) (Duda & Hart, 1973; Ballard & Brown, 1982).

The images we usually wish to measure are simpler in organization that this, and precedence can usually be ignored. We will start by making the assumption that boundaries revealed in the edge image all lie in nearly the same image plane, and should be connected to form a simple tesselation or set of feature outlines.

Sometimes is it practical to connect gaps between line segments marking boundaries by curve fitting. With the simplest case of straight lines, each end point would be tested against all other available end points to which a straight line could be drawn (without crossing another feature). Each possible line would be assigned a score based on a combination of the length of the line (the shorter the better if it is really filling a gap) and the orientation of the lines it would connect (the more closely it matches both of them, the more likely it is their continuation). The extension of this method to lines other than straight ones is clear, but may become quite complicated mathematically. The next easiest curves to fit are quadratic.

The greatest complication comes from the fact that once one line is drawn in, it may block others that would have to cross it. A scheme that begins with the line segment having the highest score, and then proceeds down the list until all end points have been accounted for is easier to program but less robust that a relaxation method that finds the combination of lines that give the best overall score.

The other problem with this method is that it makes the fundamental assumption that all line ends should be connected, and that they are paired. In fact, in real structures there

Figure 4-16: Contouring of noisy images is particularly applicable to X-ray maps obtained in the SEM. By collecting more "dots" and using a median filter, edges are better defined and can be located with a gradient operator: a) original; b) filtered; c) boundary contours.

may be some dangling ends, or branches by which one end might connect to two or more others (Hueckel, 1971).

Another approach to achieving continuity is to find the local maximum of the image containing the edge magnitude function (for instance, the Sobel). Rather than setting an arbitrary limit on the magnitude needed to represent an edge, or on the fraction of all pixels that could be along edges, this method will narrow regions where the edge function gives a high magnitude to a single pixel width along the "ridge" of the brightness. In regions where the magnitude is less, there is still a local maximum to be found, representing a lower ridge. Finding local maxima is closely related to finding the skeleton of the distance image, which will be dealt with in Chapter 6 under the context of segmentation of binary images, and indeed it uses the same algorithms and programs. It is reasonably fast, and depending on the operator used to produce the edge magnitude image in the first place, can be moderately resistant to noise. It is perhaps the most flexible technique.

The zero crossing of the LOG function is guaranteed to have a continuous path. Another method guaranteed to produce continuous lines is generating contour lines. While there is no reason to expect that any particular brightness value will produce continuous lines, since individual pixels may not take on that value, the mean value theorem assures us that somewhere between pixels there will always be whatever intermediate value we seek. It is most helpful to imagine the image (either the original image or the magnitude of the edge function) as having an elevation or height proportional to the pixel brightness. Then the image is equivalent to a conventional map, and contour lines of selected brightness are analogous to isoelevation contours marked on topographic maps.

These must by their nature be continuous. If there are two regions at different elevations, then there must be somewhere between them a location along the contour line at any intermediate elevation. When this is applied to edge images, it produces continuous boundaries, but of course the mathematical location of the boundary need not fall exactly on any pixel coordinates. The boundary lines can pass between pixels, and may either be located by interpolation, or may be arbitrarily assigned to the nearest pixel. The choice has to do primarily with how the information needed to define the features and boundaries are stored, which will be discussed shortly.

As an example, the image shown in Figure 4-17 is a microscopic view of printed letters, in which there is quite nonuniform illumination although the letters are visually distinct. By selecting a value for contouring using the "minimum perimeter sensitivity" criterion discussed in the next chapter the contour lines shown are obtained to outline the features. This is superior to the result of applying a gradient operator (a Kirsch transform, described above) and then selecting the brightest pixels, which both introduced background noise and leaves breaks in the edges.

So far we have dealt almost exclusively with techniques that are applied to the entire image, dealing with all of the pixels present. There are two other methods that deserve mention that do not work globally, but locally (Ballard & Brown, 1982). Each requires a starting point, which can often be arbitrarily selected. They are called region growing and edge following.

Edge following (Figure 4-18) should ideally produce the same result as finding the local maximum, or ridge in the brightness values of the image containing the magnitude of the edge function. Beginning at any point along an edge (possibly but not necessarily found as a maximum in the edge function image), this algorithm searches all neighboring

Figure 4-17: Outlining of features by contouring produces a continuous boundary line: a) original image (printed letters on paper); b) the pixels with brightness values between those of the features and background produce a noisy boundary; c) gradient operator also gives an irregular outline; d) contour outline drawn at the brightness value marked on the histogram is continuous.

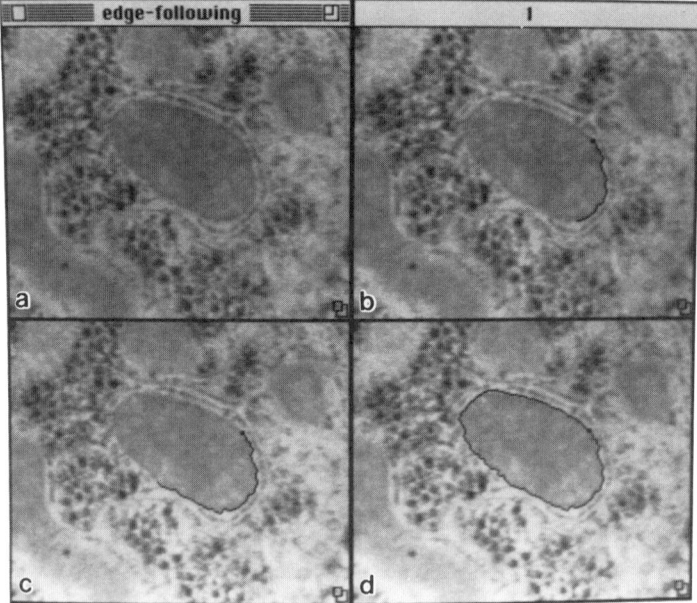

Figure 4-18: Diagram of edge following: a) original image (biological section in TEM) in which the same grey value appears in many structures; b) edge being drawn from a marked starting point; c) further drawing progress; d) completion of the outline.

Segmentation of Edges and Lines

pixels to select the one giving the path of the edge. This can be done by applying any of the various edge functions to the neighbor pixels, and selecting the one producing the maximum response. This procedure is then repeated, like an insect crawling along the ridge. If it meets the edge of the image, or its own prior path, then the current line is ended and a new starting point is selected.

Deciding upon a stopping criterion for the process is not easy. Arbitrary limits on the magnitude of the edge response to begin a line, or the amount of boundary per unit area of the image, are entirely analogous to similar restrictions on the magnitude or area percent of a discrimination threshold applied to the image containing the magnitude of the edge response. Edge following is a fundamentally serial rather than parallel operation, and the time needed to execute it depends on the complexity of the image. Also, it is sensitive to noise and does not fail in a graceful way. For these reasons it is little used in general image analysis.

Other segmentation methods

Region growing works on the basis that pixels within an object should share some property, and so perhaps belongs in the next chapter. But because the method is in some respects similar to edge following, it is discussed here instead. Starting with any interior pixel in the image, each neighboring pixel is examined to decide whether it is part of the same feature. The criteria that can be applied are quite varied. One possibility is that the point is not an edge (as defined by the magnitude of its edge function response). More often, the criteria use the brightness value of the pixel. If the difference between the new pixel and the neighbor is less than an arbitrary limit, the pixel is considered to belong to the same feature. Other criteria, including comparisons based on texture, are essentially identical to the global application of the same parameters prior to thresholding, as described in the next chapter. The process continues until no neighbor pixels can be found, whereupon the feature is deemed to be complete and a new one is begun. Again, the process continues, usually until all pixels have been tested and assigned.

Because the comparison is local, it is possible for features to vary considerably in brightness across their width or height, so long as the local variation is small. This accommodates shading of objects due to nonuniform illumination or surface curvature, for instance. However, like edge-following, region growing is serial, and is usually a much less efficient method than some of the other techniques for edge finding.

An apparently opposite technique to region growing proceeds from the entire image down rather than from the individual pixels up. A large region is tested for uniformity, for instance by comparing the brightness values for the pixels. If it is found to be nonuniform, then it is subdivided (Bright, 1987). This can be done intelligently by finding an internal boundary, but in the most basic version of the approach simply separates the original region into halves or quarters. Each of these is then tested, and perhaps subdivided again (Figure 4-19). There is also a way to compare adjacent regions and decide whether they should be combined. The final result of this "split and merge" is a collection of features selected as having some uniform brightness character that can be distinguished from surrounding regions. This method is often used in machine vision applications, but not in more general image analysis applications.

Note that in many of these methods the boundary between regions will usually lie between pixels, rather than occupying a pixel. Indeed, this is a rather basic difference between the results of the techniques described in this chapter and in the next. Some segmentation methods produce features that can exactly touch each other with a boundary that is mathematically present between pixels, but has no physical width. The others

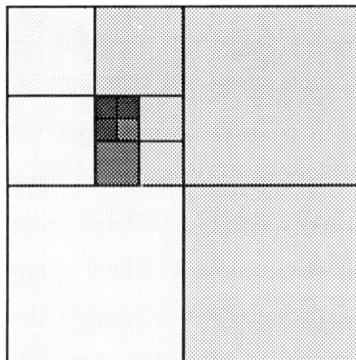

Figure 4-19: Top-down split and merge. Each region is subdivided if its brightness values vary.

produce a boundary that is itself actually an object. That is, it has width (usually a single pixel) and separates the objects on each side.

In either case, it is necessary to decide how the boundary and feature data are to be stored. When the pixels in features are stored (as discussed in the next chapter), the image is always of known size (depending on the image size) and is readily accessible for measurement. However, the identification of features and the determination of the boundary (for instance to measure the perimeter) around them must be performed when it is needed.

On the other hand, boundary representation (generally called "b-rep") produces a list of boundary lines, with information on how they are connected. If each object has a distinct boundary and objects do not touch or overlap, then b-rep curves automatically identify objects and give perimeters straightforwardly. The number of parameters needed to represent each boundary depends on the complexity of lines that are allowed (whether they are straight lines or curves, for instance). The b-rep lines are mathematical representations of the boundaries and generally pass between rather than occupying pixels. The list of such lines is as long as it needs to be for any given image, and to determine whether a particular pixel lies within any given object (for instance to measure its brightness) requires either solving equations or drawing the lines and converting them to a pixel-based representation.

Most current image analysis methods utilize a pixel-based rather than a boundary-based representation of objects. This partly results because of the great efficiency and wide applicability of the discrimination methods discussed in the next chapter. It is also true that most of the things that we wish to measure about objects, such as area and density, are most directly obtained from a pixel-based representation. Some parameters, such a perimeter, can be gotten more easily from the boundary representation, and when a pixel-based object is measured the boundary must be constructed and its length summed.

There are a few specific applications that demand a b-rep treatment, however. One is serial section reconstruction, discussed in Chapter 12. The representation of a large body of data (outlines of interesting features in a large number of parallel planes) would require massive storage. Drawing views of the three-dimensional arrangement of these features from any viewpoint, and the real-time rotation of such views on the computer screen, is greatly facilitated by simply storing the edge data, which is usually simply a list of points that form polygonal representations of the outlines. Furthermore, these points are usually directly obtained as the user traces the outlines of features with a tablet or other manual digitizing device.

Another important use of b-rep is the measurement of the widths of traces and metallization on integrated circuits, and the exposed layers of photographic emulsions used to produce them. These are examined either by light microscopy or scanning electron microscopy, with the latter gaining in importance as devices become smaller and higher magnifications are required. The boundaries of the structures as composed almost exclusively of straight line segments. We need not discuss here the complications of defining just what the "width" means for a structure whose sides are sloped and perhaps irregular. Depending on the method of imaging, using either secondary or backscattered electrons, it is possible to perform mathematical modelling of the processes of generation, emission and detection to develop relationships between the signals generated and the composition and physical dimensions of the stripes.

Finding the edges in the image analysis sense is a classic example of the power of boundary representation. In many situations, the surfaces upon which the stripes to be measured are present are carefully aligned so that the edges run in a vertical or horizontal direction on the image, and the viewpoint is precisely normal to the surface. The means for relaxing the latter criterion lies in rubber-sheeting of the image. The former makes it possible to use somewhat simplified derivative functions to find the boundaries, although in actual fact it does not improve either speed or accuracy of the resulting computation.

A derivative is performed on the brightness information, typically in a single direction normal to the presumed stripe orientation but averaged over several rows of pixels (probably several scan lines, depending on the means of image acquisition). The second derivative produces a profile whose extrema can be taken to mark some reproducible point on the physical profile, and the distance between maxima is taken as the width. Changes in profile, or any other changes in the structure, composition, or imaging conditions may significantly alter the results. But in most cases the measurement is more concerned with assuring consistency (precision) than in accurately measuring the actual dimensions. This is discussed further in Chapter 10.

This is a b-rep method because the boundary position is used directly, and is interpolated to a fraction of a pixel width by using the derivative of the brightness values, and then fitting a smooth function (polynomial, Gaussian, Lagrangian, etc.) to the extrema peaks to locate their exact center from the digitized values. It seems likely that better results could be obtained using the Hough method described below because more of the information in the image would be used, but most commercial or in-house constructed systems use the existing one-dimensional approach. This is probably due more to limited knowledge of image analysis methods than to any concern about available computer power, speed of obtaining results, or standards testing of results to compare the various methods.

Another application that is best performed using b-rep data is the fusion of stereo pair images. This is discussed in detail in Chapter 11. It is not a measurement operation, but a reconstruction of a three-dimensional structure from two two-dimensional images. This works better with boundaries than with pixels because it is the boundaries that best represent the edges of objects or surfaces, and there are fewer of them to match up between the left- and right-eye images. This is also the way in which the human visual system achieves fusion of stereo pairs, and the speed and ease with which we do so is

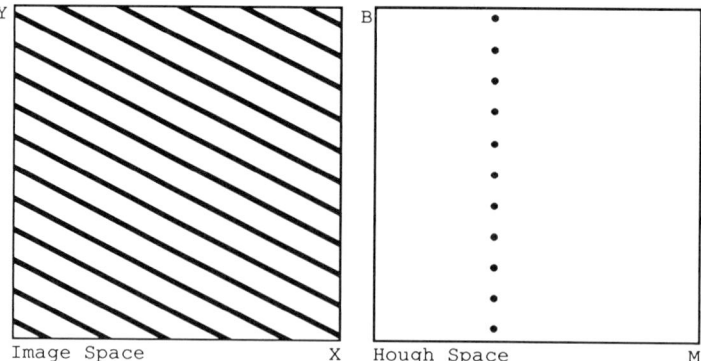

Figure 4-20: Example of a linear Hough transform

further testimony to the power of boundary representation (which, it has already been pointed out, is primarily the way our neural image processing hardware functions).

The Hough transform

The Hough transform (Hough, 1962; Duda & Hart, 1972) is a particularly powerful technique for extracting information about lines from an image. The lines can be boundaries, or any other structure present. As an initial example, we will consider finding straight lines in an image, such as the markings on a diffraction grating that is partially obscured. The technique is applied by defining a space in which the lines can be defined. A straight line requires two parameters, the slope M and intercept B in the expression $y = Mx + B$. Accordingly, the Hough space is a plane whose axes are M the slope and B the intercept of all possible lines. Usually, the Hough transform is constructed in another image memory, with some practical limits on M and B to allow covering the range of interesting lines.

Hough space is initially empty, because we do not know what lines may be present in the sample. This information is provided by the original image. For each point in the image, a series of M and B values represent all of the lines that could possibly be drawn through the point. Graphically, these lines all pass through the point and so have values of M (slope) and B (intercept) that are directly related. They form a straight line in the Hough transform plot. In other words, points in the image are mapped into lines in the Hough space that represents straight lines in the image (Figure 4-20). Each of the points in the original image produces a different line in Hough space. But they are not all equivalent. Each is weighted according to how likely it is that the original point actually lies on a line. This could be simply the value of one of the edge-finding functions at that point, such as the Sobel operator discussed above.

As points are mapped from the original image to lines in Hough space, weighted in brightness according to the magnitude of their probability for being part of the lines we seek, the result is a superposition of many lines that produces maxima in the Hough plot. Each maximum corresponds to a single line, because the inverse of the Hough transformation also produces a line from a point. When the transform is complete, the maxima in the plot identify the M and B values which will give the unique lines in the original image space which are represented.

The maxima can be found in a variety of ways, including discrimination as discussed in the next chapter, and may be selected as significant either based on their maximum value (the sum of the weights of points that voted for them, encoded as the brightness of the Hough image), or by taking some number of points starting with the brightest. In any case, the location of the points can be determined by fitting to much better accuracy than the size of one pixel in Hough space, and the lines are located with much better accuracy than the size of pixels in the original image space.

The Hough transform is not limited to finding straight lines. It can be used for any kind of structure that is defined by simple mathematical relationships (Ballard, 1981). For the case of circles, a particularly useful application is locating and measuring electron diffraction patterns (Figure 4-21). These are produced in the transmission electron microscope by diffraction of electrons from planes of atoms in the sample. Measuring the radius of the rings yields the atomic spacing of the planes, according to the Bragg law $n \lambda = 2 d \sin \theta$, where λ is the wavelength of the electrons (known from the accelerating voltage used in the microscope), n is an integer (usually 1) and θ is the half-angle of deflection of the scattered beam. In the case of the electron microscope, the distance from the specimen to the film where the pattern is recorded is so great that the angle and its tangent are essentially equal, so the result is that the radius of the circles made by electron diffraction spots is linearly related to $1/d$–spacing, with a proportionality constant that is usually determined for a given microscope by measuring a known standard material such as a gold foil. The total intensity in each ring is also proportional to the density of atoms on a particular set of planes in the crystal structure.

Electron diffraction patterns from unknown materials are difficult to measure accurately for several reasons. First, they are often not uniform continuous rings because the sample may consist of a very limited number of grains, particles or crystallites of material which each produce a single spot on the film. Along any line drawn across the pattern, some rings may not contain any spots and so cannot be easily measured (Bright & Steel, 1988). This is particularly true for rings corresponding to planes that contain a low density of atoms, yet these planes may be very important for describing or understanding the structure (particularly when they represent minor phases present in the material).

Secondly, it is not easy to locate the center of the rings. Any preferred orientation in the particles produces incomplete ring arcs whose center cannot be easily located. The direct beam is many orders of magnitude more intense than the diffracted spots, and overexposes the film so that lateral spreading produces a very broad spot. Usually this is blocked by a beam stop, so that there is no center recorded on the film at all.

 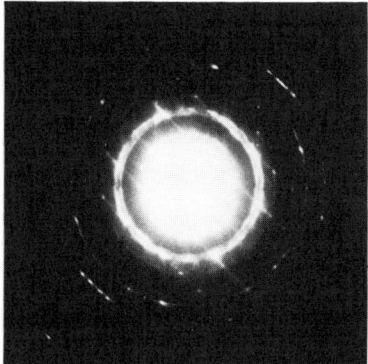

Figure 4-21: Example of a selected area electron diffraction pattern from asbestos fibers.

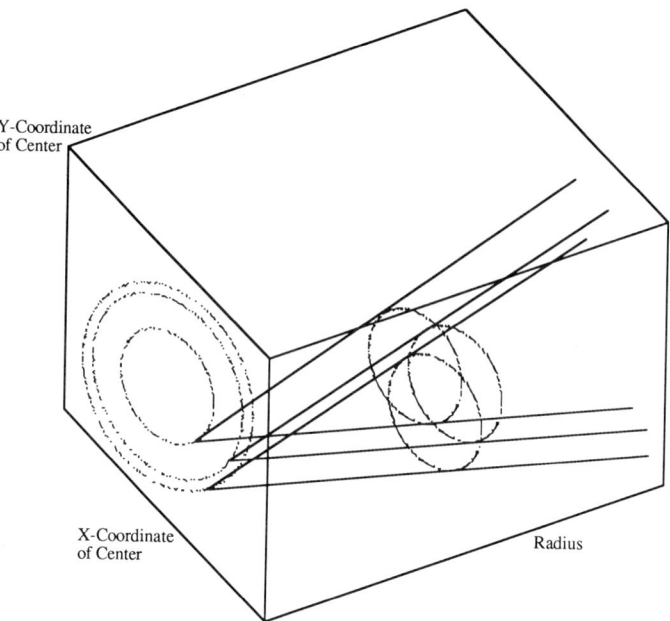

Figure 4-22. Three-dimensional Hough transform space for a circular transform to fit points on electron diffraction patterns.

In addition to locating the true center of the pattern, it is harder still to obtain a plot of integrated ring intensity as a function of diameter or d-spacing. Linear profiles of intensity vs. position along a diameter differ seriously from the integrated intensity from a spotty ring. Obtaining a good plot of integrated intensity vs. d-spacing is very important in many applications because it allows spectrum analysis methods to be used to accurately locate peaks in the plot. Small peaks that would otherwise be missed may reveal the presence of minor phases, twins or superlattice lines. Small shifts in peak position reveal changes in lattice parameter. Both kinds of information are routinely sought using bulk X-ray fluorescence methods. Making them available from electron diffraction patterns would be particularly useful in the examination of a number of materials of current interest, which are heterogeneous on a fine scale and require electron microscopy for structural analysis and interpretation.

Using the Hough transform to locate circles in the image requires three parameters for each circle (the x and y coordinates of the center, and the radius), so Hough space is three dimensional. Each point in the original image votes for all of the circles that could pass through it, in proportion to its original intensity (in practice, this may either be the absolute intensity, or preferably the intensity relative to local background in the diffraction pattern image).

The result is that each point in the diffraction pattern generates a cone in Hough space, as shown schematically in Figure 4-22. The apex of each cone lies on the original spot in the diffraction pattern, which is the zero-radius X,Y plane in the transform. The intersections of the cones produce lines, and the intersections of the lines produce spots. The spots represent the circular features (rings) present in the original pattern, and their total weight represents the integrated intensity of the ring (Russ et al., 1989).

Segmentation of Edges and Lines

In practice, the analysis of the circular Hough space transform proceeds by locating the center as the brightest spot of the projection of the entire space onto the X,Y plane. This can be interpolated to less than the dimension of the pixel spacing used to represent the pattern. Then the summed weights along the line normal to the X,Y plane at this point is the desired intensity vs. radius (or d-spacing) spectrum, as indicated in Figure 4-23.

The circular Hough transform has also been used to measure the radius of curvature of pot sherds from an archaeological site, using an image analyzer to view the sherds, as will be discussed in Chapter 7. The linear Hough has been used to locate fault lines in aerial photographs. It is a computer-intensive method that has great power in situations that can be defined beforehand, and is much under-utilized at present.

Mention was made earlier of the problem of incomplete delineation of edges and lines. The Hough transform can also be used to fill in gaps in edges and to locate and link edge points in an image. The Hough space used for this is two-dimensional, in ρ and θ. Only points with a large edge magnitude value are used. The θ value comes from the angle of the gradient (for instance the Sobel or Roberts' cross operator's arc tangent function. The ρ value is then calculated for the point as $x \cos \theta + y \sin \theta$, using the x,y coordinates of the pixel location. Hough space is then incremented by the magnitude of the edge gradient. Spots in the resulting Hough space represent the lines that link and complete broken edges (O'Gorman & Clowes, 1976)

Touching features

The final subject to be mentioned in conjunction with boundaries between features is dealing with objects that touch somewhere along their periphery, such as an array of marbles. When part of the boundaries of these objects are recognizable in the image, the boundaries can be completed in several ways. For known shapes, such as the marbles, a Hough transform (in this example, it would again be a circular Hough) could be used to find the boundaries. In other cases, local fitting of mathematical functions such as

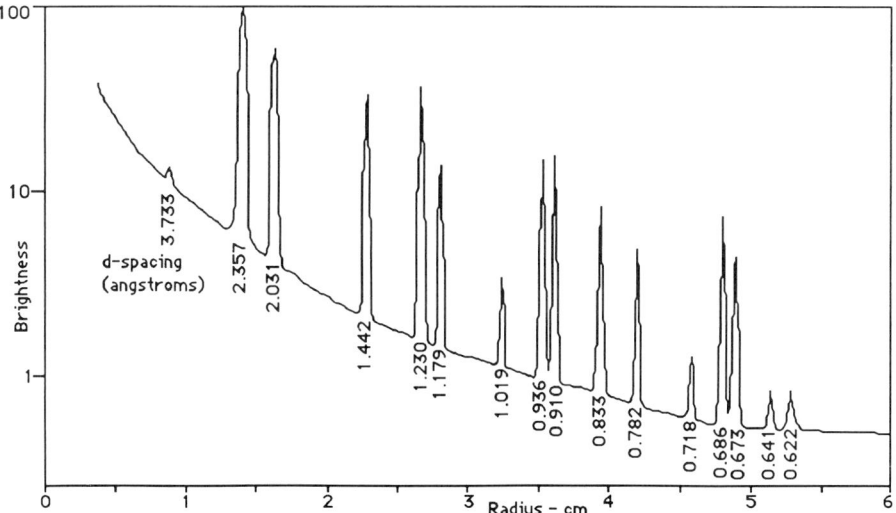

Figure 4-23. Integrated (circular) density along the central axis in Hough space for the electron diffraction pattern from gold (Figure 2-3), with the d-spacing of the peaks calculated from the measured radius marked.

polynomials can be used to extend boundary lines across gaps. These methods are usually rather image-specific, and hard to generalize.

Another method for segmentation of partially touching boundaries will be discussed in Chapter 6. This deals with the editing of binary images, and is applied once the image has been reduced to a pixel-based representation of objects. It is called convex segmentation, or segmentation by watersheds, and produces the desired separation of touching features provided that they are locally convex and only partially touching.

Manual outlining

It was mentioned earlier that manual outlining methods produce a boundary representation of features directly. The edge recognition capabilities of the human eye are employed to locate the boundaries of the features of interest and to associate them individually with features, ignoring internal discontinuities, interpolating where noise blocks an edge, and so forth. This is easy for humans to do. We see things in terms of their edges, which is why cartoons work (people don't really have black outlines around them, after all).

People are also usually reasonably adept at pointing. A variety of pointing devices are available that allow the direct entry of coordinates to the computer. These include devices such as graphics tablets or digitizing tables, on which photographs of images can be taped, or sometimes projected from either the top of bottom. In this case the human directly "draws" on the image with a pen, while the coordinates are sent to the computer. In other arrangements, the location of a marker or cursor controlled by the user is superimposed on the image as it is viewed. This may be done either optically or electronically. In the optical case, a bright point of light attached to a movable cursor may be viewed by a lens and superimposed so that it appears to lie on the specimen being viewed in a light microscope or similar arrangement. The electronic superimposition of cursors on video images is a newer technique, now quite common. In principle it offers lower resolution (limited by the number of points and lines in the video image), but if the images themselves are considered to be adequate for measurement then the precision of cursor position is the same.

Coordinates provided by these devices are either absolute or relative, but in either case the software that tracks the users motions constructs a series of lines joining the points, either as they are discretely marked by the operator or as a time series during continuous motion. When continuous point recording is used to follow tracing, the spacing of points may either be taken on a uniform time basis (which will make the spacing a function of drawing speed), or with uniform spacing as the cursor is moved. The latter is generally preferred provided that the resolution of the motion is adequate to define the intricacies of the feature outlines.

A recurring problem with manual outlining and digitization of coordinates is that any jitter in the drawing hand produces an increase in the irregularity of the outline and a corresponding increase in the length of feature perimeters. It is possible to smooth the boundary by weighted averaging of neighboring points, or by curve fitting to the points. This is rarely entirely satisfactory unless there is some independent information available about the boundary's shape and regularity. Another approach requires that the sensitivity of the hardware and software to incremental motion must be adjusted to the scale of the features being traced and the steadiness of the user's hand. Widely or irregularly spaced points may conceal local roughness.

Another common problem is that people tend not to draw on lines very well. In many cases they draw just outside the boundary line, making objects larger than they really should be. This tendency varies from person to person, and also seems to correlate with the relative brightness of features and background (that is, which is brighter and which is darker, and by how much). It is one of the reasons that manual entry is subject to greater errors and variation than automatic methods.

The boundary representation obtained in this way is distinct for each feature, and directly usable for some of the measurements to be discussed in Chapter 7, such as the perimeter and area. It can be stored as a sequence of absolute coordinates or as a series of vector directions for each side of the polygon. For some purposes it is desirable to convert it to a pixel based representation, and in this case the absolute coordinates are much preferred. However, edge following and some other methods produce b-rep tables that are not absolute and require additional logic to convert to a pixel image of the features.

The conversion process is identical to the vector to raster conversion process used in many aspects of computer graphics. It is most easily accomplished by writing directly to memory. The logic of finding the pixels along an edge or line requires deciding between 4- or 8-neighbor connectedness. Low level algorithms exist for drawing lines with minimum "aliasing" or "jaggies" due to the stair-step arrangement of pixels. A more complicated problem is filling in of feature areas. This requires knowing which side of the line is the inside of the feature, and also dealing with internal holes. Starting at a point and flooding the interior of the feature requires the ability to follow any direction. It is usually done by filling in one direction at a time while maintaining a "stack" of pending branch points. When each fill direction is completed by encountering a boundary line, the next branch point is fetched and the process repeated. New branch points are added to the stack whenever they are encountered. The process is iterative, and can be time consuming for complicated shapes.

References

D. H. Ballard (1981) *Generalizing the Hough Transform to Detect Arbitrary Shapes* Pattern Recognition 13, 111-122

D. H. Ballard, C. M. Brown (1982) Computer Vision Prentice Hall, Englewood Cliffs, NJ

V. Berzins (1984) *Accuracy of Laplacian Edge Detectors* Computer Vision, Graphics, and Image Processing 27 1955-210

D. S. Bright (1987) *An object finder based on multiple thresholds, connectivity and internal structure* Microbeam Analysis 1987 (R. H. Geiss, ed.) San Francisco Press, 290-292

D. S. Bright, E. B. Steel (1988) *Automatic Extraction of Regular Arrays of Spots from Electron Diffraction Images,* J. Microscopy, in press

K. R. Castleman (1979) *Digital Image Processing* Prentice Hall

R. O. Duda, P. E. Hart (1972) *Use of the Hough Transformation to Detect Lines and Curves in Pictures* Comm. Assoc. Comput. Mach. 15 11-15

R. O. Duda, P. E. Hart (1973) Pattern Classification and Scene Analysis Wiley, NY

E. C. Hildreth (1983) *The detection of intensity changes by computer and biological vision systems* Computer Vision, Graphics and Image Processing 22 1-27

P.V.C.Hough (1982) *Method and Means for Recognizing Complex Patterns* U. S. Patent 3069654, Dec. 18, 1962

M. H. Hueckel (1971) *An Operator Which Locates Edges in Digitized Pictures* J. Assoc. Comput. Mach. 18 113-125

D. C. Joy (1987) *A model for secondary and backscattered electron yields* J. Microsc. 147, 51

A. I. Konomopoulos (1982) *An approach to edge detection based on the direction of edge elements* Computer Graphics and Image Processing 19 179-195

D. Marr (1982) Vision, W. H. Freeman, San Francisco

D. Marr, E. Hildreth (1980) *Theory of edge detection*, Proc. R. Soc. Lond., B207, 187-217

F. O'Gorman, M. B. Clowes (1976) *Finding Picture Edges Through Collinearity of Feature Points* IEEE Trans. Computers C-25 449-456}

W. K. Pratt (1978) Digital Image Processing, John Wiley, New York

A. Rosenfeld, A. C. Kak (1982) Digital Picture Processing, Academic Press, London

J. C. Russ, T. Taguchi, P. M. Peters, E. Chatfield, J. Ch. Russ, W. D. Stewart (1989) *Automatic computer measurement of selected area electron diffraction patterns from asbestos minerals* in Advances in X-Ray Analysis vol. 32 (C. S. Barrett et al., ed.), Plenum Press, New York, 593-600

J. C. Russ, D. S. Bright, T. M. Hare, J. Ch. Russ (1989) *Application of the Hough transform to electron diffraction patterns* Journal of Computer Assisted Microscopy 1, 3-38

K. C. A. Smith, B. M. Unitt, D. M. Holburn, W. J. Tee (1977) *Gradient image processing using an on-line digital computer*, SEM 1977/I, SEM Inc., Chicago, Ill, 49-56

I. Sobel (1970) *Camera models and machine perception* AIM-21 Stanford Artificial Intelligence Lab., Palo Alto CA

G. J. Yang, T. S. Huang (1981) *The effect of median filtering on edge location estimation* Computer Graphics and Image Processing 15 224-245

Chapter 5

Discrimination and Thresholding

There have been several references in the preceding chapters to the use of brightness discrimination to select pixels belonging to features of interest. This is a widely used method of converting a grey scale image to a binary (black and white) one, illustrated in Figure 5-1. Discrimination with threshold values is much more efficient than any edge following or region growing method (as discussed in the previous chapter) because it works on the entire image at once. Hence the time required is fixed, regardless of the complexity of the image, and very short. Also, the resulting binary image is a pixel-based representation of features of interest, and is easier for most measurement operations than the boundary representation that results from the location and identification of edges.

Brightness thresholds

In the simplest version of brightness discrimination, two threshold levels are established (either manually or automatically as will be discussed below). All pixels whose brightness values lie between these thresholds are considered to be part of the features of interest, and all others are considered to be background. In other words, figure and ground are separated.

Of course, the brightness range used for the thresholds need not be continuous. In principle multiple discrete ranges could be used, but since the resulting binary images can be combined anyway (as discussed in the next chapter), this really adds no new capabilities, and most image analysis systems do not provide capability for multiple thresholds. In some cases, in fact, older systems in which the discrimination was performed by analog voltage comparators on the incoming signal, so that the grey image was not digitized or stored, but converted directly to a binary one, had only a single discriminator threshold, and all darker (or brighter) pixels were selected.

Manual setting of discriminator thresholds is most often accomplished interactively, by the user turning knobs or manipulating other similar devices (including moving pointers on a video display with a mouse or other device), while observing the resulting binary image. With the manual/interactive method, it is the user's responsibility to select threshold values that properly delineate the features of interest. Of course, this presumes that the user knows what features are of interest, a point to which we shall return below. And all of these thresholding techniques that set global values for the entire image assume that there is no non-uniform illumination or that it has been corrected as discussed previously.

We are also not making the assumption here that this discrimination will only select features of one type. It may still be necessary to distinguish between several different classes of features that have the same brightness, but differ in other respects. Figure 5-2 shows an example: the graphite flakes in the cast iron are similar in brightness to the

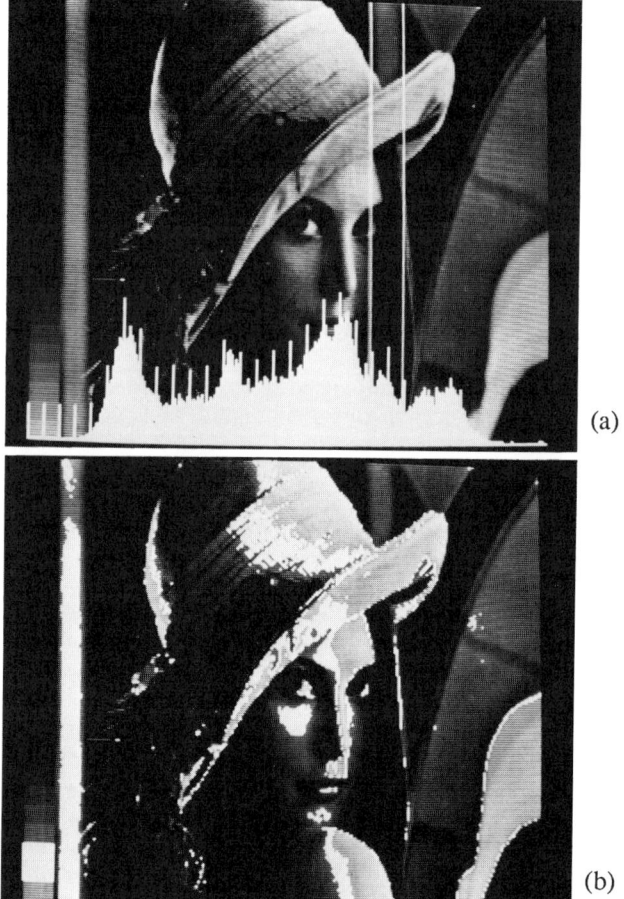

Figure 5-1: Girl image with brightness histogram (a) and overlaid binary image made by selecting pixels with brightness between two threshold levels (b).

nodules, and both are selected by discrimination. They can be separated later based on measured size or shape.

Thresholding an original grey scale image, in which the features of interest are different in brightness (either lighter or darker) than other features or background represents a very simple, almost trivial example of feature discrimination. That it is often adequate to prepare an image for counting or measurement of features is more an indication that only relatively simple tasks are usually attempted than a promise that all image analysis problems can be so easily handled. With transmission illumination, opaque objects that are well dispersed in a transparent medium are perhaps the classic example of this type of application. In materials science, many composite materials examined by incident (reflected) light also produce images of this type, because the phases of different composition are quite distinct in brightness or color. This is particularly true when trying to count or measure pores in solids, for example. The brightness difference is also found in many scanning electron microscope images, especially ones formed with backscattered electrons, because the production of these electrons is proportional to the atomic number of each phase.

In biological specimens the inherent contrast in images is much more problematic. Many structures in tissue sections have very similar compositions (overwhelmingly organic) and scatter or absorb light similarly. Sometimes the creative use of chemical stains can produce useful color or brightness variations that can be utilized for discrimination. Developing suitable staining techniques is not easy. However, if a color stain can be found that will selectively mark features of interest, it is then possible to use a complementary color filter to produce a monochrome (grey) image in which the background is light and the features dark. In remote sensing images (satellite photographs), the use of infrared or other false color images is another example of this same principle.

It is very important to understand that discriminating features with brightness thresholds need not be performed on original, as-acquired grey scale images. The smoothing operations discussed in Chapter 3 as a means of reducing noise, or background levelling methods that subtract one image from another, produce a new or derived grey scale image from the original one that can be discriminated to produce rather similar results to those for the original (hopefully better in that the uniformity of regions in the image is better). For the smoothed images, this seems natural and logical. The smoothing was applied to modify grey scale values only slightly, and the image has much the same appearance as the original. Similarly, background levelling by subtracting (or dividing) one image by another was intended to modify the brightness values of pixels specifically so that similar features in different locations would have the same brightness, so that discrimination by brightness thresholds could successfully be applied.

Thresholding after processing

Many of the edge-finding algorithms described in the previous chapter also produce a new grey scale image for which discrimination can be used. When threshold discrimination is applied to an image produced by applying a Sobel gradient operator (for instance) to an original grey image, the situation seems different. We are not selecting

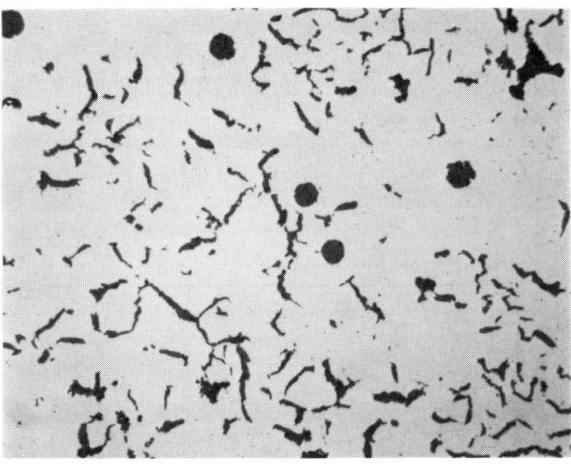

Figure 5-2: Cast iron with graphite flakes and nodules.

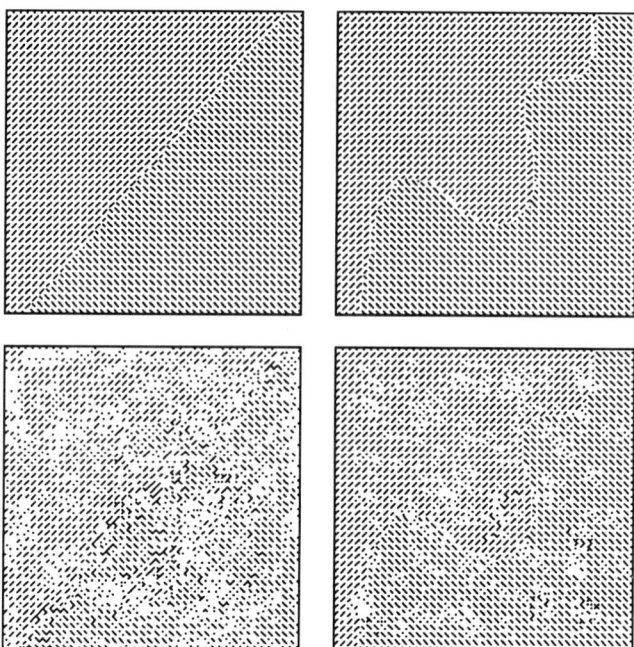

Figure 5-3: Segmentation by texture (O'Callaghan, 1974). The boundary is harder to find when the difference between the regions is less, when the boundary is not straight (top), and in the presence of noise (bottom).

features in the same sense as before. Instead, what we are selecting is regions (pixels) whose gradient, or rate of change of brightness, is high (or low, depending on the threshold settings). This may not seem intuitive at first, but in many cases it allows the selection of edge regions (or background regions) in the image.

There are other operators that can also be applied to an image before discrimination to produce particular effects or extract particular types of information (Bright & Steel, 1986). This permits selecting particular features from the image. For example, texture (Figure 5-3) is parameter which we will define in Chapter 10, which for the moment we may take as a measure of the irregularity of the brightness of neighboring pixels within a feature. If very rough particles (such as pollen grains) are dispersed on a smooth substrate (such as glass), the average brightness of the features and the background may be quite similar. But the glass is smooth and the particles are rough. This shows up in the image as a texture difference in which the pixel-to-pixel variation in brightness is large. There as several ways to extract this information, one of which is the Laplacian operator. It calculates the mean difference between each pixel and its 8 neighbors. The absolute value of the result will be small on the smooth surface and larger on the particles. The derived grey image that results from the application of this operator thus has different brightnesses for the background and the features of interest, and they can be discriminated using a threshold setting.

The Laplacian is not necessarily the best operator for this purpose. In many cases it would be preferable to find the greatest difference between a pixel and its immediate neighbors, or the greatest range of values in the location, rather than the mean difference.

This operator can either be constructed specifically for the purpose, or the equivalent operation can be performed by using the rank operator described previously. The rank operator stores in place of the original pixel, either the brightest or darkest value present for the pixel or any of its immediate neighbors. It was used in an earlier chapter to get an estimate of a varying background for image levelling purposes. In this case, we can use it to obtain a derived image which is then subtracted from the original, or two derived images (minimum and maximum) which are then subtracted as shown in Figure 5-4. The absolute value of the difference is the desired maximum local contrast value.

If there is reason to expect or to be interested in a texture running in a particular direction, then the derivative operator used in either the Sobel or Kirsch function can be applied in that single direction, and the brightness of the resulting image used as a measure of that texture. This is often useful when scratches or striations (either natural or resulting from a manufacturing process) on surfaces are of concern.

Also, as will be discussed in Chapter 7, there are other measures of texture that utilize the brightness variation over a greater distance than the nearest neighbors. The terms contrast and texture may be applied to two different parameters. Contrast is used for the mean difference between a each pixel and its immediate neighbors, and texture is the slope of the plot of mean brightness difference versus distance.

There are numerous other measures of texture, which were initially developed for the interpretation of satellite photos. Haralick et al. (1973) defines fourteen measures of image texture calculated as sums of terms as illustrated below, based on the difference in brightness values between each pixel within a feature and its neighbors. In these expressions, the array $P(i,j)$ contains the number of nearest neighbor pixel pairs (in 90 degree directions only) whose brightnesses are i and j, respectively, and R is a renormalizing constant equal to the total number of pixel pairs in the image. In principle this can be extended to pixel pairs that are separated by a distance d, and to pairs aligned in the 45 degree direction (whose separation distance is greater than ones in the 90 degree directions). The sums represent the entire feature. Haralick applied this to rectangular regions, but it is equally applicable to pixels within irregular outlines.

The first parameter shown is a measure of homogeneity using a second moment. Since the terms are squared, a few large differences will contribute more than many small ones. The second one shown is a difference moment, which is a measure of the contrast in the image. The third is a measure of the linear dependency of brightness in the image, obtained by correlation.

$$f_1 = \sum_{i=1}^{N}\sum_{j=1}^{N} \left(\frac{P(i,j)}{R}\right)^2$$

$$f_2 = \sum_{n=0}^{N-1} n^2 \left\{ \sum_{|i-j|=n} \left(\frac{P(i,j)}{R}\right) \right\}$$

$$f_3 = \frac{\sum_{i=1}^{N}\sum_{j=1}^{N} \left[i \cdot j \cdot P(i,j)/R\right] - \mu_x \cdot \mu_y}{\sigma_x \cdot \sigma_y}$$

In these expressions, N is the number of grey levels, and μ and σ are the mean and standard deviation, respectively, of the distributions of brightness values accumulated in

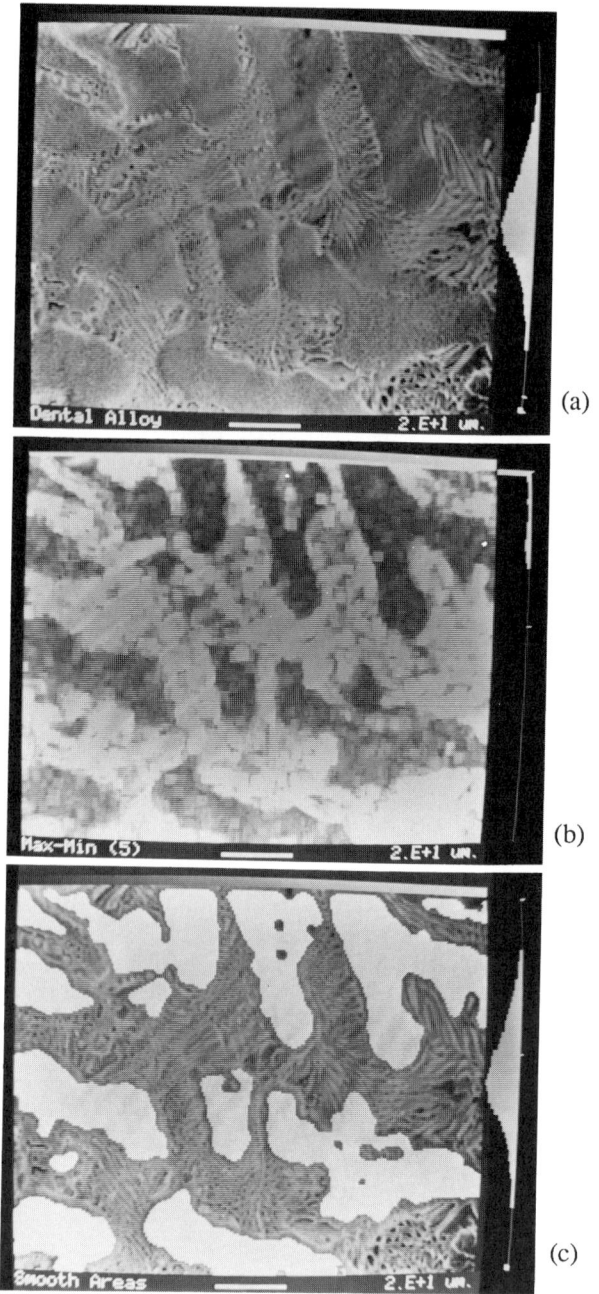

Figure 5-4: Application of a Rank operator to distinguish the smooth regions (single phase metal structure) from a visually rough one (a eutectic structure consisting of lamellar plates of two phases). a) original image; b) the difference between the 5x5 Max and 5x5 Min rank filter; and c) the region selected by setting the thresholds for the high texture region.

the x- and y-directions. Additional parameters describe the variance, entropy, and information measure of correlation of the brightness values. Haralick has shown that when applied to large rectangular areas in satellite photos, these parameters can distinguish water from grassland, different sandstones from each other, and woodland from marsh or urban regions.

Some of these are obviously easier than others to calculate for all of the pixels in an image. The resulting values can be scaled to create a useful derived image that can be discriminated with thresholds. In any given instance it sometimes requires experimentation with several texture operators to find the one that gives the best separation between the features of interest and their surroundings.

Regardless of which operator or operators are used, the key purpose of this operation is to replace the original grey scale image with a new, derived image in which the brightness represents some different parameter in the image than feature brightness. There are other examples that can be cited. For instance, consider applying a Sobel operator but using the gradient orientation rather than its magnitude to produce the derived image. The result would enable separation of regions with striations running in different directions. It often takes a practiced eye to look at an original image and determine what it is that distinguished the features of interest from their surroundings, and then to choose an image operator that will respond to that difference to produce an image suitable for discrimination with thresholds. Few systems are yet capable of applying artificial intelligence methods (which will be discussed in a later chapter) to automatically determine which parameters vary significantly between two regions of the image.

Selecting threshold settings

The most direct and primitive method for setting thresholds is simply allowing the user to look at the binary image as he or she changes the settings. When the features look right (are well separated from background and hopefully from each other) then the binary image is used for subsequent measurements. Unfortunately, this method may not give

Figure 5-5: Light microscope image of metal alloy with histogram of brightness shown at the right (white at top, black at bottom). Four distinct phases are evident.

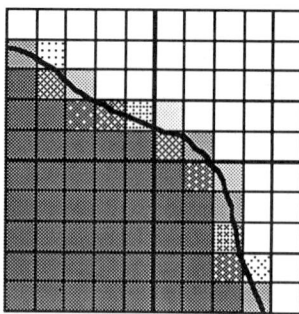

Figure 5-6: Boundary pixels average the internal and external brightness.

very reproducible results. When different features are present in different image fields, changes in the shape or location cause the operator to select different settings, and this may bias the measurement results.

It is always helpful to have the brightness histogram of the image available when setting thresholds. The histogram is a count of the number of pixels in the image that have each possible value of brightness, as shown Figure 5-5. The size of a peak in the histogram is a direct measure of the area fraction of the corresponding phase on the image. The ideal form of a brightness histogram has a peak corresponding to the pixels in the features which is well separated from the peak(s) due to background or other features which are not currently of interest. When this is the case, setting the thresholds around the peak to produce a good binary image for measurement is quite simple. Remembering the settings (either with respect to the peak position, or as absolute numerical values) allows reproducing the same setup for subsequent images of the same type.

In many cases, the situation is less ideal. Even if the feature peak is well separated from surrounding pixel brightness, there is still generally a background in the histogram due to pixels along the feature boundaries. Because the actual boundaries of the physical objects in the image do not correspond exactly to the pixels, some pixels straddle the interface as shown in Figure 5-6. The measured brightness of these pixels is intermediate between the brightness of the features and of the surroundings, and the particular value depends on the chance alignment and the fraction of the pixel that falls in either region.

This means that all possible intermediate brightness values between the levels of the features and the surroundings may be represented, and in the histogram this shows up as a continuous background between the peaks. Wherever the threshold is set in this region, it will include some of those boundary pixels in the features and exclude others. Moving the threshold toward the peak corresponding to the features will exclude more pixels and make the objects smaller, and vice versa. What is required is some consistent way to set the thresholds.

One thing that may help is obtaining a brightness histogram that excludes these pixels near edges. To accomplish this, it is first necessary to introduce the idea that the brightness histogram need not be taken from the entire image. A brightness histogram can be obtained from any set of pixels, and in fact will sometimes be usefully taken from the pixels within a single feature. This can be done by using the binary image of the feature or features as a mask, whose pixels mark the locations of interest. Then the brightness values of the original grey image are read only at these pixel locations and used to construct the histogram. This is usually done with the original (or perhaps a smoothed or background levelled) image, which may not be the image from which the binary itself was obtained.

Since it is possible, as described in the previous chapter, to use one or another of the gradient operators to produce an image that marks the edges and near-edge pixels, this image can be discriminated to mark all pixels with neighbors that do not vary significantly. Using this binary image as a mask allows the creation of a histogram of just the smooth regions of the original grey images. Usually this will eliminate the boundary pixels and produce histogram peaks representing the interior of objects. From such a histogram (Figure 5-7), it is sometimes easier to locate threshold brightness levels which will separate the peaks consistently and so demarcate the different phases or regions (Kanpur et al., 1985).

Since in the general case, there is no a priori way to know how many distinct phases will be present, or whether the phase(s) of interest are dark, light, or some intermediate shade of grey, it is up to the operator to indicate the approximate range of values to be discriminated. Often it is desirable that this be done carefully on the first of a series of images, after which the machine will perform similar discrimination (and measurement) on the remainder of the set, perhaps as they are acquired from a microscope with an automated stage. It is in this last context that the methods to be described have evolved, although they are equally suitable for other applications.

The need for automatic thresholding

Errors in the threshold settings will obviously cause bias in any measured parameters used for stereological interpretation. This applies both to global parameters such as area fraction, total length of boundary lines, etc., as well as feature specific data including area, perimeter, and so forth, for each separate object in the field of view (Russ, 1986). Using the same settings on different images from the same or similar samples may be inappropriate if the overall brightness level or contrast changes, or if the sample preparation (thickness of a transmission section, degree of etch for a surface, etc.) varies. This can contribute a major source of error in the final results.

Automatic thresholding can be accomplished in several ways (Rosenfeld, 1979; Russ & Russ, 1988; Rigaut, 1988). None is perfect in all cases, nor do they necessarily agree

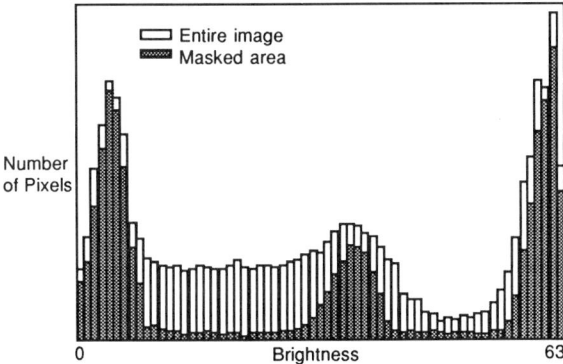

Figure 5-7: Brightness histogram of just the uniform regions (dark) in an image, showing improved peak definition as compared to the histogram for the entire image.

with each other or with the careful settings of a skilled operator. Nevertheless, interest in improving the algorithms used for automatic threshold selection continues to be high, to improve reproducibility and perhaps accuracy of measurement using otherwise highly automated systems.

We will not deal here with local ("adaptive") thresholding techniques, in which the image is subdivided into sections (ultimately to its individual pixels) and each is uniquely thresholded (Bright, 1987). This is a computationally much more demanding approach than finding an optimum setting for the entire image, and is most typically applied to images such as those from satellite mapping or reconnaissance, in which there is no expectation of uniform illumination across the image. In microscope work, it is usually (but certainly not always - see Bright & Steel, 1986) permissible to expect that each field will be uniformly illuminated, or that the nonuniformity can be measured directly and stored for subsequent levelling of acquired images, or that it can be computed from the image itself using a low-pass filter or by mathematical fitting.

Automatic methods

The following four automatic methods are considered separately below, but for clarity and comparison are defined in the following list (Russ & Russ, 1988). In all cases, if either threshold setting is at the black or white extreme limit, it is not altered by these automatic methods. The methods are applied to one or both threshold settings using the values set by the operator (using the first one or several images) as a first approximation. There is an arbitrary limit set on how far the automatic adjustment routine can depart from the nominal setting. This limit is especially important if there may be fields that do not contain the phase of interest, or are entirely different than the initial setup field, as without it any automatic method may seek out a completely different phase for measurement. An arbitrary limit of ±7 brightness levels out of 64 or ±15 out of 256 is usually suitable for this purpose.

1. *Histogram minimum method:* Each threshold is adjusted to the nearest minimum in the brightness histogram. The rationale for this approach is that the minimum between two peaks in the histogram (each peak presumably corresponding to a phase in the material) is relatively easy to define, and is the point at which peaks belonging to each phase are most recognizably separated (Prewitt & Mendelsohn, 1966; Wall et al., 1974; Weszka, 1978).

2. *Minimum area sensitivity:* Each threshold is adjusted to the point in the histogram at which the sensitivity of the discriminated area to changes in the threshold setting is minimized. This is equivalent to finding the minimum in the slope of a plot of area versus threshold settings. It can be rationalized as being the point at which a small error in the setting makes the least difference in subsequent area measurements (Weszka, 1978).

3. *Minimum perimeter sensitivity:* Each threshold is adjusted to the point at which the sensitivity of the boundary length to changes in settings is minimized. This can be visualized as finding the point at which a plot of the slope of the total perimeter around features is minimized as a function of threshold setting(s). The rationale for the method is that the pixels near the edge of objects vary in brightness between the two extremes, and thus give rise to "noisy" or irregular outlines as threshold settings vary. This method finds the settings that produce the "smoothest" outline. Unlike the minimization of area sensitivity (number 2, above) this method not only cares how many pixels are included in the image as the threshold setting changes, but whether or not they touch existing features.

Discrimination and Thresholding

4. *Fixed percentage setting:* One threshold setting is held fixed at the extreme (either black or white as appropriate) and the other is adjusted to include a fixed percentage (typically about 15%) of the area of the image. This is appropriate for images which have been processed to locate edges (e.g., with a Laplacian, Sobel or Kirsch transform). These edges demarcate the outlines of the features, and will most often be filled to produce solid profiles for measurement (Moik, 1980).

Note that in the first three methods, it is possible for both threshold settings to be adjustable (i.e., the selection of a phase at an intermediate grey level), but experience suggests that if one setting is initially at either the black or white limit, then it should remain there (the selection of the darkest or lightest phase present). Usually when the phase of interest consists of the brightest or darkest pixels present, no brighter (or darker) objects will appear in subsequent images. Further, there may be clipping in the detector or electronics so that pixels whose "true" brightness would exceed the range of the analog to digital converter are limited to its minimum or maximum value. This represents no problem unless measurements such as density or texture are to be made, in which case it is important not to exceed the measurable dynamic range.

Some examples will be presented to clarify the operation of the methods and compare their performance. The methods described here have been implemented on a microcomputer which also performs the image acquisition, processing, measurement and analysis functions.

Histogram minimum method

Figure 5-5 shows a light microscope image of a metal alloy in which several phases with different grey scale brightnesses are evident. Each phase has a distinct peak in the brightness histogram. For the white phase, the upper level threshold is set to maximum. The lower threshold must then be placed between the brightnesses of the white and light grey phase.

Binary images resulting from different placements are shown in Figure 5-8. Since the histogram has a minimum between the peaks, it is possible to use this as a threshold setting. These cannot be used for other images from a different region of the same sample when the illumination is slightly altered because the peaks will have shifted in relative size (because the image contains different proportions of the phases) and in absolute brightness, but a minimum is still present and can be selected for automatic discrimination.

Figure 5-8: Binary images produced from the image shown in Figure 5-5, using the different threshold settings shown on the histogram for each. Note the changes in feature size, shape and area fraction. In this and other printed binary images, the black areas are the pixels selected by the threshold settings.

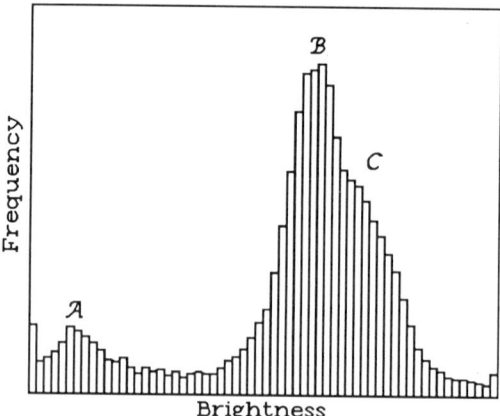

Figure 5-9: Example brightness histogram with overlapped phase peaks (B & C) with no minimum, and minor phase peak (A) far from and smaller than the major peak.

The minimum need not be found by simply looking for the lowest channel. The minima generally contain relatively few counts (few pixels), and are thus subject to statistical fluctuations. It is sometimes better to fit a parabola over several channels, up to the maxima on either side (which, because they have many more counts, are much smoother and better defined). Fitting over the range between the peaks seems preferable to an alternate method of finding the peaks (by fitting) and then arbitrarily setting the threshold midway between them. This method has been recommended (Castleman & Melnyk, 1976) based on the greater statistical confidence with which the peaks can be located, but does not take into account the fact that the minimum will not be midway between the peaks if they are not symmetrical or of the same size, or if their relative sizes change in different image fields.

Several problems can occur with the use of the histogram minimum in particular cases (Figure 5-9). First, if the peaks are not far enough separated (B and C in the figure), there may be no minimum, although in principal it should be possible to assume a peak shape and deconvolute the histogram to find the respective peak and phase areas. In practice, if two phases are so close in brightness that their histogram peaks are badly overlapped, it is not likely that any threshold setting will produce measurable profiles for one phase or the other. In this case, the use of a gradient or other local edge-finding operator is preferred, and this will produce a modified image with a very different brightness histogram.

Of more practical concern is the situation shown for peaks A and B in the figure. When the peaks are either very different in size, or far apart with a broad, flat minimum area between, or one peak is at the limit of brightness (black or white) and shows the effects of clipping, finding a consistent minimum is difficult, and small fluctuations in the image and histogram will seriously alter the result. A good example of this problem is analyzing a minor phase, which represents only a few percent of the area of the image. This is apt to vary widely from field to field, so that the peak is sometimes virtually absent. The histogram minimum may then wander, and the resulting data will be biased. In general, when more of the minor phase is present its peak will be larger and the minimum will shift toward the matrix peak, producing larger feature outlines for the minor phase in the discriminated image.

Minimum area sensitivity method

The second method finds the points in the histogram at which a small change in the threshold setting makes the least change in the area of the resulting binary image. This is not quite the same as finding the minimum point of the original histogram, because while a local minimum has a zero slope mathematically, if it is a very narrow minimum the slope on either side may be quite large. The sensitivity is actually calculated as the difference in the histogram values on either side of the threshold setting, summed in absolute value. This will select a broad minimum, if there is one, but will also locate regions of low slope that are not minima. An example is shown in Figure 5-10, along with the results from the histogram minimum method. The image is the same one as in Figure 5-5, with the positions used to set thresholds around the light grey phase selected by each method.

The plots show that the histogram minimum method (Figure 5-10a,b) selects threshold settings of 55 and 30, respectively, while the minimum area sensitivity method (Figure 5-10c,d) selects settings of 59 and 34. Figure 5-11a,b shows the resulting binary images, which are similar but not identical (the area fractions shown differ by nearly 2%). In both cases, there are broken lines of pixels which lie between the threshold settings but are not in the light grey phase, but rather along the boundary line between the white phase and a darker phase. These can be easily removed by applying an "opening" to the image (an erosion followed by a dilation - see Chapter 6) before feature measurement; this has not been done in the figures, and does not eliminate the differences in area fractions.

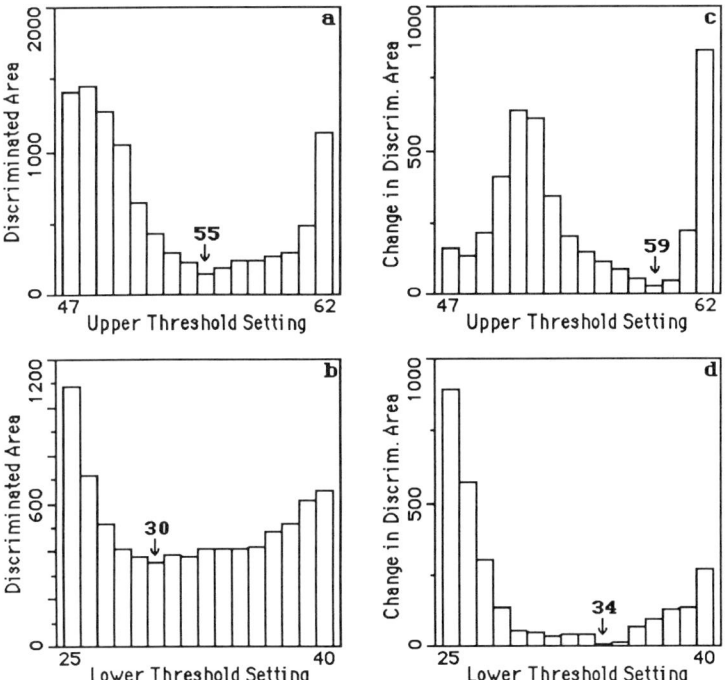

Figure 5-10: Plots showing discriminator settings for light grey phase in Figure 5-5: a, b) discriminated area vs. settings; c, d) area sensitivity (magnitude of change of area vs. threshold settings). Minima are marked.

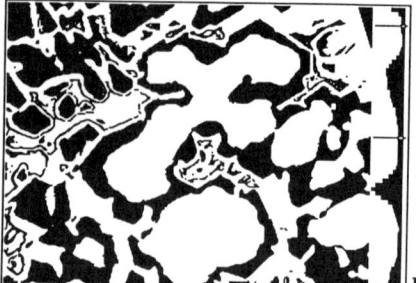

Figure 5-11: Binary images produced by optimum settings from Figure 5-10: a) minimum in brightness histogram (area fraction = 41.45%); b) minimum area sensitivity (area fraction = 39.69%).

The chief difficulty with the area sensitivity method is that it frequently falls into false local minima because the calculated sensitivity sometimes varies rather abruptly from one threshold level to another. It is also possible, although not common, for the method to select the very top of a broad peak for threshold placement (see the local minimum at the left of Figure 5-10c). Finally, while the method seems to often agree with the settings made manually by an experienced operator, when it does so, the perimeter method usually does as well, and when the area method differs, the perimeter method is usually in closer agreement.

Minimum perimeter sensitivity method

The third method handles well most of the circumstances that cause problems for the first, and often agrees with the second, albeit with more computational requirements. The length of the perimeter in the image is quite time consuming to calculate. For an isotropic image, a useful short cut is to simply count the number of transitions across the threshold setting(s) along scan lines in the image. In stereological notation this is N_L (number per unit length), and is proportional to the length of boundary surrounding the discriminated phase. Since the length of the scan lines is constant, the number of transitions is directly a measure of feature perimeter, both internal and external.

When the feature boundaries are smoothest, the number of transitions changes very slowly with changes in threshold setting. This is a reasonably efficient measurement tool; changing the threshold setting through the permitted range and counting the number of transitions produces a plot of image complexity versus setting whose minimum slope point gives the most stable, and hence preferred setting. It is possible to count these transitions simultaneously for many different threshold settings, to produce a table in which the minimum can be found. The agreement in threshold settings observed with by this method and manual settings by skilled users suggest that the operators may use the criterion of achieving smooth feature outlines.

Figure 5-12 shows an illustration of the application of this technique to the same phase and image as in Figures 5-10 and 5-11. Since the amount of perimeter varies with the setting of both upper and lower thresholds, the result is a two-dimensional array of values, which has a well defined minimum at an upper threshold setting of 56 and a lower threshold setting of 31. This produces the binary image shown in Figure 5-13, which has a fractional area close to that in Figure 5-11a, although the settings are not at the minima in the brightness histogram.

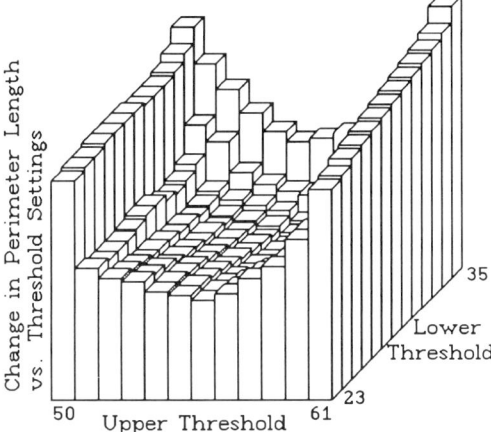

Figure 5-12: Plot of change of perimeter in binary image versus settings of upper and lower level thresholds, for image in Figure 5-5.

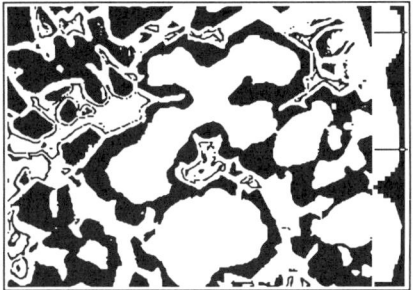

Figure 5-13: Binary image from optimum settings from Figure 5-5 (area fraction = 41.11%).

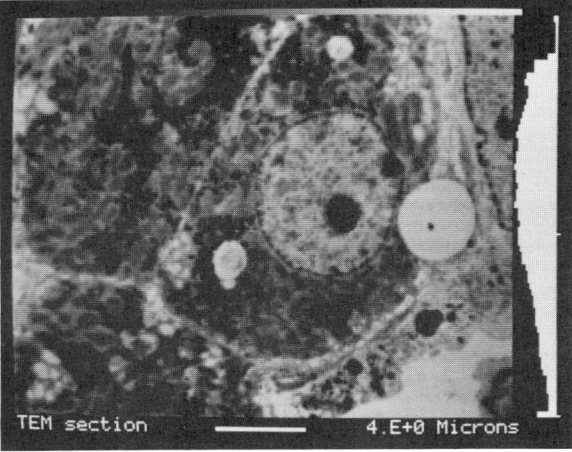

Figure 5-14: Grey-scale image of stained a biological thin section (TEM), with its brightness histogram.

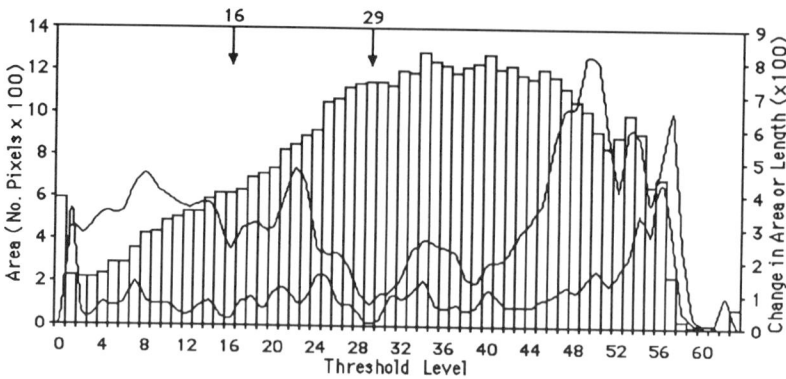

Figure 5-15: Brightness histogram (bars) for TEM image with lines showing magnitude of area (lower) and perimeter (upper) sensitivity to variation in upper threshold setting. Minima are marked.

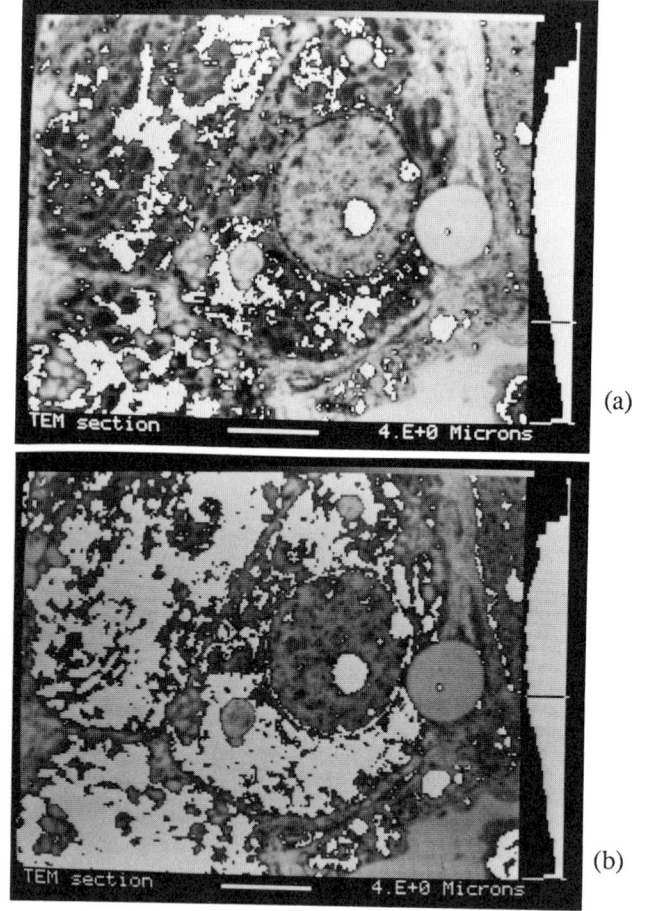

Figure 5-16: Binary images obtained with the settings from Figure 5-14, superimposed on the original grey scale image: a) upper threshold at 16; b) upper threshold at 29.

Discrimination and Thresholding

A more difficult test of the method arises with images such as Figure 5-14, a biological section viewed in TEM in which the dark (stained) regions are of interest. The histogram shows no useful minimum at all; the small rise at the black end, which appears to make a shallow minimum, is due to clipping of very dark pixels from dense, heavily stained regions.

The minimum area sensitivity and minimum perimeter sensitivity methods both find two optimum settings for the upper threshold, at 16 and 29 (Figure 5-15). Figure 5-16 shows the resulting binary images, superimposed on the original grey scale image. The higher setting corresponds to the sharper minimum in the plots, and produces an image that agrees visually with the extent of the dark area in the original image. The location of this setting is on the broad hump of the histogram distribution, and would not be easily selected without these algorithms.

In practice, the minimum perimeter criterion seems to be reliable, moderately fast (requiring a few seconds per field), and applicable to a fairly broad range of grey-scale images. Its chief drawback is that it makes the assumption that feature outlines *should* be smooth. For some features this is clearly not the case, and indeed if the fractal dimension of the object outlines are to be measured, the resulting data may be biased if this criterion is used for threshold settings.

Reproducibility testing

To evaluate the reproducibility of measurements on features discriminated using the first three methods, a test image (Figure 5-17) was produced with black paper cutouts on a grey background. This was digitized ten times, with widely varying overall levels of illumination so that no fixed threshold setting was capable of discriminating the features in the various acquired images. Each of the three automatic methods (histogram minimum, minimum area sensitivity, and minimum perimeter sensitivity) was applied to set thresholds and discriminate the features, which were then measured to determine the area, perimeter, and length (longest projected dimension). In addition, for the large feature with the "zigzag" edge, a fractal dimension (Chapter 6) was determined by repetitively measuring the perimeter and coarsening the image, and constructing a Richardson plot.

Note that the rectangle and "E" shaped features are oriented so that their sides lie at a slight angle to the lines of pixels in the image. This is a worst case for reproducibility,

Figure 5-17: Example of a binary representation of test image used for repetitive acquisition, discrimination and measurement.

Table 5-1. Measurement results for features in Figure 5-17 (ten repetitions)

Discrimination Method:	Histogram Minimum			Minimum Area Sensitivity			Minimum Perimeter Sensitivity		
Object	Area								
Zigzag	6474.4	±	161.12	6550.7	±	127.20	6552.2	±	119.65
Circular	4399.0	±	38.25	4417.2	±	33.18	4414.9	±	35.13
"E"	1794.4	±	48.73	1823.2	±	33.48	1818.7	±	41.26
Rectangular	1739.6	±	27.82	1750.2	±	22.45	1752.5	±	21.98
Object	Perimeter								
Zigzag	766.77	±	14.12	764.80	±	13.26	764.59	±	12.89
Circular	257.96	±	3.07	259.14	±	2.69	258.18	±	2.43
"E"	353.79	±	4.49	353.64	±	4.13	354.91	±	3.01
Rectangular	170.89	±	2.34	172.31	±	2.31	171.70	±	1.49
Object	Length								
Zigzag	113.9	±	0.98	114.0	±	0.78	114.2	±	0.83
Circular	78.1	±	0.54	78.5	±	0.50	78.4	±	0.40
"E"	67.5	±	0.67	67.8	±	0.40	67.6	±	0.56
Rectangular	57.2	±	0.40	57.5	±	0.50	57.6	±	0.40
Object	Fractal Dimension								
Zigzag	1.330	±	0.027	1.323	±	0.019	1.317	±	0.013

since pixels along the edge have varying intermediate levels of brightness due to averaging, and will fall within or outside the features depending on the threshold setting.

Table 5-1 shows the results (mean and standard deviation) for ten repetitive measurements (the area, perimeter and length values are in pixels). Several observations can be made:

1) By any of the methods, the more regular features (the smooth rectangular and circular features) have much lower variation in both area and perimeter than the irregular features (the zigzag and "E" features). The variation in both the area and the perimeter values is roughly proportional to the amount of perimeter for each feature, consistent with the interpretation that it is the inclusion or rejection of pixels along the periphery of each object that produces the variation in these measurement values.

2) Generally, the histogram minimum method produces the largest variation in repeated measurements. The minimum perimeter sensitivity method produces, not surprisingly, the most consistent perimeter measurements. However, it also does about as well as the minimum area sensitivity for area measurements. The length measurements show no significant differences in standard deviation between any of the techniques.

3) There is no obvious consistent difference in the mean values of perimeter or length determined by the minimum area or perimeter sensitivity methods, but both give slightly larger dimensions (as well as lower variability) on the average, than the histogram minimum method. This is consistent with the difference in area measurements, which are also consistently smaller with the histogram minimum method than those from either the area or perimeter sensitivity methods.

4) The fractal dimension of the zigzag feature is lowest, and has the lowest standard deviation, when the perimeter sensitivity method is used (as expected). However, the difference among the methods is not large.

Fixed percentage setting

Gradient search methods are widely used for feature detection in reconnaissance images, as a way to locate feature edges. They are also useful for microscope images in which phase contrast is low, or phase brightness is nonuniform (Russ & Russ, 1984). The application of a Laplacian, Sobel, Kirsch, Marr-Hildreth or other edge-finding operator to a grey scale image produces another image in which brightness is a function of the gradient of brightness in the original image (Castleman, 1979; Pratt, 1978; Rosenfeld & Kak, 1982). Discrimination of this image then produces feature outlines (Moik, 1980). These may be filled to obtain solid features for measurement by including all pixels that are within the outlines; this also removes any internal lines or structure. Measurements requiring the actual brightness values (e.g., integrated optical density or texture) can be referred back to the original grey image using the binary image for pixel selection. Figures 4-11 and 4-12 showed an example of the application of a Kirsch operator, along with the features in the binary image ready for measurement after thresholding and filling the outlines.

The selection of which of these automatic threshold setting methods to use must be made with some knowledge of the type of image being acquired, the features of interest in the images, and the range of variation to be expected in a series of images. None of these techniques is necessarily more "correct" than another, or than manual methods. However, the computational requirements are small, and the improvement in consistency of the measured results is substantial, especially when the original image can vary in overall brightness or contrast.

Color images

For color images, thresholding offers more complicated and flexible possibilities. As discussed in an earlier chapter, color images are generally stored in three image planes, one each for red, green and blue. However, it is not usually convenient to perform the thresholding with these directly. Sometimes separate thresholds for the R, G and B images can be set, each using the same methods as for monochrome images, and the results combined with either a logical AND (selecting the pixels that lie within all three binary images) or a logical OR (selecting the pixels that lie within any of the three binaries).

In other cases, it is possible to use some image processing operation(s) to obtain a monochrome image from the three color planes, and then perform the thresholding on it. In rare cases this may be done with the same kinds of operators applied to monochrome images (for instance, a texture or gradient operator), usually involving adding the results from individual planes.

But one of the most important reasons for using color images in the first place, rather than simply acquiring monochrome images based on brightness variations, is to use the color information. This usually implies that features of interest can be distinguished by a color change. In terms of the way that color images can be encoded (discussed in Chapter 2), this is best seen when the HIS (hue, intensity, and saturation) method of coding is used rather than RGB (red, green, blue). It is possible to convert from one to the other using a linear transformation as described in the earlier chapter.

To accomplish this, an operator that reads the RGB planes and creates a monochrome image in which the pixel value is the hue of the original can be used (Figure 5-18). The resulting image is not easy to interpret visually, since it encodes hue, the orientation of the color vector on the color wheel varying from blue through green and yellow, to red, as brightness. It is at first tempting to think of this as the same image as could be obtained by acquiring a monochrome image of the same scene through a color filter, since the filter would make some colors dark and pass others. There is an important difference however (even assuming we could find a suitable filter whose absorbance varied so linearly and gradually over the entire range of viewable colors). The difference is that the hue as defined above and calculated from the RGB information, has been separated from image brightness and saturation. A physical filter would not be able to make such a separation. Yet it can be very useful to have the hue, or color vector as it is also called. Discriminating this image permits selecting features based on color without regard to brightness, and generally overcomes problems with shading, varying specimen density, nonuniform illumination, and so forth.

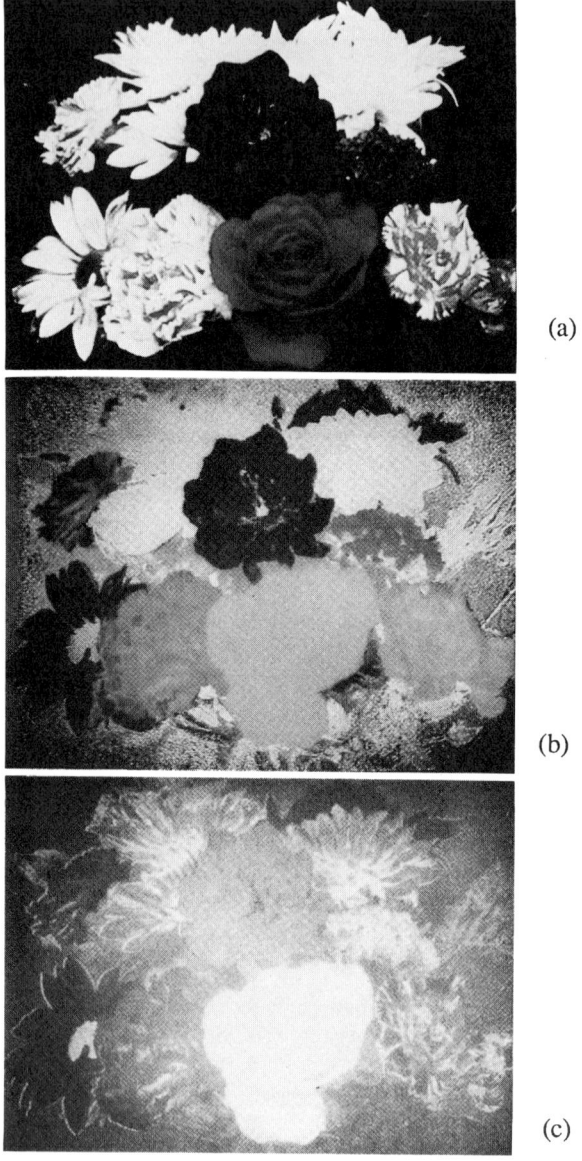

(a)

(b)

(c)

Figure 5-18: Example of hue (b) and saturation (c) images from an original photograph (a) (Data Translations, Marlboro, MA).

Discrimination and Thresholding

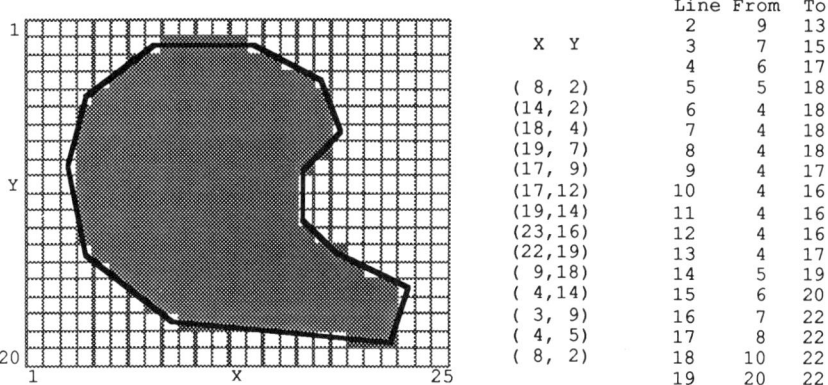

Figure 5-19: Diagram of chord encoding and boundary representation.

Analog electronic devices are available which will accept a standard RS-170 color video signal and separate out the red, green or blue signal as a 1 volt analog output which can be digitized. This makes it possible to use a monochrome digitizer to store the R, G or B plane for possible manipulation, processing, discrimination and measurement. When the hue or saturation signal is desired for image analysis work it is generally necessary to perform a full color digitization and storage, and subsequently process the image to extract the information separately.

Sometimes, other monochrome signals are desired from a stored full color (RGB) image. One fairly common and very useful example is to produce a monochrome image corresponding to the intensity of a particular wavelength of light (color). As an example, consider the aurora borealis (Figure 3-19) in which the various colors of red and green result from the excitation of specific molecules in the high atmosphere. This is exactly equivalent to obtaining a monochrome image using a black and white camera with a suitable narrow pass-band filter, but it may be more convenient to acquire the color image and subsequently process it to extract each of several different colors (different wavelengths, corresponding to different elements) in order to measure their spatial relationship in the single image. This can be done using a similar processing operator to that used for hue, with different coefficients.

Generally speaking, there are comparatively few image analysis systems that make full use of the color information in the images, even though they may be able to acquire and display full RGB images. The displays of these systems are primarily used for false coloring (using lookup tables) to show subtle variations in monochrome images, and the acquired images are generally reduced to a single monochrome image, most often the intensity or brightness image, before measurement.

Encoding binary images

The binary, or black and white image produced by discrimination of an original grey (or color) image is a pixel-based representation of the features to be measured. It requires much less storage space than the original grey image, 1 bit per pixel rather than 8 or more bits. There are other encoding methods that can reduce this even further, and also facilitate some of the measurements which will be made. The two principle methods are run length encoding (also known as chord encoding), and chain code or boundary representation (Figure 5-19).

Figure 5-20: Features with joins and branches.

Run length encoding considers the binary image as a series of scan or raster lines, and keeps track of the positions where a transition occurs from white to black, or vice versa. To avoid confusion about whether white represents features and black represents background, or the reverse, we will here use the terminology ON for a pixel that is within the selected features, and has a binary value of 1 in the image, and OFF for a pixel that is part of the background or surroundings, and has a value of 0. Display and hard copy devices may use either white on black or black on white (or some other color combination) to display this information.

Along each scan line, the position of the start of ON pixels is recorded by the line number and the position of the pixel. This requires two numbers. Then the position of the transition back to an OFF pixel requires only one number, since the line number is the same. The difference between these last two values is is just the length of the chord across the feature on the current scan line. So for each such chord, three numbers are needed: the line number, position of the start, and position of the end. Since in general most chords are many pixels long, this will reduce the storage requirements for the binary image. However, that is not its principal reason for use. Chord encoding also facilitates making the comparisons needed to decide which chords are part of the same feature, since only chords whose line numbers differ by one and whose left or right ends overlap or touch can be considered to be contiguous. A more rigorous definition of contiguity will be given below.

Note that chord encoding may occasionally increase the storage requirements for the image (the worst case is for a checkerboard image in which each pixel has 4 neighbors in the 90 degree positions which are opposite in value; for such an image the storage requirement grows from 1 bit per pixel to three numbers, each 8 or more bits in size, for every two pixels). This is usually only encountered for very "noisy" images with many small features. The list of chords in a chord- or run-length-encoded format does not directly give any information about the features present, since along a single scan line there may be intersections with several different features, or several intersections with one feature (if it is not convex), and these may be interspersed in complicated ways. Because many of the measurements we wish to make require the identification of individual features, it is often convenient to allow space in the table of chords to write in a feature number for each chord, when it is determined by a suitable algorithm. This depends on the definition of contiguity that is chosen, which will be discussed below.

What the algorithm for sorting the individual pixels or chords into features composed of touching chords must do is follow any joins or branches that occur in the features (Pavlidis, 1982). In the example shown in Figure 5-20, the first feature (U-shaped) presents a "join" as it is scanned in horizontal lines from top to bottom. It appears to be two features until they join together at the bottom of the U. Conversely, the second feature is known to be a single feature immediately, but when the two sides branch apart it is necessary to recognize that the chords in each side belong to the same feature. The

Discrimination and Thresholding

"X" and "O" shapes features represent more complex situations in which there are both joins and branches.

Of course, it is not unusual to encounter features that are intertwined in various ways, perhaps with one lying entirely within the U or even O shaped space in another, and the logic must keep these separate even when the chords occur mixed together on the same scan lines, and in the table.

Given a chord-encoded table, the identification of individual features is straightforward in logic, but intricate in terms of bookkeeping. Each chord is examined in turn (Figure 5-21). A search through previous chords on the preceding scan line is made to see if the chord touches any previous ones, and if so it is assigned the same feature identification number. If not, then a new feature number is assigned. If a chord touches the edge of the image, it is generally given a unique number or flag, and all previous chords with the same number have the identification changed. If a chord touches two chords in the previous line which have the same feature identification number, then it represents the closing of an internal hole, which can be counted if desired. If the two previous feature numbers were different, then a join between two arms of a feature has been encountered. One of the numbers is used for the chord, and all references to the other number are changed. The procedure is carried out for all of the chords in the table. The time required depends on the complexity of the image, the number of features and joins and branches, but is usually imperceptibly short. Sometimes this is performed as a separate operation, and in other cases some of the measurement operations described in Chapter 7 are carried out at the same time.

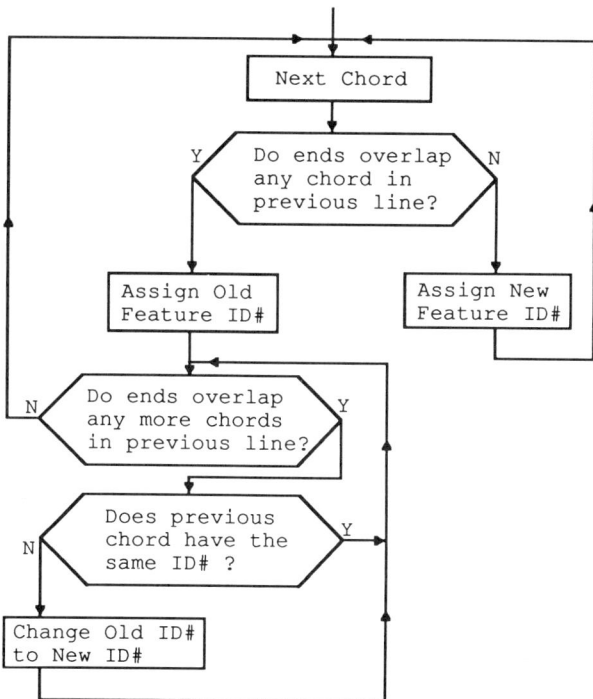

Figure 5-21: Flow chart for feature identification.

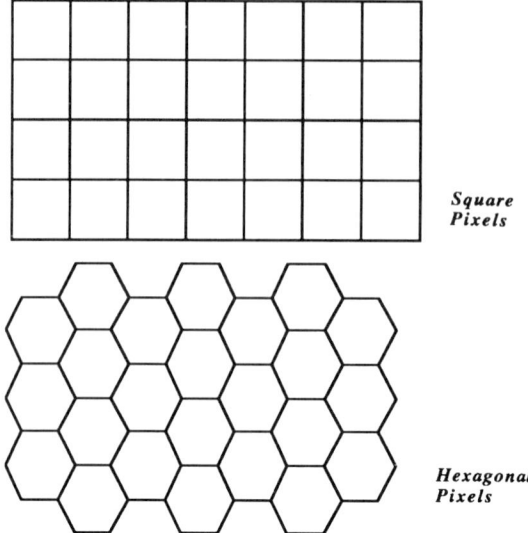

Figure 5-22: Square and hexagonal arrays of pixels.

Contiguity

For pixels in a square array, there are two possible definitions of contiguity which define how pixels within a single feature touch each other. For a hexagonal array (Figure 5-22) this confusion does not arise. In a hexagonal array, each pixel has six neighbors, all equidistant from the central pixel. A line of pixels can be formed in any of three directions, and a line of OFF pixels separating features can likewise run in any of the same three directions.

For a square array of pixels, the situation is more complicated. If the criterion chosen for pixel touching is that any of the 8 closest neighbors can touch to continue a feature, then lines of single pixels can run in any of eight directions. However, this means that OFF pixels which are separated by such a diagonal line of touching ON pixels must be considered to be separated, even though they also touch along a 45 degree direction. In other words, if the ON pixels which comprise features are allowed to form connections with 8 neighbors (called 8-connectivity), then the background of OFF pixels will be restricted to form connections with only the 4 orthogonal neighbors (4-connectivity), as shown in Figure 5-23.

Another way to look at this problem is to consider the inverse of a binary image. If all of the ON pixels are changed to OFF and vice versa (in the same way as making a photographic negative), then with a hexagonal array (6-connectivity) all of the connectivities in the features and the background will remain the same, but with a square array, if the rule of 8-connectivity is selected for the original features and 4-connectivity for the background, then after reversing the image both features and background may have different connections than they did before. This creates some problems for dealing with holes in features and some logical combinations of images. The problem only arises when irregularities along feature edges, or the separation between objects reach sizes of the order of single pixels. For comparatively large, fairly smooth objects that are well separated the choice of connectivity rules makes no appreciable difference.

Discrimination and Thresholding　　　　　　　　　　　　　　　　　　　　　　　　　　*123*

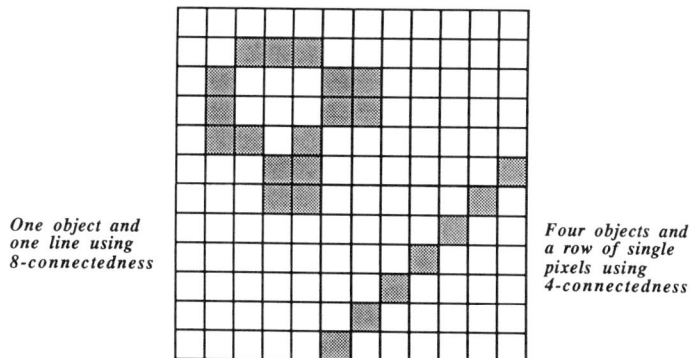

One object and one line using 8-connectedness

Four objects and a row of single pixels using 4-connectedness

Figure 5-23: An ambiguous feature and line.

Note that choosing 4-connectivity for pixels within objects does not solve this problem. It simply results in allowing 8-connectivity for the OFF or background pixels. In most cases, when a square pixel array is used (and it usually is, because hexagonal pixels create other problems, especially for the hardware of image acquisition and display, and ways of encoding images), an 8-connectivity rule is applied to ON pixels within features, and the corresponding 4-connectivity rule is used for OFF pixels in background. Sometimes, in performing specific binary image editing operations (such as filling holes) it is advantageous to temporarily switch these rules, and it is important when doing so to keep careful track of what is happening.

The other method besides chord- or run-length-encoding that can be used to compact binary images is chain code (Figure 5-24). Each feature obviously has a periphery, and in principle it is only necessary to keep track of the positions of the edge pixels. Actually, chord encoding only records the positions of the outermost pixels, as well. But chain coding does this in a way that ties together all of the information about one feature. Chain code starts with any pixel on the edge. For practical reasons having to do with the algorithm that finds these points, the initial point is usually in some consistent position, such as the leftmost pixel in the topmost line. Then the direction of the next touching

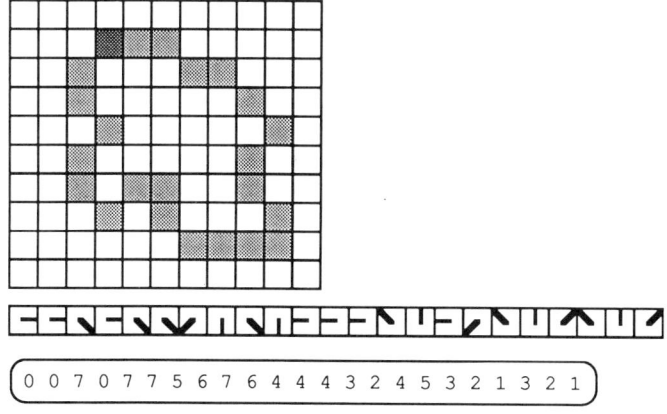

0 0 7 0 7 7 5 6 7 6 4 4 4 3 2 4 5 3 2 1 3 2 1

Figure 5-24: Chain code example.

pixel along the edge is noted. If 8-connectivity rules are in use, this can be one of eight directions. If 4-connectivity allows only the four orthogonal neighbors to be considered, then only four possible directions are available. In either case, a single digit suffices to indicate which way the boundary goes. The situation is repeated for each pixel along the periphery until the initial point is repeated. This boundary code completely specifies the feature, and from it the original pixel-based representation can be re-created if necessary.

From chain-code, some parameters for features are very easy to calculate. For instance, the length of the perimeter is simply the sum of the number of links in the chain. If 8-connectivity is used, a more accurate value is obtained by counting the number of orthogonal and diagonal steps separately and multiplying the latter by the square root of two before adding, to correct for the fact that the diagonal neighbors are slightly farther away. A simple counting procedure should produce Length = $No + \sqrt{2} \cdot Nd$, where No is the number of orthogonal links in the chain and Nd the number of diagonal ones. In the three-dimensional or "voxel" case which will be dealt with in later chapters, we could easily add $\sqrt{3} \cdot Nb$, where Nb is the number of "body diagonal" links (the length of the body diagonal of a cube, assuming we have cubic voxels, is $\sqrt{3}$).

In fact, this is not the best answer. For calculating the distance between points in a square lattice (in 2-D) or a cubic one (in 3-D) the simple use of chain code will produce a biased result as compared to the exact distance determined by the Pythagorean equation with the end coordinates. To minimize the mean error, a different set of coefficients can be used (Smeulders & Beckers, 1989). In 2-D it is $0.948 \cdot No + 1.340 \cdot Nd$, and in 3-D it is $0.877 \cdot No + 1.342 \cdot Nd + 1.647 \cdot Nb$. Averaged over all orientations, the expected error with these equations is 2.5% for the 2-D and 3.5% for the 3-D case.

The same problems associated with choosing 4- or 8-connectivity apply as for chord-encoding, and chain code can also be used for hexagonal pixel arrays, with 6 possible directions at each step. Chain code is less easily produced than chord encoding, but it is even more compact (only a single digit is needed, occupying at most 3 bits, for each edge pixel after the first). Also, it is immediately feature-specific because all of the links in each chain refer to a single feature, and all of the information on the feature outline is present in the chain. However, it is not used as widely as chord encoding.

There are several reasons for this. Probably the most important from a practical point of view is that it is more difficult to obtain. Many systems have very low level algorithms in machine code or even in hardware to produce the chord-encoded tables, while much more elaborate algorithms are needed to produce the chain code. Also, while perimeter and a few other measurement parameters are easy to obtain from chain code, these can all be accomplished with only slightly more difficulty from chord encoded tables, and the reverse is not true. Calculating area, or going back to the original grey image to determine mean brightness or other parameters, is much more difficult from chain code because it is necessary to reconstruct the full pixel image first.

Also, while chain code seems to offer an extremely simple and compact means of describing a feature, it has some specific problems. For instance, if a feature touches the edge of the image, this is immediately recognizable in the chord-encoded table because the coordinate of either the line number, or the start or end point of the chord, will be equal to the limit for the image. With chain code, it is first necessary to decide how to handle the code. If a line of links along the image edge is used to connect the chain and continue around the feature, then it is necessary to flag the data somehow so that this is not misunderstood to be the actual perimeter of the feature. Some measurements can be made using features that touch edges (for instance, the total area fraction and total specific surface area), but others, such as the size or shape of the individual feature cannot be measured. The measurement algorithms discussed in Chapter 7 deal with this problem, but the data must warn the calculation program that the feature did touch the edge. Chain code offers no simple way to do this.

Also, what about features with internal holes? We may want to measure the internal as well as the external perimeter, and certainly the area of the feature must take into account the missing holes. With a chord-encoded table this is straightforward; with chain code it is not. The outer boundary can be used to designate the region within which other boundaries are to be sought, and the perimeter around internal holes can be chain coded, of course, but somehow the chain must be labelled to indicate which other feature it is inside. There may be several such holes inside a feature and even features within holes. This complexity is not well handled by chain code, and the data must generally be converted back to a pixel-based representation before calculations or measurements are performed. Overlapping features produce even more serious problems in chain-coded format.

The preceding chapter warned that boundary representation is generally harder to use for image measurements than pixel-based representation. Chain code is a special case of boundary representation, in which the boundary is considered to run through rather than between pixels. It therefore ends up having many of the limitations of both techniques, and is used only for specific circumstances.

The most prominent one is simply applications that generate chain code directly. The common example is measurement of objects that are traced by hand using some kind of digitizing tablet. This can be done by placing maps or photographs on such a device and then using a stylus or cursor to manually outline the perimeter, either continuously or by marking discrete points. The same results may be obtained by marking points along the periphery of objects viewed on a video screen using a light pen, mouse or other pointing device whose location is electronically mixed onto the viewing screen, or by using optical

Figure 5-25: Manual entry and pointing devices (light pen, graphics tablet, touch pad, mouse).

mixing to combine an image viewed in a microscope with a view of the pointing device or tablet. In all of these situations, the pointing device generates coordinates which are fed to the computer (Figure 5-25).

Graphics or drawing tablets work in several ways; a few examples will suffice. One common type uses wires embedded within the tablet to carry a radio-frequency pulse along either the X- or Y-direction. This pulse is picked up by a small coil of wire in the stylus or cursor, and the time delay provides a measure of distance which is converted to position. The X- and Y- directions are each sampled alternately, hundreds of times a second, and a continuous stream of coordinates is produced as the stylus or cursor is moved. Usually there is also a button on the cursor or a switch on the stylus sensitive to pressure to indicate whether the user wants the coordinates to be saved. In another variation on the same theme, an ultrasonic transducer in the stylus generates sound waves that are received by other transducers at two ends of a bar across the top of the drawing area. The time delay gives the radial distance from each corner, from which the X,Y location can be solved by triangulation. In another type of tablet, pressing the stylus down causes a contact or capacitative coupling between two layers of material. In the case of direct contact, the resistance between electrodes at the top (or side) of each sheet gives the distance. In the case of capacitative coupling, an electronic circuit oscillates at a characteristic frequency. In either case, the distance of the stylus from the edge of the tablet can be determined.

Mice work differently. Whereas the tablet offers absolute coordinates, a mouse measures only relative motion. Underneath the mouse there is usually a ball that rolls on the table surface (this is not always true - some mice use optical sensors that move over a special grid). As the ball rolls, it produces signals which indicate direction. Each pulse corresponds to a known distance, and the computer can count them to determine the total motion. A mouse turned upside down so that the ball can be spun by hand is called a trackball. Another device is a light pen, used to point at video images. This does not emit light, but rather senses it with an internal photodiode. As the electron beam in the cathode ray tube passes the point where the light pen is positioned, it produces an abrupt increase in the brightness of the phosphor, which then decays gradually until the beam returns on its next raster. Electronics measure the time delays since the start of the raster frame and the start of the current line to determine the position of the pen. The chief difficulty with light pens is that they tend to get in the way of viewing the image, and that they lose sensitivity in dark regions.

With these and other equally imaginative devices, the resolution of position determination of the marker is generally on the order of 0.1 mm or less, often better than the ability of the user to position the stylus or follow the periphery of a feature. The steady stream of coordinate values in the case of free-hand drawing, or the discrete values that may also be sent, are a series of perimeter locations, in order, around the object. They may either be stored as absolute coordinates or converted to chain code. In either case, they have the advantage of being feature specific and ready for measurement. Chapter 7 will include some algorithms for measuring features using this kind of data.

Jitter while drawing around features will increase their measured perimeter. Smoothing of the boundary points or setting a threshold value requiring a minimum motion of the input device before another point can be registered are two methods used to deal with this source of error.

It is often not easy to convert these coordinates to a pixel-based representation because they may exceed the limits of resolution of the pixel-based binary image memory. Dividing down the coordinates to fit the image memory and display reduces resolution. This may be acceptable for some purposes, such as the measurement of area, but not others, such as boundary irregularities. Most binary images have the same size as the acquired grey scale image, typically 512 pixels wide and rarely more than a thousand. But with resolution of 0.1 mm or better, and a tablet with a size as great as 6 feet (these large drawing tables are chiefly used for maps), the potential coordinate range is quite large.

References

D. S. Bright, E. B. Steel (1986) *Bright-field image correction with various image-processing tools* Microbeam Analysis 1986 (A.D.Romig, Jr., W. F. Chambers, eds.) San Francisco Press, 517-520

D. S. Bright (1987) *An object finder based on multiple thresholds, connectivity and internal structure* Microbeam Analysis 1987 (R. H. Geiss, ed.) San Francisco Press 1987, 290-292

K. R. Castleman (1979) *Digital Image Processing*, Prentice Hall, Englewood Cliffs, NJ

K. R. Castleman, J. Melnyk (1976) *An Automated System for Chromosome Analysis: Final Report*, Document 5040-30, Jet Propulsion Lab., Pasadena, CA

R. M. Haralick, K. Shanmugam, I. Dinstein (1973)*Textural Features for Image Classification* IEEE Trans. Syst. Man. Cybern., SMC-3 610-621

J. N. Kanpur, P. K. Sahoo, A. K. C. Wong (1985) *A new method for grey-level picture thresholding using the entropy of the histogram* CVGIP 29, 273-285

J. G. Moik (1980) *Digital Processing of Remotely Sensed Images* NASA publication SP-431, 277

J. F. O'Callaghan (1974) *Computing the Perceptual Boundaries of Dot Patterns* Computer Graphics and Image Processing 3#2, 141-162

T. Pavlidis (1982) *Algorithms for Graphics and Image Processing*, Computer Science Press, Rockville MD

W. K. Pratt (1978) *Digital Image Processing* Wiley, New York

J. Prewitt, M. Mendelsohn (1966) *The Analysis of Cell Images* Annals of the N.Y. Academy of Sciences 128, 1035-1053

J. P. Rigaut (1988) *Automated image segmentation by mathematical morphology and fractal geometry* Journal of Microscopy 150 21-30

A. Rosenfeld (1979) *Some experiments on variable thresholding* Pattern Recognition 11, 191

A. Rosenfeld, A. C. Kak (1982) *Digital Picture Processing* Academic Press, London

J. C. Russ, J. Ch. Russ (1984) *Image processing in a general purpose microcomputer* J. Microscopy 135, 89

J. C. Russ (1986) *Practical Stereology* Plenum Press, New York

J. C. Russ, J. Ch. Russ (1988) *Automatic discrimination of features in grey-scale images* Journal of Microscopy 148 263-277

A. W. M. Smeulders, A. D. Beckers (1989) *Accurate image measurement methods (applied to 3D length and distance measurements)* Proc. 1st International Conf. on Confocal Microscopy, Academisch Medisch Centrum, Amsterdam

R. J. Wall, A. Klinger, K. R. Castleman (1974) *Analysis of Image Histograms*, Proc 2nd Joint Int'l Conference on Patt. Recog., IEEE 74CH-0885-4C, 341-344

J. Weszka (1978) *A Survey of Threshold Selection Techniques* Comp. Graph. & Image Proc. 7, 259-265

Chapter 6

Binary Image Editing

The binary pixel-based representation that results from discrimination of a grey scale image may not perfectly delineate all of the features present. This may be due to noise or other imperfections in the original image, to inadequate processing methods, or to something inherent in the image itself such as touching objects. It may also happen than the original image covers an area that is larger than the region which we desire to measure. A simple example is the determination of the number and size distribution of spots on a leaf produced by aerial spraying. It is comparatively easy to obtain images of one or more leaves from which the spots may be measured, but the image will generally be rectangular or square, and the leaves are irregular in shape. Counting should be performed only within the leaf area. Similarly, some microscopes have a limiting aperture for illumination that is more or less circular, and measurements should be performed only within that region. Many other examples ranging from counting of cell colonies within a petri dish to trees in orchards viewed in aerial photos present a similar problem (Figure 6-1).

Normally, these kinds of situations are dealt with by manipulating the binary image, which is called here image editing to distinguish it from image processing (which deals with grey scale images). Image editing also starts with one or more binary images and produces a resultant image, again usually a binary (1 bit per pixel) image. Most of the types of editing operations are performed directly on the images themselves, where the

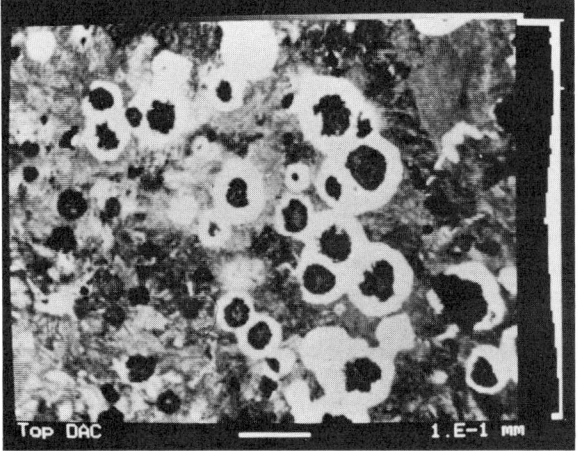

Figure 6-1: A cast iron, in which only the graphite (black) regions inside ferrite (white) are to be counted and measured.

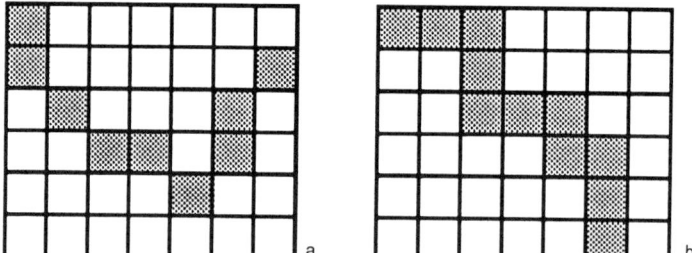

Figure 6-2: Connectivity rules: a) an eight-neighbor connected line will not stop flooding with an 8-neighbor rule; b) a four-neighbor connected line will stop flooding.

spatial arrangement of pixels is most directly analyzed. In some cases, use of chord-encoded tables are used, but they are generally used in an intermediate step and produced as required, and afterwards the binary image is again constructed. Since virtually all computer-based image analyzers can display binary images, it is usually possible and instructive to watch the various steps in editing an image to obtain the desired result. Once a suitable sequence of operations has been established, it can often be applied automatically to subsequent images as they are acquired and discriminated.

Manual editing

The most direct method of editing an image is, of course, for the user to mark on it using some pointing device (mouse, light pen, drawing tablet, etc.). This allows setting pixels to ON or OFF as desired, and can be used to connect broken features or separate touching ones. It can also be used to draw in an outline for the region within which measurements are to be performed, or to point to particular features which are to be measured.

All of these operations are straightforward, as the pointing devices usually produce coordinates (as described in the previous chapter), and there is usually a cursor or other mark on the binary image showing where the device is currently pointing. It takes only a little practice to learn to draw on a video screen displaying an image. Sometimes it is useful to allow free-hand drawing, and in other cases it may be preferable to allow the discrete entry of points which are then connected with a straight line.

There is one other mode of drawing that has some uses. It is generally called flooding or filling, and it consists of writing the selected color (ON or OFF in this case) starting at the current locations of the cursor and growing outwards in all directions until a boundary of the same color is met. This allows filling any closed shape, however intricate. Flooding in this way can be used to fill holes in features, by placing the cursor anywhere inside the hole and flooding with ON, or to erase features, by placing the cursor anywhere on the feature and flooding with OFF. When a square pixel array is used, this immediately raises the problem of what the connectivity rules are. If the usual rules of 8-connectivity for pixels within features (and 4-connectivity for surrounding pixels) are used, then flooding with ON pixels (to fill voids) must follow any possible 4-neighbor paths, while flooding with OFF pixels (to erase noise) must follow any 8-neighbor paths (Figure 6-2). This creates complexity for the programs which do the flooding, and potentially for the user (Pavlidis, 1982).

Note that this problem cannot be overcome by inverting (negating) the image. It may at first seem that if all ON pixels are changed to OFF and vice-versa, that a single

Figure 6-3: A line in a hexagonal pixel always bounds background regions.

flooding routine could be used for either features or surroundings, but this is not the case as was pointed out in Chapter 5. It is still necessary to have both 4- and 8-connectivity rules because the surroundings and features are differently defined with square pixels. With hexagonal cells (6-connectivity) the problem does not arise, of course (Figure 6-3).

While manual or interactive editing of images is certainly a necessary capability to allow the user to apply independent knowledge about the image, such as what region should be measured or which features are of interest, it is usually desirable to eliminate as much of this user input as possible. Not only does it reduce variability due to user error or bias, but it also is generally much faster to use more automatic methods, and permits the image analysis system to work with minimum operator intervention. This can be important in situations where semi-skilled operators use a system to perform quality control work, for instance.

Combining images

One principal class of operations with binary images is the logical combination of two or more to produce a single image as a result (Russ & Russ, 1986). The Boolean logic operators which are available are AND, OR, EX-OR (exclusive OR), and NOT. They can be used in any combination to produce the desired logical statement. Figure 6-4 shows examples of the principal operations.

NOT is simply the negation or inverse operator. It replaces all OFF pixels by ON and vice versa. It is usually used in conjunction with the other operators, and is order

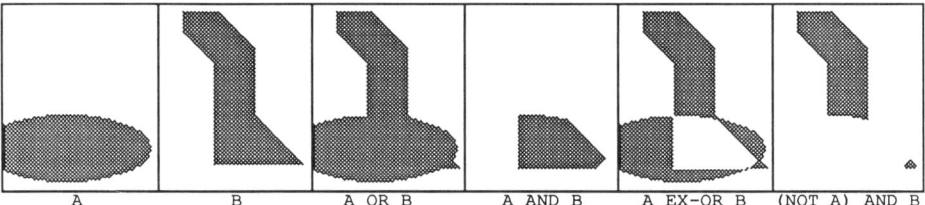
Figure 6-4: Examples of Boolean logic used to combine two regions.

dependent (for instance NOT [A AND B] is a very different statement than [NOT A] AND B).

The logical AND statement combines two images by setting to ON each pixel which is ON in both of the original images. This is a way of finding features or portions of features that are present in both images, not in only one of them.

The logical OR statement is different from AND in that it combines the images by setting to ON each pixel which is ON is either of the two original ones. This can be used to combine features whose pixel representations are obtained by two different procedures (for instance, different brightness threshold ranges for discrimination), or to add together portion of features that require different processing to permit discrimination. We will encounter an example of the latter when considering automatic filling of holes in features.

Exclusive-ORing of two images produces a result in which pixels are turned on when they are ON in either original image but not in both. The statement A EX-OR B is equivalent to A AND (NOT B). In fact, many of the Boolean logic statements that can be

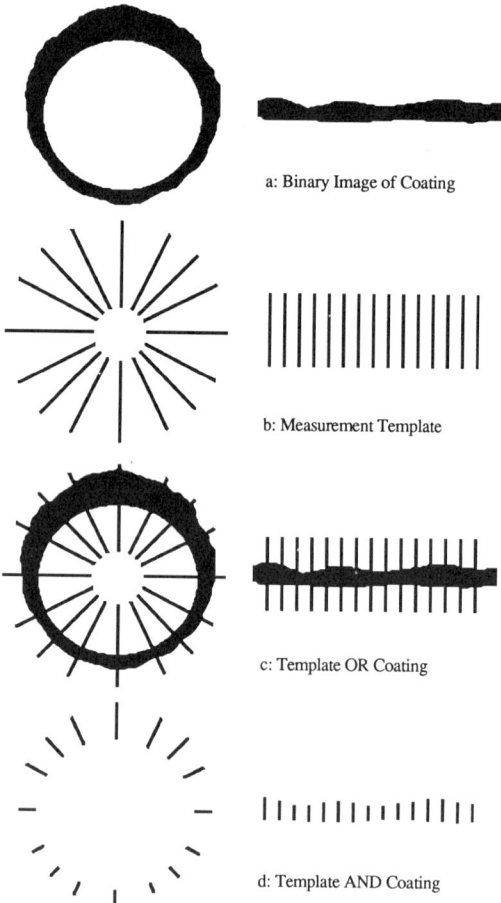

a: Binary Image of Coating

b: Measurement Template

c: Template OR Coating

d: Template AND Coating

Figure 6-5: Example of using Boolean AND to overlay a measurement template on binary images of coatings, to obtain line segments for measurement of coating thickness and uniformity.

used to combine images can be written in many different ways, while producing identical results.

Boolean or logical image combination can be used for many different purposes. One of the simplest is restricting the field to an aperture or other reduced outline. A binary image, which is often referred to as a mask when it is used for this purpose, can be obtained of a microscope illumination aperture (for which it can be acquired once and stored) or for the leaf area used as an example given earlier (for instance by choosing appropriate threshold settings to discriminate the leaf from its surroundings). This binary image or mask can then be ANDed with the image of the features to be counted, which will eliminate anything outside of the mask area. Then the counting or measurement operation can be applied to the features which remain, to obtain the desired result.

ANDing of images with specific templates is useful prior to performing measurements of specific features and dimensions. The most common example of this is the measurement of coating thickness and uniformity. Assuming that the original image of a cross-section through a coating can be thresholded (or can be processed to permit discrimination), the binary image will show the coating. As indicated in Figure 6-5, this may be either a coating on a flat surface, or on the inside or outside of a wire or pipe.

Since the overall configuration of the coating and part is known beforehand, it is practical to set up a template consisting of lines, as shown in the Figure. When these are ANDed with the binary image of the coating, a series of line segments remain. These can be easily measured (as discussed in Chapter 7) to characterize the mean thickness of the coating, and the variation in that thickness with position or direction.

Another example of combining image combinations to select features for measurement arises in scanning electron microscopy. Pores or indentation on a surface often appear darker than their surroundings, but can be confused with other discolorations and surface stains which also appear dark. The distinctive aspect of true pores or pits is that they are surrounded with a bright edge, produced by the nature of secondary electron generation from the surface. One binary image can be created of all dark regions (pores and others) by discriminating with an appropriate set of brightness thresholds. A second can be created from the white outlines, using a different set of threshold values. This second image can then be filled (as will be described below) to produce a mask that covers all of the actual pores, but whose features are larger than the pores and which may touch for closely-spaced pores. If this mask is ANDed with the first image, the result is the image of just the pores, ready for counting and measurement.

Boolean logic for image combination is a powerful tool that is used whenever several criteria are needed to select the features of interest. This is especially true when color images are used, since binary image masks may be created from each of the separate color images (e.g., red, green and blue). We will frequently refer to these logical combinations in subsequent parts of this chapter as part of other image editing operations.

Perhaps the most familiar application of this method is used with X-ray maps from the SEM. By setting the X-ray spectrometer to select the energy or wavelength of particular elements, and recording the individual pulses or the rate of such pulses as a function of beam position as it scans across the sample surface, an image is obtained showing the spatial distribution of each element. This image must usually be smoothed because of the limited counting statistics, and may also require image editing using a plating or dilation operation to fill in gaps in the image. But it is usually possible to obtain adequate maps showing the distribution of several elements of interest.

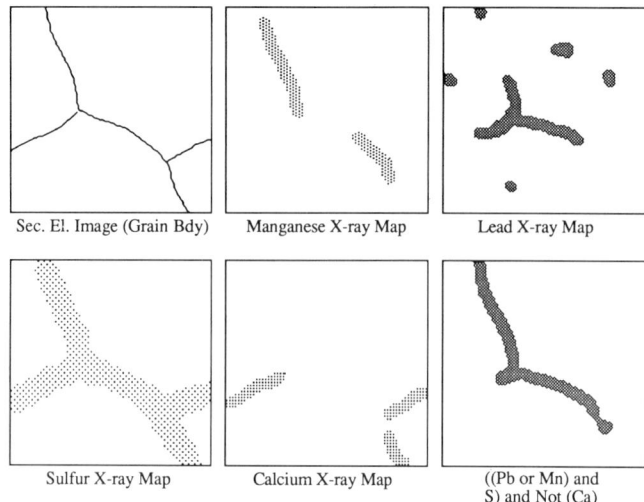

Figure 6-6: Illustration of application of Boolean image combination to X-ray maps.

These may then be combined to delineate particular compositional structures of interest (Russ & Russ, 1984). The example in Figure 6-6 shows schematically the use of four maps for a free-machining steel. The original SEM image (formed using secondary electrons) shows only the grain boundaries in the metal. Maps of the elements Manganese, Lead, Sulfur and Calcium all appear to show some enrichment along the grain boundaries, but they do not directly give information about the presence of manganese and lead sulfides. A representative combination of these individual binary images might use the Manganese and Lead images together (an OR operation) to select the metals of interest, as well as requiring the presence of Sulfur (an AND operation) and the absence of Calcium (a NOT AND operation). The entire statement describing the resulting image would then be "pixels selected as ((Manganese OR Lead) AND Sulfur) AND NOT (Calcium)".

Another area of application for Boolean logical combination of images arises in images of minerals and ores. For ore processing, it is important to know what fraction of the commercially important particles in the ore are "locked" within grains of other minerals, which may require special processing to break them up and free the desired grains. This is accomplished in the same way as shown for the pores above. A binary image of all the grains of interest is created. This is then masked with a filled image of the second phase grains, using an exclusive-OR function. The result is to show only the "liberated" grains which are not inside the other grains. The features in this image can then be measured, as described in the next chapter, or they can be subtracted from the original image to allow measurement of the locked grains.

Neighbor operations

Another important class of editing procedures involve neighbor relationships (Levialdi, 1983; Rosenfeld & Pfaltz, 1966). Many of these operations form the basis for a field known as mathematical morphology, which uses the notation of set theory and is concerned with operations that respond to feature shape (Matheron, 1975; Serra, 1982; Giardina & Dougherty, 1988). We will not use that notation or terminology, and will

Binary Image Editing

include a number of methods and examples that go beyond the usual definitions of the field. But the debt which these methods owe to that discipline is considerable.

A very basic example is known as erosion and dilation, plating and etching, or growth and decay. It serves a number of useful functions including smoothing feature outlines, joining broken or discontinuous features, and separating touching ones.

In its simplest form, erosion consists of examining each binary pixel and changing it from ON to OFF if it has any neighbors that are OFF. As in earlier discussions of neighbors and connectivity, it is necessary to specify which neighbors are considered. We usually use all 8 neighbors in the case of square pixels, and 6 in the case of hexagonal ones.

Erosion or etching reduces features all around their periphery. It is a global operation, working on the entire image at once. Features with narrow protuberances will have them

Figure 6-7: Results of erosion on binary image of dendrites in a metal alloy, using neighbor coefficients of 1 and 3, and after 1, 2 and 4 cycles of application.

removed by etching, and features connected by a narrow strand will have it removed. The names applied to the method are descriptive of this reduction, as though the features had been dipped into an acid and etched from all sides, or worn down by some eroding process.

Plating or dilation is the converse of etching. If the features were dipped into a chemical plating bath, material would be added on all sides. Mathematically, this is done by changing any OFF pixel to ON if it has at least one ON neighbor. This causes features to grow (or dilate) in size, which will fill in small breaks in features, internal voids, or small indentations along the feature surface.

In actual practice, more flexibility in plating and etching operations can be achieved my modifying the rules for adding or removing a pixel. Instead of removing a pixel if any of its neighbors is OFF, which removes all pixels which touch the feature's outside periphery, it is often useful to count the number of neighbors that are OFF and only remove the central pixel if more than some preset number of neighbors are OFF. Different test values from 1 to 8 produce different effects. Very high numbers result in the removal of isolated pixels or very small features, and very sharp small protuberances, but leave the balance of the feature outline alone. Very low numbers act much the same as the original form of erosion described above. Typically, values from 1 to 4 are most used because they produce both a general erosion and a smoothing of feature outlines (Figure 6-7).

The same thing is true for plating or dilation. If the number of ON neighbors is tested before an OFF pixel is set to ON, then control over the growth process can be achieved. Again, values between 2 and 4 are most often used to produce smoothing of feature outlines (Figure 6-8).

Erosion and dilation (also known as plating and etching, or shrinking and growing) operations may be used together (Figure 6-9). This recovers most of the original feature size while producing a smoothed shape. The sequence of erosion followed by dilation is called an "opening". The initial erosion removes small features which may represent noise, and also sharp protuberances from the feature outline. The subsequent dilation does not restore the small features, which have permanently disappeared, but it does fill in any small indentations in the outlines. The overall result is that the feature size (for the moment we will use its area or number of pixels) is restored to nearly the original value, while the shape is modified to become more rounded and smooth.

The name "opening" comes from the tendency of this sequence of operations to open up gaps between touching features, or to open up cavities near the periphery of features so they are not enclosed. If the same test coefficients are used for the etching and the plating, the original feature area is not exactly recovered because the initial erosion eliminates some of the original outside pixels, and subsequent dilation has fewer points adjacent to the outside periphery to test. Consequently, the feature area is slightly reduced. However, the choice of the same test coefficients for both operations is generally preferred because each coefficient (1, 2, 3, 4...) has some tendency to produce geometric distortion of the feature shape, for instance toward a diamond or square. Using the same coefficient tends to restore a more similar shape to the smoothed feature. Figure 6-10 compares the effect of three cycles of erosion followed by three of dilation, with varying coefficients.

Multiple applications of erosion and dilation (referred to as the "depth" of the operation) produce more extreme smoothing or shape modification, and more complete

Binary Image Editing

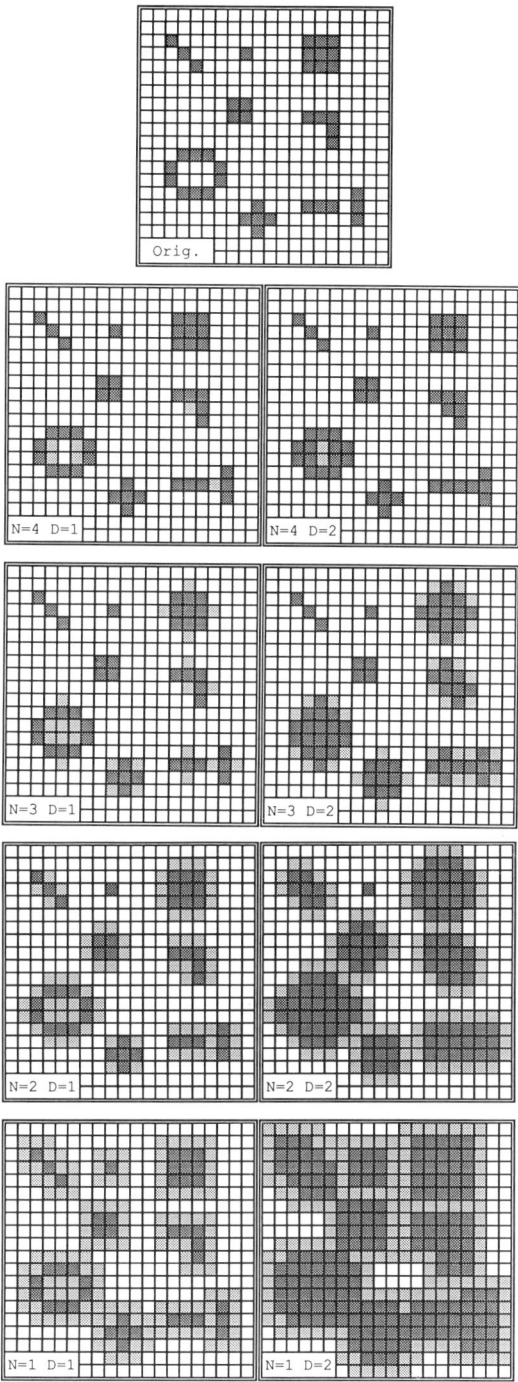

Figure 6-8: Effect of 1 and 2 cycles of dilation with neighbor rules from 1 to 4. Light shading indicates pixels added each time.

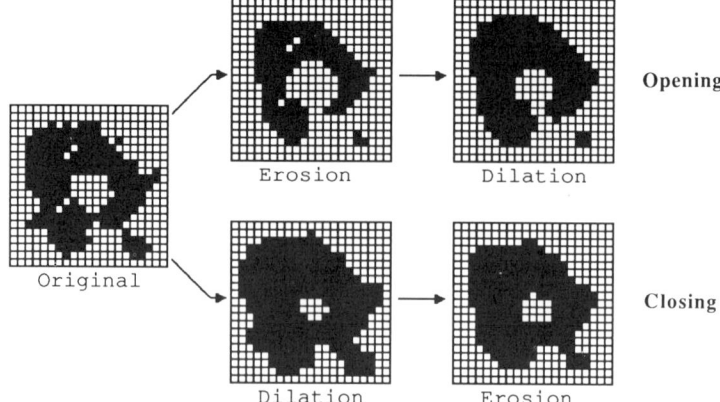

Figure 6-9: Opening and closing operations on binary pixels.

removal (in opening) or joining (in closing) of features, as shown in Figure 6-11. When more than one application of erosion (followed by an equal number of dilations) is applied, it is not necessary to keep the test values always the same. For instance an alternating sequence of using 3 and 4, or 0 and 1, as test values will result in an octagonal pattern, which is a reasonable approximation to the circle that represents the isotropic ideal of uniform response.

When the sequence is reversed, that is when dilation is applied and then erosion, it is referred to as a "closing". In some respects this is the opposite of an opening, because

Figure 6-10: Openings (depth = 3) applied to dendrite image using various coefficients.

Binary Image Editing 139

small features are not erased, small voids in features are filled in, and breaks or gaps in features are joined. However, while the plating operation tends to increase the size and hence the perimeter of objects so that more pixels are tested for removal in the subsequent step, the resulting outline is also smoother which limits the increase and so the difference in feature area after a closing cycle is much smaller than for an opening. Closings, like openings, may be carried out to more than one iteration (greater "depth"), and the coefficients may either be the same for both operations, or may be changed for each iteration. Both openings and closings, like the basic erosion and dilation operations, are used to "clean up" binary images by removing small features or feature irregularities which are presumed to be due to noise or other imaging or object imperfections (Figure 6-12).

Dilation can be used to fill in spaces in broken-up or "noisy" images, but in the process will increase the size of features. One such situation that is often encountered is

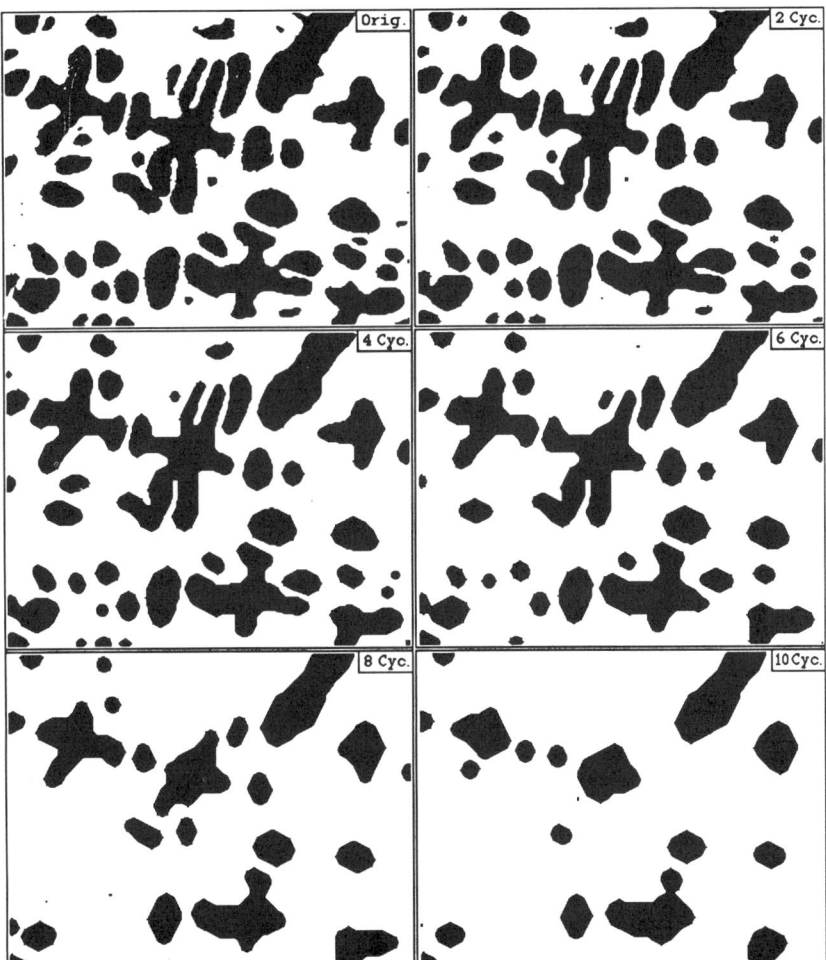

Figure 6-11: Openings applied to the dendrite image using a coefficient of 3, and varying the depth up to 10 cycles.

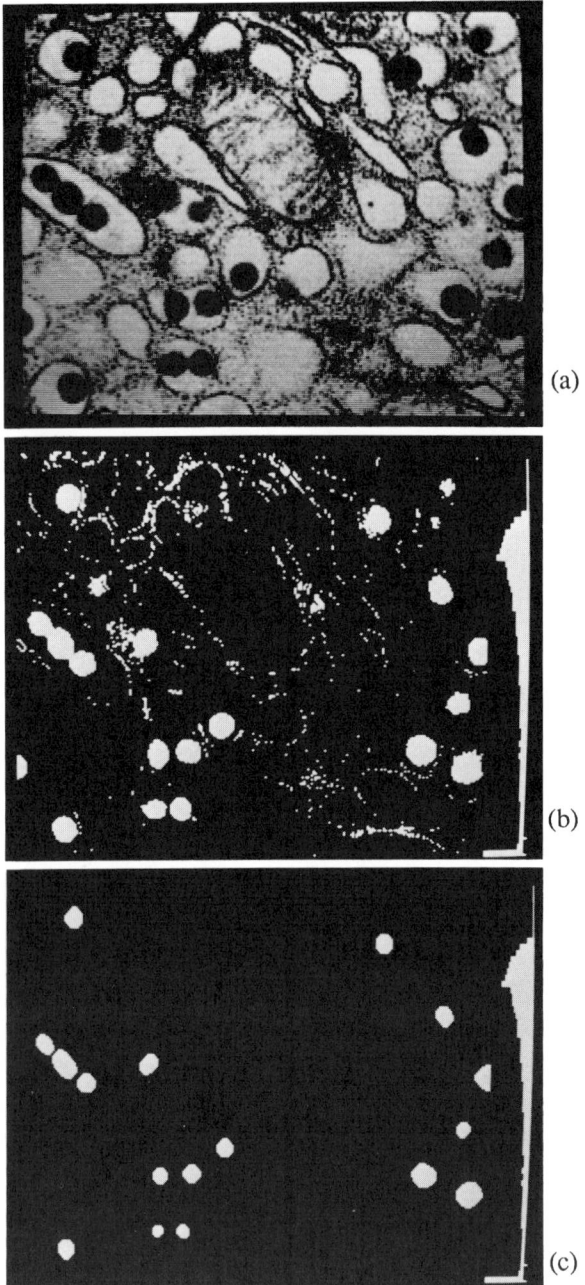

Figure 6-12: Use of an opening used to remove noise and separate regions in TEM image of biological section: a) original image - note that both the lysosomes and some membrane regions are stained equally dark; b) binary image obtained by thresholding; c) opening used to remove noise and separate lysosomes.

with X-ray maps obtained in the SEM. The individual dots in a low-intensity image are spaced apart, and most lie within features also visible in another more conventional image. When dilation is used to merge the individual spots, the result is a set of blobs that are larger than the original features, but correspond only to those that contain the element for which the map was made. The use of an AND function to combine the dilated X-ray map image and the higher-definition feature image results in the selection of just those features containing the element, as shown in Figure 6-13.

It happens that a very convenient way to actually apply erosion/dilation logic is not to count the number of neighbors around each point, but to use the ON/OFF status of the neighbors to construct a binary number as shown in Figure 6-14.

With eight neighbors, this produces a string of eight bits which have a numeric value from 0 (00000000, or all neighbors OFF) to 255 (11111111, or all neighbors ON). This number than serves as an index to a table, called a "fate table", which indicates whether the central pixel is to be turned on or off. The use of such a fate table makes it possible to respond to any possible pattern of neighbors, and this opens that way to other kinds of editing operations.

Skeletonization

One of the most interesting specialized uses of erosion is skeletonization. The skeleton of a feature is also known as a medial axis transform, and we will encounter another way to generate a skeleton shortly in the context of distance maps. But a very efficient way to generate a skeleton is to use the fate table method (Pavlidis, 1980; Russ, 1984). The skeleton consists of the lines of pixels that mark the midline of the feature. When the feature has branches or projections, so does the skeleton, and in fact a count of the number of branches or nodes in the skeleton is the most common way to describe the amount of branching in the object.

Skeletonization is also sometimes applied to the background or surroundings (by negating the image) in order to break the image up into so-called "zones of influence" around each feature (Figure 6-15). These can sometimes be used to separate touching features, as follows. First, erode the original image until all features have separated. Depending on the shape of objects and the extent of their overlaps, this may take quite a few iterations, and in some cases is not possible. Furthermore, if the features are of very different sizes some will have disappeared before others separate. We will return to these subjects shortly. For now, let us assume that the erosion is well behaved and the features separate leaving small islands. Now invert the image and skeletonize it. The result will be a series of lines which tessellate the image into zones around each island. This is sometimes called the "skiz" of the original image. Negate this image and combine it with the original image (before erosion) using a logical AND. This will cause lines of separation to appear on the original image that separate the touching features while keeping all other portions of their outlines, size and shape intact.

The size distribution of features in the negative image of the skeletonized background also gives information about the distances between neighboring features, which is sometimes of interest. We will later encounter another way to determine this information, in Chapter 8.

There are many other applications for which the skeleton is useful. Most are rather specialized. To produce the skeleton of an image for all features at once, we can use the specialized erosion technique mentioned above. The fate table is constructed so that

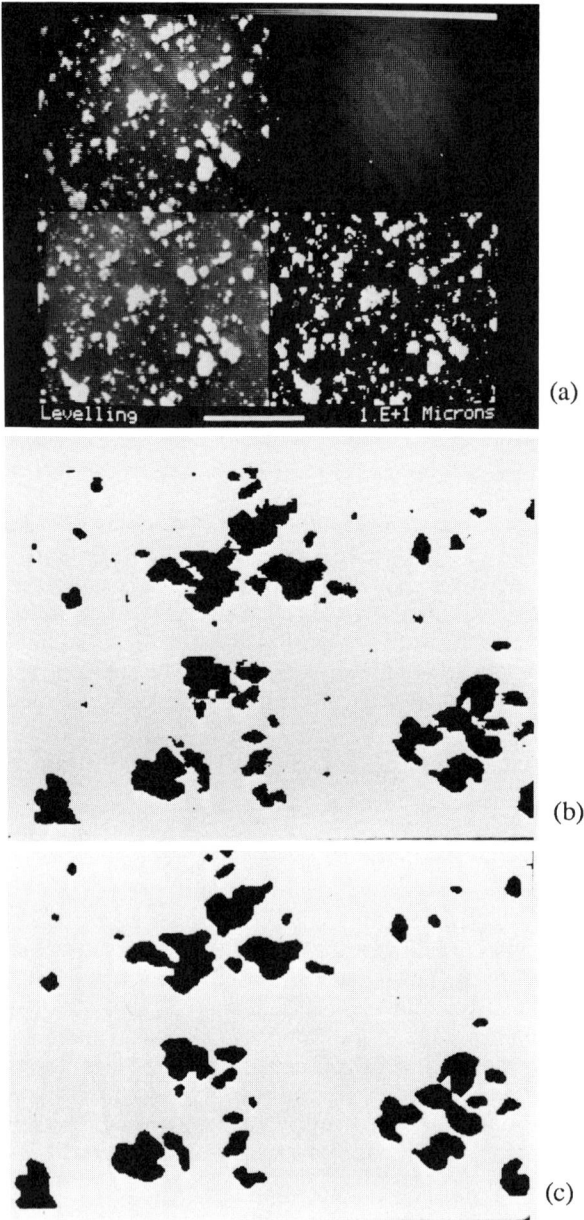

Figure 6-13: SEM image of paint pigment particles: a) original image, showing levelling used to permit thresholding, b) portion of the binary image, c) same after application of an opening, d) sulfur X-ray map image, e) regions in (b) selected by dilating dots in (d) followed by AND to show just the sulfur-containing particles.

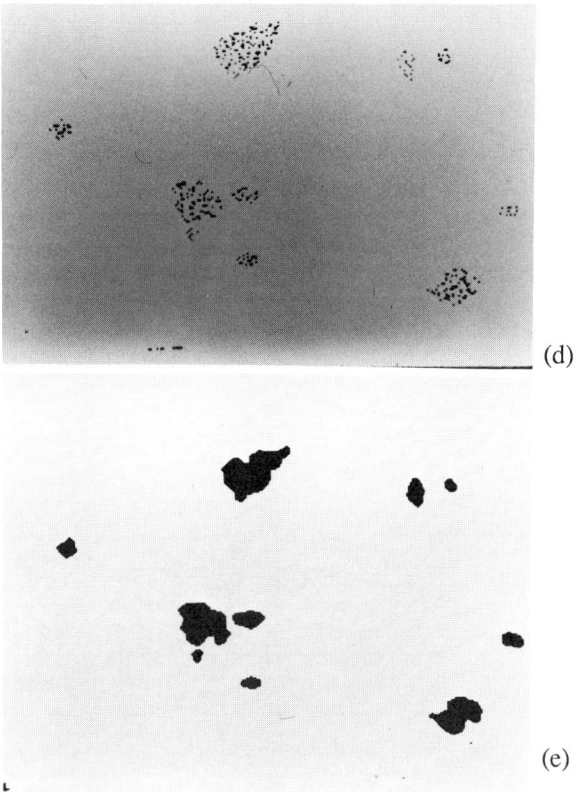

(d)

(e)

pixels with fewer than 8 neighbors, but whose neighbors are themselves contiguous are removed, because they must lie on an exposed periphery of a feature. However, pixels whose neighbors are not contiguous cannot be removed because doing so would create a break in the continuity of the feature. Furthermore, it is necessary to have a series of fate tables corresponding to each of the different directions in which connectivity is allowed. As we have seen so often, the particular rules depend on whether we allow 4- or 8- neighbor connectivity rules. Since each fate table requires only a single yes or no entry for each neighbor pattern, even with 8 such directions to consider, the entire set of possibilities will fit into a single byte (8 bits). With 256 possible neighbor patterns, this means that the entire table for skeletonization occupies only 256 bytes in the computer.

The erosion using this fate table removes only one layer from the outer periphery of the features with each iteration (consisting of 8 directional tests, applied to each pixel in the image). It must be continued until no further changes take place in the image. The result is then the desired skeleton (Figures 6-16, 6-17). It represents a simplified feature shape that can often be used for measurements that intentionally ignore some of the subtleties and irregularities that may be present.

There are several possible further steps that can be performed on the skeleton to obtain additional information. For instance, the branches of the skeleton can be disconnected by finding pixels with three or more neighbors and removing them and their neighbors. This removes the nodes where branches meet. These nodes can be counted as

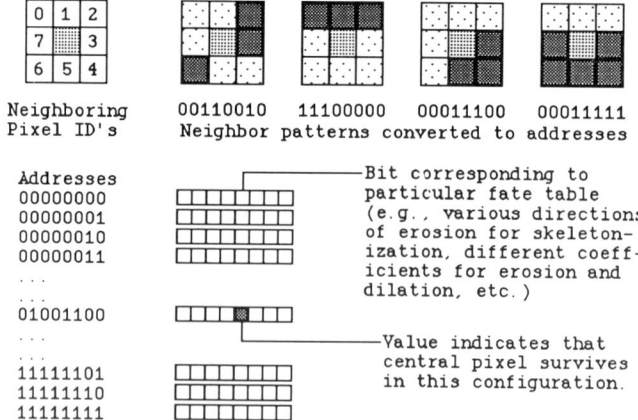

Figure 6-14: Pixel identification and construction of an address for a fate table.

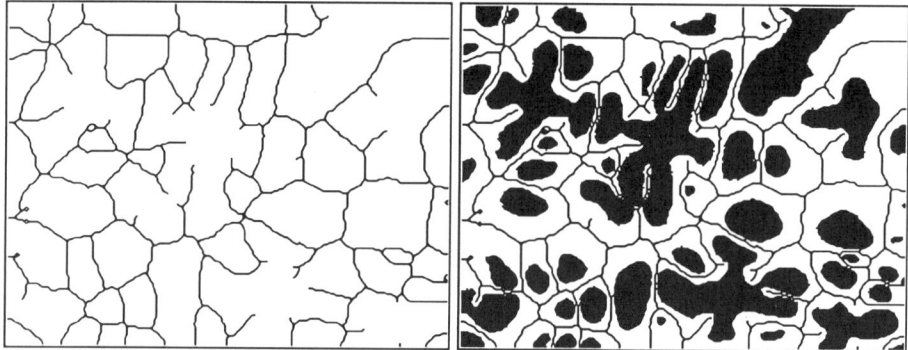

Figure 6-15: The "Skiz" of the dendrite image in Figure 6-16, by itself and superimposed on the original to show the zones around each feature.

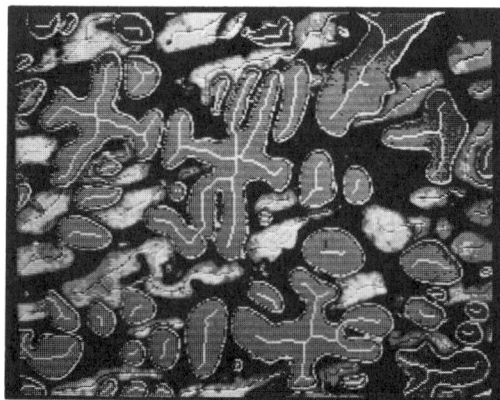

Figure 6-16: Grey scale image of dendritic alloy structure with skeleton superimposed.

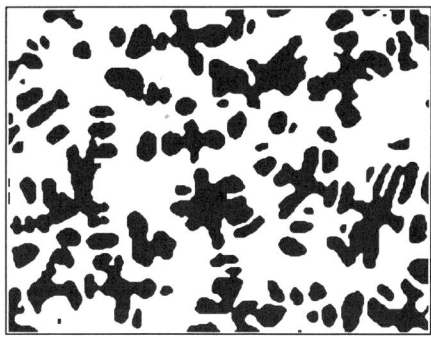
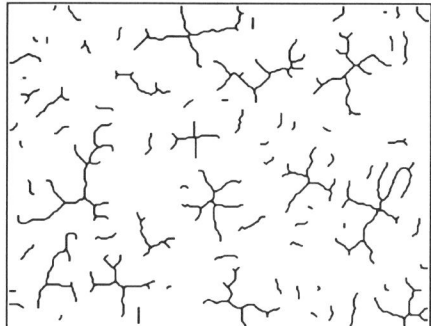

Figure 6-17: A binary image of dendritic features in Figure 6-16, and the skeletons of the features before removal of nodes.

will be described shortly. Their removal also leaves isolated branches whose length and orientation angle can be measured subsequently to characterize the original image. For instance, this method has been used to study the length and orientation of soybean roots, after acquiring a video image of the roots spread out on a white cardboard surface, and skeletonizing the discriminated binary (Figure 6-18).

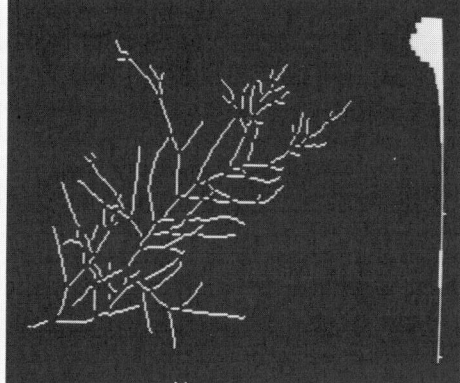
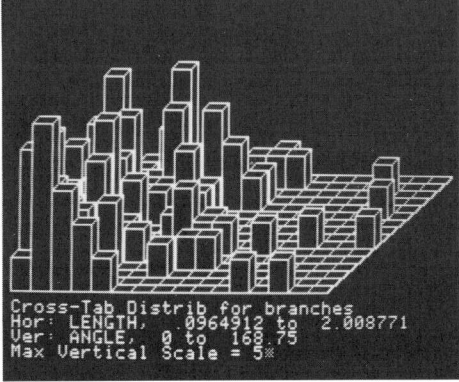

Figure 6-18: Cedar branch - image, skeleton and histogram of length and orientation of disconnected pieces.

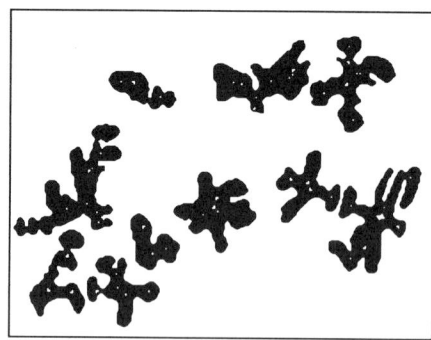

Figure 6-19: The image from Figure 6-16 showing only those features whose skeletons had more than n branch points (marked by white points).

Some systems make it possible to count the number of nodes in each skeleton directly (the number of pixels with three or four neighbors), and assign this number to the parent feature. If the system logic does not allow this, it can be accomplished by a more indirect route. To measure the number of nodes within each of several objects in an image, the following sequence of steps can be used: First, skeletonize the original binary and then remove the nodes. Combine this image with the image of the skeletons with the nodes intact using a logical AND. This produces an image of just the nodes. Now negate this and AND it with the original image. This leaves holes in the original features corresponding to each node. The number of internal holes within the features then gives the desired value, namely the number of branch points (nodes) in the skeleton of each feature, which is a measure of the feature's shape. The method for counting internal holes was described as part of the discussion of numbering a chord-encoded table with identifying numbers for each feature (when a chord touches two chords with the same feature number in the preceding line, it must indicate a join ending an internal hole).

The operation to count the number of end points of the skeleton is carried out in essentially the same way. Just as nodes can be recognized as points with three or four neighbors, so end points are those which have only a single neighbor. These can either be counted for each skeleton and assigned to the parent feature, or a sequence of steps involving erosion of the skeleton using a fate table that removes only end points, followed by and exclusive OR with the original skeleton to recover just the end points, followed by an exclusive OR with the original feature to leave holes, followed by counting the holes, can be used (Figure 6-19). Other systems directly count the nodes, end points and branches in each feature, and thus provide feature specific data that can be used to describe shape as discussed in Chapter 7. Figure 6-20 shows an example of these "topological" parameters.

Some of these operations seem to involve many steps, but most of them are virtually instantaneous, and even the more time consuming ones such as skeletonization require only a few seconds. It may also not be clear how such multi-step operations are to be devised to answer particular questions which can be asked about images. There is no easy answer to this question. It involves experience and experimentation, and careful phrasing of the original question in an unambiguous way. Once the question is asked, there is often a comparatively simple way to answer it.

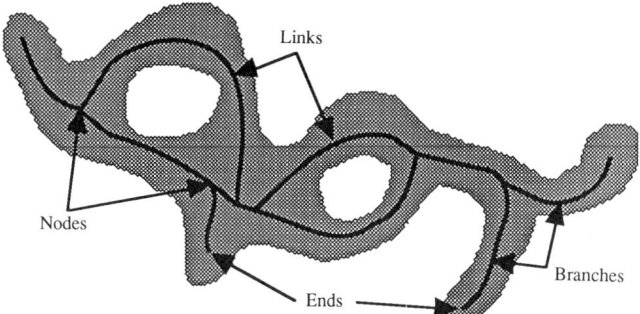

Figure 6-20: A feature with its skeleton, containing 4 Ends and Branches, 6 Nodes, 2 Loops, and 7 Links.

Skeletonization is also useful to reduce broad features such as grain boundaries that contain second phases or have been widened by etching, so that they produce just narrow lines (Figure 6-21). This allows better measurement of the "grain size" of the principal phase, which will be discussed in detail in Chapter 8. In fact, just counting the number of nodes (and knowing the area of the image in mm^2) in the skeletonized grain boundary image provides the "ASTM Grain Size Number" directly, by calculating

$$G = \frac{\log_e\left(\frac{\text{Nodes}/2 - 1}{\text{Area}}\right)}{\log_e(2)} - 2.95$$

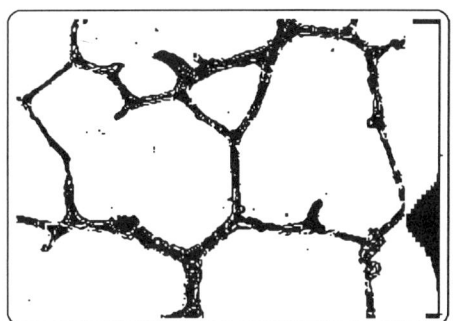
Combined light and dark regions of grain boundaries, made by imaging and then inverting intermediate grey level grains.

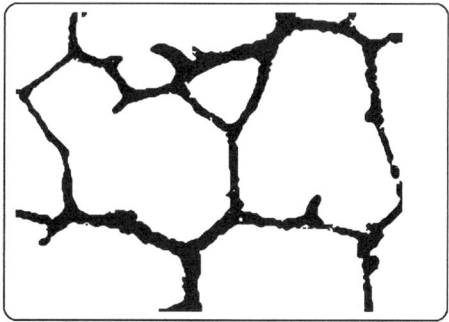
Grain boundary image produced by Plating (Coeff.=5) image after negating binary image of medium grey grains.

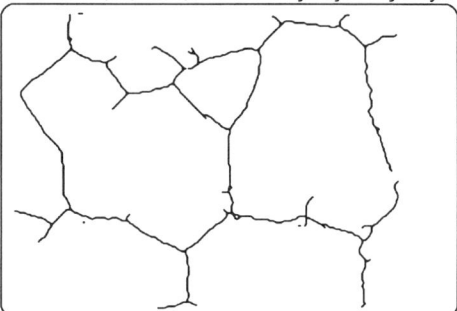
Skeletonized grain boundary image.

Figure 6-21: Skeletonization of the grain boundary image.

This method has an advantage over others such as measuring the mean intercept length within the grains (Chapter 8), in that it varies very little when the image degrades due to imperfect polishing or etching to delineate the grain boundaries. The nodes (known as triple points to the metallographer) generally etch well, even when the connecting lines do not. As much as 20% of the boundaries can disappear without significant effect on the grain size calculated from the number of nodes.

Another way to obtain a skeleton of features uses a grey scale image as an intermediate step. Each pixel in the grey scale image is assigned a brightness level corresponding to the distance from the pixel to the nearest edge of the feature (points outside the features are set to zero). In this image, called a Euclidean distance map, the central ridge of brightness is the skeleton, or "medial axis transform" (Danielsson, 1980; Shapiro et al., 1981; Bright & Newbury, 1986). It can be found using grey scale operators that locate maxima. This is somewhat more precise than the sequential erosion method for small features, because it does not depend on which direction the first erosion was performed. And for large features it is superior because the distance to the edge is measured in the exact direction to the nearest point, rather than in a 45 or 90 degree direction. However, the practical difficulty of computing these distance maps has led to short cuts that introduce exactly the same approximations, and in addition the method requires the use of a grey scale image memory, and so the most common approach is still the direct erosion of the binary image.

Measurement using binary image editing

Erosion of binary images can be used as a measurement tool, to obtain a distribution of feature sizes (Ehrlich et al., 1984). If the number of features present is counted, and then the image is eroded and they are counted again, the change in the number of features present reflects two processes. One is the disappearance of small features which do not survive the erosion. The second is the separation of touching features. If the same operation is repeated until the entire image has been erased, the plot of the change in the number of features at each step gives a distribution of their sizes. The dimension is not specified exactly, but is more or less related to the length (longest dimension) of the feature. The interpretation of the plot is somewhat confused by the separation of touching features, which increases the count, but the method is nonetheless often used, and produces results that can be used to compare populations of features that are similar in shape and in the degree of overlap (Figure 6-22).

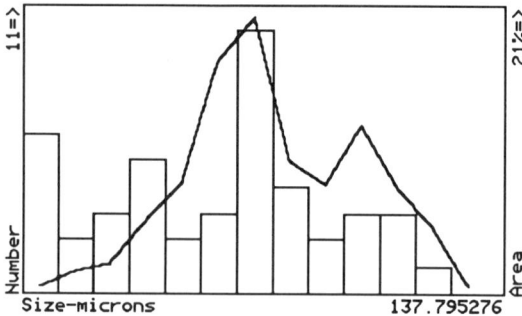

Figure 6-22: Size distribution information for the dendrites image obtained by plotting the number of features removed as a function of the number of erosion steps. The plot also shows the fractional area removed in each step.

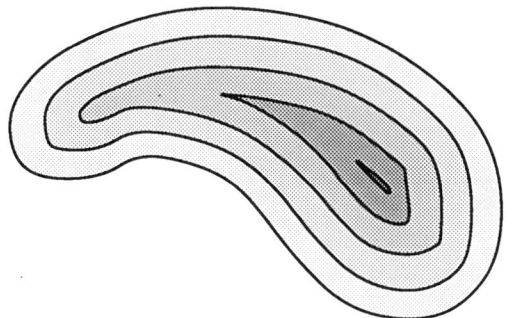

Figure 6-23: Example of concentric masks used to count features and produce a gradient plot

A similar operation can be performed using plating. If the number of objects is counted as a function of the number of plating cycles, it gives the neighbor distances between objects. This is particularly useful to measure the distances between features that may be of different size or arranged in clusters. A typical example is the distance between spots of mildew on leaves. It is also possible to plot the area fraction of a binary image as a function of dilation. This departs from a linear increase as overlaps occur, giving a measure of the tendency of features to cluster in the image. It has also been used to describe the size of pores and pore necks in binary images of sintered ceramics, soils, or other porous specimens (Chermant et al., 1981; Scrivener & Pratt, 1987). Another application is to quantify the degree of clustering of dislocations in material as revealed by etch pits on a polished surface viewed in the light microscope.

Finally, sequential erosion and dilation operations have been employed to estimate the fractal dimension of feature outlines. The concept of a fractal dimension is discussed later in this chapter. A measure of the roughness of the periphery of features, it shows up as a decrease in perimeter as the outline is smoothed by erosion and plating, or as the area of the pixels added or removed in these operations (Flook, 1978).

Erosion removes layers of pixels from the outer periphery of features present in the image. If the original image is EX-ORed with the eroded one, the result (which is sometimes called the "custer" of the feature) is just the removed layer of pixels. This can be used for several purposes that require the periphery. The first one we shall consider is the determination of gradients of some property as a function of distance from the surface of features.

If the binary image with the outer layer(s) as shown in Figure 6-23 is saved as a mask, then it can be used to measure the brightness of pixels in the original image (which can be converted to density or other parameters as will be discussed in the next chapter). The layer thickness can have any value convenient, as this can be controlled by the number of iterations of the erosion. The images can also be used as masks to combine with other binary image(s) discriminated to show other features, such as inclusions in metals or nuclei in tissue, etc., to select just those in that outer layer. These can be measured in any of the ways to be discussed in the next chapter. Usually, the area of each layer must also be used to normalize the measured data, since the areas vary as they progress in from the outer boundary.

The process of erosion and combination to produce a mask, and its use to select pixels or features for measurement, can be repeated until the original features disappear.

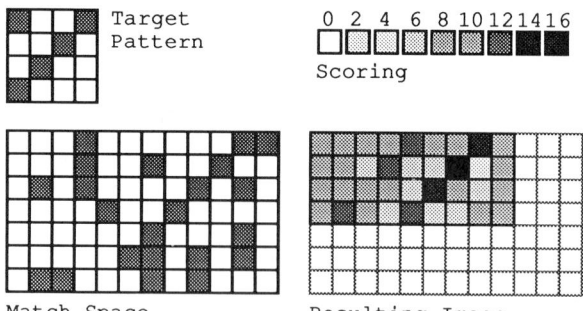

Figure 6-24: Template matching: the matrix of values shows the brightness of the grey image produced by counting matches of the template with the image fragment shown.

A plot of the data obtained during the measurement process gives a characterization of the radial distribution of whatever parameter has been determined within the original features. This characterization deals with features of any shape, including ones that are not convex, because it works in from the outer surface and does not depend on finding a center and measuring out radially from that point. It has been applied to heterogeneous biological specimens such as organs, cross sections of muscles, etc. as well as meteorites, and even distributions of economically important oil deposits as a function of distance from rivers or mountain chains.

Erosion and other fate table operations are not the only transformations that come from mathematical morphology. Another, more general concept is sometimes referred to as the "hit or miss" transformation. This consists of setting up any pattern of binary pixels, of any size and shape, and passing it over the binary image. The pattern can contain pixels that must be ON and others that must be OFF. Wherever the image matches the pattern, the central pixel (or any other single reference location, since the patterns need not be symmetric or have an obvious "center") is set ON, and in all other places it is set OFF. This provides a way to find any kind of feature that may be desired. In principle, it could be used to read printed text, although problems in pixel alignment and character variations make it rather finicky in practice. Also, it would be time consuming to match a template for each possible letter in the alphabet to all locations of an image.

A more general approach to the same task is called template matching (Vollath, 1986). As before, an arbitrary pattern of pixels is set up to be matched (the template), and may contain some pixels that must be ON and others that must be OFF. It is then positioned at each location on the binary image and a score is counted up of the number of pixels in the image that match the template conditions (either ON or OFF). This score is then stored as a brightness value in a grey scale (not binary) image. The result is a brightness-coded map of where the target patterns or ones somewhat similar to it are found. The brightness image can be analyzed by finding each of the maxima, comparing their values to some preset acceptance level, and using the locations to identify the positions of the target pattern, as shown in Figure 6-24.

This provides a fairly straightforward feature recognition tool that can be used to look for many kinds of objects in an image. It does not deal well with representations of the target feature that have been rotated with respect to the template, of course, and we will

Binary Image Editing

see in Chapter 9 a different approach to object recognition that overcomes his limitation. But for many applications (especially quality control work) it is particularly efficient and useful.

Covariance

Template matching was performed by sliding a particular pattern across the binary image and scoring how many pixels matched at each location. The same operation can be performed using the image itself as shown in Figure 6-25. Of course, the match is perfect when no offset is introduced. But as the image is shifted with respect to itself, the number of pixels that match will generally drop. If the features are not random in shape or alignment, the result is a nonsymmetric rate of decline in the fraction of pixels that match. This value can be graphed or converted to a grey scale image to reveal the anisotropy in the original image (Fabbri, 1984) as shown in Figure 6-26.

If the original image has a repetitive structure in it, then the number of matching points will rise when the shift distance and direction match the spacing of the structure. The result is a series of peaks that reveal the structure and spacing, as shown in Figure 6-27.

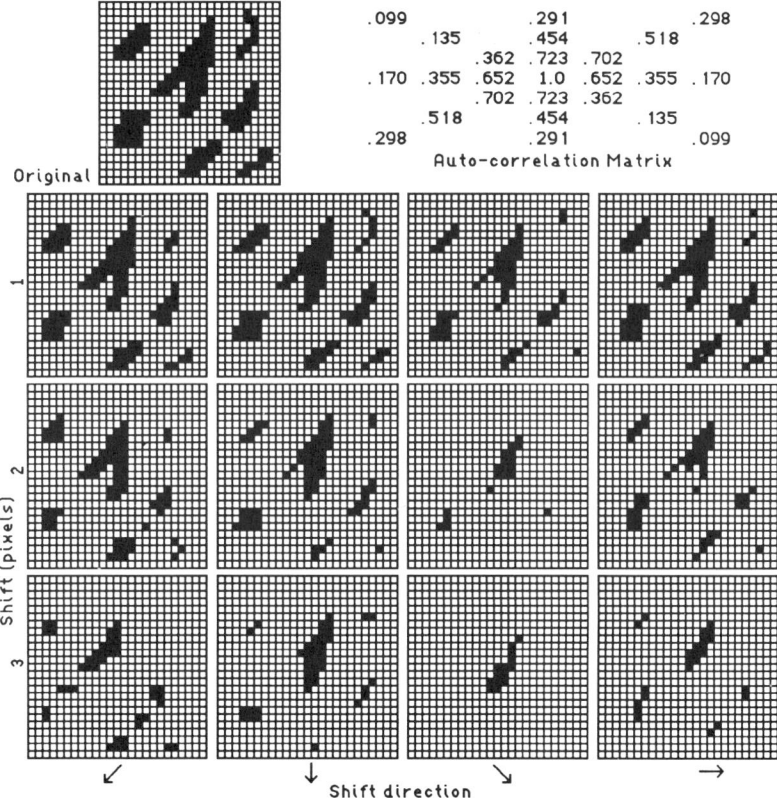

Figure 6-25: Autocorrelation performed by shifting a binary image and counting the overlaps.

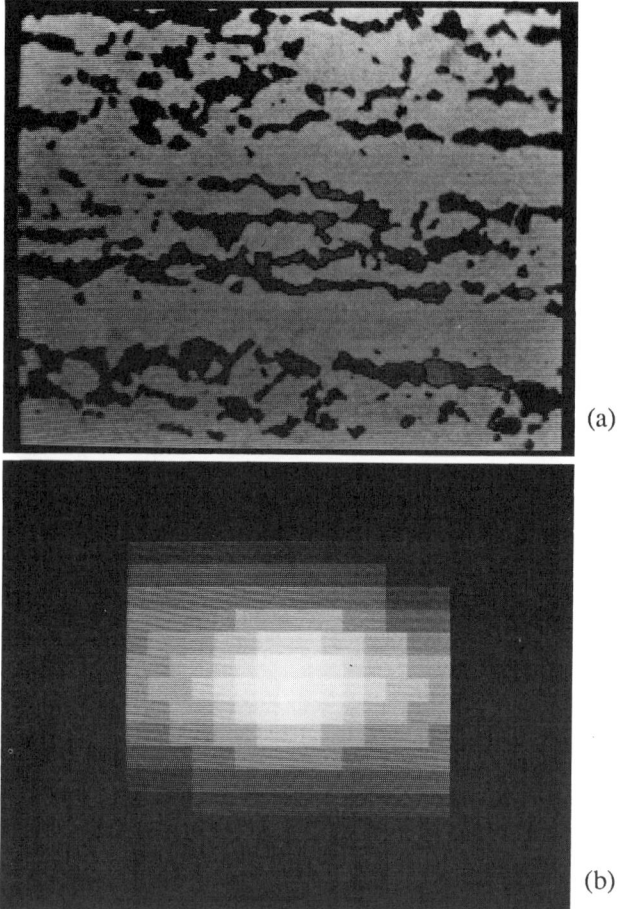

Figure 6-26: Anisotropic structure of brass (a) with its autocorrelation matrix shown as a grey scale image (b).

When this repeat is present, it sometimes allows improved quality images to be constructed by image averaging (Saxton, 1982). Chapter 3 showed that averaging together a large number of images could improve the image quality by reducing noise, or statistical fluctuations. If the same structure is repeated within a single image, then averaging is also possible by superimposing the many small images of the same feature. For this purpose, the autocorrelation spectrum is usually obtained with the grey scale image rather than the binary image. Once a repeat distance and direction have been established, then the image fragments can be averaged together. In many cases, rotational averaging of symmetrical structures as often occur in biological macromolecules or crystalline lattice images can also be used to increase the number of images to be added. The result is a considerable improvement in the visual appearance of the image, showing details that may not be visible in any single one of the many repeated images.

Care must be taken to assure that the apparent repeat is real, however. Averaging together different structures that happen to appear with near-regular spacing, or rotational superimposition of nearly symmetric features, will also produce images with improved appearance and reduced noise, but of course the results are an artefact of the processing and not real structure. It should also be recalled from the earlier chapter that global averaging of this kind can also be accomplished in frequency space, for example with a Fourier transform of the image. Then the repeat distance and direction show up as a bright spot in the magnitude image, and filtering to reduce other components of the transform image can produce a smoothed repeating image when transformed back to the spatial domain.

Watershed segmentation

Mention has been made several times in this and preceding chapters of the desirability of separating touching features so that they can be measured separately, and some ways of accomplishing this under certain circumstances. The most general way to do this is called segmentation by watersheds (Beucher & Lantejoul, 1979; Lantejoul & Beucher, 1981). Like the template matching operation, it makes use of a grey scale image to store the results of operations on the binary image.

The first step in this technique is to create a distance map, in which the brightness of the grey pixel is a measure of how far the corresponding binary pixel is from the outer periphery of the feature. Ideally, this would the the distance to the nearest point on the periphery in any direction (the "Euclidean" distance map mentioned earlier). In practice, it is usually sufficient and involves much less programming to replace the circle of uniform radius with an octagon, or even a square. In other words, the distance to the periphery is found in the four orthogonal or eight orthogonal and diagonal directions, and the shortest one selected as the distance. Its value is then saved as the corresponding pixel brightness.

There are two ways to construct this distance map. The first method uses sequential application of morphological operators (Serra, 1982). The features in the image are sequentially eroded, and after each iteration all remaining pixels have their corresponding brightness value incremented by one. When the process is complete (the image is completely eroded), the desired distance map has been constructed. This method works for hexagonal pixel arrays as well as for square ones (Jernot, 1982; Coster & Chermant, 1985), but is comparatively slow, and furthermore the number of iterations grows as feature size increases.

In the example shown in Figure 6-28, the original binary image (lower left corner) is successively eroded in four steps (first column). At each stage, isolated clusters of points

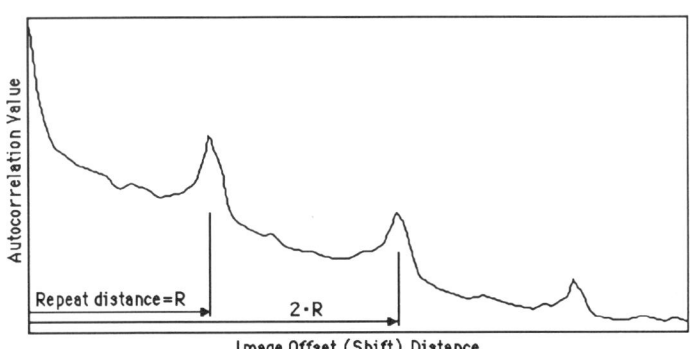

Figure 6-27: Autocorrelation plot for a repeating structure.

which disappear entirely at the next step are found and saved (middle column). Starting from the final stage, these ultimate eroded sets are thickened. The process of thickening is similar to dilation, but is modified to eliminate points that would cause new connections to form. At each stage in the thickening process, the ultimate eroded points from the corresponding stage in the erosion process are added into the growing image (right column), and dilation is limited to those points which are present in the eroded image.

The thickening is iterative, involving (for the case of 8-connectivity) twelve steps (one for each orthogonal neighbor direction and two for each diagonal neighbor direction), and continues until no further changes occur. Figure 6-29 shows the rules for the 90 and 45 degree directions for both 8- and 4-neighbor connectivity. The central pixel is filled only if the neighbors marked X are already present, and if the ones marked O are absent. The other neighbor positions are "don't care" ones which need not be examined.

The key to the method is finding all of the ultimate eroded points, which are the final, separated islands last removed by iterative erosion. It is important to understand that

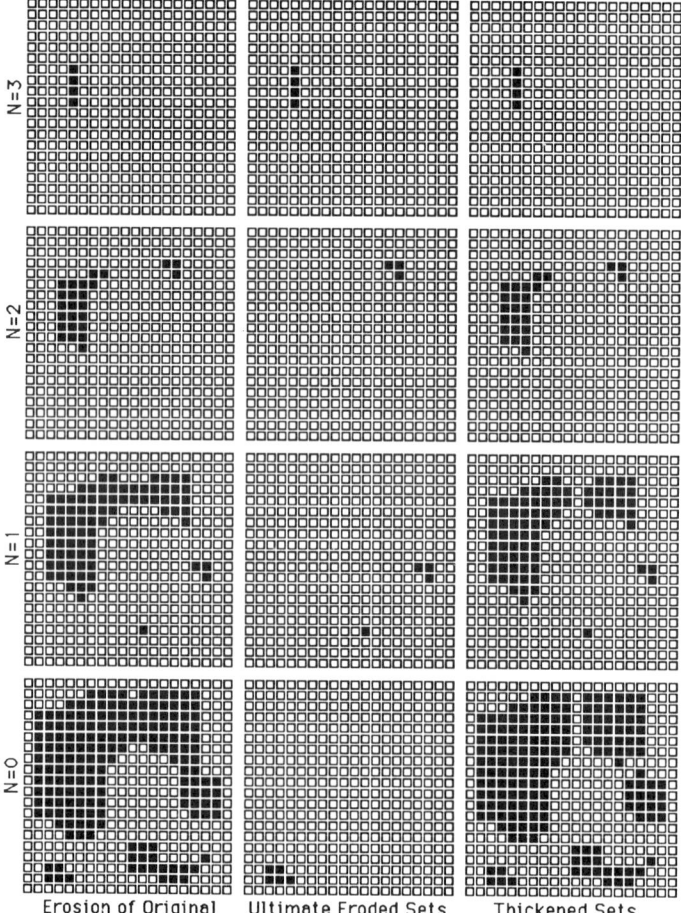

Figure 6-28: Procedure for segmentation by watershed method (see text).

Binary Image Editing

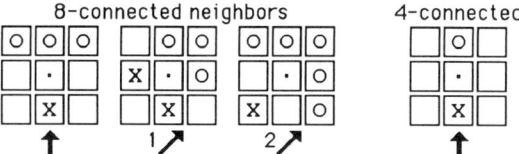

Figure 6-29: Dilation rules for conditional growth (other directions by rotation). X = required, O = prohibited, blank = don't care. For 8-connectivity, one test for orthogonal neighbors and two for diagonal neighbors are needed. For 4-connectivity only a single test for orthogonal neighbors is required.

these do not appear at the same time in the erosion sequence, because in general the features are not of the same size. The ultimate eroded sets can be arbitrarily large (e.g., a long fiber of single pixel width) if the narrowest direction is small enough to permit the entire feature to be removed in the next erosion step.

Thickening of the ultimate eroded points in reverse sequence of their removal, with the application of a rule to prevent pixels from being filled in if they cause a connection to be formed between any other pixels or were not part of the original image, then allows the recreation of the original features with lines of separation.

This apparently simple series of steps is beset with several important practical problems. First and foremost, the process is iterative. Each of the erosion and thickening steps must be repeated until nothing further changes in the image, and there are particular object shapes (for instance the use of thickening to reconstruct a long, narrow fiber or projection) which may take many repetitions. This is particularly true with the thickening process used by Jernot, which produces growth in each direction on only one out of every six applications of the procedure to the entire image (for a hexagonal pixel array).

Second, since the method requires the entire binary image at each stage of the erosion process to be available during the thickening operations (both to define new ultimate points to be added to the growing image, and to limit the thickening to existing pixels), it either requires massive storage or many repetitions of the erosion operation. Compact storage of the information can be achieved in the form of a distance map in which pixel brightness represents distance to the edge. This effectively stores many eroded binary

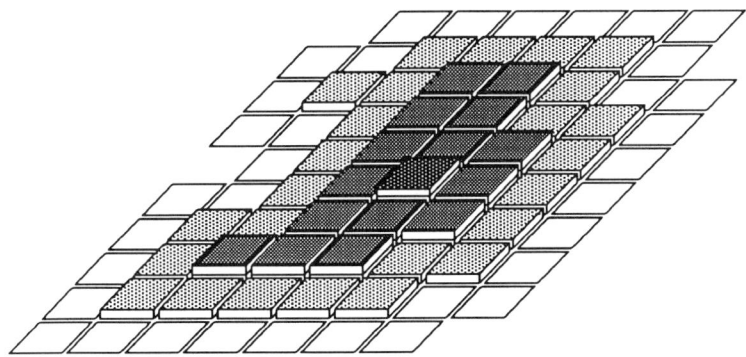

Figure 6-30: Height map in which the brightness of each pixel is proportional to its distance from the edge of the original object in the binary image.

images in different brightness planes of a single grey image. In principle, it is possible to obtain this image by iteratively applying the sequential erosion operator. As each outer layer of pixels is removed from the binary image of the objects, all remaining pixels in the grey image store are incremented. The result is the required height map.

A second method which is more efficient requires just four passes across the image (Russ & Russ, 1988). Initially the brightness values for each pixel that is ON in the binary image is set to a maximum value, and to zero elsewhere. Then starting from the top left and going through the image, line by line from top to bottom and from left to right, we impose a rule that no point may be more than one higher than its lowest neighbor. This is then repeated line by line going from bottom to top and right to left. Regardless of feature size, shape or complexity, two complete repetitions of this procedure, in both directions through the image, produces a distance or height map as shown schematically in Figure 6-30. This requires a fixed time.

If comparison to all 8 neighbors is used, the brightness "peaks" (brightness represents distance to the edge in this image) are diamond shaped in profile, and if 4 neighbors are used (only comparisons in the four orthogonal directions), square shaped profiles result. By restricting the brightness difference for horizontal and vertical neighboring pixels to 2 and for diagonally neighboring ones to 3, an closer approximation to an isotropic circle results (the ratio 3/2 is close to the ideal value, the square root of 2) and the peaks are octagonal in shape. After completion of the height map, the values are divided by two, rounding down to help compensate for the small difference between 3/2 and $\sqrt{2}$. A similar algorithm can be used for hexagonal pixel arrays, to produce hexagonal shaped peaks, nearly as good an approximation to the circle as an octagon.

The distance map can be used for several purposes. For one thing, the ridge of local maximum brightness is the medial axis transform of the image (the line of points equidistant from points along the periphery), and is another way to obtain the feature skeleton. In general, this method gives a slightly different result than the selective erosion one, and is more sensitive to irregularities in the periphery which lead to short branches from the main backbone of the skeleton.

The maximum brightness within each feature is also a measure of the width of the feature - the smallest caliper diameter through which it will pass, or the radius of the largest inscribed circle. This is otherwise an extremely difficult parameter to measure, and as we will see in Chapter 7, we often resort to estimates for it.

But our present purpose for the distance map is to use it for segmentation of touching objects. The segmentation occurs along watersheds in the distance map. These are best understood by imagining the brightness value of each pixel to be a physical altitude, so that the bright points in the centers of features become mountains. The watersheds are the valleys between the mountains. When the peaks are of the same height (meaning that the two features which touch are of the same size), the division point is midway between the two peaks. but when one feature is smaller than the other, the peak is smaller and the watershed lies closer to the smaller peak. The method makes the implicit assumption that the features are both actually convex, so that they should be segmented, and also assumes that the degree of touching or overlap is sufficiently small that there is a valley between the peaks at the center of each feature in the brightness-coded distance map.

Now we must see how to find these watershed lines. In the distance map, the plateaus (all points which are not dimmer than any neighboring point and are adjacent to

Binary Image Editing

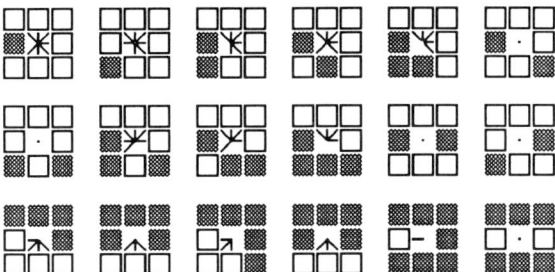

Figure 6-31: Representative neighbor patterns with the directions from which the central pixel is filled shown by the central vectors.

neighbors with the same criteria) form the nuclei for subsequent feature growth by thickening. However, they lie at different heights (representing the centers of features of different sizes). This is taken into account automatically by the thickening routine described below, so it is possible to immediately place all of the plateaus into the final binary image. This is accomplished by setting the high bit of the byte for that pixel in the image memory. In other words, the lowest n bits (usually 7) represent each pixel's brightness (distance from the nearest edge), while the highest bit indicates whether the pixel is part of the resulting segmented binary image. This makes the recovery of the final image from the grey image storage a simple matter of discrimination. It also permits watching the bright features grow during the thickening process.

Thickening is carried out in steps, starting at a height (brightness) value one less than the highest plateau. At each height, pixels having at least that value are candidates for thickening. They will become part of the binary image (by having their high bit set) if they do not create a new junction between any neighboring pixels. This operation requires applying rules of logic much like those used in skeletonization but using a different fate table. A table of the 256 possible configurations of the 8 neighbors around each pixel is stored in memory. At each position in the table, there are 8 directions of growth of the pixel considered, and for each a single bit defines the pixel's fate. Patterns of neighbors which would be connected if the central pixel were filled in are not permitted; all other patterns are filled. Figure 6-31 illustrates several examples.

This fate table is applied to each pixel in the image whose brightness is at the specified value of height or distance, and it is repeated sequentially for the image in each of the eight possible growth directions. In practice, most pixel patterns are filled according to the rules for several directions, and hence most pixels at each height level

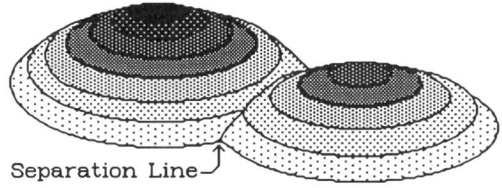

Figure 6-32: Separation of two touching features along the "watershed" line between their peaks in the distance map.

158 Chapter 6

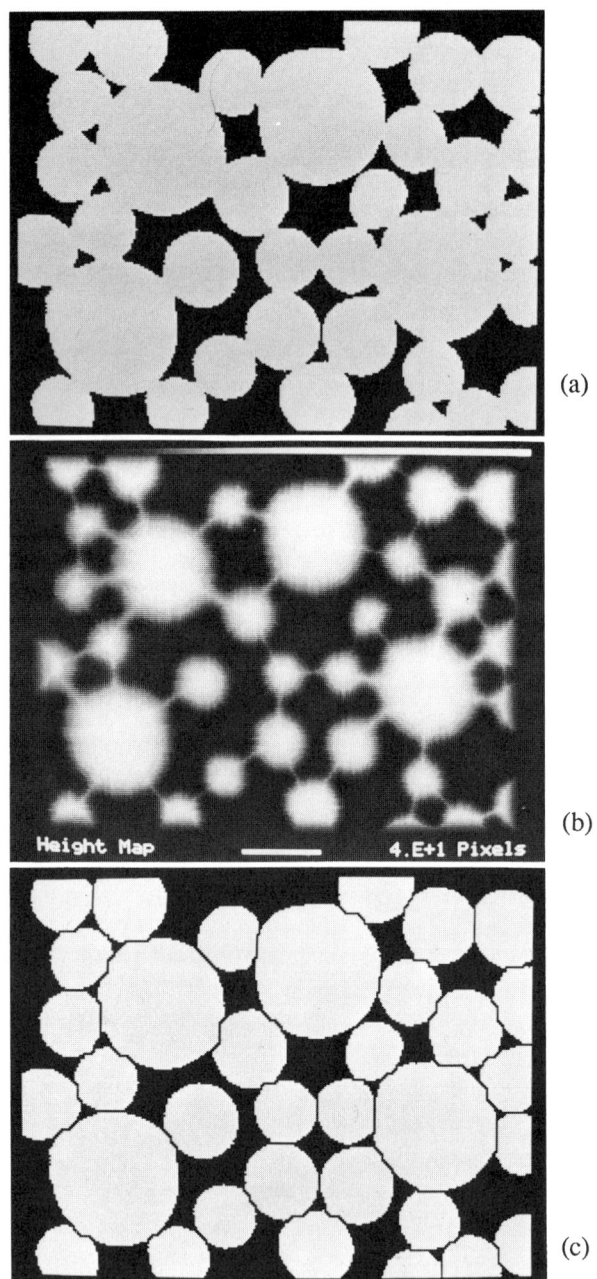

(a)

(b)

(c)

Figure 6-33: Example of a binary image with touching features (a), the distance or height map in which pixel brightness represents distance to an edge (b), and the segmented binary image produced by the watershed method (c).

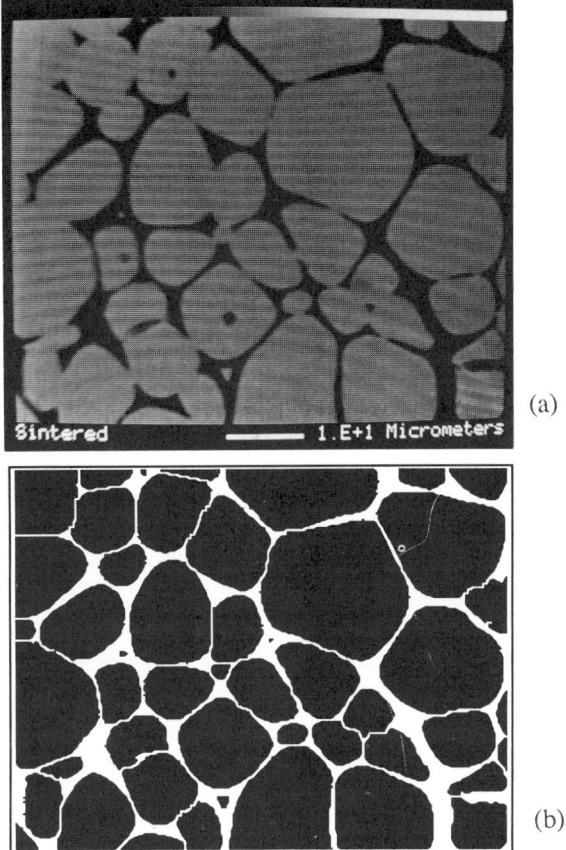

Figure 6-34: Application of watershed or convex segmentation to a polished section through sintered metallic particles: a) original image, b) segmented features.

are filled within the first few passes. The process must be repeated until no changes take place (there is no further growth). This is accomplished along with speeding up the process, by keeping a set of flags for lines in the image having any possible candidate pixels. Initially, these flags mark all lines which have pixels at the proper brightness, but as most of these are filled in, the flags are cleared so that no pixels on the line need to be considered again. This is much more efficient than the iterative erosion/dilation method in which each pixel in the entire image is repeatedly tested, and growth occurs in only one direction at a time. When all of the flags have been cleared, the process is complete at the current brightness level. After the first attempt at growth in each direction, subsequent growth is always adjacent to previous growth. The speed can be further improved by maintaining a list of all points at or above the specified height that are eligible for growth.

The operation is then repeated at each next lower brightness level, until it ends. The resulting binary image is then recovered and may be compared to the original, from which it differs only in the removal of pixels along lines to separate touching objects. The lines of pixels removed are defined by the junctions of cones in the height map, each cone built around one plateau or ultimate eroded set, and growing from a base on the

Figure 6-35: Some natural fractal shapes a) leaf with jagged edge, b) agglomeration of soot particles, c) computer modelling of "real" appearing landscapes, d) artistic representation (Hokusai - The Wave).

original binary image. The segmentation takes place along the "watershed" lines at the junctions, as shown schematically in Figure 6-32. The lines of segmentation run along specific angular directions. For the case of hexagonal pixels, orientations in multiples of 60 degrees occur. For the diamond or square structuring elements, the lines can run at any increment of 45 degrees, and for the octagonal case, multiples of 22.5 degrees are found.

Figure 6-33 show a representative example of the application of this segmentation method. The original image contains features that are broadly convex, of different sizes, and touch each other at numerous points. This type of image is commonly encountered in the examination of sintered materials, but the principle is equally applicable to many other types of images, from microscopic and macroscopic sources. The figure shows the grey-scale distance map; the central plateaus (bright regions) are the nuclei for thickening. After segmentation, the features are separated and suitable for automatic measurement. Figure 6-34 shows the application of this technique to a sintered metal specimen in which the individual features vary both in size and shape.

(c)

(d)

This procedure is much faster than the sequential one using erosion and dilation. Each pixel in the example image is visited by the program an average of 11.2 times, compared to over 1600 for the sequential method. For less complex images that cover less area or consist of smaller objects, the time required is even less. The iterative sequential method requires time that is linearly proportional to the number of pixels in the original binary image.

Mosaic amalgamation and fractal dimensions

Another binary image editing method that is used in conjunction with a measurement operation is mosaic amalgamation. To see its purpose, we must tolerate a brief aside to discuss the concept of a fractal dimension. In nature, most boundaries are not straight lines forming a polygon, or simple smooth curves, but are instead rough and irregular (Figure 6-35). Furthermore, the amount of roughness and irregularity that we can see is generally limited by our image resolution, and if we could increase the magnification, the amount of perimeter we would see in the image would increase.

Figure 6-36: The west coastline of Britain at two magnifications, one at a scale 80 times larger than the other.

The particular type of crinkled-up behavior that is of most interest to us here has a peculiar form called self-similarity. This simply means that at any magnification at which we view the line or surface, it looks the same. Whatever measurements we can make to describe the roughness and its scale will be independent of the scale. This imposes a significant constraint on shapes of features. This effect was first noted by a somewhat eccentric British mathematician named Richardson. He measured the borders of several countries and landmasses, but the best known example is the west coast of Britain (Scotland, England and Wales) as shown in Figure 6-36.

If the length of the boundary (the coastline) was determined by swinging along a map of the country with dividers set at some arbitrary distance, say 1 kilometer, a total length was obtained. Then if the divider distance, or stride length, was reduced, and the same operation repeated, the measured length was greater because the measurement was able to follow more of the irregularities of the coastline. Repeating the operation with finer and finer steps (and maps of appropriate scale) would cause the length to continuously increase, so that in effect, the length of the coastline would be expected to become infinite at a fine enough scale. Furthermore, Richardson noted that over a considerable range of stride lengths, and for a variety of borders, both natural and man-made, the slope of the plot of measured length versus stride length was constant on a log scale, as shown in Figure 6-37.

An example of the effect of increasing perimeter without increasing area is shown in Figure 6-38. This type of feature is called a Koch Island, after the mathematician who proposed them (Koch, 1904). All the features are drawn with the same area; you are asked to imagine that in each of the different images you have improved your image resolution to be able to define smaller details. The perimeter on each successive feature is increased because of the patterned irregularity that is introduced on each straight line segment of the boundary. In fact, the amount of perimeter increases by a factor of 1.5 from each feature to the next, and there is no limit (at least down to atomic dimensions) to how far we might extend this process, so the length of the perimeter is undefined. The mathematicians who considered shapes like these were horrified at the behavior of such lines and shapes because without smooth, continuous derivatives, they could not be easily handled by conventional techniques.

Mandelbrot (1982) has developed these ideas further into a useful concept, and Feder (1988) has reviewed still more practical applications and interpretations based on fractal

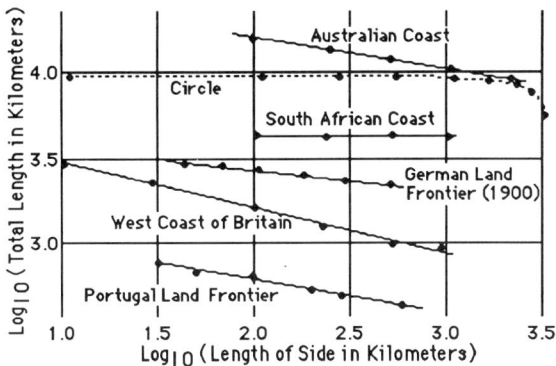

Figure 6-37: Richardson's plots (1961) of the length of various geographical boundaries versus the distance used for measurement (the length of the side of the polygon used to fit the boundaries).

dimensions. If the increase in measured length with improvement in measuring resolution is uniform (a straight line on a log plot, as shown in Richardson's data), the feature is said to be self-similar. It has been shown that many natural objects, ranging from the microscopic to large, have this character at least over a substantial range of distances. The consequence is that it is not really possible to state what the amount of boundary or surface area is. For the Koch island shown in Figure 6-38, the fact that the perimeter increases by a factor of 1.5 for each halving of the measurement distance has led Mandelbrot to describe this particular boundary line as having a dimension not of 1 (a straight line) or 2 (a plane) but 1.5. In other words, the line has a fractional dimension which reflects its ability to fill a plane, and hence it is called a "fractal."

Depending on the nature of the irregularity which we might introduce on the Koch island, the rate at which the perimeter increased with each step could vary. Figure 6-39 shows several such curves. In each case, the area of the feature would stay the same as each straight line segment in the feature boundary was replaced by the new curve, and then the sequence repeated at a finer scale, and so on, but the rate at which the perimeter would increase is given by the values shown. These values are related to the slopes of the Richardson plot that one would obtain for these features. Higher fractal dimensions reflect the fact that as the substitutions of finer scale irregularities are made, the line spreads out faster over the plane. A fractal dimension approaching 2.0 would cover the entire plane, while one close to 1.0 would remain nearly a line.

Of course, in real features the irregularity is not so regular. It is possible to apply a random pattern to increase the perimeter along edges, with a mean fractal dimension, and

Figure 6-38: Development of a Koch island with a perimeter whose fractal dimension is 1.5.

Figure 6-39: Irregularities that can be substituted on the Koch island, with different fractal dimensions.

Mandelbrot has shown some examples of this technique that produce features with stunning realism (Lucasfilm has extensively developed and utilized these methods, to produce realistic but un-earthly landscapes for movies). The concept of self-similarity simply means that on the average, the increase in boundary length is uniform as resolution is increased (for instance by working at progressively higher magnifications).

Generating a fractal boundary that is "random" can be accomplished rather straightforwardly by using the first type of generator shown in Figure 6-39, while varying the distance by which the new points on the line are offset. The outlines shown in Figure 6-40 began with polygons formed by generating a random series of points on a circle. Each side was then divided into 2 parts and the midpoint displaced either in or out by a random distance, proportional to the length of the side. This process was repeated 6 times, so that the side ultimately became a series of 128 short segments. The total feature has a fractal dimension that is easily determined by constructing a Richardson plot (obtained by summing the length of the perimeter for each of the halving steps). The mean displacement of the midpoints can be varied from zero up to half the length of the line segment whose midpoint is being displaced, to produce fractal dimensions from 1.0 (a Euclidean polygon) to about 1.25, and the shapes created are very similar in

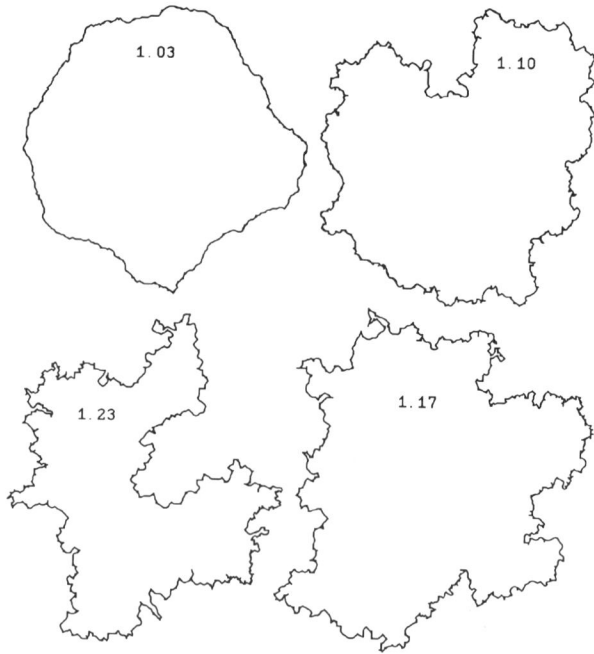

Figure 6-40: Random fractal outlines generated as described in the text, with varying fractal dimensions.

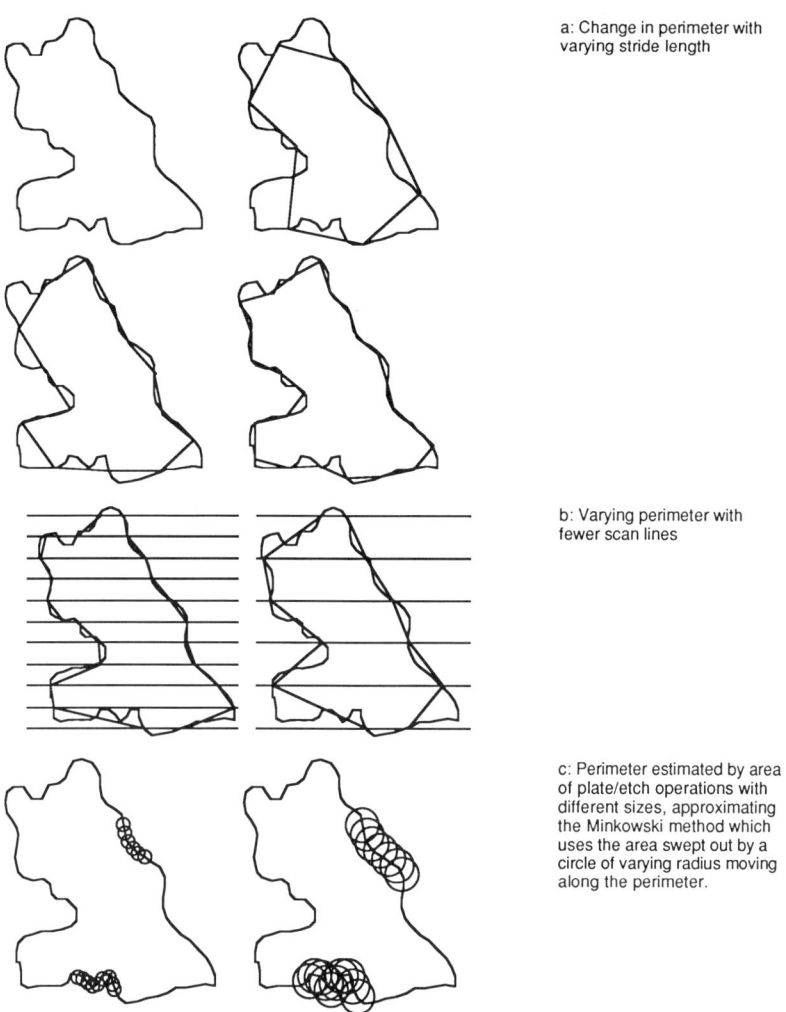

Figure 6-41: Ways to measure a feature's fractal dimension.

appearance to those of some real objects (cornflakes or dust particles or islands, for example).

For some applications, it is satisfactory to visually compare the "roughness" of real objects (typically viewed in the microscope) with generated outlines having known fractal dimensions, to determine an approximate value. This may be adequate for comparative purposes, but measurement of the dimension using a Richardson-type plot is desirable since it also provides confirmation of the self-similarity of the object's outline, established by the linearity of the plot.

There are also several ways to obtain a Richardson plot (Kaye, 1986) as summarized in Figure 6-41. For features represented by their boundary (either chain code or a series of coordinates from manual data entry), it is straightforward to calculate the perimeter as a count of the links in the chain or as a polygon formed by the points, but not using all of

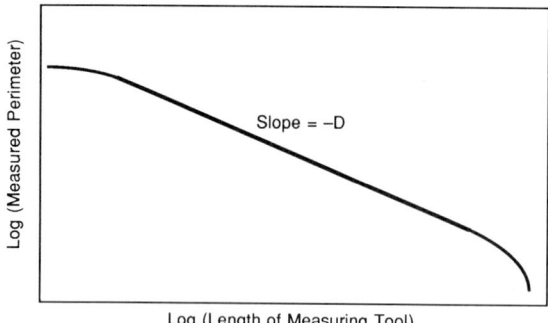
Figure 6-42: Schematic Richardson plot for a particle outline.

the links or points. If every second point is used, the polygon will miss some irregularities and the perimeter will be shorter. This can be repeated with every third point, and so on, to form a Richardson plot of the log of perimeter versus the log of the side length (the mean spacing between points) from which the fractal dimension can be obtained as described before.

When the points are not equally spaced, it may introduce some error into the result, but this is usually a minor factor, and in addition most continuous drawing routines for the entry of boundary points produce coordinate streams that are fairly uniform in spacing. It is necessary to take into account the final non-uniform polygon side that completes the boundary, if one is present. But when many points are used in the polygon (and a typical manual outline may contain several hundred) it is usually adequate to estimate the side length by dividing the perimeter by the number of points (vertices). It is also sometimes useful to calculate the perimeter using all possible starting points when skipping intermediate points, and to average the perimeter values from the various determinations.

For pixel-based feature representation, it is also possible to estimate the fractal dimension by using chords from every second, third, etc. scan line in the image. This is quite fast, but it is orientation sensitive and different results may be obtained for different rotational alignments of the same feature, unless it has a truly random outline shape.

To apply this method to real structures, such as particulates whose surface area may be important (as a substrate for chemical reactions or a site for chemical diffusion, for instance), measurements are made using a series of stride lengths and the total dimension is plotted. Usually, the perimeter thus measured is not a perfect integral multiple of the stride length, and some partial length is left over at the end. The resulting real number (integral number of strides plus a fraction) times the length is then the perimeter. If this is plotted against the stride length (λ), we obtain a Richardson plot as shown in Figure 6-42. The plot deviates from ideal behavior at short stride lengths because of the finite resolution of the image being measured, and at large stride lengths which become a significant fraction of the size of the profile (Kaye, 1984)

It is necessary to adopt a convention for dealing with cases along the periphery when swinging the dividers from one point can intersect several points along the boundary. For instance, we might decide in that case to use the point farthest from the original, and continue the process from there. Once the log plot has been obtained, the fractal dimension of the boundary is obtained from m, the slope of the linear portion of the plot as

$$dimension = 1 - m$$

Because we shall also be applying the concept of fractal dimensions (and Richardson plots) to surfaces (in Chapter 10) which have a dimension between 2 and 3, it is most useful to deal with the fractional part separately from the integer. For a similar plot of log total surface area (vertical scale) vs. the log of the area element used to measure it, the slope will still be $-m$, but the dimension will be $2 - m$. This is handled most simply by writing the dimension as $1.D$ or $2.D$ where the value of D ($= -m$) is the decimal fraction, between 0 (the Euclidean limit) and 0.999... The upper limit cannot actually exceed 1; for the Koch islands shown above, if the generating element had an increase in length so great that the fractal dimension was 2.0, then the resulting shape would spread without limit across the plane. Also, the curve would have to be self-intersecting, and the boundary would no longer unequivocally separate the inside (feature) from outside (surroundings). Quantitative comparison of the roughness of different classes of features can then be made using this parameter.

The measurement of perimeter by hand, as described, is very tedious. Computer-assisted determination of the fractal dimension of an outline may be carried out in several ways. The most straightforward is available when the feature outlines are available as a series of point coordinates around the perimeter, as discussed in Chapter 4. It is easy to sum the perimeter of the polygon using all of the points, and then to repeat this with every second point, and then every third, and so forth (Schwarz & Exner, 1980). It is possible in principle to perform the same operation with raster-scanned images, for instance by using the ends of each chord in a run-length-encoded binary image of the feature as the points in the polygon. The problem with this approach is that the points are not equally spaced, and the result may be biased if the feature outlines are not isotropic.

Another approach to determining the fractal dimension uses erosion followed by EX-ORing with the original image to get just the periphery of the features. If this is repeated after various numbers of iterations of erosion, it is possible to construct a plot of the area of the eroded layer versus the number of steps. This plot will also depend on the scale of the local roughness along feature outlines. The problem is that the value is determined for the image as a whole rather than for individual features within the image. Also, the total size of features decreases with each erosion step and this introduces some bias in the results. The latter problem can be overcome by using dilation after each cycle of erosion to restore the same approximate size to the features. This is then a process of sequential opening, and a measure of how quickly the smoothing of the feature outlines proceeds.

A more direct method of constructing a Richardson plot measures the feature perimeter (as discussed in Chapter 7) as the image resolution is progressively coarsened (Kaye, 1978; Russ, 1986). The coarsening is begun by replacing each block of 4 pixels (2 on a side) by a solid block in which the pixels are either all on or off. If the original block has 0 or 1 pixels on, it is set entirely off, and conversely if 3 or 4 pixels are on, all are set on. For the case where two pixels are set and two are not, there are 6 possible configurations. Three of these are arbitrarily chosen to produce blocks that are entirely on, and the other three result in all pixels being turned off. The result is an image with a more blocky appearance, and a shorter perimeter (but, on the average, the same area). Figures 6-43 and 6-44 show an example of this process.

The process, called "mosaic amalgamation," can then be repeated. The actual routines are much the same as those used for etching and plating. Either a series of 3x3, 4x4, 5x5, etc. blocks can be used from the original image, or for greatest simplicity, the same

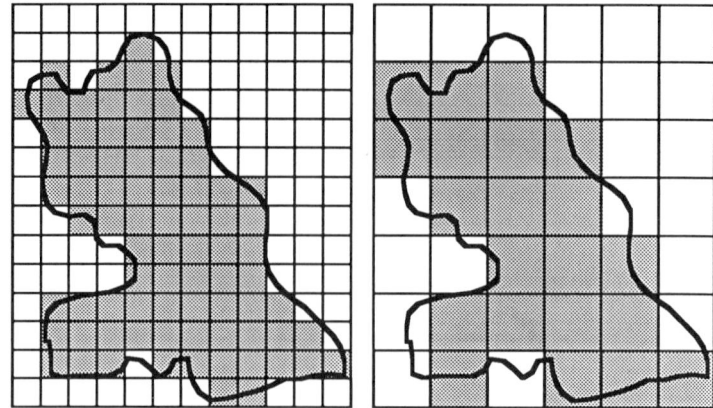

Figure 6-43: Mosaic amalgamation - change in perimeter with varying pixel resolution.

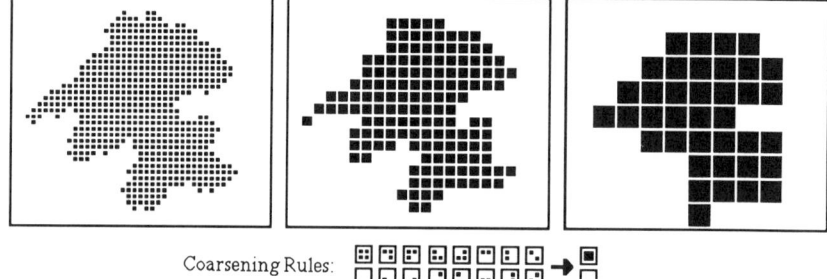

Figure 6-44: A feature made up of pixels, and the result after it is coarsened twice using the rules shown. The reduction in perimeter after several repetitions can be used to determine the fractal dimension of the outline.

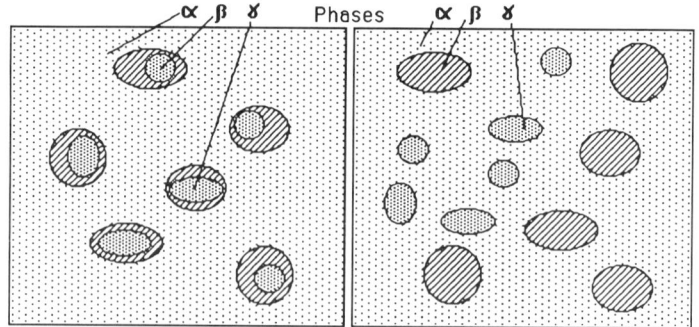

Figure 6-45: Two multiphase materials with extremes of contiguity between phases α, β and γ.

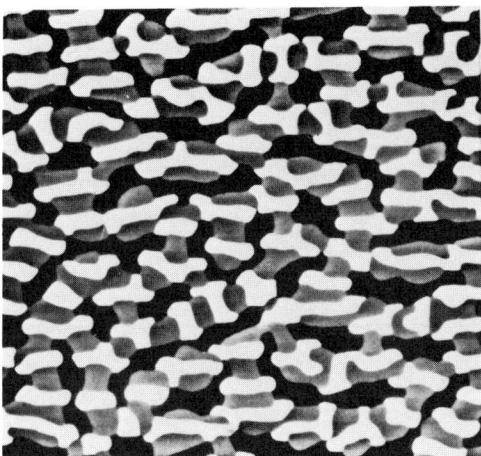

Figure 6-46: A three-phase alloy structure used for contiguity analysis (SEM image).

rules just used for the 2x2 block can be re-applied using 4x4 blocks in which the original 2x2's are treated as single pixels. Repeating this 4 or 5 times gives enough points for a good determination of D. As for the polygonal outline, this only proceeds from the finest resolution of the original image to larger dimensions, as there is no information available at a finer scale. The stride length is usually taken as the edge dimension of the block, starting with one pixel and proceeding to multiples thereof. This is slightly biased for the usual case of a square grid of pixels, since some of them are diagonally adjacent, and hence are spaced 1.4 times as far apart, but this merely shifts the plot sideways and does not change the slope. A hexagonal array would be preferred to keep the neighbor distance uniform, just as in the plating and etching situation, but is rarely used because raster scan hardware is much simpler when it uses square pixels.

Contiguity and filling interior holes

An occasionally important, but rather subtle parameter in image analysis is "contiguity." This describes the neighbor relationships between phases (Davy, 1981). Figure 6-45 shows two extreme conditions. As an example of measuring this property, we have performed contiguity measurements on a backscattered electron image (Figure 6-46) showing a three-phase directionally solidified eutectic alloy (Ball & McCartney, 1981). However, the method is equally appropriate to light microscope images, biological samples, etc.

Each of the three phases is well separated in this case, and binary images of the phases can easily be obtained. The three images shown on the left of Figure 6-47 represent each of the phases, identified as black, grey and white from their appearance in the image (in this case, the brightness is related to average atomic number of each phase). The examples shown in the figure were obtained by discriminating the image, and then using an "opening" (sequential etching and plating) to smooth the outlines and remove the narrow lines along phase boundaries in the grey image. In a more difficult case, the usual processing and editing "tricks" might be required to obtain single phase binary images. In any case, the following procedure is then used to measure the area fraction (volume fraction) and contiguity for each phase. The area fractions are read directly from the image. Representative values are 38.7% for the white phase, 35.8% for the grey

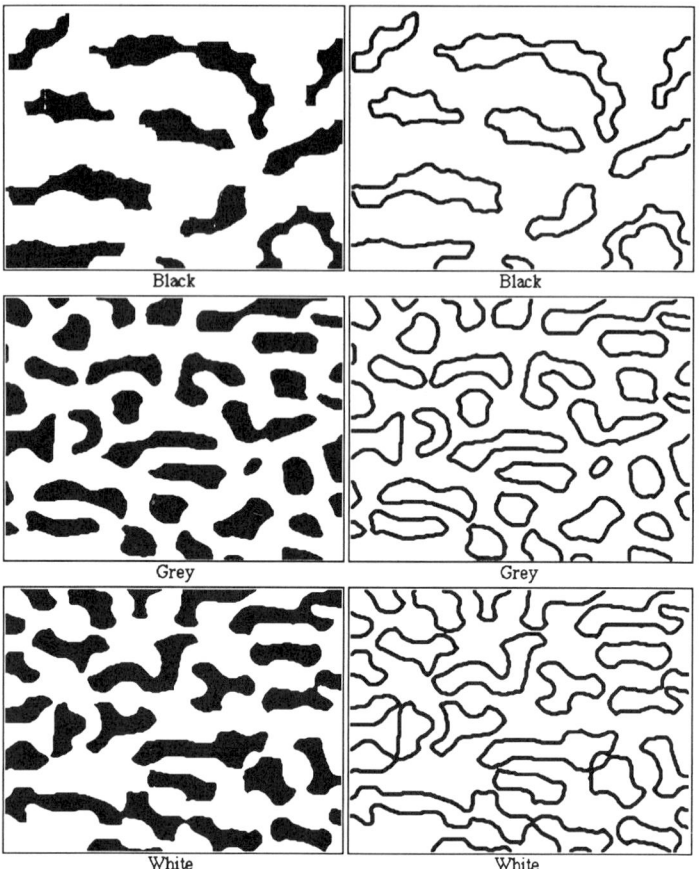

Figure 6-47: Binary images of individual phases in alloy, and the outlines around each phase obtained by plating and Ex-ORing with the original image.

phase, and 25.3% for the black phase, in this image. To obtain quantitative contiguity values, we treat each of the three binary images to obtain an outline image. The image is plated and Ex-ORed with the original to obtain an outline as shown on the right side of Figure 6-47.

Then each pair of outlines are combined with the AND function, and the relative area of the result used as a measure of the common boundary. For the three-phase alloy in this example, there are three combinations to be made (White/Grey, Grey/Black, and White/Black) as shown in Figure 6-48. The numerical results are tabulated below. The numbers have no direct meaning since they depend on the thickness of the outlines, but they can be used in ratios to determine the contiguity between phases.

Grey	0.2003	Black	0.1246	White	0.2048
White	0.2048	Grey	0.2003	Black	0.1246
AND	0.1191	AND	0.0566	AND	0.0607

Contiguity analysis tells us how much of the boundary of each phase is devoted to each neighboring phase (Moore &Wyman, 1963). The fraction (W/B)/(W/B+W/G), for

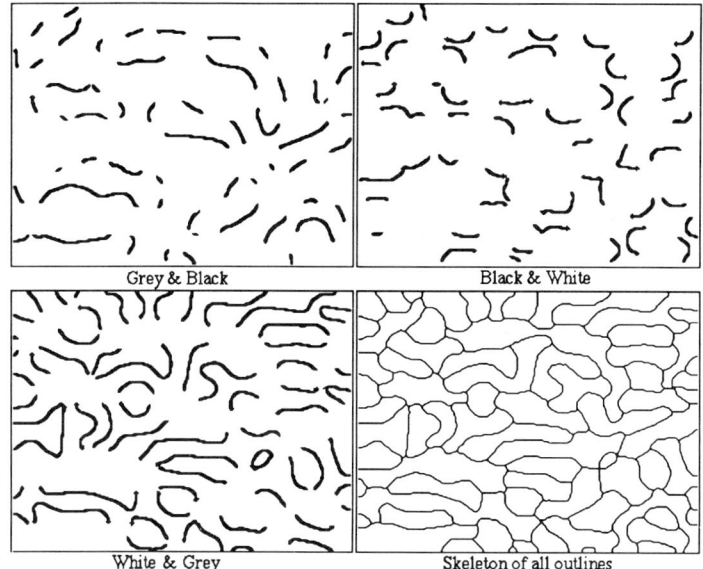

Figure 6-48: AND combinations of pairs of phase outlines, and the skeleton of the OR of all phase outlines

example, gives the percentage of contiguity between white and black phases. There are 6 such combinations here, as listed below.

Phase	Contiguous with:	White	Grey	Black
White			66.2%	33.8%
Grey		67.8%		32.2%
Black		51.7%	48.3%	

This tells us, for instance, that next to a white phase region, is it much more likely to find a grey phase than a black one, whereas the black phase is about equally likely to be adjacent to white or black. This has important consequences in materials science (bonding, diffusion, etc.), as well as the life sciences (communication between organelles, etc.). This method can be used after individual features in the image are identified (as discussed in Chapter 7) to count the number of adjacent neighbors for each grain, cell or other region in a tessellated image. For instance, Figure 49 shows a cross section of cells in the cornea of a human eye. The majority of cells have six neighbors, but some have 4, 5, 7 or 8. The frequency of occurrence of these patterns has been linked to age and disease. Similar analysis of the growth of diatoms shows that 5- and 7-neighbor cells occur in pairs to maintain an overall hexagonal tesselation.

Besides the concept of contiguity, there are many other possible positional relationships between objects. Imagine that two classes of objects have been reduced to features in binary images. The features in image A may lie entirely outside of those in image B, or they may touch then, or they may overlap them. The first and third conditions can be ascertained directly from a logical AND of the two images, while the second condition requires an additional dilation of one image or the other after it is found that the features do not overlap, to see if they will after the dilation (the same procedure could be repeated to ascertain how far apart they were).

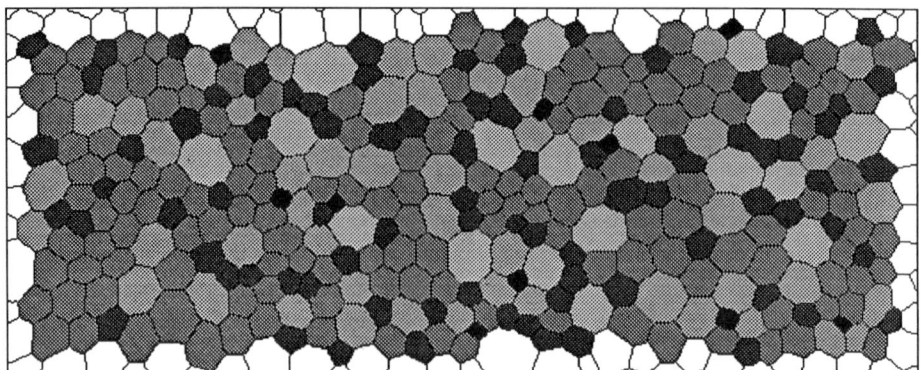

Figure 6-49: Cross section image of human cornea (courtesy Dr. Barry Masters, Emory Univ. School of Medicine, Atlanta, GA). In this processed image the cells are brightness coded by number of neighbors, from 4 to 8.

When the features of image B lie within those of image A, either adjacent to the outside or not, a similar series of operations can be used. In fact, it may be simpler in that case to negate image A and use the tests listed above. However, the concept of "within" is less precise than we may need. In Chapter 7 a more complete set of relationships is presented, including touching as well as overlapping, and distinguishing interior voids from the "outside" of the feature. Like contiguity, these relationships can be measured using a combination of erosion and dilation with Boolean logic for combining images. The only additional requirement is an image of the voids within the features.

This can be gotten as follows: Invert the original image, and produce from it a new chord-encoded table of the features present. Number these features just as was previously described for the original features. In this process, the fact that at least one feature (the general background or surroundings of the original features) touches the edge of the image field will be discovered. Remove all of the chords for any features that touch the edge from the image. This leaves the chords for the holes (regions that do not connect to the edge of the image must be within features). Reconvert the chord table to a pixel-based binary image representation, which is then the image of the holes. This image can be used as a mask to select objects within the holes. When multiple images with different features are obtained from a single field, there can be a considerable hierarchy of particle position relationships, usually referred to as parent/child relationships based on which features lie within the outlines of others. Charting and expressing these data is often more complex than obtaining the information.

The hole image can also be combined with the original image to produce filled objects without any holes, if that is desired (Pavlidis, 1979). That is the most common method used to obtain such parameters as solid area and external perimeter, as will be discussed in the next chapter.

The use of the same logic as was applied to the inverse image can also be used with the original, if it is desirable to remove any edge-touching features from the image. The problems of measuring feature-specific parameters on edge-touching features, and the various solutions to the problems, are discussed in the next chapter as well.

References

Ball and McCartney (1981) *The measurement of atomic number and composition in an SEM using backscattered detectors* Journal of Microscopy 124, 57-68

S. Beucher, C. Lantejoul (1979) *Use of Watersheds in Contour Detection* Proc. Int'l Workshop on Image Processing, CCETT, Rennes, France

D. S. Bright, D. E. Newbury (1986) *Euclidean distance mapping for shape characterization of alloy grain boundaries* Microbeam Analysis 1986 (A. D. Romig, W. F. Chambers, ed.) San Francisco Press 521-524

J.-L. Chermant, M. Coster, J.-P. Jernot, J.-L. Dypain (1981) *Morphological Analysis of Sintering* J. Microscopy 121 89-98

M. Coster, J-L. Chermant (1985) *Précis D'Analyse D'Images* Éditions du Centre National de la Recherche Scientifique, Paris

P. E. Danielsson (1980) *Euclidean Distance Mapping* Computer Graphics and Image Processing 14 227-248

P. Davy (1981) *Interpretation of phases in a material* J. Microscopy 121 3-12

R. Ehrlich, S. K. Kennedy, S. J. Crabtree, R. L. Cannon (1984) *Petrographic Image Analysis: 1. Analysis of Reservoir Pore Complexes* J. Sedimentary Petrology 54 # 4 1365-1378

A. G. Fabbri (1984) *Image Processing of Geological Data*, Van Nostrand Reinhold, NY

J. Feder (1988) Fractals Plenum Press, New York NY

A. G. Flook (1978) *Use of dilation logic on the Quantimet to achieve fractal dimension characterization of texture and structured profiles*, Powder Technology 21, 295-298

C. R. Giardina, E. R. Dougherty (1988) Morphological Methods in Image and Signal Processing Prentice Hall, Englewood Cliffs NJ

J. P. Jernot (1982) Thése de Doctorat és-Science, Université de Caen

B. H. Kaye (1978) *Sequential mosaic amalgamation as a strategy for evaluating fractal dimensions of a fineparticle profile*, Report #21 Institute for Fineparticle Research, Laurentian Univ.

B. H. Kaye (1984) *Multifractal description of a rugged fineparticle profile*, Particle Characterization 1, 14-21

B. H. Kaye (1986) *Image analysis procedures for characterizing the fractal dimension of fineparticles*, Proc. Particle Technol. Conf. Nürnberg

Koch, von, H. (1904) *Sur une courbe continue sans tangente, obtenue par une construction geometrique elementaire* Arkiv für Matematik, Astronomie och Fysik 1, 681

C. Lantejoul, S. Beucher (1981) *On the use of the geodesic metric in image analysis*, J. Microscopy 121, 39

B. Lay (1985) *Morpholog, an image processing software package* IEEE Workshop on Computer Architecture for Pattern Analysis and Image Data Base Management, p. 463-469

S. Levialdi (1983) *Neighborhood Operators: An Outlook in Pictorial Data Analysis* (R. M. Haralick, ed.) Proc. 1982 Nato Advanced Study Inst., Bonas, France, Springer Verlag New York, F4, 1-4

B. B. Mandelbrot (1982) *The Fractal Geometry of Nature*, W. H. Freeman, San Francisco

G. Matheron (1975) Random Sets and Integral Geometry Wiley, NY 1975

G. A. Moore, L. L. Wyman (1963) *Quantitative Metallography with a Digital Computer: Application to a Nb-Sn Superconducting Wire* Jour. Res. NBS 67A 127-147

T. Pavlidis (1979) *Filling Algorithms for Raster Graphics* Computer Graphics and Image Processing 10 126-141

T. Pavlidis (1980) *A Thinning Algorithm for Discrete Binary Images* Computer Graphics and Image Processing 13 142-157

T. Pavlidis (1982) *Algorithms for Graphics and Image Processing*, Computer Science Press, Rockville MD

A. Rosenfeld, J. L. Pfaltz (1966) *Sequential Operations in Digital Picture Processing* J. ACM 13, p. 471-494

J. C. Russ (1984) *Implementing a new skeletonizing method* J. Microscopy 136 p. RP7

J. C. Russ (1986) Practical Stereology Plenum, New York

J. C. Russ, J. C. Russ (1984) *Enhancement and Combination of X-ray maps and Electron Images*, Microbeam Analysis 1984, San Francisco Press, 161

J. C. Russ, J. C. Russ (1986) *Automatic editing of binary images for feature isolation and measurement*, Microbeam Analysis 1986 (A. D. Romig, W. F. Chambers, ed.) San Francisco Press, 505

J. Ch. Russ, J. C. Russ (1988) *Improved implementation of a convex segmentation algorithm* Acta Stereologica 7#1 33-40

W. O. Saxton, T. L. Koch (1982) *Interactive image processing with an off-line minicomputer: organization, performance and applications* J. of Microscopy 127 69-83

H. Schwarz, H. E. Exner (1980) *Implementation of the concept of fractal dimensions on a semi-automatic image analyzer*, Powder Tech. 27, 207

K. L. Scrivener, P. L. Pratt (1987) *Observations of Cements and Concretes by Backscattered Electron Imaging* Proc. RMS 22 # 3 167

J. Serra (1982) *Image Analysis and Mathematical Morphology*, Academic Press, London

B. Shapiro, J. Pisa, J. Sklansky (1981) *Skeleton Generation from X,Y Boundary Sequences* Computer Graphics and Image Processing 15 136-153

D. Vollath (1986) *Some fundamental considerations to precede image analysis* Bull. Mater. Sci. 8/2 169-182

Chapter 7

Image Measurements

There are two basic types of measurements that can be performed on images. The first is determining global or scene parameters, and the second involves measuring values for each individual feature that is present.

Reference areas

The first of the global parameters is the area fraction which the selected pixels cover. Notice that this only has meaning when dealing with a pixel-based representation of the selected features, since boundary representations must be re-converted to pixels to obtain this value. The area fraction is determined simply by counting pixels. A simple application of this is shown in Figure 7-1. Wood blocks are cemented together with adhesive which has been colored. When the blocks are broken apart, the quality of the bond is assessed by measuring the area fraction on which the colored surface shows. Two times the area fraction of the blocks is the amount of the fracture that occurred in the bond rather than in the wood. In the examples shown, one pair has a 36% area fraction and the other a 12% value, indicating a much stronger bond.

It is important to note that the fraction need not be based on the total number of pixels in the image, but that some other reference area can be used (Cruz-Orive & Weibel, 1981). In the previous chapter, it was pointed out that a binary image mask could be used to represent a reduced region of interest within which measurements were to be performed (examples of a microscope aperture and a leaf were given, and many more ranging from a particular cell seen in the microscope to land masses in satellite photos can be imagined). Figure 7-2 shows a leaf within which the number and area fraction of fungal spots are to be determined. When this is done, it is also natural to use the area of the mask as a reference for the area fraction measurement, and in fact for many of the other measurements which will be described here. This is usually called the reference area, and when multiple images are measured, the total reference area is summed to describe how much of the sample has been examined.

Area fractions can also be multiplied. For instance, the product of the area fraction of nuclei or mitochondria in a particular type of cell, times the area fraction of those cells in the entire tissue section, gives the area fraction of the selected organelles in the tissue section (Figure 7-3). In the next chapter we will see that this also gives the volume fraction of the organelles. The reference area need not even be planar, for instance in the case of counting the area density features on cell surfaces revealed by freeze-fracturing (Ting-Beall et al., 1986).

The area fraction gives a direct value for the extent of the feature(s) selected, but tells nothing about how they are organized. The next global parameter to consider is therefore the number of features. One method for counting features has already been described

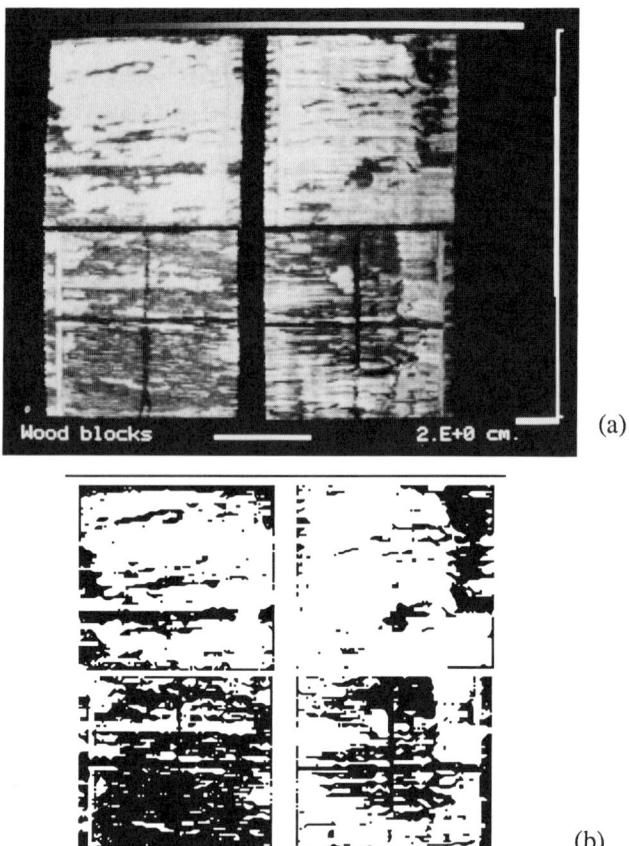

Figure 7-1: a) Grey scale image of wood blocks; b) Binary image discriminated to show fracture through adhesive.

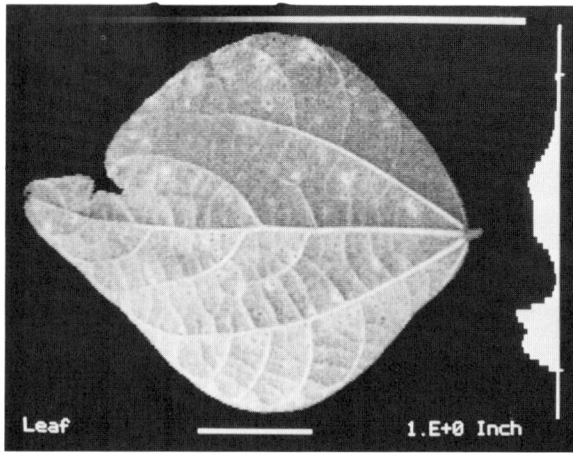

Figure 7-2: Leaf grey-scale image.

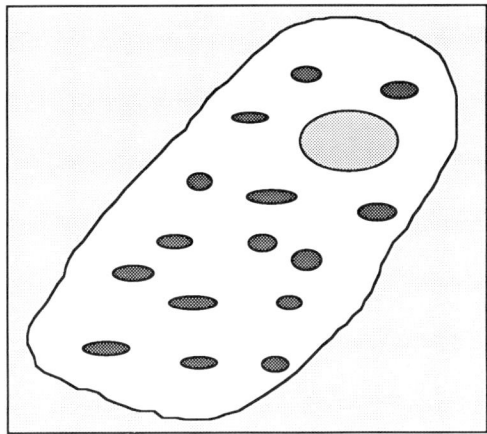

Figure 7-3: The area fraction of dark mitochondria would normally be referred to the cell, a non-rectangular reference area, or perhaps the cytoplasm, the portion of the cell excluding the lighter grey nucleus.

(numbering them using the chord encoded table). When this is done, it is necessary to correct for the possibility that some features will intersect one or more of the edges of the image frame (Figure 7-4). The usual method for doing this is to count all of the features that lie entirely within the frame, or which intersect two of the edges (for instance the bottom and right side). Features which intersect the other two sides (top and left side in this example) are not counted. If a feature touches one of the forbidden sides and another side, it is ignored. This gives an unbiased estimate of the total number of features per unit area of the entire image.

Another way to count features, which is identical when applied to the entire square or rectangular image, is to count each feature based on where its initial or final point lies. For instance, if the initial point (the left end of the first chord in the feature) is counted, then it is exactly equivalent to the situation described above, because features which intersect the bottom or right edge will have such points, while features which intersect the top or left side will not (the numbering process will recognize that the feature touches the edge and that there is no initial point).

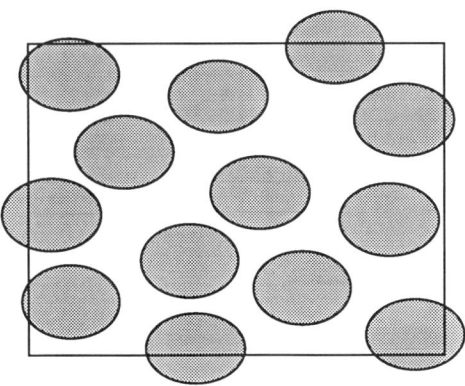

Figure 7-4: Edge-intersecting features.

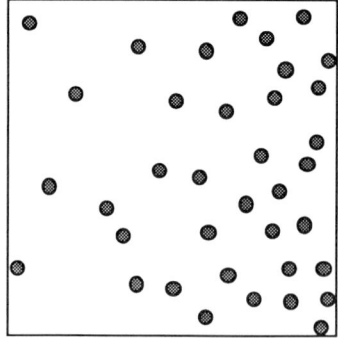 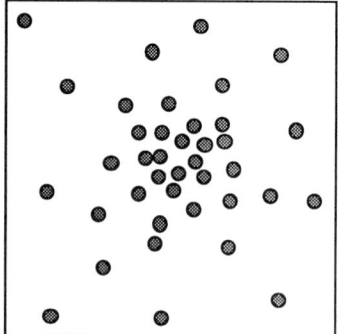

Figure 7-5: Gradient of number of objects vs. X position and vs. radial distance.

When the reference area is not rectangular, the latter method still works properly. The initial point (or final point, which gives equivalent results) can be tested to see whether it is included within the reference area, and the feature counted if it is and ignored if it is not. If the number of such objects is divided by the reference area, an unbiased result for the number of objects per unit area is obtained.

Of course, it will usually be necessary to convert from units of number of features per pixel to number per square cm. (or square mile or square micron, depending on the image magnification). This conversion is usually performed last, after all of the calculations are carried out based on pixels. The image magnification must be known, and is usually established by acquiring an image of some standard sample containing marks of known spacing (a stage micrometer) or objects of known size. In some cases, the magnification can be calculated from independent knowledge of the viewing optics, position of the spacecraft, etc. The problems inherent in combining information from several images obtained at different magnifications will be dealt with in the next chapter.

Given the number of objects and the total area which they represent (from the area fraction), it is already possible to report a mean size parameter based on the area. We still do not know anything about the shape or distribution of the features, however. Spatial uniformity (Figure 7-5) can be estimated by subdividing the image into several segments, often 10-20, consisting of vertical or horizontal slices, or concentric rings of equal area, and then determining the area fraction of ON pixels in each segment. This is exactly identical to the use of arbitrary binary image masks to select subregions of the image, and in fact may be accomplished in that way by creating the masks as they are needed. It also means that other series of masks can be used, as the example in the previous chapter of concentric rings produced by sequential erosion and Boolean logic. If the masks are not of equal area, then dividing by the area fraction to obtain number per unit area in each will normalize that variation and still permit a straightforward report of uniformity as a function of position.

To describe the size of features in another way, we may use the mean intercept length. In a classical stereological sense (as will be discussed in the next chapter on interpretation), the intercept length is the length of a chord across a feature produced by any random line drawn on the image (Figure 7-6). In practice, most images have the features themselves randomly arranged and oriented, so horizontal or vertical lines such as the scan lines in the raster are effectively random with respect to the features. Consequently, the mean intercept length is just the same as the mean chord length,

Image Measurements

immediately available by averaging the values in the chord-encoded table, or even more efficiently by dividing the total area of the binary image (the number of ON pixels, which is the sum of the lengths of all the chords) by the number of chords in the table. Of course, this assumes that no partial chords (those which touch an edge of the image field) are included in the table.

The mean intercept length gives a direct way to estimate the mean size of objects (depending on the type of image and the type of objects). Performed on the negative of the image (the background or surroundings of the features) it permits estimating the mean spacing between features. Chapter 8 will discuss these interpretations in more detail.

A rather surprising parameter than can be determined globally from the binary image, and in fact from its chord-encoded table, is the amount of boundary in the image and its net mean curvature. This needs a bit of explanation. The boundary between features and surroundings is a line in the image plane. We will see in Chapter 8 how it is related to the amount of boundary surface in a solid. The boundary lines run in all possible directions, and are intersected by the scan lines. From geometrical probability, it can be shown that the number of intersections per unit length of the scan lines (P_L) is related to the length of boundary per unit area by $B_A = \pi/2 \cdot P_L$.

Boundary curvature

Furthermore, the boundaries are not in general composed of straight lines (Figure 7-7), and the amount of curvature of the boundary can be estimated by counting the number of places where the boundary becomes tangent to any particular direction, which in our case can be taken for convenience as the scan direction. There are two kinds of tangency, internal and external. The process of numbering the features in the chord table already recognizes the first chord in a new feature (an upper, external tangent contact) and each time that a chord does not have a touching chord in the next scan line (a lower, external tangent contact). It also finds joins where two chords in the prior line are touched by a single chord in the current line (a lower internal tangent contact) and places where two

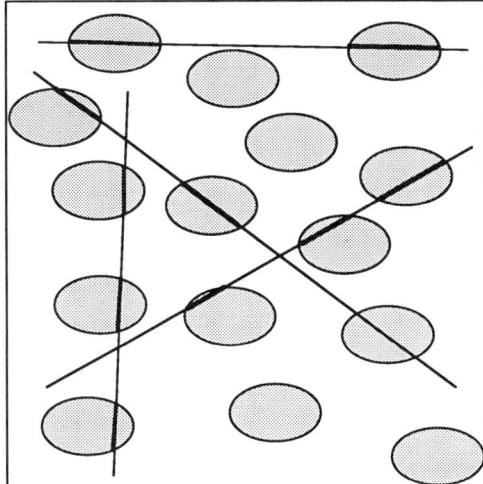

Figure 7-6: Schematic diagram of intercept length determination. Lengths of line segments in and between features are measured.

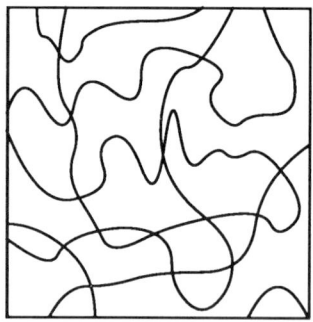 

Figure 7-7: Curved boundaries showing different amounts of curvature.

chords in the current line touch a single one in the prior line (an upper internal tangent count). If the number of external and internal tangencies are noted as T^+ and T^- respectively, then the total boundary curvature in the plane is proportional to $(T^+ - T^-)$.

It is also possible to count inflection points where the curvature of the boundary line changes from convex to concave, and to use this parameter to describe the shape of features, but it is considerably harder to incorporate this into automatic programs. The use of boundary representation such as chain code makes it more practical, but some smoothing is generally desirable to suppress minor roughness or noise. A method to determine the total curvature of lines in the image is discussed later in this chapter, in conjunction with the measurement of the curvature of individual features. A very different approach to shape as a global parameter requires analysis of the shape of the distribution of intercept lengths through the phase of interest (You & Jain, 1984). This is primarily a comparative tool, but has been used in some studies of porosity and permeability.

Another class of global measurements involves the original grey image. We have already seen that a brightness histogram can be obtained for the entire image, or through any reference mask (Russ et al., 1985; Zeineh & Kyriakidis, 1987). Given a calibration curve relating brightness to some other parameter such as density, it is easy to convert the measurement to these units. If the calibration relationship is not linear (and in general it is not, as logarithmic brightness vs. density relationships are more common), it is not correct to first determine the mean brightness and then convert it to density. Instead, each brightness value in the histogram should be converted to density and weighted by multiplying by the number of pixels with that brightness value. This operation is performed for each brightness value and the results summed. Finally, the total is divided by the total area (number of pixels) to obtain the integrated mean density.

Similar calibration curves can be established for other brightness-related parameters, such as drug dosage in autoradiograms, vegetation cover in false color (infrared) satellite photos, etc. In X-ray tomograms (Chapter 13) the brightness can be converted to an X-ray absorption coefficient, and if several images with different radiation energies are used, this can be further converted to elemental concentration.

In these examples, the brightness data were obtained from the original grey image, perhaps with processing to smooth noise or level nonuniform background. It is also possible to use the same technique to make measurements on images processed in other ways. For example, Chapter 4 described the use of a Sobel or other gradient operator to determine the orientation of the gradient, which can be coded as brightness in a grey image. A histogram of this image will represent the directions of gradients in the image, and can be used to get a mean value or a degree of nonuniformity (preferred orientation), possibly resulting from illumination or preparation variables. Likewise, brightness images can be used to encode various texture or other parameters which can be studied using the histogram method.

When feature specific measurement are discussed below, we shall also note that some very useful parameters can be determined from processed or derived images. For example, the linear Hough transform produces a grey scale image in which the "features" that can be discriminated and measured are spots whose locations identify the linear structures present in the original image.

Most of the measurements described above are actually carried out using a pixel-based representation of the image. Boundary representation requires additional calculations to find the intersections of the boundary lines with test lines, or changes in directions, etc. It is usually simpler to convert the image from vector to raster representation to obtain a pixel representation for these measurements.

Feature measurements

Although global measurements are extremely fast and efficient ways to obtain useful data about the image, more information is available by measuring the individual features present. Sometimes this is more information than we need and it is simply reduced to a single mean value, but in other cases the distribution of values can itself be characterized, or compared to another data set from a different sample. The interpretation of the data, both statistically and from a stereological point of view, is taken up in the next chapter (Underwood, 1970; Weibel, 1979; Russ, 1986). Here we shall describe the various types of measurements that are available (Hare et al., 1982), although by no means are they all used or useful for any individual image.

For convenience, the parameters will be divided into four groups. The first covers measures of size, the second deals with various aspects of shape, the third is concerned with describing feature position, and the fourth includes parameters that utilize grey scale brightness information (including color). Within each of these categories, there are a variety of individual parameters that can be measured, or calculated from others that are measured directly.

The most obvious measure of size is the area of a feature, simply the total number of pixels within it for the case of a pixel-based representation, or the area within the boundaries if boundary representation is used. The area is, of course, converted to the appropriate real units of measure (square microns, acres, etc.) as was described before. When boundary representation is used instead of pixel-based representation, it is fairly straightforward to calculate feature area. The coordinates of the boundary points are the corners of a polygon, and are usually expressed in Cartesian (X,Y) coordinates. By convention, the final coordinate pair in the boundary list X_n, Y_n is equal to the first X_0, Y_0 for a closed feature. Then the total area of the feature is simply

$$\text{Area} = \frac{1}{2} \sum_{k=0}^{n-1} (X_k \cdot Y_{k+1} - X_{k+1} \cdot Y_k)$$

When the feature outlines are obtained by manual means, such as tracing on images with a graphic tablet or drawing directly on the video screen with a mouse, it is commonly observed that the areas are too large. Human bias usually causes the user to draw around the feature rather than along the actual boundary, and the amount of error is generally proportional to the difference in brightness between the feature and its surroundings. Furthermore, the error in area is proportional to the length of the perimeter, with either a pixel or boundary representation method.

A linear measure of size is often more useful than the area, and it is common to convert the measured area to either the equivalent circular diameter or the equivalent spherical diameter. These are calculated simply as

$$D_{Circ} = \sqrt{\frac{4}{\pi} \cdot Area}$$

$$D_{Sph} = \sqrt[3]{\frac{6}{\pi} \cdot Area}$$

Both of these formulae make an assumption about shape. The first one assumes that a circle is a reasonable description of the feature. The second makes the further assumption that not only can the feature reasonably be described as a sphere, but in addition that it is viewed in projection (see Chapter 2) so that the shadow or profile of the spheroidal shape is present in the image. This will not be true if the image is a planar section cut through a solid matrix, in which case the features in the image will in general be smaller than the full extent of the three-dimensional objects. Chapter 8 will discuss these geometric models in more detail.

Parameters such as the equivalent spherical or circular diameter are often described as derived parameters, while the area itself is a measured or fundamental parameter. This reflects the fact that the latter are directly obtained from the image data itself, and the former are calculated afterwards. It is possible either to perform the calculations of derived variables "on the fly" as the fundamental parameters are obtained from the image, or later using files of stored fundamental data. The first method has the advantage that the data are presented almost immediately for the user (the time needed for the calculations is imperceptible compared to the other operations carried out on the image). The latter has the advantage that specific derived parameters can be constructed and evaluated at leisure after the measured data are complete and the user can reflect on the interpretation. Systems are available offering either method, and some offer both.

In addition to the feature area, it is often interesting to measure the filled area or the convex area (Figure 7-8). These are not so immediately obvious as the area itself. The filled area is the total area of the feature after all internal holes or voids have been filled (as described in Chapter 6). It is determined in the same way as the area, by counting pixels. The convex area (also known as the taut string area) is less simple. For a convex feature it is the same as the filled area, but for a feature with indentations or concave irregularities it is greater. Imagine a string wrapped around the feature and pulled taut. The string will bridge over the bays and indentations. The convex area is the area inside the string. This can be obtained in either of two ways, one using a pixel-based

Figure 7-8: Illustration of a feature with its filled and convex area and perimeter.

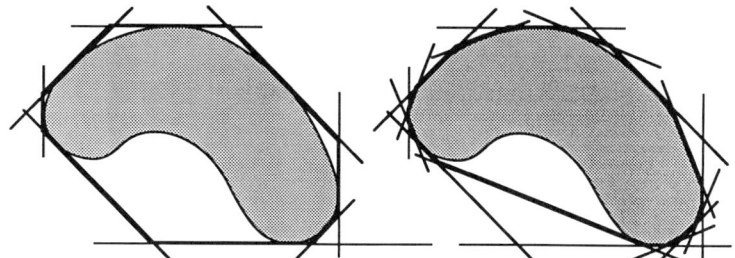

Figure 7-9: Diagram showing the convex boundary constructed from four and eight caliper diameters. With more directions, a closer approximation to the taut string boundary is obtained.

representation and the other using boundary representation. For the former, a series of plating operations are performed using fate tables that permit growth only in one of eight possible directions, and only for pixels that have neighbors ON in the selected direction. This will eventually fill the feature out to an octagon, whose area can then be determined by counting pixels. (For hexagonal pixels the feature will become a hexagon, of course.) The chief problem with this method is that features that are quite close together and concave may intersect as the plating process proceeds, giving rise to erroneous results. Also, the method is iterative and therefore slow (and requires an amount of time proportional to feature size and concavity).

Perimeter points

The boundary-representation method is closely tied to the measurement of length discussed below. The extreme left and right points on the feature can be readily determined by comparing the ends of the chords in the chord-encoded table. The extreme top and bottom can be obtained in the same way.

By rotation of coordinates, the X and Y Cartesian axes of the image can be converted to X',Y' using the relationship shown. In this rotated coordinate system the extreme X' and Y' coordinates can also be determined by simple comparisons. Only the periphery points need to be examined, and they are the ends of the chords.

$$X' = X \cos \theta - Y \sin \theta$$
$$Y' = Y \sin \theta - X \cos \theta$$

By rotating the coordinate system to a reasonably large number of different angles, the coordinates of the extreme points form the corners of a bounding polygon that closely approximates the taut string boundary (Figure 7-9). From the coordinates of the polygon corners, the polygon area can be directly calculated as

$$\text{Area} = \frac{1}{2} \sum_{k=0}^{n-1} (X_k \cdot Y_{k+1} - X_{k+1} \cdot Y_k)$$

and used as the convex feature area. In practice, 8, 16 or 32 rotations are commonly used. As will be seen in the discussion of the feature length, there is little improvement in accuracy to be gained from using more points.

Another size parameter of interest is the perimeter. For the case of boundary representation, either chain code or a string of discrete points entered from a drawing tablet, the perimeter is directly obtainable by adding the lengths of the steps along the

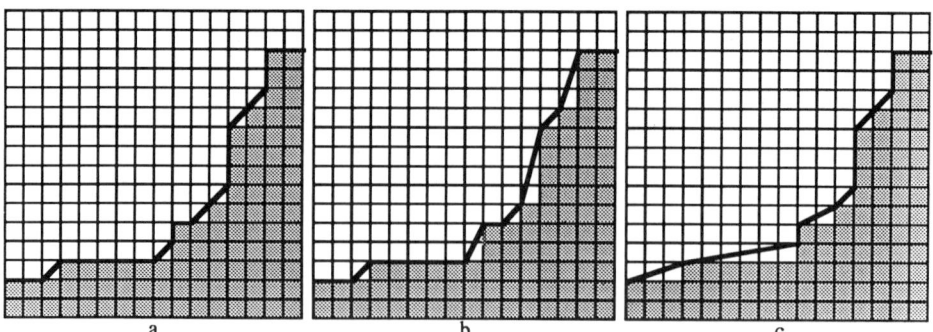

Figure 7-10: Variation in boundary perimeter with orientation (b,c) when using the Pythagorean distance, and comparison to "chain-code" distance (a).

boundary. For chain code, this is the number of orthogonal (0 or 90 degree) steps plus 1.414 times the number of diagonal (45 degree) steps. For a series of discrete points, the total perimeter length is the sum of the Pythagorean distances between the points, or

$$\text{Perim} = \sum_{k=0}^{n-1} \sqrt{(X_{k+1} - X_k)^2 + (Y_{k+1} - Y_k)^2}$$

Of course, this method is also used to determine the taut string or convex perimeter, using the coordinates of the polygon corners already described in conjunction with the convex area calculation. The taut string perimeter is less than the external perimeter for concave features.

For pixel-based feature representation, the perimeter is obtained through an intermediate step that uses boundary representation methods. It must be summed as the total of lengths of boundary lines connecting each of the chord ends. However, if the feature is not convex, it is necessary to keep track of which chord ends should be connected to others in the next or preceding scan lines when branches or joins occur. The length along the scan line of chords at the start and end of the feature, and at the branches and joins, must also be included. This requires some careful bookkeeping in the program.

Of greater concern is a bias that can result if the simple Pythagorean distance between successive end points is used. A diagonal line connecting an end point at (for instance) (10,100) to one at (11,110) would have a length of $(1^2 + 10^2)^{1/2} = 10.050$; if the same step occurred in a different orientation, we would have eleven chord ends, at (for instance) (100,10); (101,10); (102,10);... (109,10); (110,11). If the lengths between these points were summed we would get $9 \cdot 1 + (1^2 + 1^2)^{1/2} = 10.414$. This difference may seem small, but over the entire periphery of a feature it causes a significant bias. The solution, in order to obtain a perimeter value that is insensitive to the orientation of the feature, is to always use the latter method of summing distances. This is exactly the same as counting the links in a chain code around the feature. Figure 7-10 shows the comparison between approximating the boundary or perimeter length with Pythagorean distances vs. chain code.

As for the area, it is often desirable to distinguish between the total perimeter (the sum of all boundary lines for a feature) and the external perimeter (the perimeter after all internal voids have been filled). The total and external perimeter, and the convex or taut string perimeter already mentioned, are all directly measured parameters. It is possible to calculate an equivalent circular diameter from the external perimeter, just as it was from the area (although for any shape other than a circle, the value will be different). This is evaluated as D_p = Perimeter / π.

Another particularly useful size parameter is the length of a feature. This is usually defined as the maximum length straight line that can be drawn between any two points on the periphery. The line itself is not restricted to lie within the feature, so for a concave feature it may cross the boundary.

The perimeter points are directly available for a boundary-representation coding of a feature, or the chord ends can be used when a pixel-based image is compacted to a chord table (run length encoding). The length could be gotten by an exhaustive search of all combinations of points. In practice, there is a way to greatly reduce the number of comparisons that must be made. If a bounding rectangle is drawn around the feature outline, the centers of the four sides can be taken as the centers of arcs drawn tangent to the opposite sides. These arcs will intersect the boundary, so that only portions of the periphery lie outside the arcs (Figure 7-11). Only the points on or outside the arcs need to be compared to find the greatest separation, and hence the length. It is easy to find out which points lie outside the arcs, as their distances from the center points of the sides of the bounding rectangle are easily computed (and in practice it is not even necessary to take the square root, but rather to compare the sum of squares of the difference in coordinates to the square of the length of the side of the rectangle).

The bounding rectangle itself has side lengths that are called the X and Y Feret's diameters. These are caliper dimensions or shadow dimensions in the two particular orthogonal directions. They are used in some calculations, including the correction for edge intersection effects that is discussed below. The values are easily obtained by first finding the extreme (minimum and maximum) values of the X and Y coordinates, and then taking their differences. This is true for boundary representation data as well as for pixel-based binary images. In the latter case, the data can be found directly from the chord-encoded table once the features have been numbered.

Length and breadth

The length of a feature in a binary image can be obtained using the same method as was used to fit a convex polygon to the feature to obtain the taut string area, as described above. This is conveniently done by using rotation of coordinates (and a pre-calculated

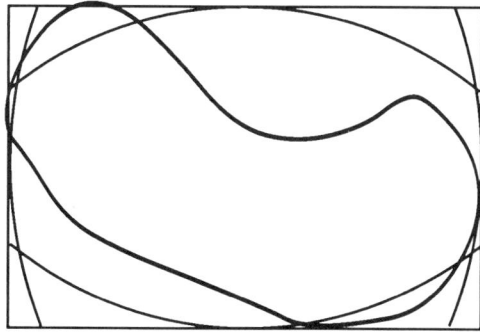

Figure 7-11: Method for locating candidate extreme points on a feature boundary.

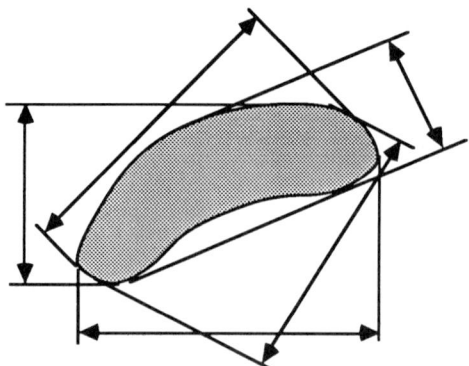

Figure 7-12: Diagram of projected, Feret's or caliper diameters in various directions.

set of sine and cosine values for the fixed angles of rotation to be used) with the ends of each chord in the chord-encoded table (Figure 7-12). As each point is converted, it may be compared to the minimum and maximum values for that rotation angle, and the extreme values saved. After the process is complete, the rotation angle whose minimum and maximum values give the greatest difference is found. This difference is taken to be the feature length.

If the feature has an actual maximum dimension that is not aligned with the rotation angles that were chosen, then the measured length value will underestimate the actual length (Figure 7-13). However, the error is not large. In the worst case it is given by the cosine of the angle of error (between the actual longest chord and the rotation angle used). Even for only 8 rotation angles (corresponding to 22.5 degree steps) this can at most produce an error of 11.25 degrees, whose cosine value is greater than 0.98 (representing a maximum error of 2% in the measured length). If 16 steps are used, the error angle drops to 5.625 degrees, with a cosine value greater than 0.995. This amounts to a measurement error of one part in 200, or typically less than one pixel for most features.

The length value is also sometimes referred to as the maximum Feret's diameter. By direct analogy there is a minimum Feret's diameter, which we will call the Breadth. This can be determined at the same time the length (and for that matter the convex area) are measured, by finding the smallest difference between minimum and maximum coordinates at any angle of rotation.

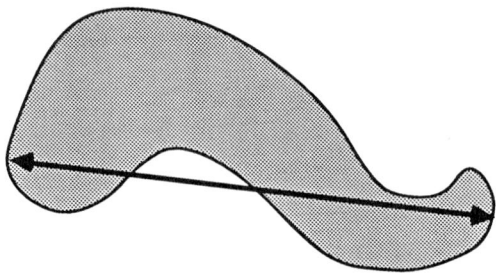

Figure 7-13: Length of a feature.

Figure 7-14: Error in the breadth for long, narrow features.

However, whereas the length value is very insensitive to the number of rotation angles used, this is not the case for the breadth. If the feature is quite long and narrow, any deviation of its actual orientation from the rotation angle produces an error that increases as the sine of the error angle (Figure 7-14). There is no practical limit to how large the error can be, since it is the product of the sine and the feature length. There is no reasonable number of rotation angles that would assure an accurate measurement. Consequently, while the breadth has some uses which we will discuss shortly, it is often replaced by a different estimate of the minor dimension of the feature, which we will call the width.

Furthermore, if the feature is bent or concave, the breadth (minimum Feret's diameter) is not a particularly good measure of how wide the feature is (Figure 7-15). If the feature can be approximated as a bent but otherwise smoothly elliptical shape, then it may be reasonable to approximate its width by the width of an ellipse of the same area and length as those measured for the feature. This is calculated simply as the quotient of (4 • Area) divided by (π • Length). This is a good example of a derived parameter. It depends on two different fundamental parameters (area and length) which are obtained independently using quite different algorithms. And it attempts to describe a particular aspect of the feature which is hard to determine in any other way (it was mentioned that the width can also be estimated from the maximum height of the ridge in the distance map, but this is computationally much more demanding).

Another approach to a representative dimension for feature size is to use the mean of the various Feret's diameters. These were obtained when the rotation of coordinates was used to locate extreme points. The difference in X' coordinates in each direction gives the Feret's or caliper diameter. The mean of these values is an effective mean diameter. However, it is a poor descriptor for non-convex shapes, while for convex ones the Circular Diameter described above is similar and easier to calculate.

From the principal size parameters so far described (Area, Perimeter, Length, Breadth and Width), it is possible to derive many others. Most of these make some assumptions about the shape of the feature, either in two or in three dimensions.

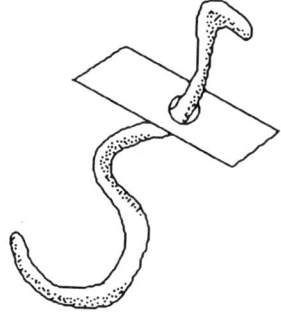
Figure 7-15: Width is an approximation to the dimension of a fiber corresponding to the smallest hole through which it will pass.

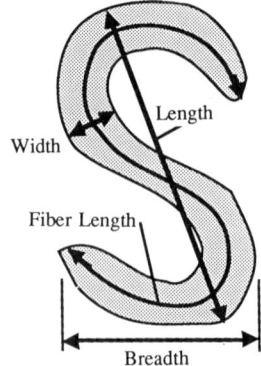

Figure 7-16: A feature showing the Length, Breadth, Width and Fiber Length.

For instance, another estimate of length for curved features (such as fibers or bent structures) is available from the perimeter. For a very long and narrow fiber, the length would be just half of the perimeter. As the width becomes significant, it is still possible to estimate a useful length dimension from the perimeter and area, assuming that the feature is shaped like a constant width stripe with circular ends. Figure 7-16 compares several of these size measures for a feature. Figure 7-17 shows the automatically generated measurement lines for a feature with a complex shape (one dendrite from image 6-16). These show the bounding 32-sided polygon, the length and breadth (and their orientation), the centroid location for the feature, a circle with the same area as the feature, and the feature skeleton.

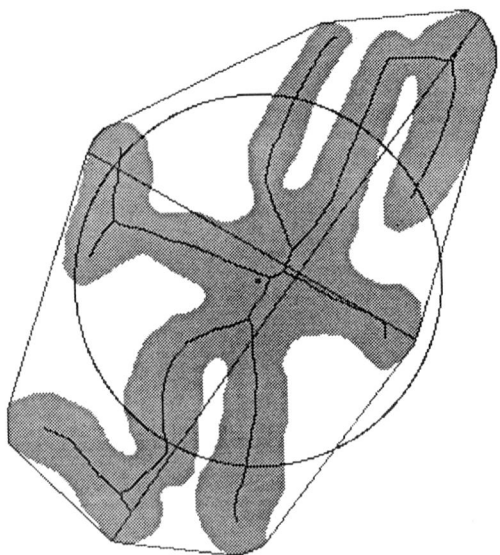

Figure 7-17: One feature from image 6-16 with automatically superimposed lines marking the centroid location, length and breadth, convex polygon, the feature skeleton, and a circle with the same area as the feature.

Image Measurements

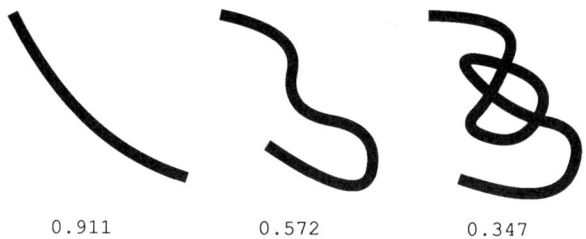

Figure 7-18: Examples of fibers with varying Curl.

Another shape model could be substituted if it were justified by a particular application, but for this model the "fiber length" can be estimated with either

Fiber Length = $0.3181 \cdot \text{Perim} + (0.033012 \cdot \text{Perim}^2 - 0.41483 \cdot \text{Area})^{1/2}$

Fiber Length = $0.5 \cdot \text{Perimeter} - 2 \cdot \text{Area} / \text{Perimeter}$

The first expression contains constants with various π terms. In one limit as the area becomes negligible, both expressions approach half the perimeter; the area term introduces a correction for the finite (if variable) width of the feature. In the other limit as the feature becomes circular, the first expression approaches the diameter. These models are not particularly sensitive to crossovers or "knots" in a long narrow fiber. Closely related is a shape parameter that describes the extent to which features of this type are curled up. The Curl (Figure 7-18) is simply the ratio of the measured Length (the maximum projected dimension of the fiber) to the Fiber Length. For fibers that are nearly straight this approaches 1.0, and becomes much smaller as the fiber is more curled up. It is also fairly insensitive to crossovers in the fiber image.

For three-dimensional features viewed as projected shadows, the length and breadth or length and width can be used to estimate volume and surface area, provided that a shape is assumed (Hilliard, 1968). If the basic shape is taken to be a prolate ellipsoid (like an American football) or an oblate ellipsoid (a discus) then the volume is

$$\text{Vol}_{\text{prolate}} = (\pi/6) \text{ Length} \cdot \text{Breadth}^2$$

$$\text{Vol}_{\text{oblate}} = (\pi/6) \text{ Breadth} \cdot \text{Length}^2$$

In both of these models, the advantage of the ellipsoid is that its parameters are easy to calculate, and by varying its length and breadth it is able to model objects ranging from fibers or needles (prolate with a large length to width ratio) to platelets (oblate with a large width to length ratio).

Figure 7-19 compares the volume of ellipsoids of revolution calculated using the length and breath (L,B), length and width (L,W), and fiber length and width (F,W) measured from projected images of a variety of regular polyhedra (Russ & Hare, 1986). Depending on the orientation of the polyhedra, different images and hence different measured dimensions are obtained. The plots show the extreme limits and range of possible estimated volume. It is apparent that very angular shapes like tetrahedra are poorly modelled by an ellipsoid, but the accuracy improves for the other shapes. For very short cylinders (disks) better results would be obtained using an oblate rather than prolate ellipsoid. The rather large range for extreme shapes is expected, but would average out for a large collection of randomly oriented objects. Generally, the Length,Width model is superior to the Length, Breadth model and the Fiber Length,

Width model is best for the long skinny shapes for which it is appropriate, as we would expect.

The surface area of the ellipsoid can also be straightforwardly calculated as

$$\text{Surface Area} = \frac{\pi}{2} \cdot \left\{ B^2 + \frac{A \cdot B \cdot \sin^{-1}(e)}{e} \right\}$$

$$e = \frac{(A^2 - B^2)}{A}$$

This is apt to underestimate the actual object surface area because any irregularities or local roughness will increase the surface area greatly while making only a small

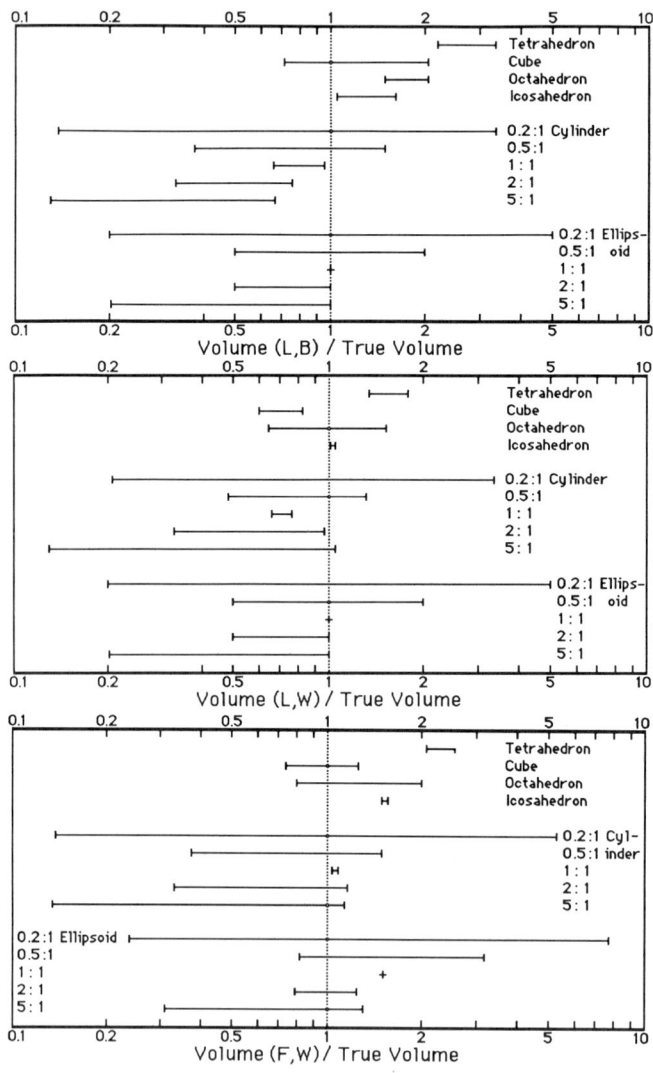

Figure 7-19: Range of errors for surface area of ellipsoids estimated from projected images of various regular polyhedra in all orientations.

Image Measurements

difference in the volume. The surface area can be estimated using a model that takes into account the shape of the projected feature outline. Shape is discussed separately below. For now, we will introduce without discussion a dimensionless "formfactor" calculated from the area and perimeter of the feature as Formfactor = 4π Area / Perimeter^2. This takes on values from 1.0 (a perfect circle) down, with very small numerical values representing very irregular shapes. If the surface area of the ellipsoid determined from the feature length and breadth is divided by this formfactor, a useful estimate of the actual three-dimensional object's surface area is obtained.

Figure 7-20 compares the surface area calculated for the same shapes used in the volume comparison above. The agreement is generally better than the volume results.

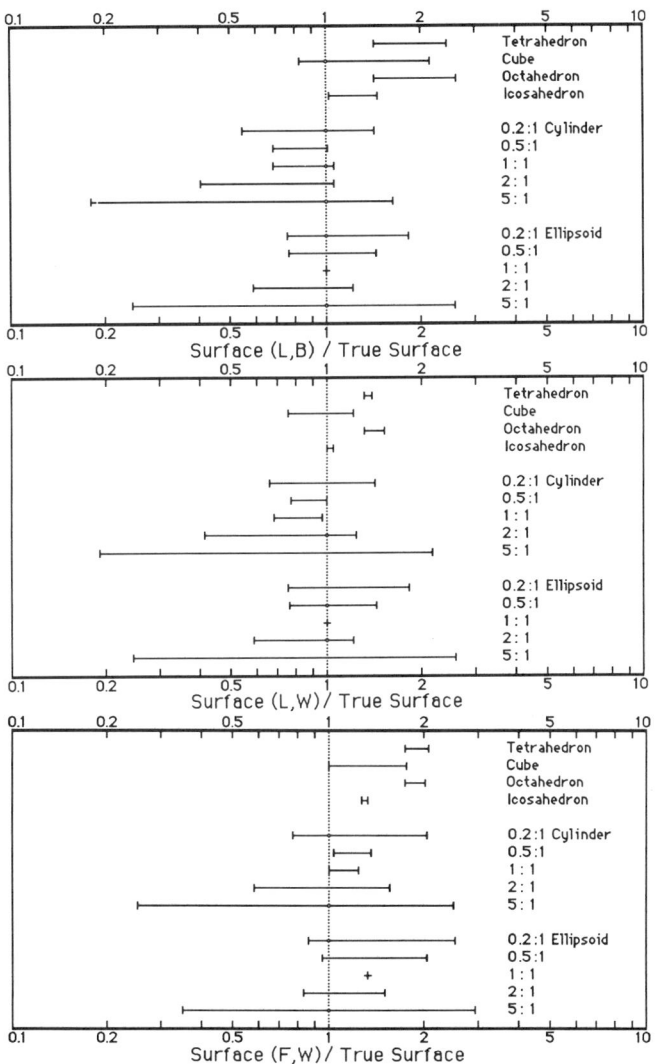

Figure 7-20: Range of errors for surface area of ellipsoids estimated from projected images of various regular polyhedra in all orientations.

Again, the Length, Width model is generally superior to the Length, Breadth model, and the Fiber Length, Width model is best for long, skinny shapes.

For projected or shadow images of objects, it is rather common to report an equivalent diameter calculated as $(4 \cdot \text{Area}/\pi)^{1/2}$, or the diameter of a circle with the same area. If the objects are randomly oriented in space, this is also the mean caliper diameter. This mean caliper diameter is also known as the Minkowski height of the object, which can be understood by imagining the objects scattered randomly on a substrate. In that case the mean height of the tops of the objects is just the mean caliper dimension.

This height can be calculated quite straightforwardly from the curvature of the surfaces and edges of the object. The relationship is $M = 2\pi D$ where D is the mean caliper diameter. The total curvature M is defined in terms of the mean local surface curvature H (which depends upon the two principal radii of curvature) and the dihedral edge angle χ as

$$M = \int H \, dS + \frac{1}{2} \int \chi \, dl$$
$$H = \frac{1}{2} \cdot \left(\frac{1}{r_1} + \frac{1}{r_2} \right)$$

A few simple examples will illustrate the meaning and usefulness of this relationship. For a sphere, there are no edges (the second integral vanishes) and the curvature is everywhere uniform so that $H = 1/r$. The first integral is just the product of this value times the area, giving $M = 4\pi r$. When this is equated to $2\pi D$, it gives the mean caliper diameter $D = 2r$, which is expected for a sphere.

For a cube, there is no surface curvature and the first integral vanishes. The edges have a dihedral angle of $\pi/2$, and the total length of edges for a cube with edge length a is $12a$. The total value for M is thus $1/2 \cdot 12a \cdot \pi/2 = 3\pi a$. Equating this to $2\pi D$ gives $D = (3/2) a$ as the mean caliper diameter. This is between the minimum caliper diameter a and the maximum value $\sqrt{3} \, a$, but is not an easy answer to obtain by analytical integration or Monte-Carlo computer modelling, both of which require rotating the cube uniformly to all possible orientations. The same method can be directly extended to other polyhedra.

For a cylinder of length L and radius R, both terms in the equation for M must be evaluated. The end faces are flat, but the side of the cylinder has a uniform curvature $H = 1/2R$ and area $2\pi RL$, giving an integral of πL. The edges have a dihedral angle of $\pi/2$ and a total length (for both ends) of $4\pi R$. The value of the second integral is thus $\pi^2 R$. The Minkowski equation for the cylinder gives $D = (L + \pi R)/2$.

Radius of curvature

Measurement of the global mean curvature of boundaries was discussed earlier. However, this technique is not applicable to the case of measuring the curvature of individual objects, nor ones viewed in projection rather than in section. Image analysis measurements of individual features in an image field generally determine a number of different parameters, some of which are measured directly (e.g., area) and some of which are calculated from them (e.g., width as a function of length and perimeter). We will examine some ways to estimate the radius of curvature of objects by this method. Our purpose is not only to develop a practical method to measure this specific parameter, but also to illustrate the variety of ways in which particular feature descriptors such as this can be measured (Chapin & Norton, 1968; Riggs, 1973; Blakemore & Owen, 1974; Agin & Binford, 1976; Shapiro, 1978; Pavlidis, 1978,1980; Aherne & Dunhill, 1982;

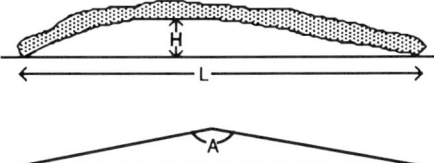

Figure 7-21: Measurement of the height *H* of a curved or bent object allows calculation of an effective angle *A* = 2• arc tan (*L*/2*H*).

Russ, 1989). The parameters used in these models are ones already defined above: area, perimeter, convex area, length, breadth, width and fiber length.

One particular context in which the need for the curvature of features arose is in the archaeometric study of materials, but it is clear that the same parameter will also be useful in many other situations. Many archaeological objects are curved in a simple but not necessarily regular way (e.g. flakes from stone tool manufacture, whose curvature reflects the technology of knapping and also changes progressively during the course of knapping, or pot sherds whose curvature may reflect the size of the original containers or, near the neck, the individual stylistic designs of the maker) We want to measure this curvature using the automatic image analyzer in order to obtain statistically useful amounts of data with minimum operator time or variability.

The curvature with which this section is concerned is difficult to determine directly, so several methods to estimate it using formulae for derived parameters will be presented. Strictly speaking, this is a size parameter with a linear dimension, but obviously it is related to the shape of the features. For irregular shapes, we seek a meaningful "average" value.

Several *ad hoc* parameters which are related to curvature have been tried in various applications. Some are only indirectly related to the actual nature of the features, but simply try to assign arbitrary values to shapes that are visually distinct. The data may be useful for automatic feature recognition but will not in general contribute to an understanding of the objects.

Among the parameters that more directly deal with actual feature dimensions, one (applied to flint flakes used for tools and spear-points) is simply the angle (Figure 7-21) between two lines fit to the mid- and end-points of the object. This was done manually (Andrefsky, 1986), but is not so easy to measure as the sketch suggests, because the placement of the points or tool along an irregular object surface is variable, and is complicated considerably by any curvature in the second direction or asymmetry in the piece. Also, it requires operator attention to each piece. The other obvious method of estimating curvature is by comparison to a series of circle templates. This calls for considerable operator judgement, and is especially complicated for objects that are not uniformly curved.

For digitized images, in which the entire projected outline of the object is available, several other parameters can be easily defined and measured. They provide more consistent characterization of the shape of the object, but are still inferior in several respect to the other methods we will introduce below.

One parameter similar to the angle described above can be obtained from the height *H* (here used to describe the convexity, or *embayment*). This is calculated as (Breadth −

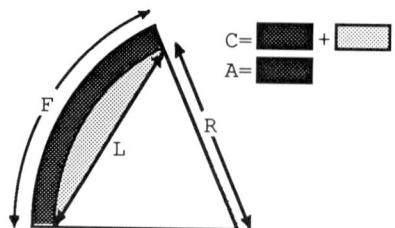

Figure 7-22: Radius of curvature or an object (dark shading) can be determined from the area of the embayment (light shading), which is the difference between the convex area and the area, and the fiber length and projected length.

Width) using the measured parameters as described above. In order to deal with objects of different length, either the ratio of Height to Length H/L or the actual angle A, twice the angle whose tangent is $L/(2 \cdot H)$, may be used.

Instead of H as measured above, it is often preferable to obtain a value for the embayment that is averaged over the entire length of the object. Instead of Breadth minus Width, we can use the area of the embayment divided by the object's length. This is obtained as (Convex Area − Area) / Fiber Length. The resulting parameter is a mean value rather than a maximum; for a uniformly curved object it will be proportional to the value of H.

A better method of describing the shape of a "bent" object is the radius of curvature. This can be obtained in several ways. For example, Figure 7-22 shows a relationship between Radius of curvature (R) and the Area, Convex Area, Fiber Length and Length (A, C, F and L, respectively). The radius of curvature to the inner edge of the object is R (the radius to the midline, if required, can be determined as R + Width/2). The area of the lightly shaded region is $(C - A)$. This equals the difference in area between the circular wedge and the triangle. The area of the wedge is $F \cdot R/2$, and that of the triangle is $L \cdot X/2$ where X is the altitude of the triangle, calculated from simple geometry as $(R^2 - (L/2)^2)^{1/2}$. By algebraic manipulation, this yields an equation

$$R = (-b \pm (b^2 - 4 \cdot a \cdot c)^{1/2})/(2 \cdot a)$$

where

$$a = (F^2 - L^2)$$
$$b = -4 \cdot F (C - A)$$
$$c = 4 \cdot (C - A)^2 + L^4/4$$

This model is easily programmed into the image analysis system to calculate radii of curvature of object profiles as a derived parameter. The problem with the model described above, or a similar one in which (Breadth − Width) is used, is that the algebraic solution involves squaring the terms, and this introduces another possible arithmetic solution that may have no physical meaning. Also, the effect of small roughnesses on the feature profiles or other irregularities produces a much larger error in the radius because of the differences between terms. For this reason, this is not always to be a "robust" method when shapes become much more irregular than the idealized arc shown in the figure.

A better approach is to first reduce the feature profile to its skeleton (or medial axis transform, a line equidistant between the edges). This can be done for the entire binary image using specialized erosion/dilation rules built into the image analysis system as

Image Measurements

described in detail in Chapter 6. This removes any contribution of the object's width or minor irregularities along the profile. Then the Length and Fiber Length (L and F respectively) obey the simple relationship $L/(2 \cdot R) = \sin(F/(2 \cdot R))$. However, this cannot be solved directly for the radius R. An iterative solution is possible, by guessing at an initial radius (e.g. twice the length), placing this value in the right side of the equation, and then solving for a new value on the left side. By repeating this operation, a stable value of R emerges. This takes longer than most of the other methods (a fraction of a second per object), but gives a more meaningful answer.

This method can also be used to measure the arc, or angular extent of the object. The angle is arc $\sin(L/(2 \cdot R))$, where L is the length and R is the radius. Figure 7-23 shows a test image consisting of 7 lines. Three have a large radius (a, b, c) and three a radius about half as large (d, e, f), while one is a straight line (g). Pairs of lines are drawn covering approximately 45, 90 and 135 degrees (a+d, b+e, c+f, respectively). Applying the measurement program to this test image produces the results shown, which demonstrate the expected accuracy from an image in which the features consist of a very small number of pixels (the radii are given in pixels, or picture points). The straight line is identified directly (zero is the reported value for an indeterminate radius and angle), and the other objects are reasonably classed by both radius of curvature and angular extent.

This method has been applied to profiles of pot rim sherds as shown in Figure 7-24. The results, compared to those obtained by hand (using visual comparison to a set of circle templates) were fairly good (Rovner, 1987). However, the method does not work well if the skeleton of the object has branches, which tend to develop near irregular ends or at abrupt bend points. The figures show some representative features with their skeletons. Automatic pruning of the skeletons has not worked well enough to solve this problem reliably, so that some manual interaction is still required, depending upon the shape and surface irregularities of the feature.

Image processing approaches

A very different method, which is used in machine vision work, can locate the center of circular (or other known shape) outlines, even if only a part of the arc is visible. It is used to guide the robotic placement of parts on shafts, and other similar applications.

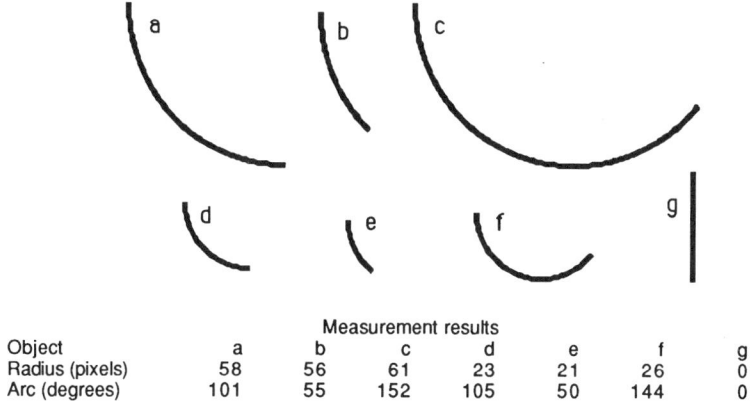

Object	a	b	c	d	e	f	g
Radius (pixels)	58	56	61	23	21	26	0
Arc (degrees)	101	55	152	105	50	144	0

Figure 7-23: Test image for curvature measurement.

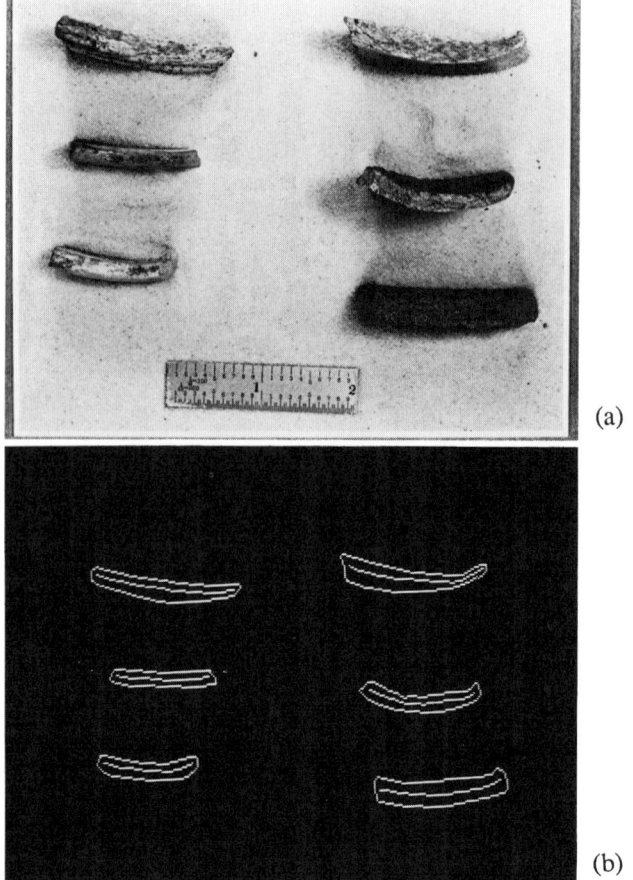

Figure 7-24: a) Image of pot sherds (prepared by setting them on edge in a pan of white sand); b) the digitized binary image of the sherd outlines with their skeletons.

Although it deals with the location of a curved feature and assumes the radius is known, which is not the main thrust of this section, it is included here because it leads rather directly to a more general method. The circle-locating routine uses image processing on the original grey scale image. A Sobel transform (Sobel, 1970) or one of the other similar edge-finding algorithms is used. The transform calculates the derivative of brightness in the X and Y directions. Normally, the magnitude of this function, calculated for each pixel in the original image, is mapped into another image plane to delineate edges. In this case, an arbitrary threshold is used to detect pixels (whose magnitude values exceed the threshold setting) which lie along the periphery of the part or feature to be located. Then, for those pixels only, the orientation of the maximum edge brightness gradient is calculated as

$$\vartheta = \arctan\left(\frac{\partial B/\partial y}{\partial B/\partial x}\right)$$

This angle gives the direction of maximum rate of change in the brightness, which should be radial to the circle outline. Of course, because of the uncertainty of lighting conditions, there is a 180° uncertainty in the direction to the center. But since the radius is presumed known, it is possible to plot the point where the center should lie at a distance R from the pixel in both directions. This is done by adding one to the brightness value at the corresponding pixel location in a new image plane. After this procedure is carried out at each point around the circle or portion thereof, the actual center will have accumulated many such dots, while the points at the "wrong" end of the direction vectors will form a very faint circle at twice the feature radius as shown in Figure 7-25. The resulting image can be measured to find the point of maximum brightness, which then locates the feature center. The same method can be employed with other shapes provided they are known, so that the distance and direction to the center can be specified as a function of the direction of the local edge derivative.

Note that this technique does not address the current problem, because it makes the assumption that the shape and size of the feature are known, when in fact that is what we are seeking to find out. However, the method is included here because it is related to curvature measurement and because it can serve as an introduction to the next method, in which circles of any radius are detected and located.

Related to this method is another approach to locating circular features and arcs, and determining the radius of curvature, which employs the circular Hough transform. The most common form of the Hough transform (Hough, 1962; Duda & Hart, 1972) is the linear transform, in which each point in the original image is mapped onto all of the points (in a two-dimensional space whose coordinate system is M and B in the equation $y = Mx + B$) which would define a line that could pass through the object. Once this has been done for all points in the image, the total value from all superimposed mappings gives peaks in Hough space corresponding to the linear structures present in the original image, and these can be found by discrimination and plotted on the original.

The same method can be used for nonlinear shapes (Sklansky, 1978). Chapter 4 discussed the circular Hough transform and showed an example of its application to election diffraction patterns. This method is quite similar to the center-finding technique discussed above in its use of incrementing points to mark the center of the circle(s). It amounts to performing the circle-locating operation with each possible radius, and selecting the one with the brightest point as giving both the circle radius and center location. This same method allows finding the center of any partial pattern of arcs. A simple but obvious example is locating the center of a tree ring pattern from a core that does not intersect the center (Figure 7-26), so that measurements of radial distance can be

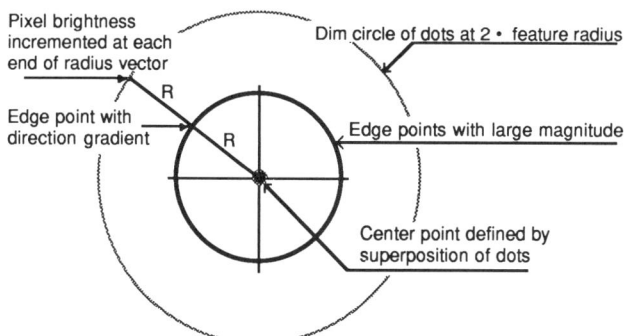

Figure 7-25: Locating the center of a circular feature from the edge pixels.

Figure 7-26: Example of locating the central point for partial concentric arcs (e.g., tree rings).

recorded. Note that the location of the center can be established to better than a single pixel dimension, by fitting to the brightness values in the image.

For individual arcs, the method and its storage requirements can be simplified by examining each planar section corresponding to a different radius in the three-dimensional circular Hough space sequentially. Each plane will consist of superimposed circles (sections of the cones in Figure 4-22) centered on each point in the feature (or its skeleton - the presence of branches has only a minor effect), whose brightnesses are added together in a grey scale image. The maximum brightness point in the image is easily determined, and the its value noted. The process is continued with increasing radius values until the maximum brightness value decreases, indicating that the central point has been passed. In the image with the maximum brightness, the location of the center is found. If the arc is not perfectly circular, or is an incomplete arc, the central region is also irregular, generally ridge-shaped, and the brightest point is the best approximation to the center of the arc. The radius is then the number of that image in the sequence. Figure 7-27 shows the plane in Hough space that locates the center for an arc (image of a pot sherd). It lies at the radius value with the maximum brightness value obtained by summing the cone sections from each point in the feature. The image (from a single pot sherd) and its skeleton are also shown. The relationship of this method to that described above for locating the center of known circles is clear.

This method is not fast, and requires dealing with the objects one at a time, whereas the earlier ones based on derived parameters can deal with an entire field of objects at once to provide a curvature value for each. This is a practical limitation that can be overcome by redrawing each identified feature singly for measurement. The Hough method also does not do well with very straight objects, since the number of iterations is

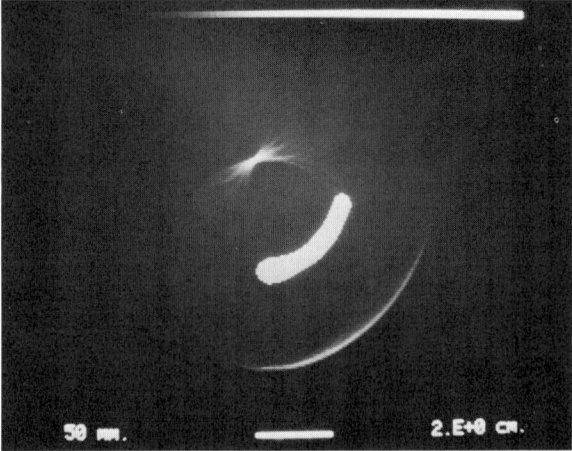
Figure 7-27: Using the circular Hough transform to locate the center point of an arc.

Image Measurements

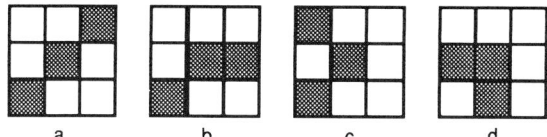

Figure 7-28: Pixel patterns showing a) no curvature; b) a small curvature to the right; c) a larger curvature to the left; d) an even larger curvature to the left because the same angular change occurs in a shorter length of the line.

proportional to radius. Furthermore the method fails if the center is not within the image field.

Counting neighbor patterns

Another rather different approach to measuring the curvature of features uses the same neighbor pattern technique referred to before in the context of skeletonization. The definition of curvature from analytical geometry is simply the change in direction per length of line. Applied to the skeleton of a single object, this can be established by counting the incidence of various local pixel patterns along the length of the object. The various patterns will include ones that are straight (have no curvature), and ones that exhibit varying degrees of curvature to the right or left. These can be counted by template matching, also known as "hit-or-miss" matching, for each pattern.

The integrated curvature of the skeleton, which can be used to describe the total curvature of the object, can then be determined from the sum of the number of incidences of each pattern. Figure 7-28 shows several examples. It is important to count the patterns in proportion to their contribution to the derivative $d\varphi/ds$, the angular deviation along the line. The numerator in the pixel-based approximation to this derivative is always (for square pixel patterns) a multiple of 45 degrees, but the denominator may be either 1 or 1.414 times the pixel spacing, depending on whether the line's orientation was initially in one of the 45 degree or 90 degree orientations with respect to the pixel grid. Notice that the total curvature obtained in this way is independent of the position and orientation of the feature, and that the sum may be acquired in any order so that pixels can be counted in raster order, and do not require progressing along the possibly irregular shape of the skeleton.

This method makes use of the same routine as described in Chapter 6 in the context of erosion or skeletonization to obtain neighbor patterns for each pixel, and simply counts the incidence of each pattern. It requires that individual features be measured one at a time and also makes the assumption that the skeleton is a good representation of the feature shape and curvature. This technique is especially good for determining the total curvature along irregular feature skeletons or other boundaries, as discussed earlier in this chapter, as it sums both the curvature and the length of the line directly. The result of adding the absolute value of all deviations and dividing by the length is a value expressed in degrees or radians per unit of length that describes how "irregular" the feature is. It is also possible to count individual patterns and produce a histogram of the frequency of local curvature values along boundaries.

However, when just the net deviation along the length is used, this method is not good for our intended purpose here of describing the radius of curvature of arcs. This is because the net angular deviation between the ends of the line is constrained to be a multiple of 45 degrees, and changing even a single pixel in the original feature that causes

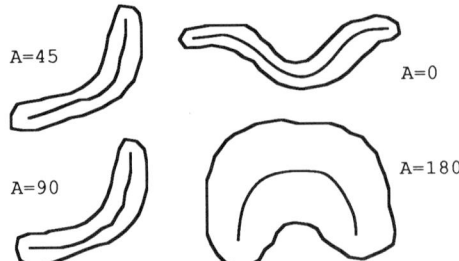

Figure 7-29: Several features with their skeletons and the net angular extent of each (in degrees).

the end-most pixels in the skeleton to shift can introduce a 45 degree change in the net curvature. Figure 7-29 shows several features with their curvature (net angular extent) values. Notice that two of the features are nearly identical but produce slightly different skeletons and curvature results that are widely different.

The radius of curvature is especially recommended as a descriptor of curvature, as it corresponds to a physically meaningful dimension for many objects. Of the various methods to determine this radius described here, the first type (derived parameters calculated from measured ones) has the advantage of speed. Even the iterative method shown operates quickly for an entire field of objects, producing a file of results within seconds or minutes that can be used for statistical analysis. For relatively simple objects, or ones with relatively simple shapes, it seems to produce adequate results. In some cases, the use of a skeleton of the features (which can also be produced for the entire discriminated or binary image at once) helps to reduce the dependency of the derived value on the details of the shape of the feature outline. This should be the method of choice if the features are simple enough to allow it to work.

The second class of methods, utilizing image processing methods such as the Circular Hough transform, are much more tolerant of incomplete or irregular shapes. The method is somewhat more demanding on the computer, since memory for the three dimensional Hough space must be available and the operations are carried out on the original grey-scale image rather than on a binary image. Also, the method can only deal effectively with a single object at a time (or at least with widely separated features). For relatively complex objects that cannot be handled by the derived parameter methods, this is probably the best choice. Certainly for another class of images such as electron diffraction patterns it is the premier technique available.

Curvature measurement from neighbor patterns (in conjunction with skeletonization) is a useful approach to determining total or absolute magnitude of boundary curvature as discussed at the beginning of this chapter, but is not usually suitable for estimating the arc length or radius of curvature of individual features.

Shape

A major category of measurement parameters describe the shape of features. This is an extremely important thing to be able to describe (Exner, 1986), and is often a key factor in being able to recognize or select objects of interest. But there are few definitions which we can draw from common experience to guide us. The radius of curvature just discussed is one way to characterize the shape of particular features, but shape factors are generally dimensionless numbers, and usually are obtained by combining size parameters in various ways. For example, the aspect ratio of an object is a more-or-less familiar concept, which we can calculate as the ratio of the measured length to either the breadth or the width. This is dimensionless, of course, and expresses a quality of feature shape that ignores some things (such as the local smoothness of surfaces) but provides a measure of how elongated the feature is.

If the image is a projected image of objects such as particles lying on a substrate, then it is likely that gravity or electrostatic forces will have aligned the objects so that their greatest length is parallel to the plane of view, so that it can be measured (footballs do not stand on end very well). If this is the case, then the aspect ratio measured as just described is correct for the three-dimensional object as well as for its two-dimensional feature image. However, if the objects in the image are randomly oriented in space, the features seen in the projected image will in general be foreshortened and the measured aspect ratio will be too low. This cannot be corrected for an individual feature because its orientation is not known. But for an entire population of objects with random orientations, the corrected aspect ratios can be estimated based on geometric probability. This gives a value for the aspect ratio of

$$\text{3-D Aspect Ratio} = 1.0 + (4/\pi) \cdot (\text{Length/Width} - 1.0)$$

There are a wide variety of shape descriptors available. Most of them are derived parameters, and some of the more common will be discussed here. Two words of caution are required: first, the names of these factors are frequently substituted or interchanged, and there is no general agreement on them, and secondly sometimes the factors appear as the inverse of the form given here. In this text, most factors are calculated so as to give values between zero and one, for convenience in plotting.

Probably the most widely used shape parameter is the "formfactor" calculated as $4\pi \text{ Area} / \text{Perimeter}^2$. We encountered this earlier as a means to estimate the surface area of irregular but approximately ellipsoidal particles. The formfactor is 1.0 for a perfect circle. Any other shape will have more perimeter for the same area, and the formfactor describes this increase. For a square the formfactor is 0.785, and for more irregular shapes it becomes much smaller. A many-petalled flower or the cross section of villi in the intestine may have formfactors of 0.05 or even less (Figure 7-30).

"Roundness" looks very much like the formfactor at first glance. It is calculated as $4 \cdot \text{Area} / \pi \cdot \text{Length}^2$, which again is just the formula for the area of a circle and gives a value of 1.0 for a perfect circle. But instead of perimeter, Roundness uses the length (longest chord) of the feature. This makes it more sensitive to how elongated the feature is, rather than how irregular its outline may be. It is not unusual to find features that vary independently in roundness and formfactor, as we will see shortly.

Another set of shape parameters deal with how convex the feature is. This can be described based on either the perimeter or the area. The ratio of taut string perimeter to external perimeter is called the convexity, and is 1.0 for a feature that has no concavities

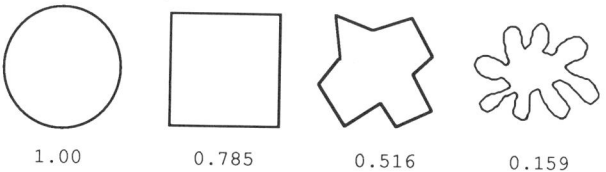

Figure 7-30: A few features with varying formfactors.

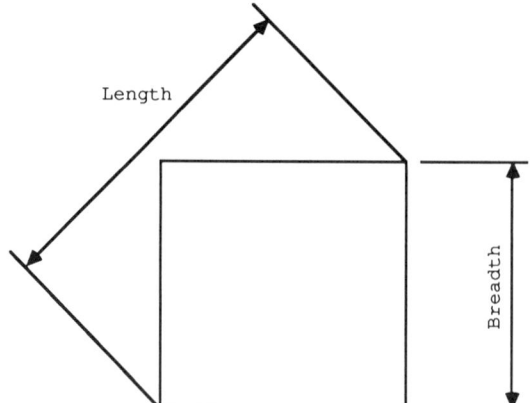

Figure 7-31: A Square with Length and Breadth marked.

or indentations around its periphery. Solidity is the ratio of area to convex or taut string area, and again this is 1.0 if the object is everywhere convex. Both parameters decrease for features which are concave, but in somewhat different ways. The perimeter-based convexity will decrease quickly for features with deep, narrow intrusions while the solidity will not, for instance.

Finally in this category, it is sometimes useful to consider how fully the feature fills the area described by its measured length and breadth. The two projected or Feret's dimensions can be used to specify two different areas. One is a rectangle with their lengths as sides (sometimes called the minimum bounding rectangle). This does not take into account the possibility that the length and breadth dimensions were not originally perpendicular to each other (they frequently are not). A somewhat better reference area to use is that of a parallelogram with sides having the length and breadth as their lengths, meeting at the same angle as the original length and breadth were determined (using the rotation method described earlier). The ratio of the feature area to this bounding parallelogram is called the "extent" of the feature.

For example, consider a feature that is nearly square with side = 1.0 (Figure 7-31). The length as defined above will be a diagonal, with dimension equal 1.414, and the breadth will be the length of one side. These meet at a 45 degree angle. The area of the minimum bounding rectangle would be 1.0 • 1.414 = 1.414, while that for the parallelogram would be 1.0 • 1.414 • sin (45) = 1.0. This is the same as the true area of the feature, giving an extent of 1.0, a more reasonable value for this case of a square feature.

A principal use of these shape factors is to compare different populations of features, and for recognition of features. We will see in a later chapter how to select the most distinctive shape parameters for any particular purpose, and how they can be used for automatic recognition. In addition to this purpose, some interpretation of feature morphology is sometimes directly available from the shape factors. For example, a plot (plots of data and the use of the correlation coefficient are discussed further in the next chapter) of formfactor versus some measure of size, such a length, for all of the features in an image will sometimes show a definite trend. If the formfactor increases with size it indicates that the features become rounder as they increase in size, and more irregular as they become smaller. This is often an indication that the smaller objects have been produced by fragmentation of the larger ones. Conversely, a trend of decreasing formfactor with increasing size often indicates that the larger objects have been formed by agglomeration of the smaller ones.

It is interesting to note in the example in Figure 7-32 that the Formfactor and the Solidity can distinguish between the more angular features in the left hand column from the more smoothed ones in the right column, while being relatively insensitive to how stretched out the features are. Conversely, the Roundness and Aspect Ratio can distinguish the features in the three rows which change in elongation, while being relatively insensitive to the angularity.

A similar series of shape factors based on moments of inertia of the feature shape have been used by Exner & Hougardy (1988). They define *Elongation* as the square root of the ratio of maximum to minimum moment, *Compactness* as the ratio of the maximum moment of inertia to that of a circle of the same area, and *Symmetry* as the ratio $(h - z)/h$ where h and z are the Feret's diameters in the directions of the maximum and minimum moments of inertia. As for the dimensonless shape factors discussed above, these can often be shown to sort out specific features, but they are not unique: an unlimited number of visually distinct shapes can be designed which are the same in one or more of these factors.

Also included here in the category of shape parameters is the number of internal holes or voids in the feature (Figure 7-33). This has a clear meaning for the two dimensional

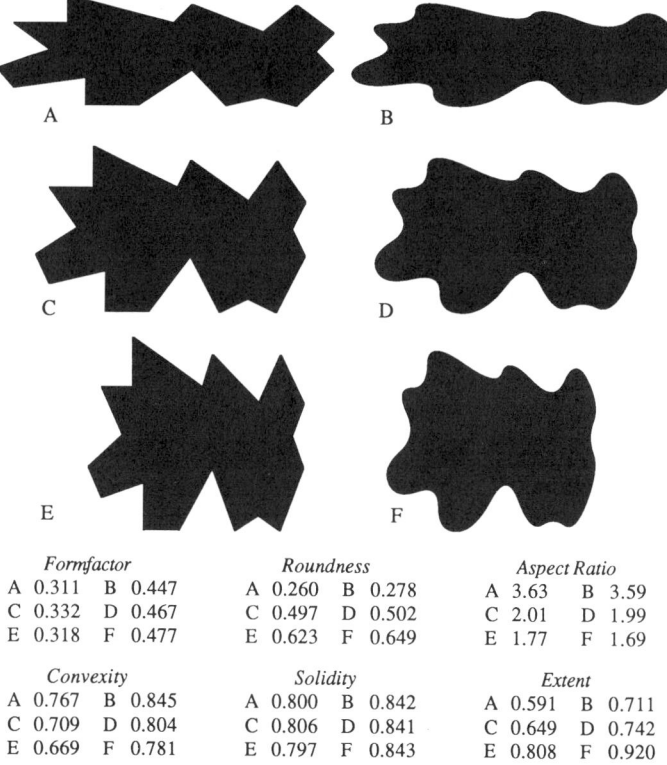

Formfactor		Roundness		Aspect Ratio	
A 0.311	B 0.447	A 0.260	B 0.278	A 3.63	B 3.59
C 0.332	D 0.467	C 0.497	D 0.502	C 2.01	D 1.99
E 0.318	F 0.477	E 0.623	F 0.649	E 1.77	F 1.69
Convexity		Solidity		Extent	
A 0.767	B 0.845	A 0.800	B 0.842	A 0.591	B 0.711
C 0.709	D 0.804	C 0.806	D 0.841	C 0.649	D 0.742
E 0.669	F 0.781	E 0.797	F 0.843	E 0.808	F 0.920

Figure 7-32: Features with shape variations compared by their Formfactor, Roundness, Aspect Ratio, Convexity, Solidity and Extent.

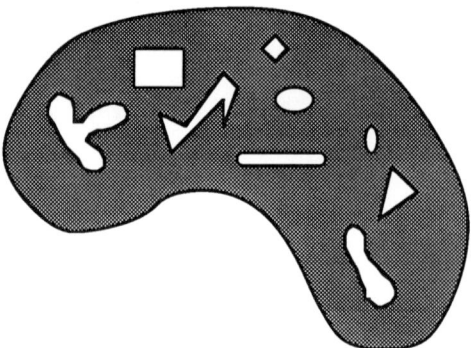

Figure 7-33: Binary image of a feature containing multiple holes. The number, dimensions, shapes, etc. of the holes can also be measured, and may be useful in comparing the parent features.

feature, but if the feature has been produced by sectioning through a three-dimensional object, it is not possible to determine the number of voids (the genus or connectivity) of the three-dimensional object from any information obtained on the two-dimensional section image. This will be discussed more in a later chapter in which multiple sections are used to study topology.

Nevertheless, the number of holes can be of use in distinguishing classes of 2-D features. These may be internal holes in the binary image resulting from the presence of another phase or type of internal object, such as organelles within a cell or islands in a lake. For the moment, we will just refer to them as "holes". Since the total area of holes can also be determined (as the difference between the filled area and the net area of the object), this means that the mean size of voids is readily obtained as well.

Likewise, from the difference between the total and the external perimeter, the net perimeter around the holes can be calculated. This plus the net hole area and number of holes allows the mean formfactor (or another shape descriptor) to be computed for the voids within each object. Because the formfactor is sensitive to the roundness or squareness of the voids, this derived parameter can sometimes be used to distinguish features which have internal voids resulting from membranes, or other structures which tend to assume a smoothly rounded shape, from features that result from an overlapping tangle of fibers in which the holes tend to be angular. This seems to be pretty much the same criterion that humans use to decide which of the two types of features they are seeing. The principal areas of application for this method are in biological specimens, at magnifications ranging from structures within the cell to whole organs.

As was mentioned before, in conjunction with the use of selective erosion to produce a skeletonized representation of features, the count of internal holes is also a tool that can be used to determine the number of nodes, ends or branches within each feature. These are also shape parameters in their own right, which describe the extent of branching. Combined with the difference between the taut string and net area or perimeter, they permit estimates to be made of the size (either length or area) of indentations around the feature periphery, in much the same way that the mean formfactor of holes was derived above.

For example, in examining cross-sections of man-made fibers, the size of the surface indentations (which control the flexibility of the fibers) can be characterized as the difference between the convex area and the area, divided by the number of end points in the skeleton. Further, the mean depth of the indentations is 1/2 · (perimeter − convex perimeter) / (number of ends). This depth is an important parameter by which the wear on the extrusion dies is judged.

Corners as a measure of shape

The idea often resurfaces that an important visual descriptor of object shape is the number of corners that the feature shows. Simplified line drawings in which all corners are preserved, but are joined by straight line segments, seem to convey much the same information as the original drawings. Sharp corners are often associated with man made objects, but in natural scenes, corners also reveal intersections between different objects or surfaces, and can be used to reconstruct three-dimensional shapes.

For this reason, it is interesting to attempt to characterize a shape by the number of corners it possesses. The difficulty is that most features observed in real images have "corners" that are quite variable. Some measure of how prominent a corner is must be used along with the means of detecting them. Two principal types of approach have been used.

The first uses the skeletonization of the feature. Depending on the skeletonization rules, and whether it is done by an iterative erosion method or by constructing a Euclidean distance map, the skeleton will generally develop branching lines that point toward corners that are sufficiently long and narrow with respect to the main body of the feature. A square, for instance, will have a skeleton consisting of four arms pointing to the corners. Counting the number of end points in the skeleton of a feature, and associating that value with the feature for analytical purposes, is rather straightforward.

The difficulty with the method is that it often does not detect places on the periphery of an object that have a large change in orientation of the boundary that takes place over a larger distance, and that the amount of local curvature required to generate a branch in the skeleton is somewhat orientation dependent. Also, local variations that we would normally associate with "noise" in the feature boundary, or are at any rate small compared to the overall feature size, may have significant local curvature to produce a branch indicating a corner that we would prefer to ignore in favor of more global ones.

The second approach offers more flexibility in interpretation. It requires that the feature boundary be converted to a chain-code representation. Then a chord is drawn between points in the chain that are some distance n apart. The value of n can be made large enough to ignore local variations that are not to be considered as "real" corners. As the position of the chord is advanced along the chain, a plot is made of some derived value such as the change in slope of the line (Freeman & Davis, 1977), the length of the line, or the net area between the line and the chain. These vary somewhat in their sensitivity to minor irregularities and in the computational effort required. But in all cases, the parameter shows a peak of significant size when a corner is encountered. This can be thresholded to locate corners, which can then be counted to characterize the shape (Freeman, 1978), or used either to construct a simplified polygonal representation of the feature.

For a convex shape, the polygon is similar but not identical to the convex polygon constructed using the Feret's or projected diameter method described earlier. However, the corner-finding method also locates interior corners and so can construct a polygonal representation of a non-convex shape. An extension of this approach describes feature shape as a sequence of numbers encoding the sequence of corners and sides (Bribesca & Guzman, 1980).

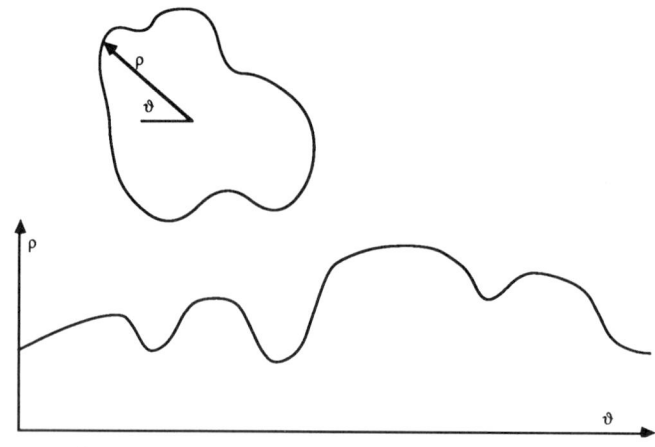

$$\rho(\vartheta) = a_0 + a_1 \cos(\vartheta) + b_1 \sin(\vartheta) + a_2 \sin(2\vartheta) + b_2 \cos(2\vartheta) + ...$$

Figure 7-34: Example of unrolling a feature profile as a plot of $\rho(\vartheta)$.

The fractal dimension is also a type of shape descriptor. It was described in Chapter 6 in connection with the techniques for determining it from successive perimeter measurements. The result is a single value that characterizes the roughness of feature outlines, provided that the roughness is of a particular type called "self-similar," meaning that the degree of roughness would appear the same at any magnification. (Examples are shown in Figure 6-39). Many naturally occurring objects and surfaces have a roughness that fits this criterion well, at least over a finite range of dimensions.

Harmonic analysis

Another very different approach to shape is available, often known as spectral or harmonic analysis (Schwartz & Shane, 1969; Ehrlich & Weinberg, 1970; Beddow et al., 1977; Flook, 1982; Kaye et al., 1983; Barth & Sun, 1985). In this method, the outline of a feature is converted to a waveform. The simplest method of accomplishing this is to use the centroid of the feature as the center of coordinates and plot the length of the radius vector as a function of angle (Figure 7-34). This function obviously repeats after 360 degrees, so that the function is endlessly repetitive. This makes it suitable for Fourier analysis, that is, the magnitude of the radius vector can be expressed as a summation of terms of the form shown in the figure.

There are a few points that need to be made about this type of analysis. First, given enough terms these coefficients will exactly reproduce the original feature outline. There is no loss of information at all. Second, the magnitude of the *a* and *b* coefficients depend on where around the periphery of the outline the zero value for the angle is selected. This is equivalent to varying the phase of the $\rho(\vartheta)$ plot, or shifting it sideways. A canonical orientation can be selected for all features if desired, for instance using the axis of minimum moment for the shape, or the maximum radius, or the length vector (these are discussed below under measures of object orientation). However, as the subsequent use of these factors virtually always combines the factors as $c_i = (a_i^2 + b_i^2)^{1/2}$, which is called the amplitude of the *i*th harmonic, and this value is insensitive to the phase, this is not really necessary.

Third, the Fourier method is very powerful for this purpose because it is extremely efficient. The low order harmonics contain a great deal of information about the shape of

Image Measurements

the feature, and the magnitude of the coefficients decreases very rapidly for higher order harmonics. In practice, only the first 20 or perhaps 30 coefficients are actually used to obtain outlines that are arbitrarily close to the original outlines as shown in Figure 7-35, at least within the available pixel resolution of the image (but for a perfectly random fractal outline all coefficients would be important). The zeroth coefficient, in fact, is just the mean circular radius of the feature, and by itself would produce a circular estimate for the feature shape. Furthermore, as more coefficients are added, the values of prior coefficients do not change. This is equivalent to saying that the harmonic terms are independent of each other in the way they describe the feature shape, or that in mathematical terms they are "orthogonal".

One difficulty that arises with Fourier analysis as described here is that the feature shape may be sufficiently irregular or indented that the $\rho(\vartheta)$ function may not be single valued or smoothly continuous. This can be handled by expressing the shape in a slightly different way, for instance as the angle of the boundary line as a function of position along the perimeter. This is also a smooth, continuous and repeating function, so the same arguments as before apply (it is, however, somewhat messier to use from a calculational point of view).

Also, it is important to mention that the sine and cosine functions are not the only set of orthogonal functions which can be used to build up a series of terms that converge to the original function. In principle, any of these other sets of functions can be used. In practice, the Fourier transform has several advantages. First, it is relatively easy to

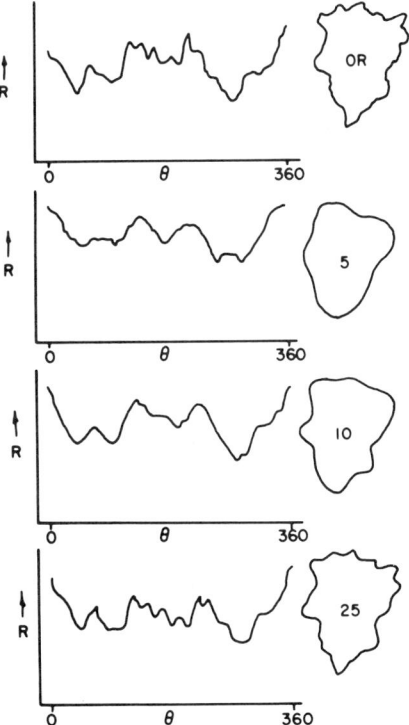

Figure 7-35: Reconstruction of feature shape using different numbers of frequency terms.

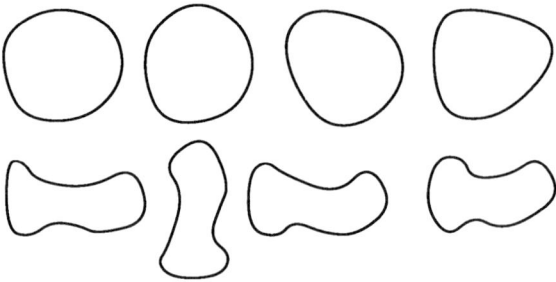

Figure 7-36: Example of two sets of chaotic shapes. Members of each set are similar in Fourier analysis, but the sets are different from each other.

calculate the terms, and in fact the FFT (Fast Fourier Transform) algorithm is a very well known and widely used tool. Secondly, for many kinds of features it can be shown that the trigonometric functions are the most efficient ones to use (requiring the fewest terms to describe the shape to any desired degree of accuracy).

This is strictly true for features that have a shape that is uncorrelated. In other words, the $\rho(\vartheta)$ function (or whatever else is used to unroll the outline) has values whose difference is not a function of the difference in angle. Most shapes which we will be dealing with are "chaotic" shapes (discussed further below) and fall into this category. However, there is a technique to find the optimum set of functions for any given shape. This involves a Karhunen-Loève transform and the solution of a large matrix for eigen functions. It is not a simple computation but needs to be done only once. For a shape that obeys the relationship above (called a "stationary" shape) the result is just the sine and cosine functions used by the Fourier method.

Once the coefficients have been obtained, they represent the shape of the feature outline. It is clear that for most kinds of objects, the details of the shape can vary somewhat. For instance, if we measure the profile of a sand grain, the coefficients of the first 20 or so Fourier terms will describe the shape, but no other sand grain will have an identical shape and so its coefficients will be different. These are called chaotic shapes (Figure 7-36). What is interesting is to ask whether the coefficients for sand grains are different from those for some other kind of particle (say coal) in a way that permits all sand grains to be distinguished from all of the coal grains (or even other sand grains from a different source).

This kind of comparison is made by plotting histograms of the coefficients for a population of one kind of particle. The initial population sample which is measured for this purpose is called a training set. The histograms show the frequency with which amplitude coefficients with particular values are found. There are N such histograms, where N is the number of harmonics used in the summation (usually 20-30, as mentioned above). The same set of histograms can be plotted for the second class of particles, from its training set. These histograms can then be compared using a statistical test such as a chi-squared test, to decide which of them are significantly different (the comparisons are made between the same harmonics, for instance the histogram of the amplitude coefficients for the 8th harmonic of class A is compared to the histogram of the amplitude coefficients for the 8th harmonic of class B).

In very many cases, it turns out that there are only a few harmonics for which the classes have statistically significant differences in distribution. In principle, this means that by acquiring the profile for a new and unknown particle, unrolling the shape to obtain an $\rho(\vartheta)$ function, and then calculating just that single amplitude coefficient, it would be possible to decide (within some probability or confidence limit) whether the particle was of type A or B. For instance, this method has been used to distinguish sand grains from different river systems (where presumably the erosional processes produce different shapes). The significant differences were found only in the 7th, 12th, 18th, and 19th harmonics (the others are not statistically different as assessed by the chi-squared test).

There are many other examples of successful classification using this method. These include the shape of species of mussel shells, identification of sediment particles dredged from bays with the river from which they came, abrasive particles from different commercial sources, discrimination of carcinogenic cells from normal ones, classification of Foraminifera as a function of the water temperature in which they grew, and many more. In the last example, it was even possible to develop a continuous scale of temperature versus the values of the selected harmonic amplitudes. The use of this kind of analysis for classification or discrimination is very general and powerful, although the calculations are not trivial. The method has been most extensively identified with sedimentation studies, to characterize particles and pores.

Shape analysis is also of interest for three-dimensional objects. This will be discussed further in Chapter 12, which deals with serial sections and the interpretation of three-dimensional structure. There are three-dimensional analogues of the various two-dimensional shape parameters, formed by combining various size measurements in dimensionless ratios. These can be calculated and used to characterize and distinguish objects. However, it is also possible to apply harmonic analysis to three-dimensional objects. Given a series of surface points with X, Y, Z coordinates, a Fourier expansion in two angles (ϑ, Φ) produces a series of coefficients that describe the contribution of various harmonics to the object shape.

However, one is naturally tempted to ask what it "means" that the distribution of coefficient values for the 13th harmonic for one class of objects is distinguishable from that for another class. Clearly, while any particular harmonic is somehow connected to the amount of irregularity that exists at that spatial repeat distance along the outline of the objects, it is not as intuitively obvious to the human observer how that is related to shape as some of the parameters discussed before, such as aspect ratio. Classification of objects using these more familiar shape factors is discussed in detail in Chapter 9 on Object Recognition. The principles shown there can also be applied to the coefficients obtained by harmonic analysis.

But this difference between Fourier coefficients and "familiar" shape parameters is perhaps just a matter of degree, since parameters such as formfactor or roundness are not all that "intuitive" either until the observer becomes familiar with them by looking at a lot of different shapes. Some researchers who use these methods reject any notion of assigning a "meaning" to the coefficients. Others make a distinction between parameters that deal with the details of the features versus ones that deal with the gestalt, or entire feature. The former kind of knowledge, to use the terminology of George Polanyi, is physiognostic while the latter is telegnostic. None of this should suggest that one is "better" than the other.

The spectral analysis method and the shape parameter method perhaps represent different means to the same end - tools to obtain quantifiable values that can be used to statistically separate different populations of objects which are somewhat variable within

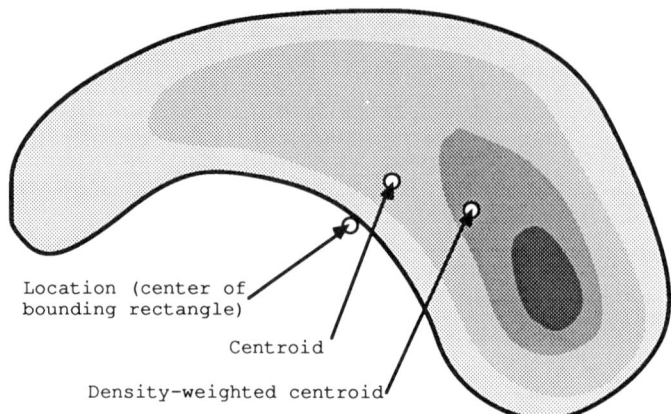

Figure 7-37: Drawing of a feature showing the centroid, density weighted centroid, and location.

each population, but are nonetheless distinguishable in the mass. In one method we give names to the parameters, and can perhaps learn to judge or recognize them with the human visual system, and in the other case they only have numbers and a more abstract quality. But in both cases the use of a sufficient number of them may serve to provide the desired classification. For the particular case of Fourier analysis, it is possible to perform an analysis that selects which coefficients are really important, and whether they give a result that approaches the theoretical limit of distinguishability or whether more terms are needed. For the shape parameters this is more difficult because they are not truly orthogonal and may not even be linearly independent in all cases.

Position

Measurement of feature position may be either within the image itself, or with respect to some global coordinate system (e.g., latitude and longitude or position relative to the corner of a microscope slide), or relative to other objects. The position is normally given in the same units as used for length, but as in the case of latitude or longitude this need not necessarily be the case.

One common way used to give the feature's position is the X and Y coordinate of the centroid of the object. This is the point at which the feature would balance if it were cut from a uniformly thick piece of cardboard. It is therefore the point at which the moment of the object is minimized, and so it represents the point through which the axis of the object (to be discussed below) passes. The centroid location is calculated by integrating the moment of the feature about any X and Y axes and dividing by the area.

$$C.\ G._x = \frac{\Sigma (X_i + X_{i-1})^2 \cdot (Y_i - Y_{i-1})}{Area}$$

The resulting X,Y coordinates give the location. For a convex shape this point will always lie within the boundaries, but for a concave feature or one containing internal voids, it may not. The centroid location can also be weighted by density (Figure 7-37).

The centroid location is a good example of a parameter that requires a pixel-based representation for determination. Attempting to calculate it from the location of boundary

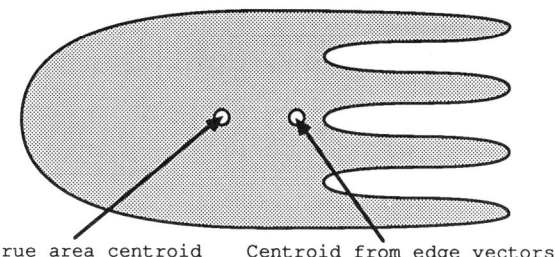
Figure 7-38: Error in centroid location estimated from boundary data.

segments in a boundary-coded representation produces a biased result in which the apparent point moves closer to the more irregular portion of the boundary (Figure 7-38).

A different descriptor for position is the center of the bounding rectangle formed by the X and Y Feret's diameters. As mentioned before, these are the projected or shadow widths of the feature in the horizontal and vertical directions. The coordinates of the center of the bounding rectangle can be very easily obtained, and often serve as an adequate measure of location. Only if the feature is symmetrical in the X and or Y directions will these coordinates be the same as the centroid. For some purposes a brightness- or density-weighted centroid may be used instead, for instance to locate spots in electron diffraction patterns, Hough transform images, or two-dimensional gels used for chemical separation.

The difference between the centroid coordinates and the center of the bounding rectangle gives one measure of orientation. If the feature is curved (like a letter C or U as shown in Figure 7-39) then the centroid will fall to one side of the center of the bounding rectangle in same direction as the feature is curved. The vector direction from one point to the other is straightforwardly calculated, and may be used as a measure of orientation. The absolute distance between the two points may be used as a measure of asymmetry, usually normalized by dividing by some characteristic dimension of the feature such as its length.

There are other candidates to be used to describe the feature orientation. One is the angle of the length chord (the longest chord connecting boundary points in the feature). When this is determined directly from boundary representation, it may have any angle as defined by the points chosen along the periphery. When the length is determined from the ends of chords by rotation of axes, the angle is limited to one of the rotation directions. This will typically be one of 8, 16 or 32 directions, and produces a quantization of orientation angles that is sometimes inadequate to show subtle differences. Usually for purposes of constructing histograms of feature orientation, this orientation angle is quite suitable.

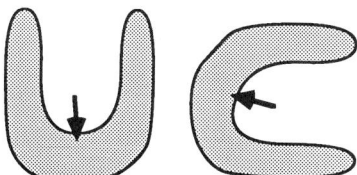
Figure 7-39: Example of orientation vector for U and C shapes.

A more precise measure of orientation angle may be obtained from the axis of minimum moment. This is the axis around which the feature has its minimum rotational moment, and it is also used for the Fourier shape coefficients described previously. It can be calculated from the arc tangent of the moments, which are determined as part of the centroid location calculation. It is most convenient to work directly with the pixel coordinates (x,y), since the centroid coordinates are not known beforehand. Six sums must be formed for the pixels in the feature, but three of these are also used for other purposes (the area and centroid location discussed before):

$S_N = \Sigma i$ (the number of pixels in the feature)
$S_X = \Sigma x_i$
$S_Y = \Sigma y_i$
$S_{XX} = \Sigma x_i^2$
$S_{YY} = \Sigma y_i^2$
$S_{XY} = \Sigma x_i y_i$

With these sums, the net moments and the angle of minimum moment are calculated as follows:

$M_{XX} = S_{XX} - S_X \cdot S_X / S_N$
$M_{YY} = S_{YY} - S_Y \cdot S_Y / S_N$
$M_{XY} = S_{XY} - S_X \cdot S_Y / S_N$
$R = ((M_{XX} - M_{YY})^2 + 4 \cdot M_{XY}^2)^{1/2}$
$\Theta = $ arc tan $((M_{XX} - M_{YY}) + R) / (2 \cdot M_{XY}))$

When orientation angles are measured, they may either be expressed in terms of an engineering angle (measured counterclockwise from the X axis, which extends to the right), or a compass angle (which starts at north, straight up, and increases in the clockwise direction). In most cases angles are expressed in degrees, but of course these values may be freely converted to and from radians as required in any subsequent calculation or analysis.

Neighbor relationships

The various position measures are most often used to show histograms that describe the uniformity, or conversely the presence of some gradient or nonuniformity in the field of view. Some of these interpretations will be discussed in the next chapter. Another use of the data is to obtain relative positional information for objects in the same image or in several images of the same field of view. This can lead to descriptions of adjacency or association (Figure 7-40).

Within a single image, or several adjacent ones, it is possible to determine the distance (and if necessary the direction) of the nearest neighbor. In rare cases, the second, third, or farther neighbors are also considered. The nearest neighbor to each feature can be found in two different ways. One is by sequential dilation while counting the number of cycles until the object touches a neighbor. This is generally awkward because it cannot be done for all features at once, is not perfectly isotropic (dilation produces squares, hexagons or octagons rather than circles) and is very sensitive to the original shape of the object.

The alternate method is an exhaustive search of the positions of all the features. Of course, this can only be done after all of the feature locations have been determined and stored in data files. There are various tricks to speed up searching, such as sorting the

Image Measurements 213

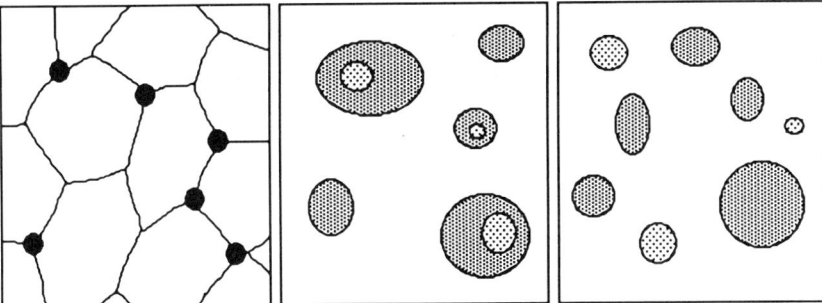

Figure 7-40: Examples of association between different features.

data in the X or Y dimension, or both, and not taking the square root of the Pythagorean sum of $(x_1 - x_0)^2 + (y_1 - y_0)^2$, but these effect only the program that implements the search. Usually, searching through more than a few thousand features begins to take a significant amount of time. We will see in the next chapter that the nearest neighbor distances can be used to determine whether the feature spacing is random, clustered or ordered.

One approach to measuring feature spacing and alignment employs the linear Hough transform. When high resolution TEM images show atomic alignment, the individual spots are fuzzy and it is difficult to measure their spacing with accuracy. Figure 7-41 shows an example, the silicon lattice near a boundary with SiO_2. The Hough transform reduces each line of atoms to a single point, and in the enlarged portion of the image it is easy to locate the maximum brightness within each point and determine the mean and standard deviation of atom spacing. This technique has been used to measure small changes in lattice parameter near dislocations and interfaces, as the result of specimen strain, or due to the presence of other elements in the matrix.

Neighbor relationships between features in several images of the same field allow a variety of relationships to exist. They are usually expressed in a data structure called a "tree", in which objects inside others are connected in such a way that they can all be related. The various positional relationships between objects are based on relative position, that is whether objects touch, or one is "inside" another. However, the concept of "inside" must be made somewhat more specific, since there are a number of possible relationships (Figure 7-42).

The first two possible relationships between two features, A and B are shown in cases 1 and 2. At this point we have introduced two of a total of four useful and different spatial relationships: *OUTSIDE* (disjoint, sharing no pixels) and *TOUCHING* (boundary pixels have each other as neighbors, using either an 8- or 4-connected definition of touching). Notice that both of these concepts are a little bit easier to deal with in a pixel-based representation than a boundary definition. It is sometimes hard to tell if boundary lines "touch".

Also notice that *OUTSIDE* means that there are no common pixels but by itself it does not distinguish case 1 from 3, which is *A is outside B but within the boundary of B*. This is how we can deal with the "hole in the doughnut" problem in which one object lies inside a void within another. It introduces the separate notion of the area of B (its pixels) and the outer boundary of B (which is also involved in the "touching" definition). It is this example and ones like it that reveal the inadequacy of common words such as inside or within to define these relationships explicitly.

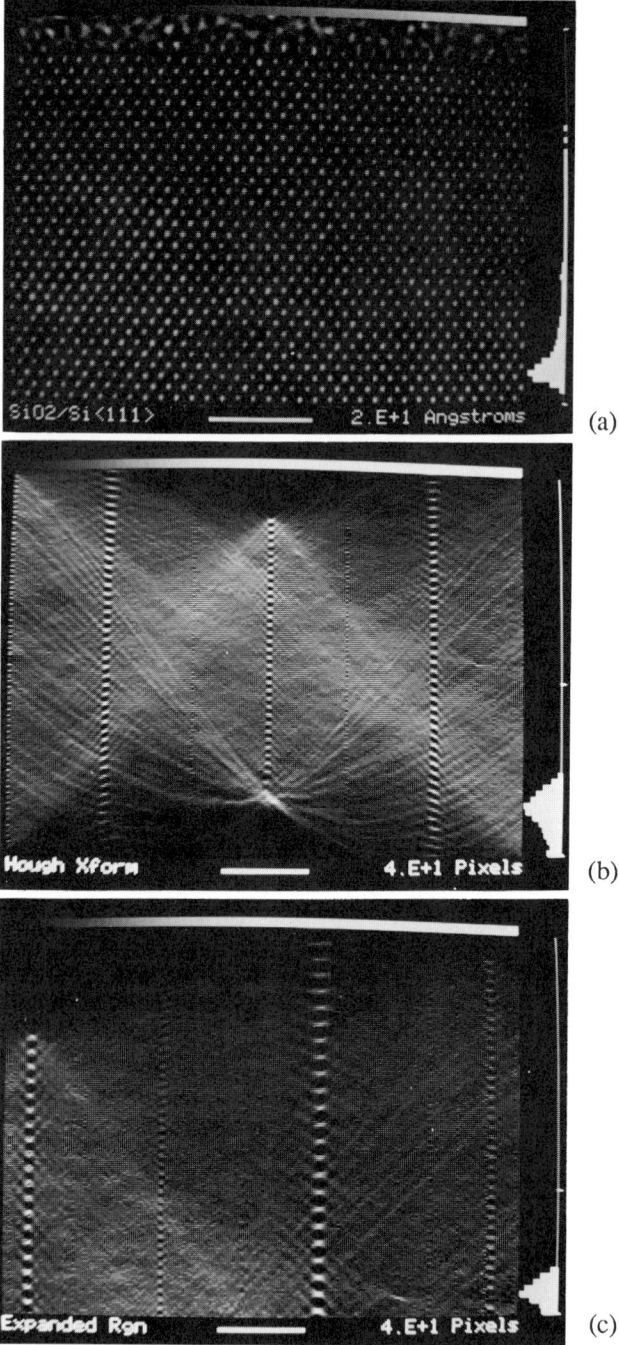

Figure 7-41: High-resolution transmission electron microscope image of Si atomic lattice near an SiO$_2$ interface (a), with the Hough transform (b) and an enlarged area (c) showing the spacing of points in Hough space corresponding to parallel planes of atoms in the lattice.

Image Measurements   215

We can now generalize the relationship operators by adding *INSIDE* (A is *INSIDE* B if all of the pixels of A are also pixels of B, the usual definition from set theory) and *OVERLAPPING* (A overlaps B if some but not all of the pixels of A are also pixels of B). Also, we should add one more feature parameter besides the area (all pixels) and outer boundary (pixels on the outer periphery). That is the inner boundary (pixels on the

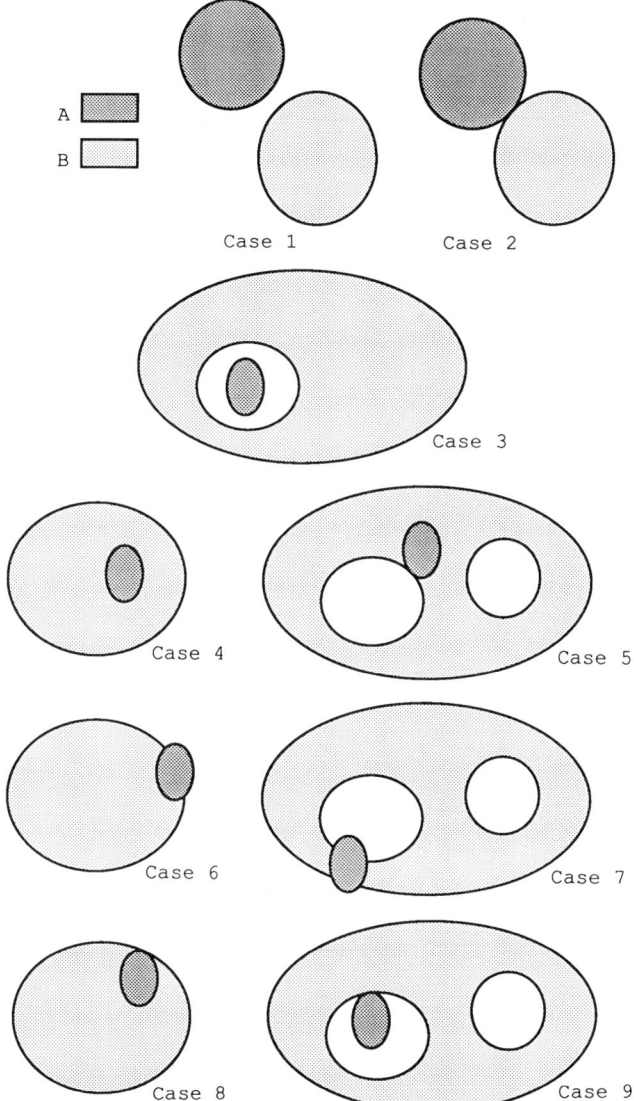

Figure 7-42: Possible feature relationships: Case 1) A is outside and not touching B, and vice versa; Case 2) A is outside B and touches B's outer periphery, and vice versa; Case 3) A is inside B; Case 4) A is inside B (and B overlaps A); Case 5) A is inside B and touching its inner boundary; Case 6) A overlaps B and its outer boundary; Case 7) A overlaps B and its inner and outer boundaries; Case 8) A is inside B and touches its outer boundaries; Case 9) A is outside B and touches its inner boundary.

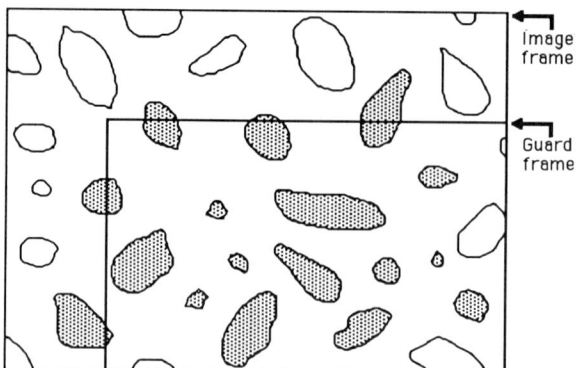

Figure 7-43: Application of a guard frame to an image. Only shaded features are measured and counted.

periphery but not the outer periphery). Then we can combine these in any order, as shown in the figure.

Of course, there are many more combinations, but this terminology provides for all the possibilities. It permits asking questions such as "for all particles outside phase B that touch phase B..." or "for all phase A regions inside phase B and not touching the outer boundary of B...", etc. In some cases, it is important to be able to select for measurement or counting features in one or another of these classes, or to measure the extent of touching. Methods for doing this using erosion/dilation logic were discussed in Chapter 6.

Edge effects

The problems of measuring features that touch the edges of the image frame have been mentioned only briefly to this point. This is closely tied to concepts of feature position. A feature that touches an edge generally would not if it were located elsewhere in the image frame. If a feature is too large to fit into the field of view, then the image magnification must be reduced to permit any useful measurements to be made. But large features that do fit in the field of view are statistically more likely to touch an edge than small ones, so that counting all features regardless of size or position would bias the results to give too few large features.

There are two approaches normally used to deal with edge-touching features. One is to define a reduced portion of the image in which counting and measurements will be made (Figure 7-43). Features which lie wholly within the internal "guard" frame, or which lie partially within this frame and do not intersect any portion of the outer image frame, are measured and counted. The reference area is that of the guard frame. In principle, the area of the guard frame should be one-quarter of the total image frame, which reduces the area and resolution of the images significantly.

A second method is to use all features in the image area that do not touch any edge, and which therefore can be measured, but to count each feature more than one to compensate for the probability that features of this size might touch an edge. This probability is simply

$$\text{Count} = \frac{W_x \cdot W_y}{(W_x - F_x) \cdot (W_y - F_y)}$$

Image Measurements

In this equation, the W_x and W_y values are the width of the image area in the horizontal and vertical directions, and the F_x and F_y values are the projected or Feret's diameters of the feature in the horizontal and vertical directions (Figure 7-44). In effect, this places a guard frame around each feature. Small features have an effective count that is close to 1.0, while larger features have correspondingly larger values.

This same correction method should be used for measurement of intercept lengths, discussed in Chapter 8, since any intercept or chord through a feature cannot be measured if it intersects the edge of the image. Consequently, the frequency distribution of intercept lengths $N(L)$ should be adjusted by multiplying the number in each size class by $L/(L_0 - L)$, where L is the length and L_0 is the dimension of the image in that direction. Since most automatic systems measure intercept lengths in either the vertical or horizontal direction, this is simply the dimension of the image. As for the adjusted count based on feature size, the correction increases the effective count for large chords (more of which will be rejected because of edge intersection) as compared to small ones. This can significantly change the mean intercept length in an image, and other parameters calculated from it.

Brightness

All of the measurement parameters discussed so far have been primarily associated with binary images (pixel-based representation) or boundary representation of features, for instance from manual entry of outlines with a graphics device. There are also important data in the original grey image, which can be measured for each individual feature in the image. This is usually done with a pixel-based feature representation because it is then easy to use the ON pixels from some discrimination and thresholding step to locate the corresponding grey pixel and read its brightness value. In doing so, it is often wise to perform an erosion of the feature masks so as to avoid the outermost row(s) of pixels on the features, where the brightness values may change due to pixels partially overlapping onto the features' surroundings.

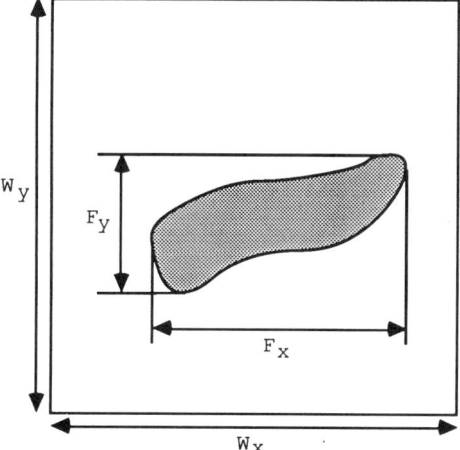

Figure 7-44: Feret's diameters and image dimensions used to calculate edge correction.

The brightness values may be used to form an entire histogram for each object, but it is hard to make use of so much data. In most cases, only the mean brightness value and perhaps the limits or standard deviation for each feature are really useful. The brightness can be converted to other units (density, dosage, etc.) using an appropriate calibration curve, and if the curve is nonlinear the density or other parameter should be converted for each pixel value and integrated over the feature rather than being calculated from the mean, as explained earlier.

Depending on the camera or other sensor used, the relationship between measured brightness and optical density may vary. For a linear detector such as a photomultiplier tube, it is simply

$$\text{Optical Density} = -\log_{10}\left(\frac{\text{Transmitted Intensity}}{\text{Incident Intensity}}\right)$$

Some detectors such as film have a logarithmic output that produces a direct relationship between brightness and density, while others, such as vidicon cameras, have an output that is not perfectly logarithmic (described as a gamma value not equal to 1.0). In some cases an input look-up table is used to convert measured brightness to optical density in these cases, before the image is stored.

Of all the various descriptions of texture described before, only a few are usually needed to characterize the feature. One is the contrast, defined as the mean brightness difference between each pixel and its immediate neighbors. The second is the texture, defined as the slope of mean brightness difference from each pixel of neighbor pixel at distances of 2, 3,... pixels away. For images of rough surfaces, this slope correlates well with the perceived surface roughness.

A particular type of rough surface is a fractal surface. As for the fractal outlines, fractal surfaces are rough at all magnifications. It happens that sections through fractal surfaces produce outlines that are fractal, with a dimension exactly one less than the dimension of the surface. This offers one way to measure surface roughness, but one that requires extensive specimen preparation. The texture (as defined above) also correlates with fractal dimension, with either diffuse light illumination or when viewing the specimen surface with secondary electrons in the scanning electron microscope. This correlation has been used to measure the roughness of particulates as well as various materials and natural surfaces. It will be discussed in more detail in Chapter 10.

Several other measures of "texture" were described in an earlier chapter. These use various combinations of neighbor brightness, and can be extended to deal with color images as well. For color images, either differences in hue, or the brightness difference of any color plane, or a combination of differences in several colors may be used. These measures are all applicable to individual features, and are most easily applied by using a pixel-based binary image representation of the features as a mask to select the pixels in the original or processed grey scale image for measurement.

References

C. J. Agin, T. O. Binford (1976) *Computer Description of Curved Objects* IEEE Trans Comput. C–25#4, 439-449

W. A. Aherne, M. S. Dunhill (1982) Morphometry E. Arnold, London

W. Andrefsky, Jr. (1986) *A Consideration of Blade and Flake Curvature* Lithic Technology 15, 48-54

H. G. Barth, S.-T. Sun (1985) *Particle Size Analysis* Anal. Chem. 57 151R

J. K. Beddow, G. C. Philip, A. F. Vetter (1977) *On Relating Some Particle Profile Characteristics to the Profile Fourier Coefficients* Powder Technol. 18 15-19

C. Blakemore, R. Owen (1974) *Curvature Detectors in Human Vision* Perception 3, 3-7

E. Bribesca, A. Guzman (1980) *How to describe pure form and how to measure differences in shapes using shape number* Pattern Recog. 12 101-112

P. G. Chapin, L. M. Norton (1968) *A Procedure for Morphological Analysis* Information System Language Studies no. 18, The Mitre Corp., Bedford MA

L-M. Cruz-Orive, E. R. Weibel (1981) *Sampling Designs for Stereology* J. Microscopy 122 235-257

R. O. Duda, P. E. Hart (1972) *Use of the Hough Transformation to Detect Lines and Curves in Pictures* Comm. Assoc. Comput. Mach. 15, 11-15

R. Ehrlich, B. Weinberg (1970) *An exact method for characterization of grain shape* J. Sediment. Petrol. 40 205-212

H. E. Exner (1986) *Shape: a key problem in quantifying microstructure*, Proc. 1st Int'l Conf on Microstructology, Gainesville FL

H. E. Exner, H. P. Hougardy (1988) Quantitative Image Analysis of Microstructures, DGM Informationsgesellschaft mbH., Oberursel, Germany

A. Flook (1982) *Fourier Analysis of Particle Shape* in Particle Size Analysis 1981-2 (N. G. Stanley-Wood, T. Allen, ed.) Wiley Heyden, London

H. Freeman (1978) *Shape Description via the use of Critical Points* Pattern Recognition 10#3, 159-166

H. Freeman, L. Davis (1977) *A Corner-Finding Algorithm for Chain-Coded Curves* IEEE Trans Comput. C–26#3, 297-303

T. M. Hare, J. C. Russ, J. C. Russ (1982) *Image measuring algorithms for a small computer*, Microbeam Analysis 82, San Francisco Press

J. E. Hilliard (1968) *The calculation of the mean caliper diameter of a body for use in the analysis of the number of particles per unit volume* in Stereology (H. Elias, ed.) Springer Verlag, New York NY, 211

P. V. C. Hough (1962) *Method and means for recognizing complex patterns* U.S. Patent 3,069,654

B. H. Kaye, J. E. LeBlanc, G. Clark (1983) *A study of the physical significance of three-dimensional signature waveforms* Proc. Fineparticle Characterization Conference

T. Pavlidis (1978) *A Review of Algorithms for Shape Analysis* Computer Graphics and Image Processing 7#2, 243-258

T. Pavlidis (1980) *Algorithms for Shape Analysis of Contours and Waveforms* IEEE Trans. Patt. Recog. Mach. Intell. PAMI-1#2 301-312

L. A. Riggs (1973) *Curvature as a Feature of Pattern Vision* Science 181#4104, 1070-1072

I. Rovner (1987) private communication

J. C. Russ (1984) *Implementing a new skeletonizing algorithm*, Journal of Microscopy, 136, RP7-8

J. C. Russ (1986) Practical Stereology, Plenum Press, NY

J. C. Russ (1989) *Automatic methods for the measurement of curvature of lines, features and feature alignment in images* J. Computer Assisted Microscopy 1 (1989) p. 39-78

J. C. Russ, T. M. Hare (1986) *Size models for real particles viewed in projected images*, Microbeam Analysis 86, San Francisco Press, 513

J. C. Russ, W. D. Stewart, J. C. Russ (1985) *Densitometric image measurements*, Amer. Lab, Apr 85, 41

H. P. Schwartz, K. C. Shane (1969) *Measurement of Particle Shape by Fourier Analysis* Sedimentology 13 213-231

S. D. Shapiro (1978) *Feature Space Transforms for Curve Detection* Pattern Recognition 10#3, 129-143
J. Sklansky (1978) *On the Hough Technique for Curve Detection* IEEE Trans. Comput. C–27#10, 923-926
I. Sobel (1970) *Camera models and machine perception* AIM-21 Stanford Artificial Intelligence Lab.
H. P. Ting-Beall, F. M. Burgess, J. D. Robertson (1986) *Particles and pits matched in native membranes* J. Microscopy 142 311-316
E. E. Underwood (1970) Quantitative Stereology, Addison Wesley, Reading MA
E. R. Weibel (1979) Stereological Methods vol. I & II, Academic Press, London
Z. You, A. K. Jain (1984) *Performance Evaluation of Shape Matching via Chord Length Distribution* Computer Vision, Graphics and Image Processing 28 185-198
R. A. Zeineh, G. Kyriakidis (1987) *Computer-aided soft laser scanning densitometry*, American Laboratory News Jan 87, 16

Chapter 8

Stereological Interpretation of Measurement Data

Stereology is a coined word, in use now for nearly three decades to describe the study of relationships between three-dimensional structure and the measurement parameters that can be obtained from conventional two-dimensional images. Many of the fundamental relationships and principles of stereology come from the mathematical field of geometrical probability which has deep roots (Buffon, 1777), but it is not essential to be comfortable with the derivations of the tools to be able to utilize them. (For comprehensive references to stereological interpretation, see DeHoff & Rhines, 1968; Underwood, 1970; Weibel, 1979; Russ, 1986).

There are two fundamental classes of measurements which we need to be concerned with, and two different categories of images to which they may be applied. Measurements are described as either global or feature-specific. The former class of values apply to the entire structure, but do not attempt to distinguish the details of how the components are organized (Rhines & DeHoff, 1986). The simplest example is the volume fraction of a phase in a complex material. This value, which can vary from 0 to 100%, is an integral result that tells us nothing about whether the phase is finely or coarsely distributed, what the shape of the particles may be, and so forth. Feature-specific values, on the other hand, attempt to describe (at least in a statistical sense) the frequency distributions of particle size and shape for the phase. Very different levels of effort are required to obtain feature-specific results, as compared to global data.

Figure 8-1 illustrates several different types of images for which stereological interpretation can be made. Stereology has classically been applied to planar sections cut through materials, to reveal a statistically random surface on which measurements can be made. This corresponds to the kind of sample preparation used for materials to be examined in the metallographic microscope, in which such planes are cut and polished. It is also encountered in many situations in the life sciences in which microtomed slices are prepared for examination in the transmission light microscope, or the transmission electron microscope. The hallmark of these images is that sections through objects in the sample do not show the full extent of the object. Instead, the features on the plane of examination are smaller.

In addition to these plane section images, there are many situations in which we view a shadow or projected image of features. Transmission images through relatively thick transparent sections, or views or particles or other objects resting on a substrate, exemplify this type of image. In these cases the projected or shadow image of each object produces a feature in the image that does show the maximum extent of the object, at least in one direction. On the other hand, surface indentations and irregularities may be hidden in this projection.

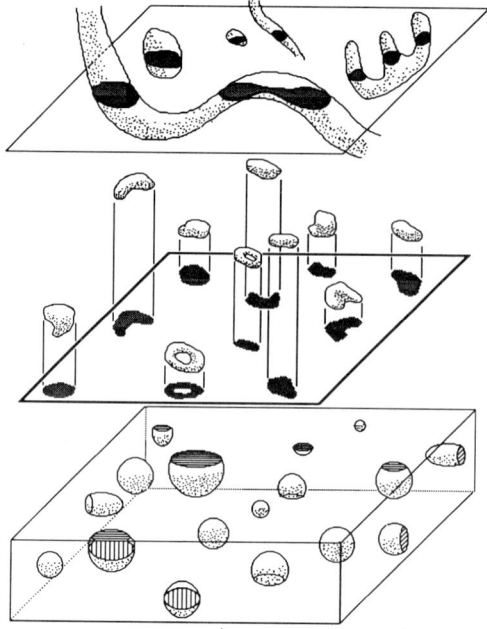

Figure 8-1: Examples of plane-section image, projection image, and finite section.

Global measurements

Measurements intended to characterize the global parameters of a sample are usually performed on planar section images. Historically, many of these measurements have been made manually or with minimum machine aids. The techniques are highly efficient, usually requiring that the operator perform a counting experiment rather than a series of measurements. The advantage of counting over measurements is not only ease and efficiency; it also makes the expected errors in the results very easy to predict and control.

The discussion that follows will use the word "phase" to denote the features or structure of interest. This word is common in the materials field, and recognizable to other users of stereological methods such as geologists. The life scientist has a more complex lexicon, which we will defer for the present. Briefly, a phase is any identifiable region or set of regions within the sample. There is no need to restrict it to a single connected structure, or isolated convex regions, or any other particular configuration.

Volume fraction is probably the most obvious of the global measurements. It has been known for more than a century that the area fraction of the phase on the plane of the image is equal to the volume fraction of the phase in the solid. This provides a way to measure the volume fraction, for instance by cutting out the images of the phase regions and weighing them, and ratioing the value to the weight of the entire picture. Estimating the area fraction of an image occupied by a phase is very prone to errors, as shown in Figure 8-2.

A somewhat easier way to get the volume fraction, also known for many years, is to draw random lines on the surface and measure the fraction of the total line length that lies within the phase (Figure 8-3). An instrument to assist with this measurement (the

Stereological Interpretation   223

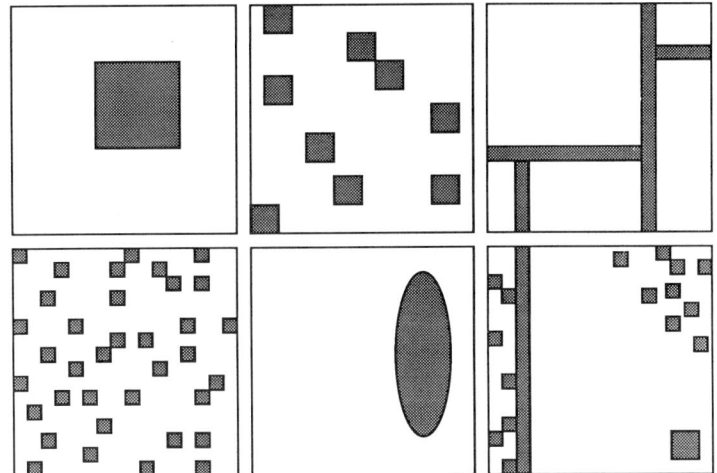

Figure 8-2: Example of area fraction of phase. The same area is organized in different ways, but covers the same 14% of the test region in each image.

Hurlbut counter) appeared many years ago, consisting of a mechanical drive for the microscope stage with a series of turns counters. The operator viewed the sample surface as it passed through the field of view, and pressed keys to engage counters corresponding to the various phases of interest. The ratio of each phase's counter value to the counter recording the total number of turns of the drive screw gave the lineal intercept fraction, and hence the volume fraction for each phase. Modifications of this setup in which the operator's eye was replaced by a photocell and voltage discriminators, to select phases based on their relative brightness, pointed the way towards the automatic systems of the present day.

An even more efficient technique for measuring the same parameter is also available. This consists of placing a grid of points on the image, and counting the number of the

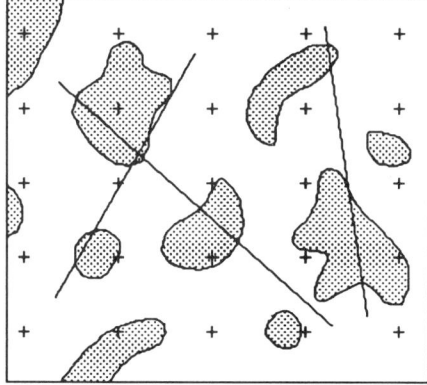

Figure 8-3: Schematic representation of a microstructure shown on a section image through a solid. Shaded areas represent one type of object of phase, surrounded by the white matrix. Superimposed points and lines are used for the measurement and counting operations described in the text.

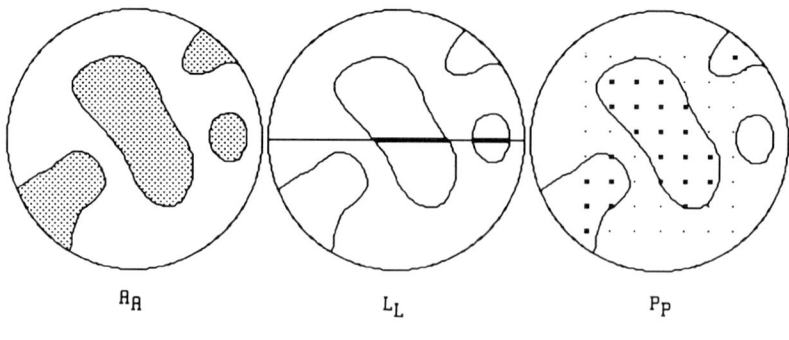

$$V_V = A_A = L_L = P_P$$

Figure 8-4: Equality of measures of volume fraction.

points that fall on the phase of interest. The fraction of the total number of points that lie in the phase is also equal to the volume fraction of the phase. Since counting is always an easier operation, and usually one with fewer human errors than measurement, this has become the preferred method (Saltykov, 1958).

Notice that it is important that the points be randomized with respect to the structure. If the structure is randomized, then the points can be arranged in a regular grid with no loss of generality. The complete description of this condition is that the structure be isotropic, uniform and random. Isotropy means that the structure is indistinguishable in any direction. Uniform means that the results are the same in any location. And random means that successive measurements are independent. Generally this means that sequential measurements (in this case the points) must be far enough apart to avoid redundancy, for instance by not being so close together that they lie within the same feature. If the structure does not meet these criteria (is not IUR) then the test probes must be made isotropic, uniform and random by an appropriate sampling strategy.

The statement of the equality of results from each of these methods, and their relationship to the volume fraction of the phase, is customarily written as shown in Figure 8-4.

This also serves to introduce the notation used for these stereological parameters. V, A, L and P are volume, area, lines and points respectively. The subscript denotes the reference quantity, and so for instance L_L would be read as line length per unit of reference line. The quantities present in the equality above are all dimensionless. There are other stereological quantities of importance that have units of length raised to some power, and so will depend upon knowing the magnification of the image. They also involve some few additional notational symbols.

Global parameters

First of these is the surface area per unit volume of a boundary, such as the interface between the phase of interest and the matrix in which it resides. This is written as S_V and has units of area/volume or length^{-1}. The subscript V indicates that volume is the denominator, and S is used to represent a surface area that is not necessarily planar.

Many properties of structures, both in materials and life sciences, depend upon boundaries. Examples include chemical transport across membranes and high temperature creep. The total amount of the boundary is therefore of considerable interest.

Stereological Interpretation

On a planar section image, the boundaries appear as lines separating the phase of interest from its surroundings. If the total length of these lines are measured, and divided by the total image area, the units (1/length) are correct. But it is necessary to account for the fact that the boundaries will not in general intersect the image plane at right angles. Statistically, this angle of intersection can range over all possible angles. Using geometrical probability to perform the integration results in the relationship $S_V = (4/\pi)\, B_A$, where B is the length of the boundary lines and A is the image area. The factor $4/\pi$ corrects for the non-normal incidence of the boundaries with the image plane.

In the preceding chapter, contiguity was determined by using dilation and image combination to measure the fraction of the total phase boundary that lies between particular phases of interest. The measurement determines B_A, but since this is directly related to S_V, the fraction of the boundary length corresponding to the selected phase boundary is also the fraction of the total surface area per unit volume. The contiguity is expressed as a fraction or percentage (Figure 8-5).

Because the determination of B_A involves measurement, it has been replaced at least for manual measurements by another technique that only requires counting. If a series of test line of total length L are drawn at random on the image, and the number of points at which the line crosses the trace of the boundary of interest are counted, the resulting value P_L (number of points per unit line length) also has units of (1/length). The integration over all possible orientations of the line with respect to the boundary surfaces again introduces a constant, giving as a result

$$S_V = 2\, P_L = (4/\pi)\, B_A$$

Because the S_V parameter is determined by counting the number of intersections of grain boundaries with test lines, it is closely related to a parameter very commonly used for materials (and particularly metals) characterization. The ASTM grain size is calculated as

$$G = (-6.6457 \log_{10} (1/P_L)) - 3.298$$

where the line length is measured in millimeters. Another definition of grain size is based on the number of grains per unit area visible on a polished surface N_A, and the image magnification M.

$$G = (3.22 \log_{10} (N_A \cdot M^2)) - 2.95$$

An alternate way to get the ASTM grain size of a microstructure from the number of nodes in the grain boundary tesselation, rather than the mean intercept length, was shown

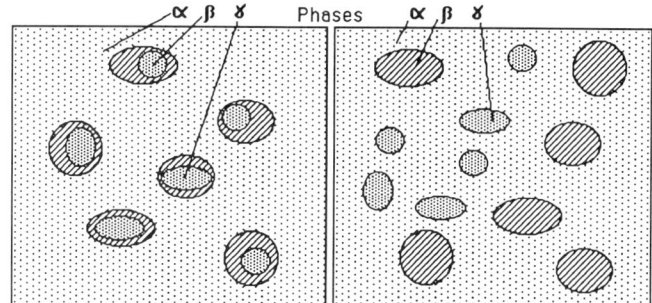

Figure 8-5: Microstructures showing two extreme cases of contiguity between the various phases.

in Chapter 6. This has the advantage of being less dependent on the perfection of delineation of grain boundaries in sample preparation and imaging (a common problem with some metals). It is an equivalent definition, based on Euler's relationship for polygons.

It is unfortunate that the name "grain size" is attached to measurements which determine either the number of grains per unit area (N_A) or the number of intersections of grain boundaries with lines (proportional to S_V). Neither of these is really the size of the grains, although they are stereologically defined parameters that may relate to structural properties (for instance, one definition of the ASTM grain size number is really a logarithm of S_V, or a measure of the amount of grain boundary surface present in a structure, and it is therefore not surprising that creep rate and other mechanical properties are related to it).

It is possible to add even more distortion to results by mis-interpreting grain size. For example, measuring the median equivalent circular diameter of the intersection area of grains cut by a plane of polish has been used as a "grain size" dimension in a recent study (Venkataraman & DiMilia, 1989) of grain growth in sintering (normally associated with an Arrhenius-type dependence on temperature). There is no stereological basis whatever for such a measure, which is not the "grain size" in any sense, and in fact is not any unbiased stereological measure of structure. A more physically meaningful method for measuring actual grain size, without bias, is discussed below.

Since both volume fraction V_V and surface area per unit volume S_V can be determined straightforwardly by counting experiments, it is often practical to do this using microscope reticles or transparent overlays on photographs, containing arrays of lines and points. One example of this uses a checkerboard grid of lines printed on a transparent plastic overlay and placed on the micrograph. The points where the lines cross are used to count P_P. Since the total number of points is known, the number of points that lie in the phase or structure of interest can be used directly to estimate the volume fraction.

Likewise, the number of intersections of the grid lines with the phase boundaries can be counted. The total line length on the grid is known, but must be adjusted to the scale of the image by dividing by the image magnification. Then the resulting value of P_L gives the desired S_V.

There is one very important restriction that must be kept in mind in making these measurements. The derivations of these equalities made the explicit assumption that the lines and planes used to sample the microstructure are randomized. But a regular grid of lines and points does not seem very random. How can this work?

The key to understanding this is that the lines and planes must be randomized with respect to the microstructure. If the structure itself is a random one, then any set of lines and points can be used. If the structure has a preferred direction or gradient, then randomized lines and points must be used. In fact, the degree of anisotropy can be determined using the same grid. If the P_L counts are determined separately from the horizontal and vertical lines, the values can be compared to see whether they are significantly different (in a statistical sense), and if so then there is preferred orientation present in the structure.

If it is known beforehand that the structure has a natural orientation, such as the arrangement of cells near a membrane or rocks in strata, it is possible to obtain unbiased results using a "vertical sectioning" technique (Baddeley et al., 1986). In this case, rather than attempting to cut entirely random surface orientations, a vertical line is positioned normal to the planar orientation, and a vertical section plane passed through that line at a random rotational orientation. In this plane, the lines used for counting must be arranged not at random, but with sine-weighting. In other words, since the plane is already at right angles to the structure, vertical lines must be fewer in number than horizontal ones. Either straight lines with this weighting probability may be drawn, or a series of cycloids (which are themselves sine-weighted) can be drawn (or used as an overlay or reticle to superimpose on the image) and intersections counted, to obtain spatially "randomized" results.

In order for the normal statistics of counting experiments to properly predict the accuracy of the results from these experiments, it is necessary that the points and lines properly sample the structure (Cruz-Orive & Weibel, 1981). In brief, this requires that they intersect different regions. The spacing of the lines and points with respect to the image is controlled by their spacing on the overlay or reticle, and by the image magnification. Ideally, there should be very few occurrences when more than one line or point fall within a single phase region. It is important in making these global stereological measurements to remember that it is much better to measure many different fields of view, making only a few counts or measurements on each, than to concentrate on one or a few fields and obtain a large number of counts. The latter method may not truly represent the structure, and the high number of total counts will be deceptive.

When proper counting precautions are observed, the expected precision of the results is proportional to the square root of the number of counts. Expressed as a relative error, one standard deviation is just $N^{1/2}/N$. This makes it possible to decide how many points should be counted in the phase of interest to obtain a result with the required precision, and also allows comparing results with appropriate confidence limits.

The precision of a V_v measurement with an automatic system is not same as for manual point counting. Although a very large number of points (the number of pixels in the image) are counted, they are closely spaced and oversample the features. The standard deviation may be more properly estimated (Exner & Hougardy, 1988) as $A_A/N_f^{1/2}$, where A_A is the fraction of pixels in the phase and N_f is the number of separate features included in the measurement. In most cases this probably overestimates the variance. The error in measurement of each phase region is itself proportional to the length of perimeter line, and hence to feature shape.

Other important global parameters that can be determined on a section image include the mean curvature of the boundary surface, mean free distance between features, and mean intercept length. These are less familiar than V_V or S_V, and will be discussed in turn.

The boundary surface described for S_V measurement may also be curved (Cahn 1967; DeHoff, 1967, 1978, 1981; Rink, 1976). Regardless of whether the boundary is organized as a single continuous sheet winding through the volume, or as a series of discrete bounds around separate objects, or some combination of the two, its mean curvature can be determined. This curvature is often related to important processes occurring within the structure, including pressure or chemical concentration differences across membranes. For some particular types of nonconvex objects, which are otherwise difficult to characterize, the mean curvature is proportional to size (e.g., the length of tubules or fibers, and the perimeter of sheets or "muralia").

The total mean surface curvature per unit volume of structure has units of (length^{-2}) and is determined by counting the number of locations along the boundary lines that

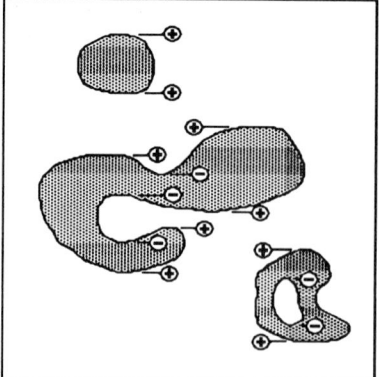

Figure 8-6: Representative microstructure for determining the integrated mean curvature from the tangent count. Points marked + and - are positive and negative tangents with a horizontal test line.

appear on the image plane that have any selected orientation. In practice, a line is swept across the image and the points where it is tangent to the boundary are counted (of course, we are making the usual assumption that the boundaries are randomly oriented). External and internal tangent points must be distinguished in this operation, as shown in the figure. The net tangent count is then used to determine the integral mean curvature per unit volume as shown in Figure 8-6.

$$M_V = \pi \cdot (T^+ - T^-)/A$$

The remaining two global parameters to be considered here cannot be determined simply by counting. They require measuring the intercept length, the length along the test line that passes through the phase of interest or through the matrix between the phase regions. The mean intercept length of random test lines (or a regular array of test lines if the microstructure is random) through features on the test plane is written as L_3. The subscript is a reminder that we are actually dealing with intersections through a three-dimensional object. Notice that the mean value does not necessarily require measuring each individual intercept, but just keeping track of the fraction of total line length in the phase regions L_L and the number of intersections with the boundary N_L. Then $L_3 = L_L/N_L$.

Notice that there will be only a single intercept for each convex object, but other more irregularly shaped objects may have any number of intersections with the test line. Regardless of the shape of the phase, and whether it is a single object or many, the value of L_3 is directly related to parameters we have seen before:

$$L_3 = 4 V_V/S_V$$

and furthermore, the mean surface to volume ratio for a phase consisting of particles, cells or grains is

$$(S/V)_{mean} = 4/L_{3\,mean}$$

If the intercepts that are measured are those of the test line with the matrix surrounding the phase of interest, we can determine the mean free path between the dispersed phase regions. This is important for correlation with mechanical properties of materials, for instance, in which particles serve as strengthening agents. Since the mean intercept length L_3 shown above is determined by measuring the fraction of the total line

Stereological Interpretation

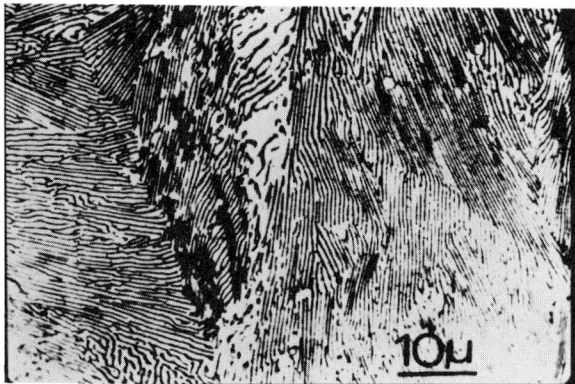

Figure 8-7: Pearlite (eutectoid or lamellar) structure in steels, viewed in the light microscope.

length in the phase, it follows that the remaining length of the test lines gives the mean free path λ, or

$$\lambda = L_{3mean} \cdot (1 - V_{V\alpha})/V_{V\alpha} = (1 - V_V)/N_L$$

where the subscript α indicates the phase of interest. Notice that the mean free distance between objects in the matrix is not the same as the mean nearest neighbor distance.

Intercept distances can also be used for other types of structures. For instance, in a lamellar structure such as the eutectic shown in Figure 8-7, a histogram of intercept lengths oriented at random to the structure, plotted with the horizontal axis organized as 1/intercept length, produces a simple linear plot from which the spacing perpendicular to the platelets can be determined. Geometrical reasoning can also be employed to realize that the mean thickness or spacing of lamella can be estimated as $2 \cdot V_V/S_V$ (Weibel & Knight, 1964) or as one half the mean spacing measured in random directions.

Measurement of contiguity was introduced in Chapter 6. Briefly, the contiguity C for particles or grains of one phase in a structure is defined as the ratio of the length of boundary between particles of that phase to the total length of boundary for the phase. It can be measured based on the length of the boundaries in an automatic system, or by counting intercepts with random lines in a manual experiment. There is a relationship between several of these parameters, namely

$$L_{mean} = L_3 \cdot (1 - C) \cdot V_V/(1 - V_V)$$

where L_3, C and V_V are as defined above, and L_{mean} is the mean size of a phase region, calculated as $4 \cdot V_V/S_V$.

Mean free path

Imagine standing in a forest (not a dense jungle or tangled scrub, but a forest of uniform and well-spaced trees). How far can you see in any direction? This is not an idle question, but one with a rich history and continuing importance. It first arose long before the word stereology (or indeed most of its concepts) arose, and was used to illustrate a question of some astronomical importance. That was the problem of why the sky is dark at night. Given an unbounded universe of stars, eventually in any direction there should be a star at some distance, however great. In that case, the cumulative starshine from all

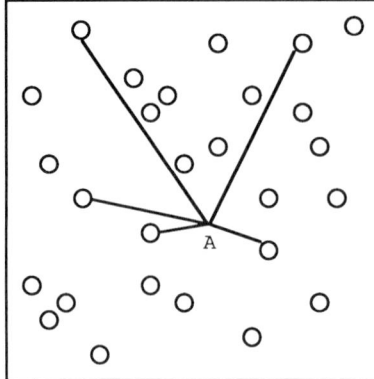

Figure 8-8: Lines of sight from an observer at point A to trees or stars.

directions should sum so that the night sky is as light as the sun (and, of course, the temperature of the earth would rise to that of the surface of the sun).

We will return to this problem shortly. Let us now go back into the forest (Figure 8-8). In some directions there are trees close at hand, but in others the distance to a tree is quite large. Still, eventually they all overlap and do not permit us to see the world outside the forest (or, indeed, to determine if there is one). The average distance is known as the background limit. It is just the ratio of the average area of the forest occupied by a tree divided by its average width or diameter. The average number of trees we can see (in whole or part) is the area of a circle with radius equal to the background limit, divided by the average area of the forest occupied by a tree.

Now, the average area of the forest per tree is just the inverse of the number of trees per unit area, or N_A in our standard notation. Further, the mean diameter of a tree can be calculated from the total area fraction of the trees and their number, as

$$D_{mean} = \sqrt{\frac{4}{\pi} \cdot \frac{A_A}{N_A}}$$

where A_A the area fraction and N_A the number per unit area are familiar quantities. Finally, we must note that for an observer standing at any point in the forest, the distance along his line of sight extends in two opposite directions, so that the total average length of a line between two trees is equal to twice the mean distance he can see. This distance is just

$$\sqrt{\frac{\pi}{A_A \cdot N_A}}$$

In the more interesting case of three dimensions, the concept of a mean free distance between objects generalizes straightforwardly. Considering the case of stars in the universe, the "background distance" is the volume occupied per star (the reciprocal of N_V) divided by the projected area of a star. If all stars are assumed to have the same size, this works out to

$$\text{Diam} = \sqrt[3]{\frac{6}{\pi} \cdot \frac{V_V}{N_V}} \qquad \text{Area} = \left(\frac{9\pi}{16}\right)^{1/3} \cdot \left(\frac{V_V}{N_V}\right)^{2/3}$$

Stereological Interpretation

so that the mean free distance between stars is

$$4 \cdot \left(\frac{2}{9\pi}\right)^{1/3} \cdot \frac{1}{V_V^{2/3} \cdot N_V^{1/3}}$$

which is just a simple expression in terms of familiar parameters V_V and N_V.

This reasoning was applied in the mid-1700s to the problem of why the sky is dark. The problem had been acknowledged in one form or another since the time of the ancient Greeks, but with the new astronomical knowledge provided by the telescope, it was reasserted. Jean-Phillipe Loys de Cheseaux and, later, Heinrich Olbers stated it clearly. Various attempts to resolve the apparent paradox were devised, such as the presence of dark or absorbing matter in space (which does not work, because it, too, would eventually be heated by starlight and re-emit the same amount of radiation). Even in this century, attempts to use the observed fact that the night sky is dark as proof of the expansion of the universe (redshifting the light from remote stars so that it is not perceived) have been inaccurate or incomplete based on thermodynamic considerations. (Harrison, 1987).

The actual explanation is simply that the speed of light in not infinite, and light from the remote stars that would brighten the night sky has not arrived here yet! This explanation was first given by the amateur scientist Edgar Allan Poe, and given a more rigorous mathematical explanation by Lord Kelvin, but even today many are not aware of it. As long as the viewable horizon is less than the background limit, we cannot see all of the stars in the universe (or trees in the forest) that are actually in our line of sight. By the same reasoning that gives us the number of trees that we can see in a two-dimensional forest, this permits estimating the number of stars that would be seen within the background limit. This is of the order 10^{46}, subject to minor adjustments if the fact that stars are not all the same size is taken into account (about 6000 stars can be distinguished by the naked eye). This number of stars corresponds to a background limit greater than the age of the universe multiplied by the speed of light.

By the way, the fact that stars are not randomly distributed in space but are clustered into galaxies, clusters and superclusters of galaxies, etc., does not invalidate this line of reasoning at all. The mean free path remains unchanged, although the distribution of the path lengths along random paths changes.

By now it should not surprise us that the same stereological relationships that apply to astronomical measurements should work in the microscopic world. The distance between precipitate particles or defects (whether they are randomly distributed or constrained by the crystalline structure of the matrix) obey the same equations.

Problems in 3-D interpretation

There are other global relationships that can be measured or derived. In fact, the richness of stereological interpretation of three-dimensional structure from counting and measurements on two-dimensional images continues to be developed by the application of geometrical probability and applied by both materials and life scientists. It is important to note, however, that there are some very fundamental and interesting parameters that cannot be determined in this way. For instance, it is not possible to look at a two-dimensional image of a plane section and determine whether the phase of interest consists of discrete particles, or whether they are connected, even to the point of forming a single network. This can be very important if the "phase" is porosity, because open and closed porosity produce very different properties.

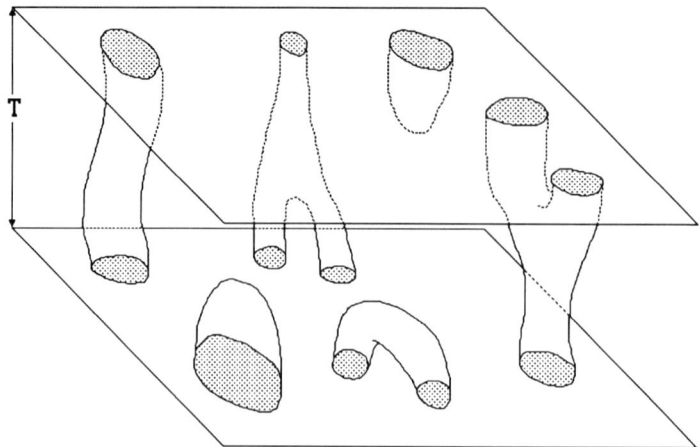

Figure 8-9: The disector - counting the new features in one plane image that do not appear in the other is complicated by possible branching behavior.

In fact, even such a simple thing as the number of objects per unit volume N_V is not determinable from this image. Two routes to determine N_V are available, one a measurement on two or more parallel image planes at different depths in the sample, and one requiring some assumptions about the shape of the objects giving rise to the features in the image.

The first method, involving the use of two (or more) planes, is called the disector (Sterio, 1984). In very simple terms, it involves examining two images and counting objects that appear in one that are not in the other. As shown in Figure 8-9, this corresponds to features that have end points in a volume of the sample defined by the product of the area of the images and the distance between them. This is N_V directly. However, it is not always easy to recognize which features in the two images are new. The figure shows examples of branching behavior that can easily mislead the observer in making the count of new features. In addition, features may enter or leave the volume being sampled by the sides, or apparently discrete objects may join outside the volume. All of these will bias the result.

The only way to limit the error from this source is to make the two planes very close together, so that no important changes occur between them. Unfortunately, this has the direct consequence that very few new objects will appear, so a very large area of material must be examined to obtain a statistically useful result. For biological materials prepared by microtoming, it is sometimes practical to carry this procedure out by examining a large number of sequential sections, to determine where objects start and end.

The disector gives a direct value for N_V without bias or requiring any assumptions about the actual three-dimensional structure. However, it requires a significant amount of effort. The disector is essentially a three-dimensional volume probe, just as planes of intersection, lines used for counting intersections, and points used for counting volume fraction have dimensions of 2, 1 and 0 respectively. In general, any unbiased measure of a global structural property of dimension n must employ a probe of dimension $(3-n)$. Besides the disector, there are some microscopy methods which are fundamentally three dimensional, and produce a continuous series of planar images. These are discussed in Chapter 12.

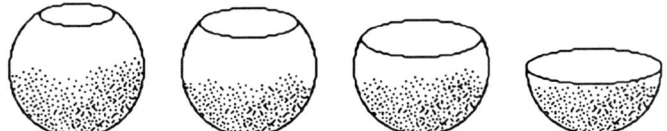

Figure 8-10: Different circular features are produced by random intersections of a plane with a spherical object.

The other approach to determining number per unit volume is much easier to carry out, but requires that a shape assumption be made about the particles in the dispersed phase. If this assumption is erroneous, the result can be quite wrong. Unlike a "design-based" method such as the disector, the model-based approach is based on the fact that for any predicted three-dimensional shape, it is possible to calculate using geometrical probability the frequency distribution of the two-dimensional features which would be seen by random sectioning. Although this method will yield a value for N_V, it also produces a complete size distribution histogram for the objects, and so leads us naturally to the subject of feature specific measurements.

Feature specific measurements

The relationship between the size of objects (the name we will adopt for three-dimensional phase regions) and the size of features (our name for the two-dimensional areas that are seen in the section images) is easiest to show for a very simple shape like a sphere (Wicksell, 1925; Cruz-Orive, 1976). The size distribution of the circles produced by random sectioning of a sphere (Figure 8-10) can be calculated analytically or modelled with a computer. In either case it produces a frequency plot as shown in Figure 8-11, with many rather large circles approaching the sphere's equatorial diameter, and some smaller circles when the section plane passes through the sphere at high latitudes. For more complicated shapes, it is possible to calculate, compute or model similar frequency histograms. This usually requires performing a large number of individual measurements, and computerized or computer-assisted methods are commonly used (Moore, 1968; Hare et al., 1982).

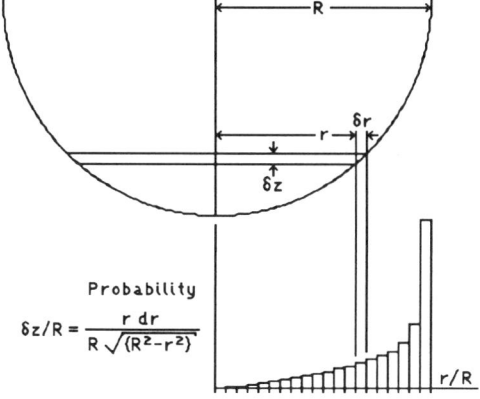

$$\delta z/R = \frac{r\,dr}{R\sqrt{(R^2 - r^2)}}$$

Figure 8-11: Construction of the frequency distribution of circle sizes obtained by random sectioning of a sphere.

Table 8-1. Matrix of coefficients used to calculate the size distribution of spheres

Profile Size Class	1	2	3	4	5	6	7	8	9	10	11	12	13	14	15	Total
1	+1.0000															+1.0000
2	–0.1547	+0.5774														+0.4227
3	–0.0360	–0.1529	+0.4472													+0.2583
4	–0.0130	–0.0420	–0.1382	+0.3779												+0.1847
5	–0.0061	–0.0171	–0.0408	–0.1260	+0.3333											+0.1433
6	–0.0033	–0.0087	–0.0178	–0.0386	–0.1161	+0.3015										+0.1170
7	–0.0020	–0.0051	–0.0093	–0.0174	–0.0366	–0.1081	+0.2773									+0.0988
8	–0.0013	–0.0031	–0.0057	–0.0095	–0.0168	–0.0346	–0.1016	+0.2582								+0.0856
9	–0.0009	–0.0021	–0.0037	–0.0058	–0.0094	–0.0163	–0.0329	–0.0961	+0.2425							+0.0753
10	–0.0006	–0.0015	–0.0026	–0.0038	–0.0059	–0.0091	–0.0155	–0.0319	–0.0913	+0.2294						+0.0672
11	–0.0005	–0.0010	–0.0018	–0.0027	–0.0040	–0.0058	–0.0090	–0.0151	–0.0301	–0.0872	+0.2182					+0.0610
12	–0.0004	–0.0009	–0.0013	–0.0020	–0.0028	–0.0041	–0.0057	–0.0088	–0.0146	–0.0290	–0.0836	+0.2085				+0.0553
13	–0.0003	–0.0006	–0.0010	–0.0016	–0.0021	–0.0028	–0.0040	–0.0056	–0.0085	–0.0140	–0.0280	–0.0804	+0.2000			+0.0511
14	–0.0002	–0.0006	–0.0007	–0.0012	–0.0016	–0.0022	–0.0029	–0.0039	–0.0055	–0.0083	–0.0136	–0.0270	–0.0776	+0.1925		+0.0472
15	–0.0001	–0.0004	–0.0007	–0.0009	–0.0013	–0.0016	–0.0022	–0.0028	–0.0039	–0.0054	–0.0080	–0.0132	–0.0261	–0.0750	+0.1857	+0.0441

When measurements of an image of a section plane through a specimen containing a range of sphere sizes are made, the result is a frequency histogram of the size of the observed circles. This observed histogram must consist of the superposition of many of the individual size histograms for all of the sphere sizes present in the structure, in proportion to the number of each size. Since the shape of the individual distributions is known, a series of simultaneous equations can be solved to recover the desired histogram of sphere sizes that must have generated the measured circle size histogram.

This is usually done by determining the matrix of coefficients that relate the number of circular features in size class i that are due to random sectioning of spheres in size class j. This matrix of values is given by

$$N_{Ai} = \Sigma\, \alpha'_{ij}\, N_{Vj}$$

This matrix can be inverted to obtain the α matrix, which can be stored and used whenever it is needed and appropriate. A 15x15 matrix of these values is shown in Table

Stereological Interpretation

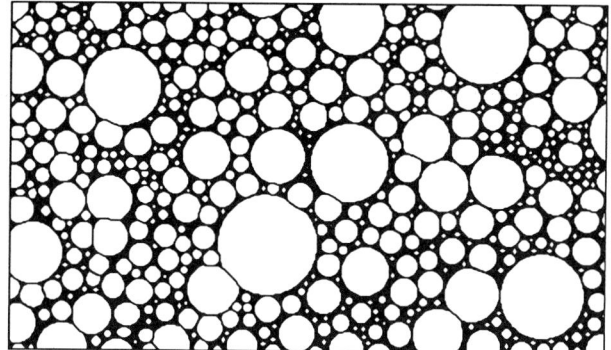

Figure 8-12: Pores or "cells" in bread showing circular features visible on a planar section.

8-1. Then the measured circle size histogram is multiplied by the matrix to obtain the number of spheres in each size class as

$$N_{Vj} = \Sigma\, \alpha_{ij}\, N_{Ai}$$

The column marked "total" can be used directly to multiply the number of features in each size class and add them to obtain the total number of objects per unit volume N_V.

The total number of objects per unit volume is easily determined by summing these values, along with the size distribution of the spherical objects. Figures 8-12 and 8-13 show an example of the application of this technique. The specimen is bread, which has been cut to show the internal pores (the cut surface was inked to obtain contrast). The circle size distribution is quite different from the sphere size distribution, as shown. The size of pores or "cells" in bread controls the "chewability" and "mouthfeel" of the bread. Pores are also important in materials and biological structures (Pohl & Redlinger, 1977).

Of course, there are a great many other types of second phase particles, inclusions or other objects for which a sphere is not an adequate model. It is possible to develop a similar matrix of values to transform the area measurements on a section image to the

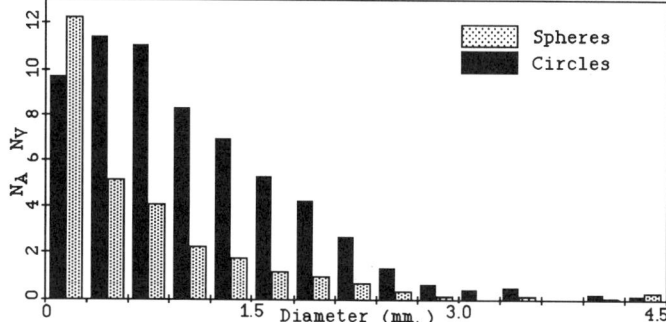

Figure 8-13: Frequency distributions for N_A (number of circles per unit image area, based on circle diameter) and N_V (number of spheres per unit volume, based on the sphere diameter) for the pores shown in Figure 8-12.

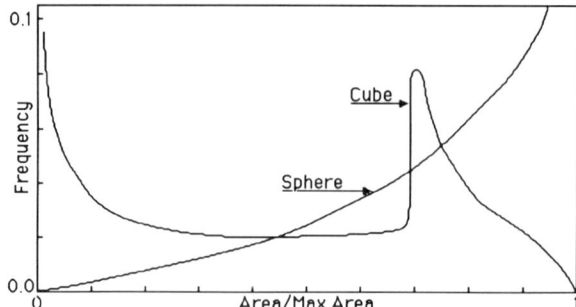

Figure 8-14: Distribution of feature sizes for cube and sphere.

volume frequency distribution for many known shapes (Figure 8-14). The problem is that only rarely are the shapes known. Sometimes it is possible to apply independent knowledge to choose an appropriate shape, for instance a platelet or cylindrical rod. More often, the actual shapes are stochastic and exhibit a natural variation that can introduce a significant error is it is ignored.

A particularly versatile and robust shape that is often used to model unknown objects is the ellipsoid of revolution (DeHoff, 1962). This is formed by rotating an ellipse about one of its axes, and can be either prolate (elongated like an American football) or oblate (flattened like a discus). Extreme variations of this shape serve for everything from rods (prolate) to platelets (oblate). Any plane section through an ellipsoid of revolution reveals an ellipse (Figure 8-15). It is possible to use measurements of the size and aspect ratio of these ellipses to estimate the frequency distribution of the size and sometimes the shape of the three-dimensional ellipsoids.

Two parameters are required to characterize the ellipse shapes. Since these ellipses are not an exact model of the actual images seen in the planar sections through objects that are only approximately shaped like ellipsoids, we would like to choose parameters that are not too seriously affected by shape. For instance, the perimeter would be a poor choice. Instead, either the area of the feature and its length (longest chord), or the length

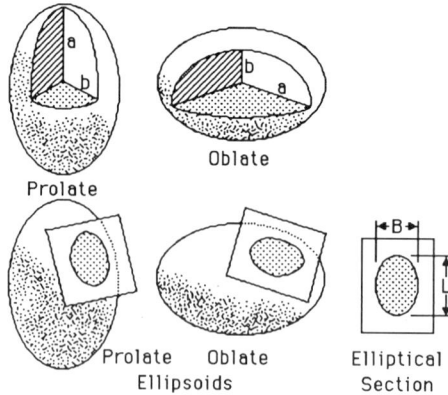

Figure 8-15: Ellipsoids of revolution, and the ellipses formed by intersections with planes.

Stereological Interpretation 237

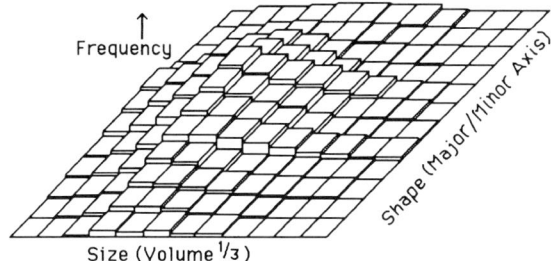

Figure 8-16: A two-dimensional histogram of frequency for size and shape can use quite different parameters (shown here, the cube root of volume and aspect ratio.

and breadth (shortest projected or caliper dimension) are most often used (although if the objects are not strictly convex, the width as described in Chapter 7 may be preferred).

If the measurements on the two-dimensional features are summed into a two-way histogram of frequency based on the two measured parameters, the resulting data can be used to obtain a similar two-way histogram of the size and shape distribution of the generating three-dimensional ellipsoids, as shown in Figure 8-16. This is done using a simple matrix multiplication, entirely analogous to the case of spheres described above.

$$N_{Vhi} = \Sigma \, \alpha_{hijk} \cdot N_{Ajk}$$

The matrix of α values is calculated beforehand, by inverting a matrix of intercept probabilities that is generated either analytically or (more commonly) by using a Monte-Carlo computer program to simulate a large number of intersections through the important range of shapes and sizes (Kendall & Moran, 1963; Matheron, 1975; Santalo, 1976; Miles & Serra, 1978; Warren, 1987). This is time consuming, but need be done only once and the matrix of values stored.

Unfortunately, while it is not difficult to carry out the calculations to determine the particle size distribution from the feature sizes, this is an example of a problem that is ill-conditioned in a mathematical sense: small variations in the measured data produce much larger variations in the calculated results. For instance, physically nonsensical negative density in some size ranges can be calculated, and in general the tails of the computed distribution (the largest and smallest objects) have rather large uncertainties. A thorough discussion of these types of errors, and the history of concern about them, can be found in Coleman (1989).

In practical terms, the problem with this approach is that the result of the matrix multiplication is a histogram of object size and shape that contains statistical uncertainties based on the number of objects counted, and that these uncertainties or errors propagate from large sizes downwards (that is, because it is possible to measure a small feature that results from a large object, but not vice versa, the number of small objects estimated to be present depends on the number of large features actually counted). If this counted number of large features, particularly for some of the more extreme shapes, is small, then it can produce a very large uncertainty in the final result. This can only be dealt with by counting more such objects, which in turn means counting many of all objects since the large objects with extreme shapes are likely to be only a small part of the total population.

Things become much simpler if it is not necessary to consider a range of shapes. If it is acceptable to model the objects by an ellipsoid of revolution having a fixed aspect ratio,

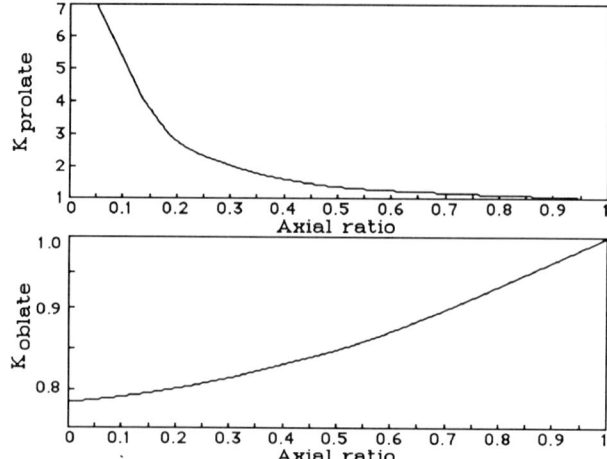

Figure 8-17: k(q) shape factors for prolate and oblate ellipsoids.

then the result is a histogram of object size derived from one of feature size that is very similar to the case for spheres. In fact, the matrix of values derived for spheres can be used with only the addition of a correction factor, so that the number of objects in each volume class N_{Vj} can be calculated from the number of features in all area classes N_{Ai} as shown.

$$N_{Vj} = \{k(q)/\delta\} \; \Sigma \; \alpha_{ij} \cdot N_{Ai}$$

The correction factor k is a function of the shape of the ellipsoids, usually expressed in terms of their axial ratio q, while δ is the size increment used for the histogram and the α factors are the same ones shown above for spheres. $k(q)$ factors have been calculated for both prolate and oblate ellipsoids, as shown in Figure 8-17. Note that this implies a serious problem: we must know (or guess) what the axial ratio is, and even whether the objects are to be modelled by a prolate or oblate shape.

No individual feature seen in the two-dimensional image can be used to obtain this information, but a statistically large number of them can be. The method is based on simple probability. If a large number of planar sections are cut through many prolate spheroids, then the mean diameter of the most nearly equiaxed sections will be about the same size as the mean minor diameter of the sections that have the greatest elongation or aspect ratio. Conversely, if the sections have been cut through oblate spheroids, the mean diameter of the most nearly equiaxed sections will be about the same size as the mean major diameter of those with the greatest aspect ratio (Figure 8-18).

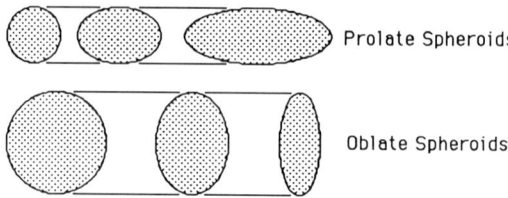

Figure 8-18: Comparison of elliptical profiles to determine if the generating shapes are oblate or prolate.

Stereological Interpretation

Because this method uses the measured length and breadth of the features, it is often the practice to measure just those parameters. The size histogram of the features is based on the length, because as discussed in the previous chapter this is a fairly robust measurement, insensitive to feature orientation. Only the most extreme shapes (perhaps the 5% with the greatest aspect ratio and the 5% with the least aspect ratio) are used to make the comparison to decide whether the objects are to be modelled as oblate or prolate. The ratio of the minor to major axis of the most unequiaxed features is used to estimate q, and the corresponding value of k is then used to modify the results from the spheroidal model.

Distribution histograms of size

The histograms discussed so far have shown frequency (number of occurrences) of features or objects as a function of some parameter such as size or shape, or in previous sections, position, orientation, etc. These kinds of histogram plots are fairly straightforward to interpret, and familiar to most researchers. The particular selections of plotting scales that have come into wide use are those that produce very simple graphs for many real samples, easily represented by a few parameters. The most obvious of all is a plot of number as a linear function of size. But the size parameter can be a linear measure, such as length or the square root of area, or an area measurement, or even a volume measurement. This choice significantly alters the shape of the resulting plot, by stretching the size scale in different ways.

It also means that the "mean size" as reported by a simple statistics test will not select the same group of features of objects, if the calculation is based on a parameter with different dimension (length, length^2, or length^3). This fact should be kept in mind when comparing different data sets. It is also important if the results from image analysis are to be compared with some other technique. Sieving, sedimentation techniques, and light scattering are all used to measure particle size, for example. They respond to different aspects of object size, and may not report data on a consistent size scale for comparison to image data.

The types of distributions discussed so far are based on a parametric model, usually some simple geometric shape such as a sphere, ellipsoid, polyhedron, etc. for which the probability distribution of linear or area intercepts can be computed beforehand (either analytically or using Monte-Carlo methods) and used to unfold the measured 2-D size distributions. If the assumed shape is incorrect, or worse still if the shape changes with size, these methods can give quite erroneous results. There has been a recent trend away from these "model" based measurements toward "design" based techniques which rely on stereological principles to define what can be measured in an unambiguous and unbiased way. The disector mentioned earlier is one such method, which can provide an unbiased measure of the number of features per unit volume. There is also a technique that can determine a distribution of feature sizes without requiring any shape assumptions, and in fact works correctly even if the shapes vary.

Unfortunately, the type of distribution which the method produces is not as familiar to common experience as the plot of number vs. size we have been dealing with. Instead, we shall employ a measure of particle volume. The technique starts by point-sampling the volume. If polished sections are oriented randomly with respect to the surface, this simply consists of placing a grid of points on the image and finding ones which fall within the outlines of features. (Of course, as discussed before, it is by counting these points that volume fraction can be most efficiently determined.) Then at each of the points lying within the phase of interest, a random line is generated at a uniformly random

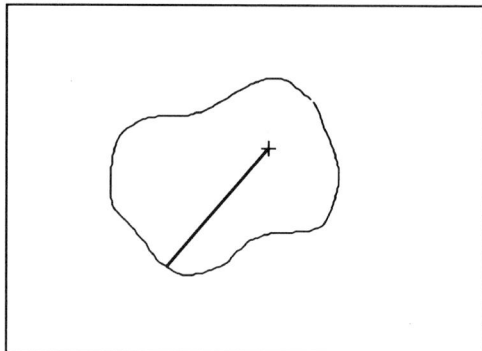

Figure 8-19: Diagram showing a randomly placed point falling within a feature, and a randomly oriented line through that point being used to measure the distance to the boundary.

angle, and the intercept length l from the original point to the boundary of the particle is measured as shown in Figure 8-19.

The surprising and quite non-intuitive result is that the volume of the particle is estimated without bias by the quantity $(4\pi/3) \cdot l^3$ (Gundersen & Jensen, 1983, 1985; Gundersen, 1986; Gundersen et al., 1988). This means that if a number of such measurements are performed, then the mean value of the volumes determined in this way is indeed the mean particle volume. If a distribution of frequency vs. volume is constructed, it corresponds to what a sieving experiment would yield. In other words, it tells us what volume fraction of the particles have volumes in a selected range, not the number of particles.

This method is extremely efficient for comparing different populations of features. Measures of only a few dozen point-sampled intercept lengths provide a very robust measure of mean size (where size now means volume), quite independent of shape. There are a few words of caution: If the feature as shown on the plane of section is concave, the intercept line may not lie entirely within the feature. It is necessary to measure the length of line within the feature, excluding portions which fall outside. If the 3-D object is concave in a way that produces multiple intersections with the section plane, the intercept distance must include all chords across the object (which requires recognition by the user that the features represent sections through the same object).

The method can be extended directly to structures that are not isotropic by the use of a vertical section plane (Baddeley et al., 1986). In this method, the section plane is cut in a direction vertical with respect to a surface, bedding plane, or other known orientation, but is randomly rotated around the normal direction to that surface, as shown in Figure 8-20.

Then, on the plane where examination and measurement are to be performed, point sampling is used as before to select points that lie within the features of interest. However, instead of choosing the orientation of the intercept lines randomly, they must be sine-weighted to compensate for the choice or orientation of the section plane. This is done by generating random numbers in the range 0...1, and then using the arc sin of this value as the angle of the intercept line from the horizontal. This will produce fewer lines with a direction near the vertical as indicated in Figure 8-21. The intercept lengths are

Stereological Interpretation 241

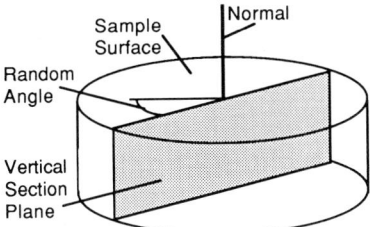

Figure 8-20: Schematic of a vertical section plane in an oriented structure.

then obtained in directions that isotropically and randomly sample the structure in three dimensions.

Interpreting distributions

Plots in which the horizontal axis is a measure of shape are more difficult to characterize, because it is not at all clear that a linear variation in any shape factor correctly distributes the data. For instance, a change in aspect ratio of 0.1 from 1.0 to 1.1 does not necessarily mean the same thing as a change from 3.0 to 3.1, and certainly corresponds even less to a change in formfactor from 0.7 to 0.8. Great care must be taken in making comparisons of frequency histograms of shape parameters. This will take the form of plotting several different horizontal scales until the shape of the histogram appears "simple" (for instance a symmetric or even a normal distribution), and then using the same scale for all data sets which are to be compared.

Nonlinear plotting scales are often used with size parameters as well, for the same reason (to make the distributions more symmetrical). One of the most common of these scales is logarithmic. For a variety of incompletely understood reasons, many real particles formed by a variety of different processes ranging from the nucleation and growth of phases in materials to the grinding of particles in milling operations, or even the wear of natural rocks in erosion or riverbeds, gives rise to a frequency distribution that is more-or-less Gaussian or normal when frequency is plotted against the logarithm of size. Many distributions of object size are found to be close to log-normal, although some (such as the size distribution of grains in materials during high temperature

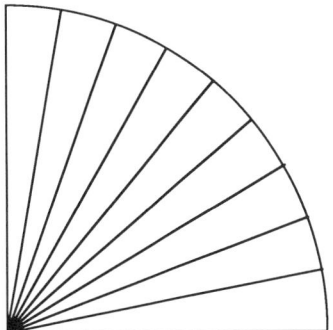 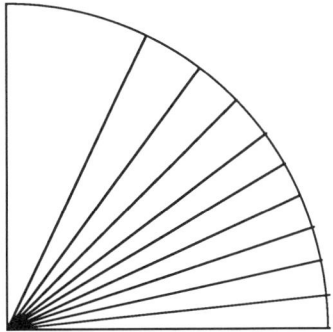

Figure 8-21: Sets of lines generated with equal angular spread Angle = π/2 • *Rnd* and sine-weighted Angle = arc sin (*Rnd*), where *Rnd* is a uniform random number between 0 and 1.

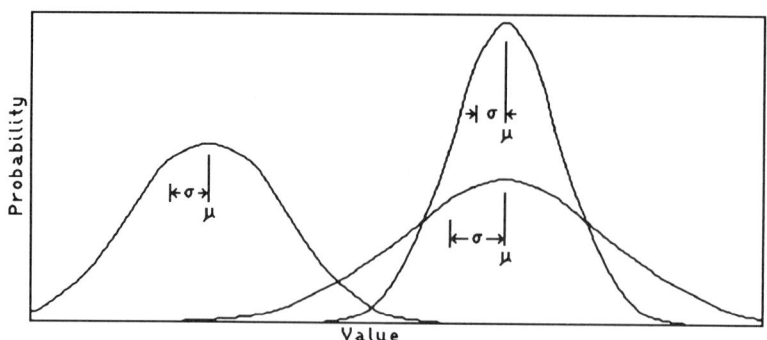

Figure 8-22: Gaussian distributions with different means and standard deviations.

treatment which causes grain growth) have been shown to deviate from this shape when examined in detail (Chermant, 1986).

In a logarithmic (or any other nonlinear scale) plot, of course the histogram data consist of bins used to count frequency that are of constant width on the plot, and so do not represent equal linear increments of size (or whatever parameter is used). For a logarithmic plot, the width of the bins is a ratio of the upper and lower limits. For instance, bins could be selected that are a factor of 1.414 wide based on equivalent circular diameter. This would mean that each succeeding bin contained the number of features whose linear size dimension increased by the square root of two, which is equivalent to a change in the area of a factor of two.

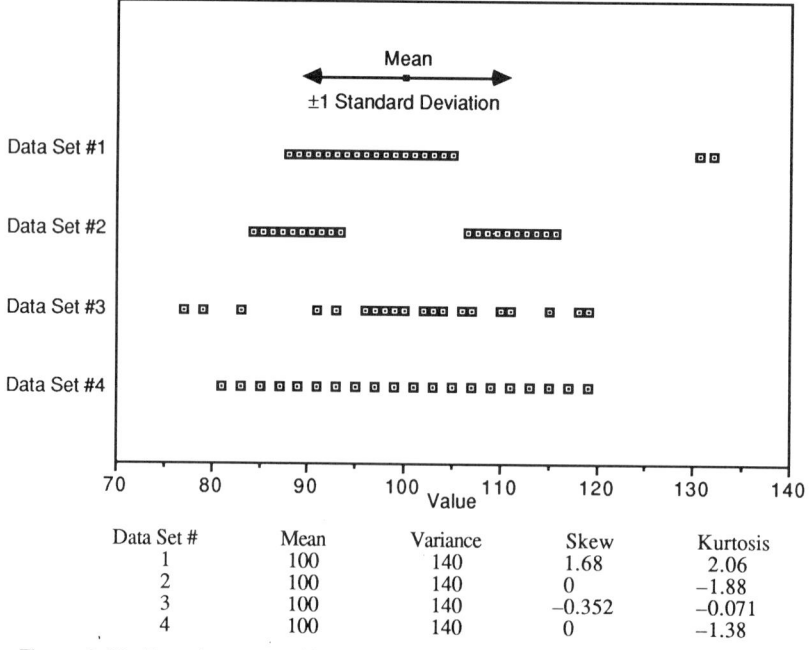

Data Set #	Mean	Variance	Skew	Kurtosis
1	100	140	1.68	2.06
2	100	140	0	−1.88
3	100	140	−0.352	−0.071
4	100	140	0	−1.38

Figure 8-23: Four data sets with the same mean and standard deviation values.

Stereological Interpretation

Gaussian or normal curves are very desirable for frequency histograms, because they can be completely described by just the mean value and standard deviation (Figure 8-22). Unfortunately, these two parameters can also be calculated and reported for any data set. Without examining the distribution, the mean and standard deviation may be misunderstood to imply that the distribution is a normal one when this is not the case. Additional statistics such as the skew and kurtosis can also be examined to verify a normal curve, but the complete plot is the best way to report the data. (Just as the standard deviation is the second moment of the distribution, the skew is the third moment and the kurtosis is the fourth. For a Gaussian distribution the skew will equal zero and the kurtosis will equal 3.0.) More information on these descriptive statistics can be found in standard texts on statistical analysis of data (for instance, see Bevington, 1969).

The inadequacy of mean and standard deviation as descriptive statistics is demonstrated in Figure 8-23. Each data set contains twenty observations, and each has the same mean (100) and standard deviation (11.8). However, the distribution of the values in each set is quite different. To a certain extent these differences can be seen in the kurtosis and skew values tabulated. The kurtosis value is the fourth moment of the distribution and reveals how tightly clustered the points are about the mean, while the skew value is the third moment and describes the asymmetry of the distribution (the mean and variance are the first and second moments of the distribution, respectively). However, these parameters are not commonly understood or used, and furthermore do not show the nature of the distribution as well as a graphical plot.

A distribution that shows a normal curve when plotted on a logarithmic scale is said to be "log normal" (Figure 8-24). Its mean is a geometric mean rather than an arithmetic one, and the standard deviation is a measure of the size ratios that correspond to the decline in frequency of occurrence. Comparison of normal distributions using standard statistical tests such as Analysis of Variance that use the mean and standard deviation are not strictly applicable to log-normal distributions.

Nonparametric tests

In fact, any comparison of distributions of data must be done with some care. If the distributions are normal, an Anova test may be an efficient way to decide if the two (or more) populations are the same or different. But in the more common case in which the distributions being compared are not both truly normal, this should not be done. Instead,

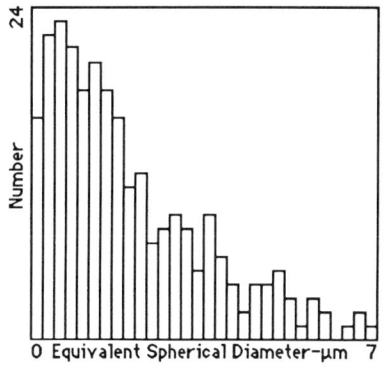
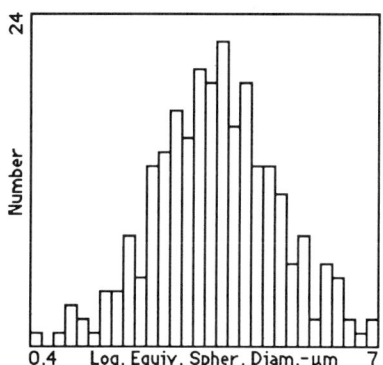

Figure 8-24: A "log normal" distribution (measured on paint pigment particles viewed in SEM image) plotted on linear and log scales.

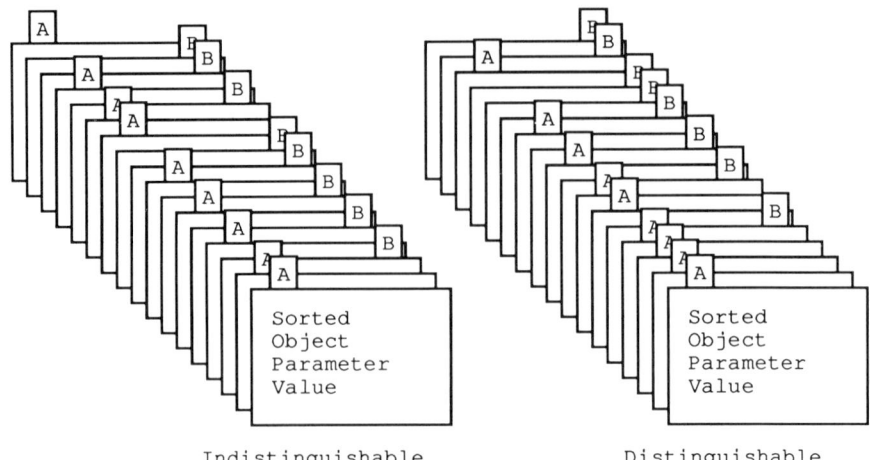

Figure 8-25: Principle of the Wilcoxon test.

a non-parametric test should be used. One of the most common of these is the Wilcoxon (Figure 8-25). This is somewhat more commonly known as a Mann-Whitney U-test, since although it was originally designed by Wilcoxon, it was extended by Mann and Whitney to deal with data sets of unequal size.

This test uses the rank order of individual observations in the two data sets. The simplest analogy is to write the measured value of the test parameter for each object onto a card, and then arrange all of the cards (for both sets of objects being compared) in order. If the cards corresponding to the two sets are randomly mixed together in order, then the populations cannot be distinguished. The binomial theorem provides a way to calculate the probability of the observed sequence of cards from the two sets occurring in the observed order, and so establishes the probability or confidence limit of being able to distinguish the two sets of objects.

To perform the test, the rank values of either population are summed (this is usually done, for convenience, with the smaller population; it need not be done for both since the

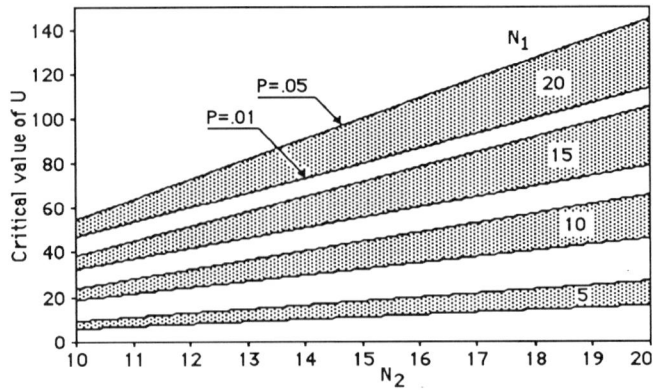

Figure 8-26: Critical values for the Wilcoxon test.

Table 8-2: Example data for the Wilcoxon test
Set 1: 27, 29, 30, 31, 31, 32, 36, 37, 41, 43
Set 2: 28, 30, 31, 31, 31, 33, 34, 34, 35, 37, 39, 40

second total is known from the total number of objects). Ties in rank order are resolved by giving each tied object the same averaged score. Then the parameter U is calculated as

$$U = W_1 - n_1(n_1 + 1)/2$$

This value is compared to tables such as the graph in Figure 8-26 to determine whether the difference between the populations is significantly different. For instance, for the example data in Table 8-2, the calculated value of $U = 57$. The critical values for populations of $n_1 = 10$ and $n_2 = 12$ objects. indicate that we cannot state with confidence that the two populations are different. If the score was smaller (indicating a more segregated order for the sorted values), we might be able to conclude otherwise: a score of 30 would give a 95% confidence ($P = 0.05$) that the two groups were different, and a score of 24 ($P = 0.01$) would give a 99% confidence that they were different.

The Kruskal-Wallis test is an extension of the Wilcoxon to deal with more than two groups (in somewhat the same way that the Anova test is a generalization of the t-test). For a total of n observations in k groups, the sum of ranks within each group is totalled as R_i and used to calculate the H parameter

$$H = \frac{12}{n \cdot (n-1)} \cdot \sum_{i=1}^{k} \left(\frac{R_i^2}{n_i}\right) - 3 \cdot (n+1)$$

This is compared to the value of $\chi^2_\alpha(\nu)$ from tables, where $\nu = k - 1$ the degrees of freedom. If this value is exceeded for a given value of α then the groups are considered to be different at that confidence factor.

The Kolmogorov-Smirnov test approaches the same problem in another way. Cumulative frequency distribution plots of the two populations are constructed on the same graph, and the maximum vertical distance between them found. This difference value D will always be some value less than 1, since the vertical axis on the cumulative plot for each population is 1.0, but it may occur anywhere along the value axis. Depending on the number of members of each population (m and n, which need not be the same), it is then possible to calculate the probability that such a difference could occur by chance. For populations with more than about 20 members, these test values can be acceptably approximated by an equation of the form $A \cdot \{N/(m \cdot n)\}^{1/2}$, where $N = m + n$. The values of the A parameter for different levels of confidence are:

P	0.10	0.05	0.025	0.010
A	1.07	1.22	1.36	1.52

Table 8-3. Representative data for Kolmogorov-Smirnov Test. Two sets of length values measured on different image features. Are they from different populations?

Set A: 1.023, 1.117, 1.232, 1.291, 1.305, 1.413, 1.445, 1.518, 1.602, 1.781, 1.822, 1.889, 1.904, 1.967
Set B: 1.019, 1.224, 1.358, 1.456, 1.514, 1.640, 1.759, 1.803, 1.872

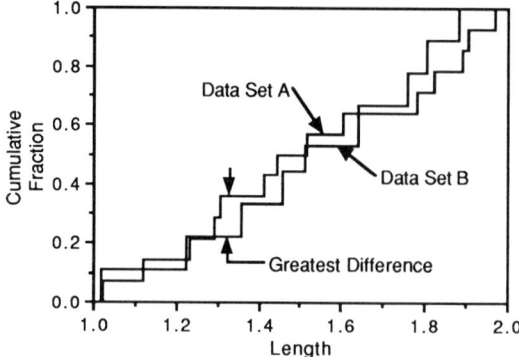

Figure 8-27: Example of cumulative plots of the two data sets shown in Table 8-3, with the greatest difference between them marked.

The data in Table 8-3 show length values from two sets of features. Set A contains 14 objects and Set B contains 9. When these data are plotted in a cumulative plot of number vs. length (Figure 8-27), the greatest difference between the two plots can be found; in this example it is 0.135. The calculated test value for $P = 0.1$ (a confidence level of 90%) is $1.07 \cdot \{23 / (9 \cdot 14)\}^{1/2} = 0.457$, and since the difference is less than this we conclude that there is no evidence (at this level of confidence) that the two sets of data are different. To conclude that at $P = 0.01$ (99% confidence) the two data sets were different, the maximum difference value between the two cumulative curves would have to be at least 0.649 for this number of observations in each set.

As interpreted here, the probabilities are based on what statisticians call a two-tailed test. This means that the question we have asked is "are the two sets different." If the question was instead "is this set greater than the other" the comparison would use a one-tailed test, and the P values (confidence limits) would be halved for a given test value. Usually in these kinds of comparisons, it is the more conservative two-tailed test that is used.

A more detailed comparison can be carried out by performing a complete chi-squared comparison between the two frequency distributions. This same test is also the most

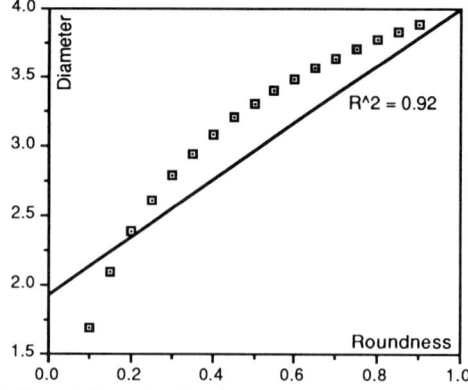

Figure 8-28: Correlation plot showing data with a nonlinear relationship.

complete way to test whether or not a distribution is normal (or log normal) by comparing it to an ideal Gaussian distribution. The comparison is performed by summing the squares of the differences between the contents of bins in each distribution, divided by the frequency in the bin of the comparison distribution. These values will be very small if the two distributions are similar. Along with the number of bins used to form the distribution, this value can be compared to standard plots or tables to ascertain the probability that the observed magnitude could have occurred by random variation.

For data that do not exhibit normal (Gaussian) distributions, correlation plots may also be quite misleading. Figure 8-28 shows an example in which the data vary in a logarithmic relationship. A normal linear correlation coefficient shows an R^2 value of 0.92, but examination of the data suggest that the relationship is actually perfect, with no out-of-order points. This kind of nonlinear behavior is especially common when dealing with the various shape factors encountered in image analysis, which are definitely not normally distributed.The tool most often used in these situations is the Spearman Rank test, which performs the correlation not using the actual values of the measured or calculated parameters, but their rank orders instead. The interpretation of the R^2 value is the same as for the "usual" correlation method, which is discussed later in this chapter.

There are other tests which are less frequently used in a typical image analysis setting, but may be provided by a computer-based statistics package (assuming one is used). The Wilcoxon Signed Rank and Kendall Rank tests are alternatives to the Spearman Ranked Correlation test which use the rank order of observations rather than the values themselves. This may offer advantages when dealing with nonlinear relationships. The Friedman test is an extension of these methods to deal with more than two groups of values. The Wald-Wolfowitz test determines whether a sequence of class occurred in a random order. More information on these specialized tests can be found in standard reference or text books such as Hollander & Wolfe (1973) or Gibbons (1985).

The reader whose statistical experience is limited should refer to appropriate texts to brush up on these crucial points. Most computer-based image analysis systems offer a variety of statistical tools to calculate the values for Anova, Wilcoxon, chi-squared and other comparison tests, and will often convert the answers directly to probabilities that the distributions are the same or different. However, it is always still up to the user to select the appropriate test(s) and properly interpret the results, and this is all too rarely done. (There is a certain fascination with any computer program that will print out a result with many digits past the decimal point, and we all tend to write it down and treat it as gospel in approximate inverse proportion to how well we understand it!)

Cumulative plots

Distribution histograms can be plotted in other ways as well. One of the most common is a cumulative plot in which each bin records the total number of all features whose measured parameter value is larger (or smaller) than the position along the plotting axis (Figure 8-29). When the measurement variable is a size, these are referred to as cumulative oversize or cumulative undersize plots, and they correspond to the usual way in which data are plotted from other techniques. This is particularly true for sieve methods of determining size distributions, in which the total weight if each screen is used (however, these plots are not of number vs. size, and should not be directly compared to image measurement data).

The cumulative plot sometimes gives a direct answer to practical questions as may arise in quality control. For instance, rice is graded by the USDA as long grain based on the maximum fraction of grains (by number) that are shorter than a fixed length. Video

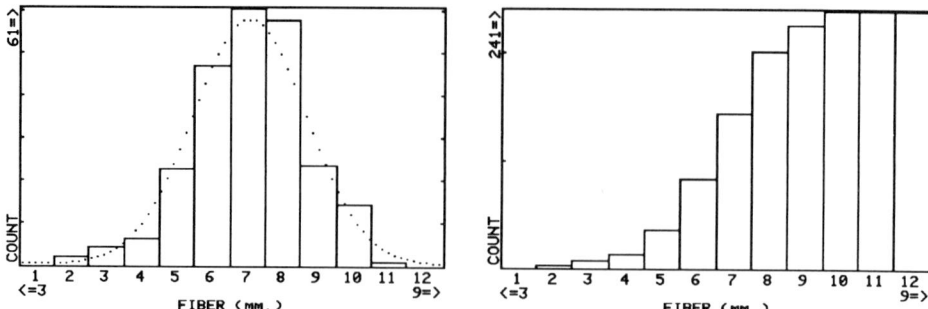

Figure 8-29: Fiber length of features plotted on a conventional and cumulative scale.

image measurement can be used to measure rice grains very easily, for instance by spreading the grains on black velvet (the texture of the cloth helps to separate grains so they do not touch). The image (Figure 8-30) is easily discriminated by selecting the white pixels, and the length of each feature can be straightforwardly determined. A cumulative undersize plot of frequency vs. length can then show directly whether the fraction of grains shorter than the cutoff value exceeds the specified limit (Figure 8-31).

Cumulative plots are also useful for normal or log-normal distributions. If the vertical scale used for frequency is converted from linear to probability values, then a normal distribution will be shown as a straight line. Since deviations from a straight line are much easier for most users to judge than are deviations from a Gaussian shape, this helps in deciding whether the data are normally (or log-normally) distributed.

A very common occurrence in these plots is a systematic deviation below the straight line for the smaller size classes. This may not be due to any actual deficit in the number of small objects, but rather in their visibility. Small objects are often undercounted because they either (a) pull out of the section surface and are lost; b) produce too little contrast to be visible, or at least easily discriminated along with the larger features; c) may not be adequately resolved by the microscope or other imaging technique; or d) may

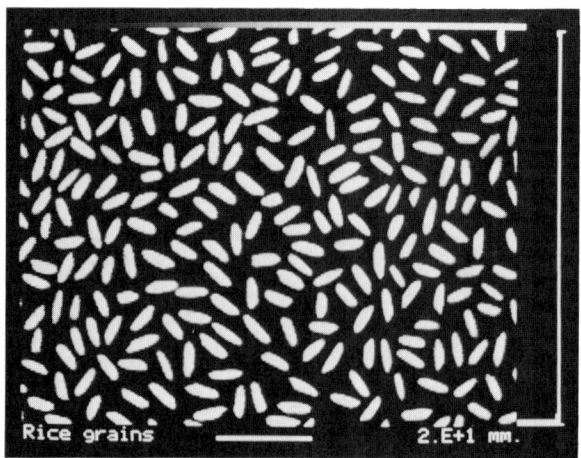

Figure 8-30: Image of rice grains.

Stereological Interpretation

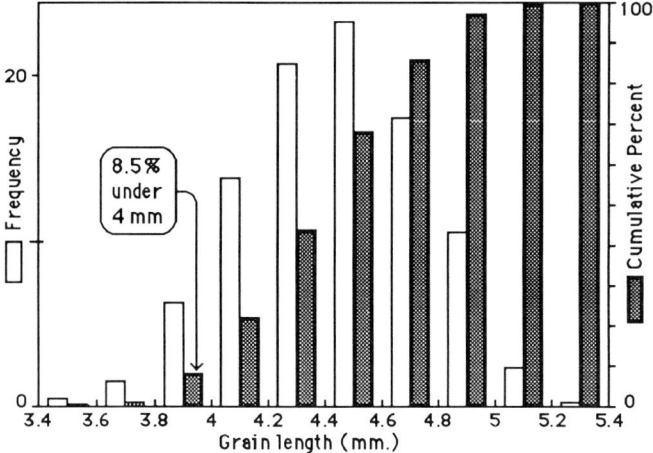

Figure 8-31: Cumulative plot of frequency vs. length.

hide under or next to larger features and cannot be distinguished. This produces the deviation shown in Figure 8-32. As a result of this known problem, it is common to count only the larger size classes and then if the data appear to be normal or log normal, to fill in the bottom quarter or third of the size range based on this assumption. On a probability plot, this simply requires drawing a straight line fitted to the data for larger sizes.

Another type of histogram or distribution plot, which we have already seen but not discussed in any detail, is the two-way plot (Figure 8-33). Of course, if multiple feature parameters have been measured, it is possible to imagine plotting any combination of them in an n-way display of frequency. However, the practical limitations of display technology make it hard to find acceptable ways to show more than two independent parameters are one time.

In a two-way plot, the number of features in each bin can be summed to produce striking graphs with a three-dimensional appearance. The chief problem with this type of presentation are that the number of counts in some of the bins may be rather sparse,

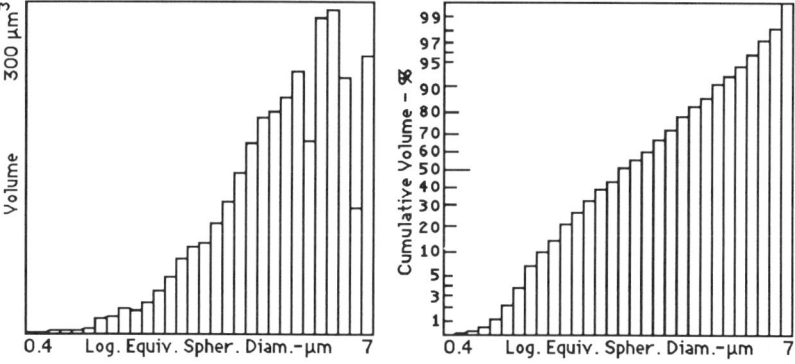

Figure 8-32: Data from Figure 8-24 showing volume plot and same with probability scale. Note the deficit at small sizes.

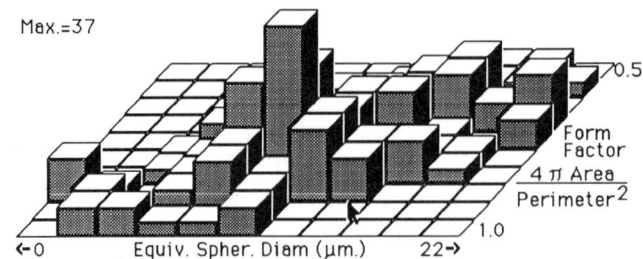

Figure 8-33: Two-way frequency histogram plot [horizontal axis = size (length); vertical axis = shape (aspect ratio)].

much more so than in the case of a simple one-axis distribution, so that the statistical counting fluctuations are distracting. This can be controlled either by acquiring large numbers of data points, or by properly choosing the bin sizes to be larger so that more counts are accumulated.

The two-way plot may also obscure some data because the columns for some bins may hide behind other parts of the distribution. If this occurs, it is usually possible to either interchange the variables and limits on the axes, or perhaps to rotate the display on the computer screen to find a vantage point from which the desired information is visible. A typical application of the two-way plot occurs when there are two measurement criteria used in a quality control situation. In the earlier example of measuring the length of rice grains only one parameter was used. For peanuts, there are two: the size of the nut and its lightness (resulting from blanching of the shelled nuts). Figures 8-34, 8-35 and 8-36 show a typical image and results. A minimum fraction of the nuts must exceed an established threshold in both values.

Up to this point, distribution or frequency histograms have consisted of plots of the number of objects with a measured value for size, shape or some other parameter within the limits of each bin. While these are perhaps the most familiar type of plot, there are others which are particularly rich in stereological information. These use one parameter to select objects (the same size or shape parameter just discussed, for instance), but instead

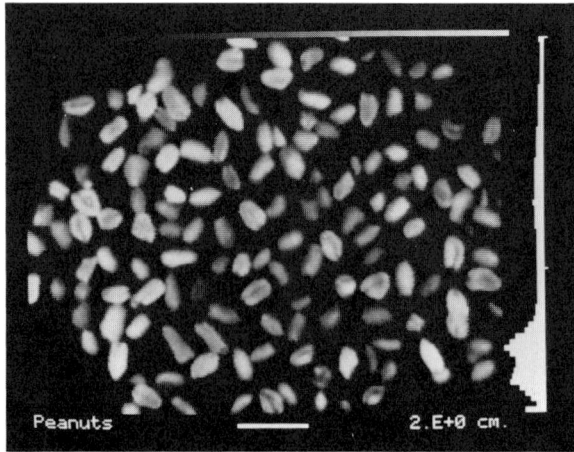

Figure 8-34: Peanuts.

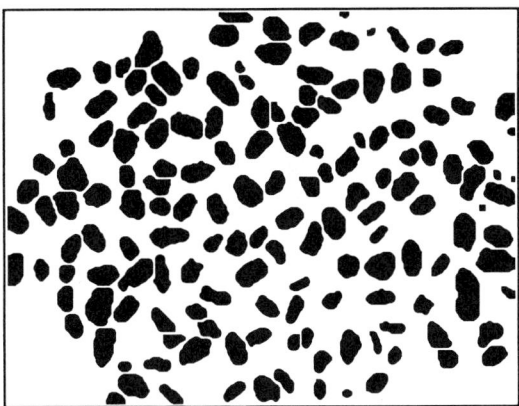
Figure 8-35: Binary image of peanuts obtained by discrimination and watershed segmentation to separate touching features.

of counting the objects they instead sum up some other value within each bin. For example, a plot of the volume of paint pigment particles as a function of their diameter produces a plot similar to the result from a sieving experiment in which the weight of each tray is determined (because weight is proportional to volume).

Plots of total volume or surface area (using the models described in the previous chapter on feature specific measurements) of particles sorted by their length or equivalent spherical diameter are especially common. As for the frequency plots, these may be presented in cumulative modes, the horizontal axis may be either linear or log, and the vertical axis may be either the value, the fraction of the total, or a probability scale. In the latter case, if the number, surface area and volume of particles in a normal (or log-normal if the horizontal axis is the logarithm of size) distribution are plotted, they produce a set of parallel straight lines. Since by definition the point at which each line crosses the 50% probability level gives the mean value, it is immediately clear that the "mean diameter" of the particles represented by this plot is different depending on whether we are talking about the number, surface area or volume of the particles (Figure 8-37). The particle with the mean volume is much larger than the mean particle based on number.

One particular type of plot uses size as a sorting variable and sums for each bin an "effective count" that corrects for edge effects on the original images. This parameter is

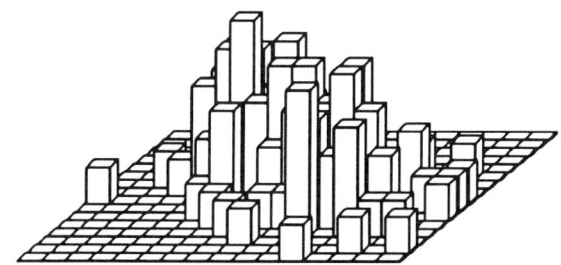
Figure 8-36: Two-way histogram of the size and brightness values for the peanuts in Figure 8-34 (some details toward the rear are obscured by information in the front).

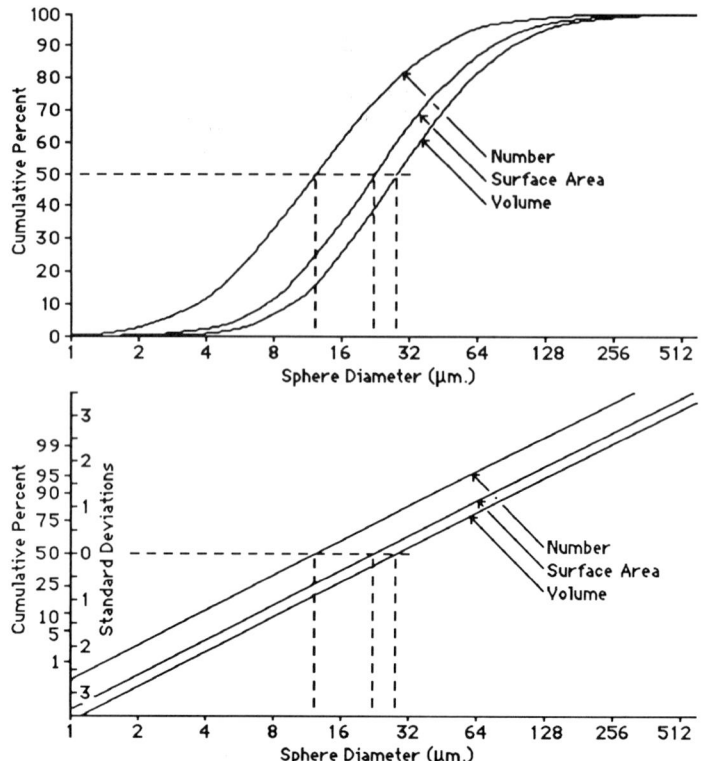

Figure 8-37: Plots of volume and surface area on log scale with probability and cumulative scales. Note the different means.

calculated for each counted object in the field, and represents the inverse of the probability that an object of these dimensions would be able to be counted in an image of the size being used. Very large features have a greater probability of intersecting an edge of the image, which makes it impossible to measure their full extent. The measurement programs recognize this, as was described in Chapter 7, and do not try to measure features that extend beyond the edge of the image region. But this would lead to an undercount of large features, while small ones would rarely touch the edge and would be correctly counted. The adjusted or effective count parameter described in Chapter 7 corrects for this probability (Russ & Hare, 1985), and makes large features count more than one in proportion to their extent. The value is calculated as

$$\text{Adj. Count} = (W_x \cdot W_y)/\{(W_x - F_x) \cdot (W_y - F_y)\}$$

where W_x and W_y are the dimensions of the image area in the x and y directions, respectively, and F_x and F_y are the projected or Feret's dimensions of the feature in these two directions. The value is close to one when the feature size (F_x and F_y) is small compared to the size of the image, but increases when the feature extends across a significant portion of the image in either direction. Summing these values in the histogram bins instead of simply counting the number of features produces a distribution corrected for edge effects in the original images.

Stereological Interpretation

Of course, measurements on features are commonly accumulated over many image fields to produce a set of data representing the specimen. If all of the image fields are of the same size and at the same magnification, no special care is needed in this process. But when different image magnifications are used, it is not sufficient to simply add the counts together. This is because at high magnification, more small features can be seen which would be missed at lower magnification, and more of the large features will be lost because they intersect an image edge.

As discussed before, the finite resolution of digital images is such that features varying in size by an order of magnitude or more cannot be accurately measured in a single magnification image. It is necessary to use images taken at appropriate magnification to show the large and small features separately, only combining the data for final presentation (Figure 8-38). But when this is done, it is necessary to know how much image area was actually examined at each magnification.

For instance, if 10 images at 100X were used to measure and count large features, and then another 10 images at 1000X were used to measure and count small features in the same specimen, the low magnification image would not show the small features and vice versa. But adding together the results from the two sets of images would not correctly represent the relative abundance of large and smaller features.

The correct way to handle this is to combine the counts in the various size bins of the histogram in proportion to the amount of area (in real measurement units) examined in each set of images. In the example above, it would take 100 pictures at 1000X to cover as much area of the section surface as each picture at 100X. If fewer than 1000 high magnification pictures have actually been counted, then the data must be multiplied by the ratio of the image area viewed at 100X divided by the area viewed at 1000X before it is added to the data from the ten 100X pictures. This method works for any combination of measurements at different magnifications. It also should be used before comparing distributions. This is why most systems keep track of the total reference area (summed area of measured images) as they are measured and added to the data base.

Plotting shape and position data

Although most of the examples cited thus far deal with some measure of size as the sorting or classification parameter, the other main categories of measurement parameters are also used. As discussed in the previous chapter, these are measures of shape, position and density. Values for any of these parameters can be selected for sorting, to

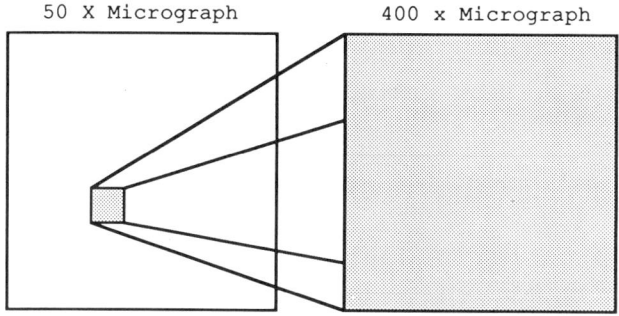

Figure 8-38: Increasing the magnification of an image by 8 times allows examination and measurement of only 1/64 as much area.

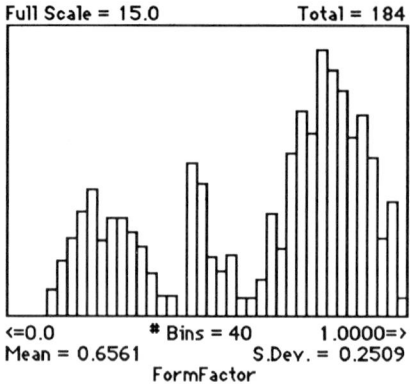

Figure 8-39: Histogram of frequency vs. shape parameter (a portion of the image appears in Figure 9-3).

produce distribution histograms showing the numerical frequency of features in the population being studied (Figure 8-39). We will see in the next chapter that shape is often a primary tool in recognition of objects, and that distributions with multiple peaks may indicate different families of objects that can be distinguished.

The main difficulty with using the various shape parameters is that they are not "linear" in the same way that size dimensions are. An increase in the value for length (or equivalent circular diameter, or any of the other size parameters) can be sorted either arithmetically or geometrically (a linear or logarithmic scale), and the results are understood and interpretable. The shape parameters are usually ratios of these size dimensions, and behave in quite different ways. It is not easy to interpret the meaning of these changes, and it means that the concept of a normal (or log-normal) distribution has little if any meaning. This is particularly important when comparing different distributions, since only non-parametric tools can be properly used.

Position parameters are fairly straightforward. A plot of the number of features as a function of their distance from the surface of a specimen, for instance, is readily understood as a gradient. It may be more important to plot the sum of the area of volume of features sorted by their distance that the number, of course, if the concern is with the amount of the phase (Figures 8-40 and 8-41). This type of plot in which a second variable is summed was described before.

Figure 8-40: Image showing a gradient of feature size.

Stereological Interpretation 255

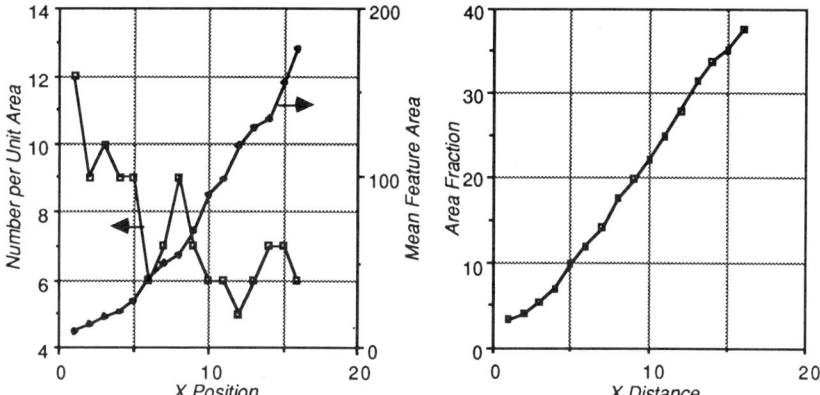

Figure 8-41: Distribution plots of feature count and area, and total area fraction vs. X position.

It may be important to characterize gradients of things other than the number or size of features. The variation of shape, orientation, or density of features with position can also be important (Figure 8-42). Finally, the gradient may not be along a straight horizontal or vertical path. One of the common situations is that variation takes place inwards from a surface. As was shown in an earlier chapter, one way to measure these variations for irregular surfaces is to make a mask of the overall shape, and then use sequential erosion to create individual masks for each band of depth. These are then used to count or measure the features within each band, from which gradient plots can be constructed.

The orientation angle of features is also a position parameter (Figure 8-43). When it is used as a sorting parameter for distribution histograms, some special considerations arise. First, the usual plotting scale covers only 0 to 180 degrees because in most definitions of orientation, either end of the object axis is the same. There are exceptions, such as using as a measure of orientation the angle of the vector between the feature centroid and the center of the enclosing rectangle, which would vary from 0 to 360 degrees. In any case, the abrupt discontinuity produced by the ends of the conventional histogram axis are arbitrary and potentially misleading. It is possible to plot these data as a rose diagram, to better show the true orientation pattern present. If the actual range of the data is 0 to 180 degrees, the resulting plot is symmetrical as shown in Figure 8-44.

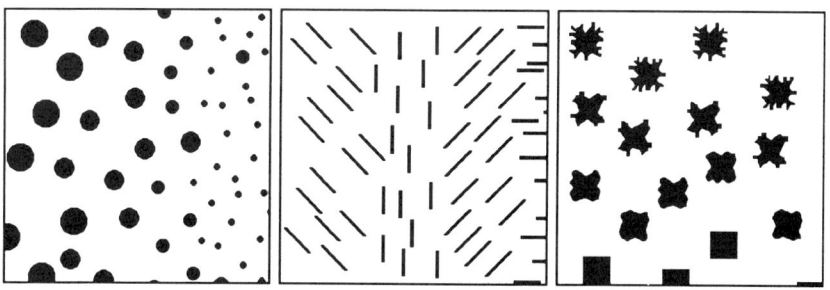

Figure 8-42: Gradients of shape and orientation.

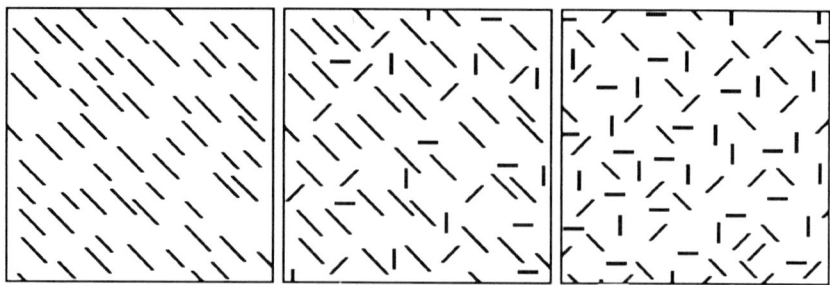
Figure 8-43: Partial and total anisotropy can be characterized by histograms.

These diagrams can also be used with a variable other than number summed in each histogram bin. An example was shown earlier in which the length of features was summed as a function of their angle. This is often used to study preferred orientation, whether of magnetic iron oxide particles used for recording media (Figure 8-45) or fault traces visible on a geological map. It reflects that fact that in both situations the importance of the features is in proportion to their length, and the magnitude of peaks in the resulting histogram better describes the directionality inherent in the structure than a simple count would. In some cases it is also useful to plot the total density of features as a function of their orientation.

When angle values are used in histograms, it may be important to remember that some measurement techniques produce angles that are quantized. For instance, if the angle of the Length is used, and this is determined from the maximum Feret's or projected diameter in any of a series of directions spaced perhaps every 5 or 10 degrees, then all angle values will be in multiples of that step. This means that the histogram bins should also be in multiples of the step value. If it is not, then bias and aliasing can occur if the number of possible angle values in each histogram bin varies.

Finally, density values may be used. This encompasses all parameters that are derived from feature brightness, including color data. It may be possible to establish a

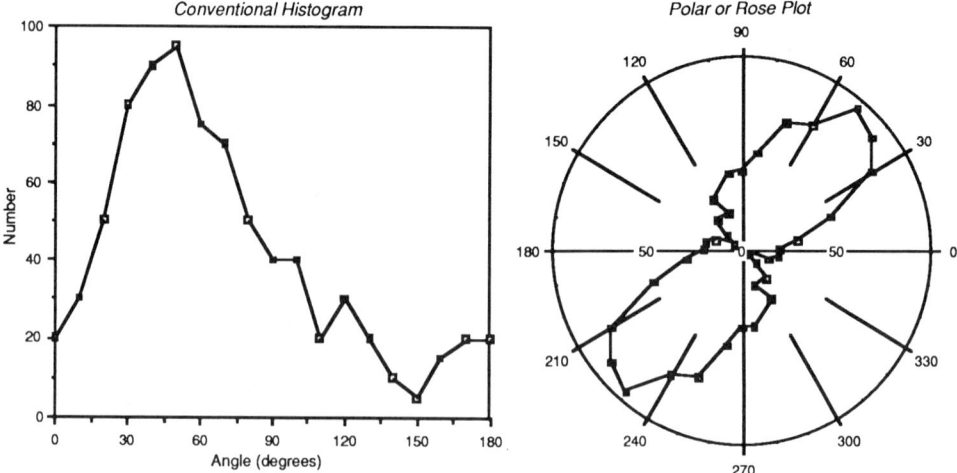
Figure 8-44: Frequency vs. angle plotted as a conventional histogram and a rose or polar plot.

Stereological Interpretation 257

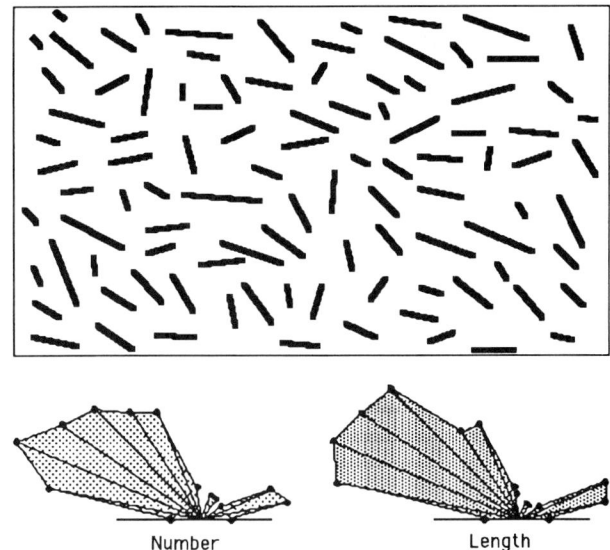

Figure 8-45: Magnetic iron oxide particles (binary image from TEM) and rose histogram of number and length vs. angle.

calibration curve that relates the brightness of X-ray maps to elemental concentration, or the density of a light microscope image to the dosage of some drug whose reaction site has been stained. These calibration curves are generally either approximately linear or logarithmic, depending on the imaging technique, but the actual details of the calibration, and whether it is possible to obtain it by calculation or whether it must be fitted to a series of measured points, does not matter in the discussion that follows.

Usually, the sorting value will be the derived parameter value, making use of the established calibration curve. This has the advantage that the units along the sorting axis have some meaning to the user. The use of cumulative histograms may be especially effective in revealing patterns of concentration, presence of different phases, and so forth.

Other plots

In the example shown above of a two-way histogram of frequency as a function of feature shape and size, it appears that not only is there a peak in the center of the distribution, but that in addition there may be some trend linking the two parameters. This is most directly studied with a correlation plot, in which the axes are the values for the two parameters and each feature is plotted as a point. Trends show up readily in this kind of plot, and can also be quantified using a correlation coefficient.

The example in Figures 8-46 and 8-47 shows an image of alumina particles viewed in the SEM. From the measured projected area, an equivalent spherical diameter was calculated as a measure of size. The shape is described by the formfactor, which decreases from 1.0 as the irregularity of the object outline increases. The result is a plot showing considerable scatter, because these particles are stochastic in nature, but nevertheless has a definite trend. The larger particles tend to be more irregular than the small ones.

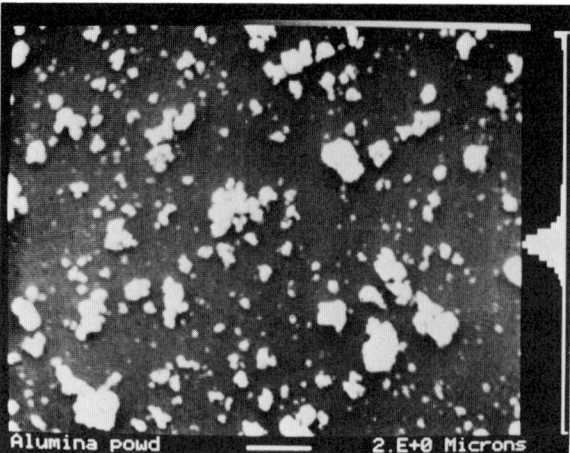

Figure 8-46: SEM image of aluminum oxide particles.

Sometimes this kind of result will simply reflect the inability of the imaging technique to resolve details on the smaller features. In this case, the trend reveals something about the origins of the particles: the larger ones are agglomerates of smaller ones. The small ones are fairly regular in shape, but the agglomerates are not (this is a fundamental property of most agglomeration processes) and so have more irregularity and a lower shape factor, on the average. The opposite trend, in which the larger particles are more regular and the smaller ones become irregular, is often an indication that the small ones were produced by fragmentation of larger ones, producing rough and irregular fragments from more smoothly shaped original objects.

A statistical measure of the degree of the correlation between the two variables in this kind of plot is provided by the correlation coefficient and the Student's t-test. The linear correlation coefficient is generally used, and consists of fitting the "best" straight line through the data and summing the squares of the deviations of the points from the line. A normal linear least squares fit line would have the form $y = mx + b$, and the optimum values of the constants m and b are calculated from the N pairs of x_i, y_i data points.

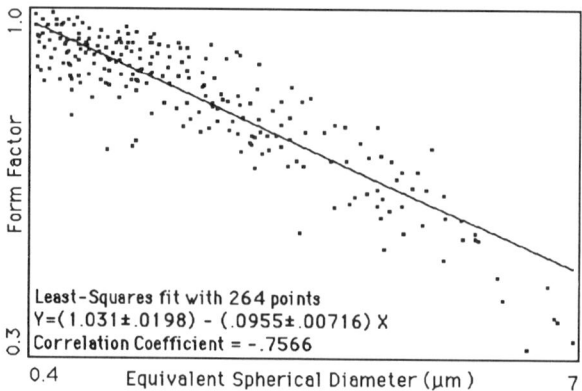

Figure 8-47: Plot of size vs shape for alumina particles.

Stereological Interpretation 259

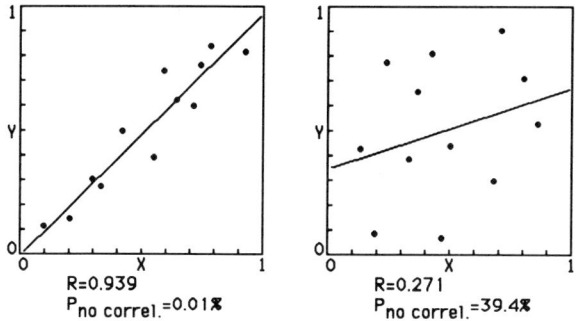

Figure 8-48: Data sets with high and low correlation coefficients.

$$m = \{N\Sigma x_i y_i - \Sigma x_i \Sigma y_i\}/\{N\Sigma x_i^2 - (\Sigma x_i)^2\}$$
$$b = \{\Sigma y_i - m\Sigma x_i\}/N$$

This makes y the dependent variable, and minimizes the vertical deviations of the points from the line (Draper & Smith, 1981). If the fit is also performed the other way (to minimize the horizontal deviations) an equation of the form $x = m'y + b'$ would result. The correlation coefficient r is then defined as $(mm')^{1/2}$. The magnitude of r can vary from 1.0 (perfect correlation in which all of the points lie on the line) to 0.0 (no correlation) as indicated in Figure 8-48. The sign of the coefficient may be either positive or negative, indicating whether the values of the two parameters increase together or whether an increase in one is accompanied by a drop in the other.

The significance of the magnitude of r depends on the number of points involved in the fit. The number of degrees of freedom in the fit is $(n - 2)$ for n points. Graphs or curves of the probability that a given value of r could arise by chance fluctuations allow the significance of a given correlation plot to be estimated (Figure 8-49).

The correlation coefficient can also be used with many different kinds of data. For instance, we used the automatic image analyzer to count the number of chips in homemade chocolate chip cookies. Every child knows that you can turn them upside down to count the chips, which melt through to the hot pan. Discrimination of the image (Figure 8-50) produces a feature for each cookie with holes that correspond to the chips, from which the number of chips in each cookie are counted (Figure 8-51). The number

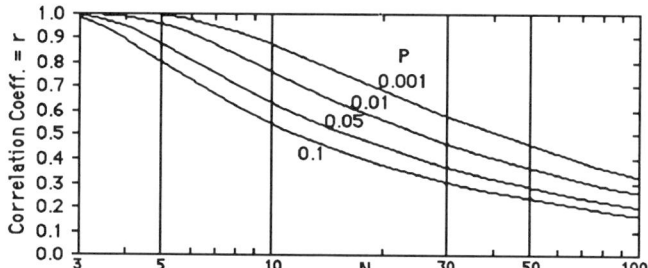

Figure 8-49: Probability of significance of values of the linear correlation coefficient r for N observations.

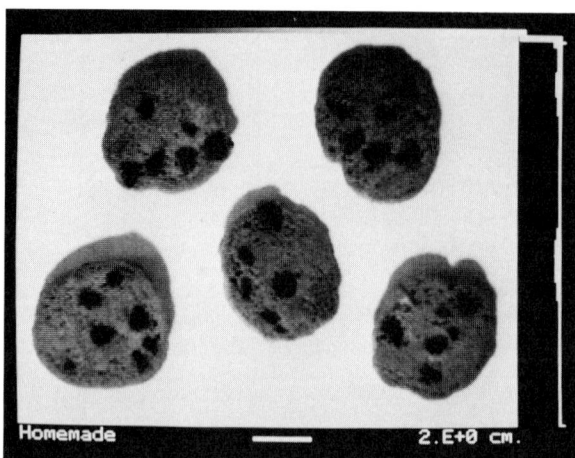

Figure 8-50: Image of chocolate chip cookies.

of chips was found to be correlated with the size of the cookies for homemade cookies (Figure 8-52), but not for "bought" cookies. This is not really surprising, since the bought cookies have fewer chips and also are subject to tighter "quality" control. Also, the process of making the homemade cookies involves dropping the dough from a spoon to the cookie sheet, and the amount of dough tends to be controlled by the number of chips.

When the parameters involved in the correlation coefficient are not inherently linear, or the trend between the two variables is not well modelled by a straight line, the value of r cannot be so simply interpreted. For the shape factor shown in the example, this is the case. We have already seen that formfactor is not a linear parameter. In other cases, the density may vary logarithmically in a plot relating density to size, or size itself may be best plotted on a logarithmic scale if a population of features has a log-normal size distribution. When transformed variable axes are used, the primary assumption used in the linear fit, namely that the deviation of all points from the line should be equally minimized, is invalid. Another problem arises when angle values are used, because angle values "wrap around" at 180 degrees. A plot using angle as a parameter may show a very distorted result that is not easily used in correlation.

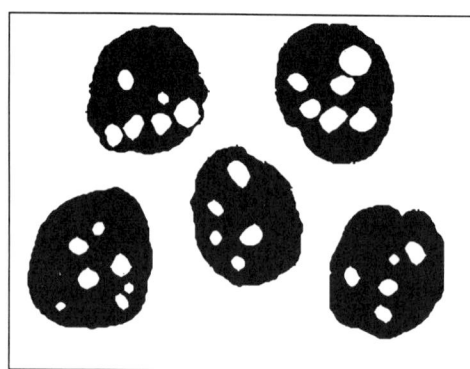

Figure 8-51: Discriminated image from Figure 8-50 in which holes correspond to chips.

Stereological Interpretation

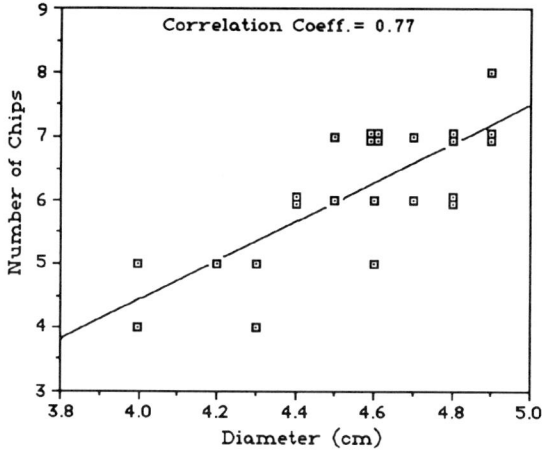

Figure 8-52: Correlation of number of chips vs. cookie size.

The most common solution to these problems of nonlinearity is to plot the rank order of observations rather than the actual value of the measured parameters. This can be done for one or both axes. When the rank order of each feature is plotted on both parameter axes, it is possible to use the binomial theorem to decide whether the data are correlated. This is similar to the Wilcoxon test discussed before.

Another situation in which stereological information can be obtained from a plot, but the data to be plotted are not those directly measured, deals with the distribution of features on the image plane. It is often interesting and important to characterize the distribution of features as regular, random or clustered. This can be important in many different fields. Astronomical studies of galaxies are finding clustering on various scales. Recall that this clustering did not affect the mean free path between stars, but it certainly can influence the distance to the nearest star. Precipitates that form in a solid metal may be regularly spaced when diffusion depletes the region around each of the element involved in their formation. Particles scattered on a substrate may not be randomly arranged if subtle forces act between them.

The tool to uncover these relationships is the neighbor distance. When particles are randomly scattered on a surface, the distance between the particles is a Poisson distribution (Figure 8-53). We saw earlier in this chapter that a mean free path between

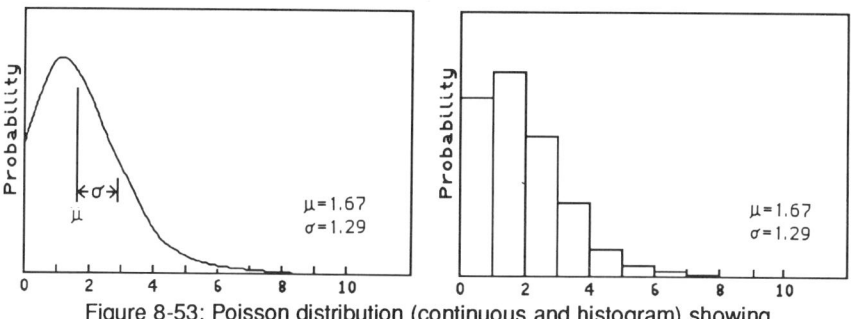

Figure 8-53: Poisson distribution (continuous and histogram) showing mean and standard deviation.

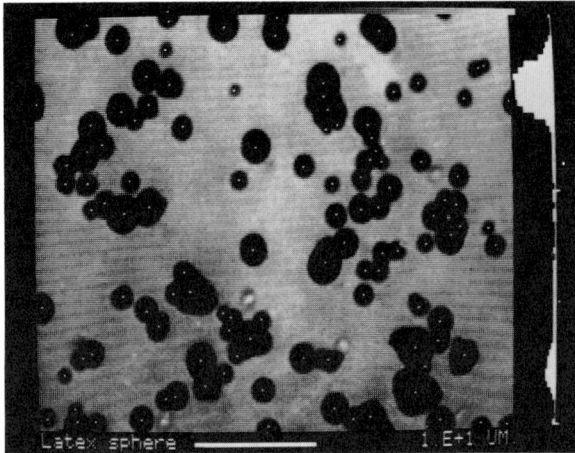

Figure 8-54: Feature centroids in touching latex spheres located by the watershed segmentation method.

features can be determined from intercept length measurements. This is a global rather than a feature-specific parameter. If the location of each feature (most typically the coordinates of the centroid are used, but other methods are also possible as shown in Figure 8-54) is determined, then it is possible to sort through all of the recorded data to find which is the nearest neighbor to each feature, and what its neighbor distance is. In fact, this same method can find the first k nearest neighbors (called a kNN search), and this technique will be discussed in Chapter 9 in connection with feature recognition (in which case the coordinates will be measured parameters other than the feature's position).

Searching and sorting are standard computer routines. They are not fast, and in fact the time required grows rapidly with the number of features, but they are much faster than humans and allow us to do things that would not be attempted by manual means. The distance between features is calculated as the vector or Pythagorean distance between the two centroid coordinates. This calculation is performed for each feature and all of the others in the field, and the shortest value obtained. A histogram of the frequency of these nearest neighbor distances is then constructed (Figure 8-55). Its shape tells us a great deal about the distribution (Schwarz & Exner, 1983).

We will assume that the features are small enough that the centroid-to-centroid distance is essentially correct, and no adjustment for periphery touching is needed. Also, the edges of the field of view will be ignored. This is valid if there are a lot of features measured, so that only a few lie near the edges where there actual nearest neighbor may not be visible. Then if the actual distribution of the features is random, the histogram of nearest neighbor distances should be a Poisson distribution. This can be characterized by its mean value, since the standard deviation of an ideal Poisson curve is just the square root of the mean. Furthermore, the mean value should be directly calculated from the number of features per unit area (Mean = $0.5/\sqrt{N_A}$). This means that if N_A is determined, there can be only one histogram that would correspond to a random distribution of features in the image.

If the mean nearest neighbor distance is significantly greater than the Poisson mean for a random distribution, and the standard deviation is less, then it is an indication that

Stereological Interpretation

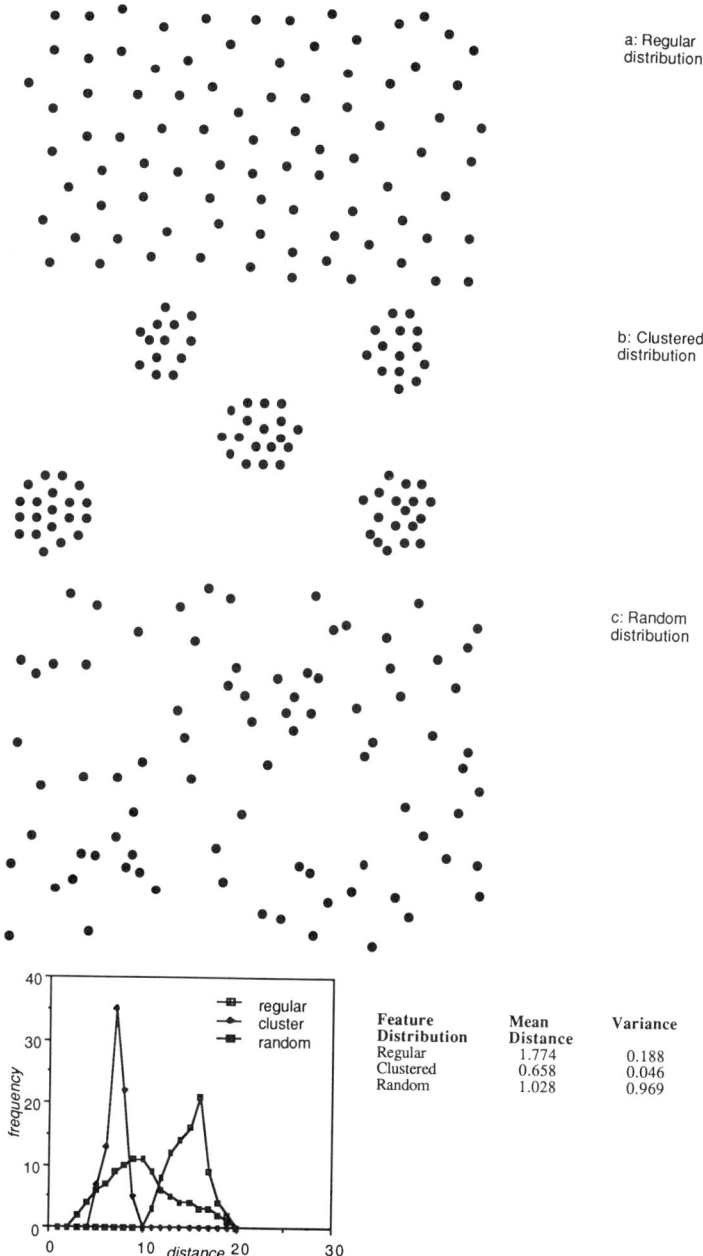

Figure 8-55: Binary images with regular (a), clustered (b) and random (c) feature distributions, their nearest neighbor distance distributions, and descriptive statistics (mean and standard deviation) ratioed to those for an ideal random Poisson distribution.

the features are regularly spaced. In the limit, for a perfectly regular spacing in a grid pattern, the standard deviation would drop to zero but the mean spacing would be twice the Poisson mean for the random case.

Conversely, if the features are clustered together (in one or more groups) the mean nearest neighbor distance will be less than for the random case. The standard deviation will also be less if the clusters are well separated and distinct. A greater standard deviation indicates that the clusters are superimposed on a background which may be either random or regular. Statistical tests can be applied to assign significance to these various cases, by comparing the values to those for a Poisson distribution and taking into account the number of points considered.

A more complete analysis that looks beyond the nearest neighbor distance uses a plot of the number of neighbors as a function of distance (Baddeley et al., 1987). For a random distribution of points, this curve rises as the cube of the distance. For a uniform or regularly spaced array, the curve rises more slowly than this, and for clusters it rises more quickly. With the entire curve, it is possible to show clustering at one scale and a random spacing or regular spacing at another scale. However, constructing such a graph for a useful number of points is not often done because of the effort required.

References

A. J. Baddeley, H. J. G. Gundersen, L. M. Cruz-Orive (1986) *Estimation of surface area from vertical sections* J. Microscopy 142, 259-276

A. J. Baddeley, C. V. Howard, A. Boyde, S. Reid (1987) *Three-dimensional analysis of the spatial distribution of particles using the tandem-scanning reflected light microscope* Acta Stereologica 6 Suppl. II, 87-100

P. R. Bevington (1969) *Data Reduction and Error Analysis for the Physical Sciences*, McGraw-Hill, New York NY

G. L. L. Buffon (1777) *Essai d'arithmetique morale* Suppl. a l'Histoire Naturelle, Paris, 4

J. W. Cahn (1967) *The significance of average mean curvature and its determination by quantitative metallography*, Trans AIME 239, 610

J. Chermant (1986) *Characterization of the Microstructure of Ceramics by Image Analysis* Ceram. Int. 12 67-80

R. Coleman (1989) *Inverse Problems* Journal of Microscopy 153 233-248

L-M. Cruz-Orive (1976) *Particle size-shape distributions: the general spheroid problem*, J. Microscopy 107, 235

L-M. Cruz-Orive, E. R. Weibel (1981) *Sampling Design for Stereology* J. Microscopy 122, 235-257

R. T. DeHoff (1962) *The determination of the size distribution of ellipsoidal particles from measurements made on random plane sections*, Trans AIME 224, 474

R. T. DeHoff (1967) *The quantitative estimate of mean surface curvature*, Trans AIME 239, 617

R. T. DeHoff (1978) *Stereological uses of the area tangent count*, in Geometrical Probability and Biological Structures: Buffon's 200th Anniversary (R. E. Miles, J. Serra, ed.) Springer Verlag, Berlin, 99

R. T. DeHoff (1981) *Stereological meaning of the inflection point count* J. Microscopy 121 13-19

R. T. DeHoff, F. N. Rhines (1968) *Quantitative Microscopy*, McGraw-Hill, New York NY

N. Draper, H. Smith (1981) Applied Regression Analysis (Second Edition) Wiley, New York NY

H. E. Exner, H. P. Hougardy (1988) Quantitative Image Analysis of Microstructures, DGM Informationsgesellschaft mbH., Oberursel, Germany

J. D. Gibbons (1985) Nonparametric Methods for Quantitative Analysis (2nd Ed.) American Sciences Press, Columbus, OH

H. J. G. Gundersen (1986) *Stereology of arbitrary particles* J. Microscopy 143 3-45

H. J. G. Gundersen, E. B. Jensen (1983) *Particle sizes and their distributions estimated from line- and point-sampled intercepts. Including graphical unfolding* J. Microscopy 131 291-310

H. J. G. Gundersen, E. B. Jensen (1985) *Stereological estimation of the volume-weighted mean volume of arbitrary particles observed on random sections* J. Microscopy 138 127-142

H. J. G. Gundersen, T. F. Bendtsen, L. Korbo, N. Marcussen, A. Moller, K. Nielsen, J. R. Nyengaard, B. Pakkenberg, F. B. Sorensen, A. Vesterby, M. J. West (1988) *Some new, simple and efficient stereological methods and their use in pathological research and diagnosis*, Acta Pathologica, Microbiologica et Immunologica Scandinavica 96 379-394

H. J. G. Gundersen, P. Bagger, T. F. Bendtsen, S. M. Evans, L. Korbo, N. Marcussen, A. Moller, K. Nielsen, J. R. Nyengaard, B. Pakkenberg, F. B. Sorensen, A. Vesterby, M. J. West (1988) *The new stereological tools: Disector, fractionator, nucleator and point ampled intercepts and their use in pathological research and diagnosis*, Acta Pathologica, Microbiologica et Immunologica Scandinavica 96 857-881

T. M. Hare, J. C. Russ, J. C. Russ (1982) *Image measuring algorithms for a small computer*, Microbeam Analysis 82, San Francisco Press

E. Harrison (1987) Darkness at Night Harvard Univ. Press, Cambridge, MA

M. Hollander, D. Wolfe (1973) Nonparametric Statistical Methods, Wiley, New York, NY

M. G. Kendall, P. A. Moran (1963) Geometrical Probability No. 10 in Griffith's Statistical Monographs, Charles Griffith, London

G. Matheron (1975) Random Sets and Integral Geometry Wiley, NY

R. Miles, J. Serra (ed.) (1978) Geometrical Probability and Biological Structure - Buffon's 200th Anniversary Springer Verlag

G. A. Moore (1968) *Automatic Scanning and Computer Processes for the Quantitative Analysis of Micrographs and Equivalent Subjects* in Pictorial Pattern Recognition, Thompson Book Co, Washington, DC

D. Pohl, F. Redlinger (1977) *Metallographic Measurement of Pore Shape and its Significance for the Mechanical Properties of Sintered Iron* Powder Metallurgy Int'l 9/4 164-168

F. N. Rhines, R. T. DeHoff (1986) *Microstructology: Behavior and Microstructure of Materials*, Dr. Riederer-Verlag, Stuttgart

M. Rink (1976) *A computerized quantitative image analysis procedure for investigating features and an adopted image process* J. Microscopy 107 267-286

J. C. Russ (1986) *Practical Stereology*, Plenum Press, NY

J. C. Russ, T. M. Hare (1985) *Measurement of edge-intersecting features in SEM images*, Microbeam Analysis 1985, San Francisco Press, 133

S. A. Saltykov (1958) Stereometric Metallography, Metallurgizdat, Moscow

L. A. Santalo (1976) Integral Geometry and Geometric Probability Addison Wesley, Reading MA

H. Schwarz, H. E. Exner (1983) *The characterization of the arrangement of feature centroids in planes and volumes* J Microscopy 129 155-169

D. C. Sterio (1984) *The unbiased estimation of number and sizes of arbitrary particles using the disector* J. microscopy 134 127

E. E. Underwood (1970) *Quantitative Stereology*, Addison Wesley, Reading MA

K. S. Venkataraman, R. A. DiMilia (1989) *Predicting the Grain-Size Distributions in High-Density, High-Purity Alumina Ceramics* J. Am. Ceram. Soc. 72 33-39

R. Warren (1987) *Microstructural modelling in stereology*, Acta Stereologica 6#1, 157

E. R. Weibel (1979) *Stereological Methods vol. I & II*, Academic Press, London

E. R. Weibel, B. W. Knight (1964) *A morphometric study on the thickness of the pulmonary air-blood barrier* J. Cell Biology 21, 367-384

S. D. Wicksell (1925) *The Corpuscle Problem* Biometrica 17 84; 18 152

Chapter 9

Object Recognition

Object or feature recognition includes tasks of very different levels of complexity. At the upper end, this includes some of the most important and interesting applications of artificial intelligence. At the low end, it can be as "simple" as OCR (Optical Character Recognition), used in processing checks, sorting mail, and closely related to the location and reading of UPC (Universal Product Codes) symbols on packaged food at the supermarket.

Consider as a beginning how we can prepare a general purpose computer-based image system to recognize characters (letters and numbers) on a printed page. In many practical situations there are tight constraints on the color, size, positioning, and especially the orientation of the characters. When this is so, a technique known as template matching can be applied.

Figure 9-1 shows an example of binary template matching. The top row contains templates for letters A thru E. Matching each against a letter E from a different font produces the results shown in the bottom row. Solid pixels are ones that match the template, and hollow ones are unmatched pixels in either the target pattern or the template. The score for each template is the net difference between the matches and

Figure 9-1: Example of template matching.

```
2 7 3 9 8 8 9 9 9 3 4 7 7 7 7      3 5 7
1 6 9 7 7 7 2 2 6 8 6 3 7 8 8      5 9 5        5 8 6 6 5 3 3 2 5 7 6 7 7
8 8 4 3 4 4 9 2 4 7 7 6 5 3 3      7 5 3        7 6 4 6 3 4 3 6 6 3 7 6 4
5 9 5 6 5 4 8 7 9 1 6 9 4 8 4                   6 4 6 3 6 6 5 7 2 5 8 3 3
9 4 7 1 7 8 6 8 3 4 9 5 1 4 8                   5 4 3 5 8 7 7 1 1 8 4 3 4
5 5 5 8 5 9 5 6 3 9 8 7 9 3 1                   4 3 5 6 6 6 6 2 7 7 2 5 3
2 5 1 7 2 5 6 3 4 7 2 6 4 9 4                   5 3 5 2 5 6 5 3 5 4 6 3 4
8 3 6 3 6 9 4 2 2 7 8 3 7 4 6                   3 5 3 5 *9* 1 0 4 5 7 5 5 5
4 5 5 8 8 4 1 9 9 7 8 3 6 8 6                   6 6 7 7 1 0 4 7 7 6 3 5 6
7 5 5 4 1 5 6 7 6 8 2 3 2 3 3                   7 6 4 2 3 6 6 5 6 3 3 5 4
5 3 3 5 6 6 2 6 7 9 9 5 8 3 1                   4 5 5 7 4 3 6 7 6 5 4 7 3
5 9 6 5 2 3 9 9 9 5 8 3 6 2 4                   7 6 7 5 1 6 7 6 6 6 5 4 0
8 8 4 7 4 5 8 1 7 7 6 6 5 9 5                   7 4 5 5 7 5 1 5 7 6 5 3 6
4 5 4 8 6 4 2 7 9 7 2 4 5 9 7                   4 4 7 5 6 1 3 8 5 4 5 6 7
7 3 7 4 8 8 2 9 3 7 3 5 5 7 1
          Image                    Target              Result
```

Figure 9-2: Single-digit representation of brightness values in an image fragment, and the result of matching a 3x3 target pattern using cross correlation. The location of best match is marked in the result (highlighted '9'). Notice that the target does not exactly match the image even at this location.

mismatches. This value is usually normalized by dividing by the number of pixels in the template. Notice in the example that the letter E is identified even though the font is different (with serifs) than the template. However, the score for some other characters may be close or even higher, and any misalignment, rotation or difference in size of the character with respect to the template makes a substantial difference in the score that can alter the identification. We will consider later what is required to deal with characters whose size and orientation can vary.

Locating features

Another application that can be described as a recognition problem is locating known features in an image. An example of this will be used in our discussion of locating matching points in stereo pair images for surface topographic measurement, in Chapter 11. In that case the pattern of brightness values that is being searched for is taken from one image, and looked for in another. In other cases, patterns that represent specific objects or parts of objects (for instance, corners) are known beforehand.

The most common method used for this task is cross-correlation, which is analogous to template matching, but is carried out using grey-scale images. A target pattern is matched to every possible location in the image (or, in some cases to every location within a specified region) as shown in Figure 9-2. At each location (i,j) the cross-correlation score is calculated as shown, where the summation reflects the size of the target, and B and T are the brightness values of pixels in the image and target, respectively:

$$\frac{\sum_{x=-1}^{+1}\sum_{y=-1}^{+1} B_{x+i,y+j} \cdot T_{x,y}}{\sqrt{\sum_{x=-1}^{+1}\sum_{y=-1}^{+1} B^2_{x+i,y+j} \cdot \sum_{x=-1}^{+1}\sum_{y=-1}^{+1} T^2_{x,y}}}$$

This value is high where bright pixels in the target and image align. The normalization in the denominator corrects for any variation in the overall brightness of different regions of the image. The correlation score is then scaled appropriately to produce another grey scale image, whose brightness is proportional to the degree of match. Bright spots in this image indicate matches between the image and the target.

This technique is sometimes used to locate edges and corners in images, by searching for independently specified patterns. It is also very commonly used to match points between two images, for instance left and right stereo pair views, or sequential images being used for motion flow purposes. In these cases, the pattern of brightness values is taken from one image and matched against regions in the second image within a reasonable search region. Sometimes, brute force matching using every point in one image and searching for its match in the second is used. Usually, matching from each image to the other is used for only the "interesting" points (selected on the basis of a large variance or gradient in the local brightness values), and conflicts are resolved by selecting the match with the best score, or by imposing restrictions such as requiring continuity in the order of points from one image to another (accomplished with a relaxation technique).

It is also possible to use this method to search for objects in images, for instance in a surveillance situation. In this case, multiple target images, representing all possible orientations or views of the object of interest, must be used. Also, the targets are often quite large and detailed. Consequently, this approach is rather computer intensive. Even though it is well suited to the use of highly parallel computers, it is not often used because of the difficulty of predicting all of the possible target patterns that might be of interest, problems of false matches and missing "camouflaged" objects, and because it does not take advantage of any intelligent interpretation of the image to locate, identify or measure features that could be of interest.

Parametric object description

At one extreme, such as the text-reading case, the objects to be recognized form a small and fully defined population, known *a priori*, and presented to the vision system in a tightly controlled format which may be two-dimensional and well aligned. At the other extreme lies the desire to have the computer "understand" natural 3-D scenes, including the ability to recognize surfaces and their hierarchical connections, for instance to permit automatic navigation or robotics handling (Roberts, 1982; Ballard & Brown, 1982; Ballard et al., 1984). Our interest here is between these limits: to be able to program the image analysis system to recognize discrete features as encountered in a typical microscope image.

By breaking the recognition problem down to simple steps, we will see that this kind of recognition can be achieved in many useful situations. In some cases, in fact, the computer can rather quickly surpass even the performance of the human who initially trained the system.

In previous chapters, a large number of measurements of individual features were presented that are available to the computer-based image analysis system. Furthermore, stereological tools allow interpretation of these parameters to describe three-dimensional characteristics of objects whose two-dimensional representations appear in the images. The kinds of parameters available, measures of size, shape, position and brightness, include many more things than the human eye normally considers when responding to images. Some things that the human eye is very poor at quantifying are also determined. How then is the human visual system so good at recognizing objects, even those presented in an unusual orientation or context? Furthermore, how can the computer system be taught to perform the same kind of recognition?

Many systems can measure a number of different parameters for each feature present and sort them in some kind of order, often displaying the results by color- or brightness-coding the features on the display in proportion to the magnitude or rank order of the

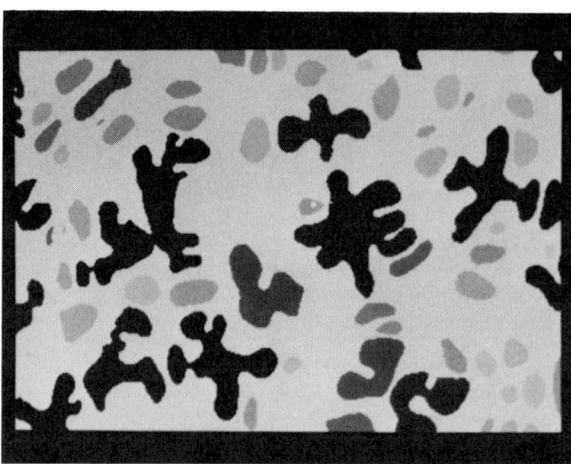

Figure 9-3: Features brightness-coded by their formfactor.

values (Figure 9-3). This is inadequate for recognition purposes because it is not clear that any single feature parameter will distinguish objects, nor if it would, where the cutoff point to separate different populations should lie.

While the image measurement programs can quantify some parameters that do not readily correspond to things the eye can determine (such as absolute brightness, or various shape factors), it is convenient and often necessary to include them. They represent a different, not necessarily better approach to something that the human visual system does very well, comparing objects. The comparison may be made between objects in a single image, or between objects currently seen and others in memory. Because the visual system is oriented more toward comparison than measurement, it also has wired-in mechanisms to facilitate these comparisons in spite of handicaps. For instance, to measure color the eye compensates for extreme differences in shading or the color of illumination, by normalizing the image based on information within the image itself.

Not all of these mechanisms are fully understood in a physiological sense. But with a computer image analysis system, the electronic sensors that respond to color can be absolutely calibrated. Light sources and shading can be controlled while acquiring the images, so that they do produce accurate color information.

The shape of objects is recognized in complex ways by the visual system. Instead of deriving numerical values as discussed in Chapter 7, the eye rotates objects to the same orientation before making comparisons and deciding that features are the same or different (Figure 1-2). Even then it is easily fooled by symmetries and reflections. Some of these optical illusions will be considered in the final chapter. A particularly interesting and revealing characteristic of the visual matching of patterns occurs when similar objects in different orientations are viewed. The length of time required to make a decision as to whether objects are the same is directly proportional to their angular deviation. We literally turn things over in our minds as we examine them. This is also true when comparing objects to our memory of them. The memory is stored in some specific orientation, and must be re-oriented to match the viewed object.

It appears that the parameters measured by a computer system include the same kinds of information available from early vision. At higher levels, the human visual system has a capability that we will not try to program into our recognition system. This is the ability to imagine what an object might look like in some other orientation. This process is not too different in principal from the use of geometric probability to construct a three dimensional solid and then measure the two-dimensional parameters that would be found in all possible orientations. But it is much more efficient, and also works better in the reverse direction – imagining what the solid might be shaped like that could produce a given two-dimensional image.

Since our experience is primarily with projected images (solid objects viewed in a transparent medium of air), this projective or imaginative capability works better for these cases than for sections through objects. Indeed, it presents a real challenge to try to imagine some section examples. Try to draw the appearance of a representative section through some complex shape(s) embedded in an opaque matrix in random orientations to test your own skills. If you have difficulty in visualizing the result, make a real sample by mixing Cheerios into fudge or noodles into Jello. The appearance of the section surfaces does not readily yield an interpretation of the three-dimensional shape of the objects.

By allowing the use of any of the measured feature parameters, even those that do not have good parallels in the human visual system, it might appear that the computer would have an advantage over the human eye and brain in feature recognition. But nature has been extremely efficient in developing a solution to this problem which we can only approach. Computer image comparisons are serial, while the human brain not only works in parallel, but can devote an extraordinary number of cells to each pixel in the image to watch for particular object parameters. It is a standing joke in human vision research that no one has yet found a "grandmother" cell, that is a cell that has been trained to recognize your grandmother, patiently and quietly examining every image that falls upon your eye, only signalling the conscious levels of the brain when she comes into view. There is probably not such a single cell, but there are certainly groups of cells that selectively recognize faces (a unique collection of features such as eyes, nose and mouth, each of which have a particular shape and color), and from there we can postulate a hierarchy of tests and matches that lead to your grandmother's image.

Our computational equivalent of the grandmother cell will start with a look at how such a cell might work, if it existed. The output of the cell is simply a logical signal, OFF or ON. The input must be a number of factors, such as 1) the object is a face, 2) the hair is white and arranged in a bun, 3) the eyes are blue and looking at me, 4) the pattern of wrinkles around the mouth is the same as one stored in memory, etc. None of these by itself is adequate to assure recognition or trigger the cell to fire. But at the same time we don't require all of the conditions to be met exactly: perhaps her hair style has changed, or she is wearing a hat and we can't see it. In other words, there must be a weighting scheme that operates. There are also some negative factors, for instance if the face has a bushy red mustache then it isn't grandmother. Some factors are more important than others, and when the total score is high enough, the cell signals that grandma has appeared. The result will not always be correct. We may miss grandma in some situations, or mistake someone else for her occasionally. But the success rate is remarkably high and the process is extremely efficient in both speed and storage requirements.

This kind of cell is actually not too different from the way a neuron is believed to work (Figure 9-4). There are computer programs that emulate these neural nets, in which

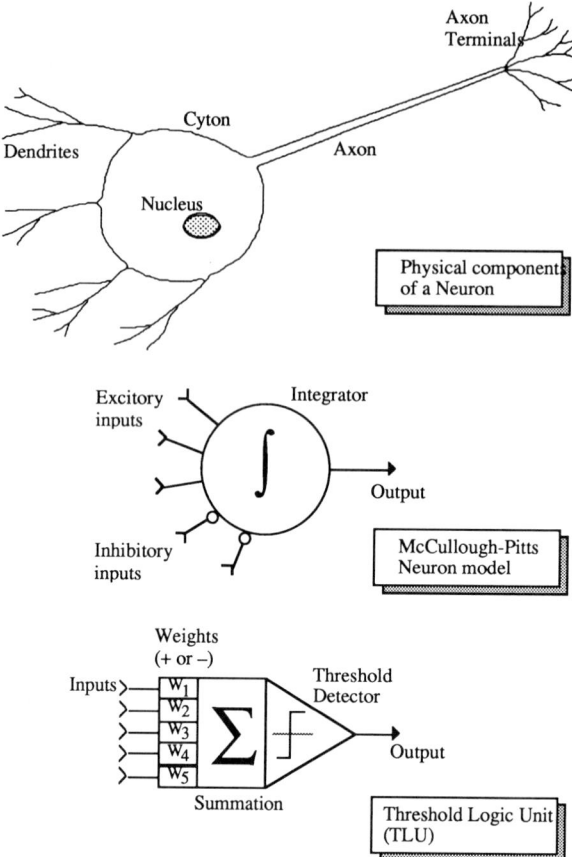

Figure 9-4: Diagram of neuron and Threshold Logic Unit

multiple inputs from other neurons or sensors are modelled, some more important than others and some weighted negatively. When the rate of pulses being received from all these weighted inputs exceeds a threshold, then the cell fires, or when the summed output of the weighted inputs exceeds a threshold, the output signal is sent on to the next logic unit. The main functional difference in "real" neurons is that the output is proportional, rather than simply being an on-off signal.

The education of a neuron is accomplished by reinforcement. Each time the cell fires (signals that it has recognized something), it increases the weights of the inputs from other neurons that caused it to trigger. In the most complete programs, the backtracking from a successful conclusion reinforces all of the intermediate neurons that contributed to the decision. This feedback method will eventually result in pathways that correspond to images of common objects being strongly weighted, to produce rapid and confident recognition.

Distinguishing populations

A suitable simplified software model for this process requires that we first identify the inputs, and in our case these will be the parameters we can measure in the image. The most efficient way to set up a recognition system is to decide which measured parameters are important in any particular situation, in other words which parameters distinguish the population of objects to be recognized from other populations. Rarely will a single variable produce a histogram as in Figure 9-5 that separates two populations completely. For statistical reasons, we would like to have the minimum number of parameters necessary, since this will permit training the system with the fewest example objects and provide the greatest number of degrees of freedom in any mathematical fits that are performed.

To find the important parameters, we can start by measuring any that we suspect may be useful on a small training set. This may require as few as 20-50 objects that are independently known to be of the target class, and a similar number of other objects. For the present, we will use an example in which the objects are either of category A (the target) or B. This can be identified for statistical comparisons by assigning a value of 1 or 2 to each object using a variable named, for example, TYPE.

Which of the measured parameters are "important" to distinguish the two categories? An efficient tool for this study is the correlation coefficient. If the parameter values are plotted against the TYPE value, then the correlation coefficient for the plot will indicate whether from a statistical point of view there is any significant correlation present. We will not normally expect to find a single parameter that can separate the type categories with high probability. But we can select the few parameters that produce the highest values of the correlation coefficient for further use.

In addition, there may be some parameters that are redundant. For example, the derived parameters Convexity and Formfactor defined in Chapter 7 both use the Perimeter. If one of them shows a high correlation with category discrimination, and if the reason for the correlation has to do with shape rather than size, then there is a good chance that others that incorporate perimeter will also show a correlation. Amongst these shape factors there will probably be only one or a few that summarize the use of shape for recognition. In other words, some of the derived factors are redundant, and we want to know which to use. This will be determined by a stepwise regression, described below.

It should be noted that the objects selected for this training measurement will not, in general, be a truly representative set. In principle, this introduces some bias into the correlations and into the choice of parameters. In practice this effect is offset by two factors. First, the human operator generally selects (either consciously or unconsciously)

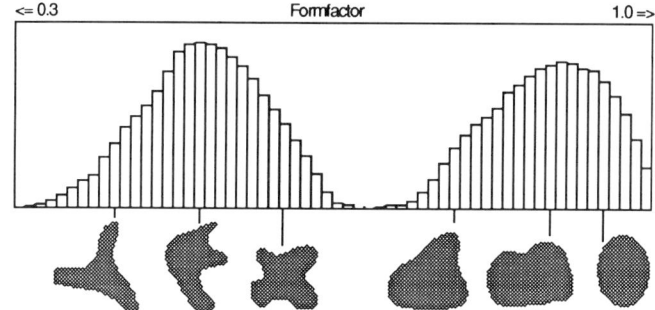

Figure 9-5: Example of two classes of objects distinguished by a shape factor, with their probability distribution functions (pdf's).

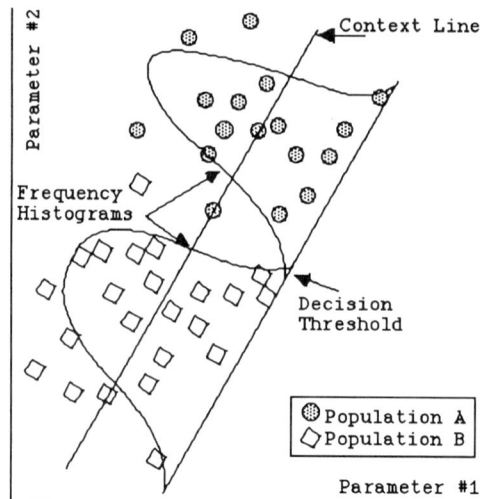

Figure 9-6: Two populations of points with a context line along which they are best separated in two dimensions.

"typical" objects that are more different in terms of whatever the significant parameters are than a truly random selection would provide. It is dangerous to count on this factor, of course. But, in many situations, if the selection is not random, it is apt to err in the direction of making the selection of useful parameters easier rather than more difficult. Secondly, we need not rely on this initial training set to establish the values of the important parameters which separate the object categories. We are at present only trying to find which parameters they are, and subsequent measurements on many more objects may later be added to develop the numeric values for the selection criteria.

The initial set of selected measurement parameters is then used to define a context line or discriminant function. This is easiest to visualize for the case of two measurement parameters (Figure 9-6), but generalizes directly to the n-dimensional case. Imagine that the values for two measurement parameters are plotted on the X and Y axes of a Cartesian graph, with a single point representing each object in the training set (some in category A and some in category B). Usually, these clusters of points will not be completely separated if projected along either the X or Y axis. If they were, then that measurement parameter would be all that was required to distinguish the two categories and provide the desired recognition.

There will be some line on the plot that represents an axis at some angle to the original axes, along which the points are maximally separated. This is called the context line. In two dimensions it can be drawn on the plot, and the histogram of values projected onto it. It is determined as the line which produces the minimum moment for the data points, or by performing a linear least-squares regression in which the category TYPE value is considered to be the dependent variable and the two measurement parameters plotted along the X and Y axes are the independent variables. It will always be the case that the separation of the two categories of objects will be greater along the context line axis than along either of the individual parameter axes.

When more than two measurement parameters are evaluated for the objects, the plotting of the values takes place in a higher dimensional space. This makes it difficult to visualize, but the mathematics of the situation is the same. There is some context line along which the points are maximally separated. This line is found by multivariate linear regression, in which the artificial TYPE variable is computed as a function of the form $a_1X_1 + a_2X_2 + ...$, where the a values are coefficients determined from the mathematical fitting operation, and the X values are the measured parameter values. The a values are the direction cosines of the context line in the n-dimensional parameter space.

The coefficients each have a precision value that can be evaluated as part of the fitting procedure. Because there are a limited number of sets of measured data (because a limited number of objects were used for training), both the coefficient values and their estimated uncertainties are subject to finite error. This error can also be estimated, and in fact helps us in deciding which parameters are not actually helping the fit.

Using a technique called stepwise regression, we can find which of the parameters are really necessary to the fit and which are not. Any parameter whose coefficient is smaller than its error is automatically dropped. The final result is a series of coefficients which are the direction cosines of the context line in n-space, where n is the number of parameters that actually need to be measured. In many cases this is only a few parameters, others having been eliminated even with individually high correlations with the TYPE value, because they are effectively superseded by other, more highly correlated parameters that make use of the same directly measured information.

It is helpful in understanding the context line to remember the original division of measured parameters into the four classes of size, shape, position and brightness. Parameters within those groups often substitute for each other, with only one or a few from each group actually being used. Generally there is a "best" choice parameter in any particular instance. For example, among the several shape parameters, it may be found that formfactor is superior to roundness for a particular application, while the reverse may be true in some other situation.

When parameters from difference classes are used, they generally represent different aspects of the objects' appearance. This is very desirable as it improves the ability to distinguish objects. In many cases, it turns out that the human user is not sure what aspect of appearance is actually distinct between categories of objects which he or she has no difficulty in distinguishing visually. After using regression to determine a context line it is sometimes possible to examine the parameters that have been selected and verify their reasonableness in terms of visual appearance, but sometimes it is not.

Decision points

Once a context line has been established, a distribution histogram of the plotted points for each measured object can be constructed for the two categories of objects which are to be discriminated. This is done by calculating the dummy variable TYPE from the equation for the context line. This is usually called the discriminant function. It is simply a linear combination of the measured variable values times the coefficients determined by the regression. Ideally it should give a value of 1 or 2 corresponding to the arbitrary values assigned to the TYPE variable before the regression was performed. In practice there will be some scatter, because the fit is imperfect and the measured values themselves have some associated error. The next task is to decide upon the value of the discriminant function that will actually be used to separate the two classes of objects (Dixon, 1983).

In principle this can be done in several ways. We could select a minimum point between the two histograms manually (Figure 9-7), test its performance with an Anova

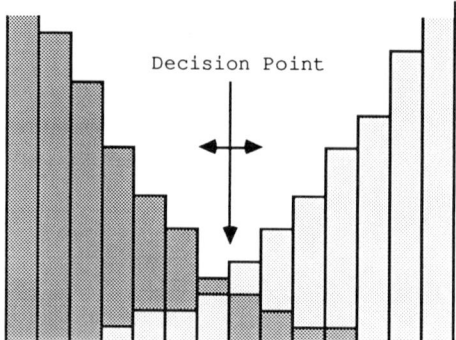

Figure 9-7: Overlapping histogram tails with a decision point at the minimum along a context line.

or other statistical test, and perhaps revise the initial estimate. Automatic search patterns can also be used to find a minimum, but this hides the fundamental and false assumption that the minimum point is the correct one. If the two populations are of different sizes, then this is the wrong criterion, and if the distributions are of arbitrary shape, simple statistical tests are inappropriate.

To determine the probability of misclassifying an object based on the discriminant function with any particular cutoff value, the area of the actual distribution histogram is used. The fraction of the total area that extends beyond the cutoff is the probability of error. In most cases, it is desirable to make the probability of misclassifying an object as either category 1 or 2 to be the same. However, this is arbitrary and there is no fundamental reason for this equality. For instance, in a quality control environment, the potential liability costs of mistakenly accepting an object as being within the target class may be quite high, while the cost of mistakenly rejecting it as being not in the target category may be much smaller. In this case, the cutoff value should be adjusted to reflect these costs, and produce error probabilities whose ratio corresponds to the relative costs.

In some cases, the shape of the distributions for categories 1 and 2 can be satisfactorily approximated by a Gaussian or normal curve. When this is so, the error probability and the degree of confidence with which the two categories can be distinguished is straightforwardly provided by the Anova (Analysis of Variance) method. This uses the difference between the distribution means and the two standard deviations to compute an F value, from which the probability is obtained.

However, in many cases the assumption of a Gaussian shape, central to the Anova method, is not justified. There are other statistical comparisons, called nonparametric tests, which can still be used. One of the simplest is the Wilcoxon (also known as the Mann-Whitney or U test when the populations being compared are different in size). This functions by considering the order of the values of the discriminant function along the context line, and the category to which each object belongs. Then using the binomial theorem to calculate the probability of this sequence occurring at random, the probability of successfully separating the two classes of objects can be assessed.

The simplest way to estimate the expected error is simply to use the area fraction of the histogram of discriminant function values along the context line (Figure 9-8). This is also nonparametric and has the advantage that it easily permits dealing with the unequal

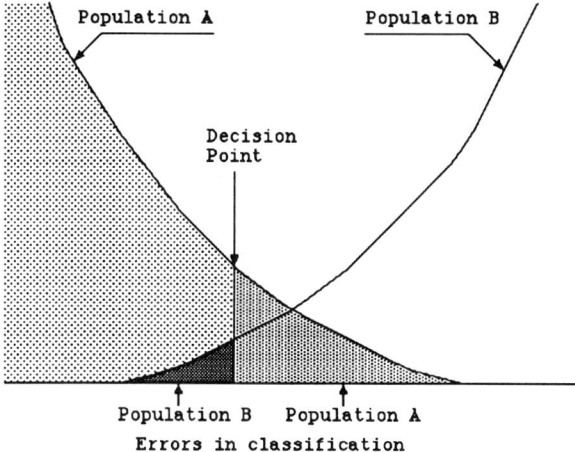

Figure 9-8: Detail of a decision point between two slightly overlapping populations, chosen based on the probability of error as defined by the normalized area of the tails.

probability case discussed above. Furthermore, it is readily recalculated as the histograms change.

The histograms may change with time because more objects will be measured. The initial training set of objects may have been non-representative of the actual population of objects, either in terms of the measured parameter values used in the discriminant function, or in terms of the frequency of objects in the two categories. As more features are measured, the shape of the histogram will be better defined (Figure 9-9), and the cutoff value may change.

In principal, the equation for the context line could change as well, or even the important measured parameters. However, as a practical matter it is much more difficult to allow these to change as data are acquired, and it is usual to consider it as fixed. This is due partly to data storage requirements. If the context line is considered fixed, then it is

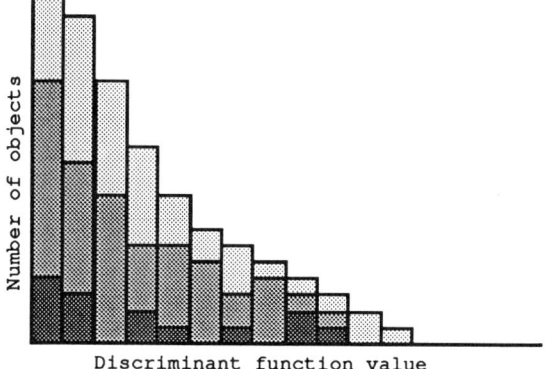

Figure 9-9: Improved statistical precision at the tail of a distribution with increasing number of observations.

only necessary to calculate the discriminant function for each object measured and add it to the evolving histogram of values, which requires minimal storage space. If the context line itself is to be recomputed then all of the individual measured parameters would be required for each object, which is often impractical. Of course, it means that the training set of objects cannot be too unrepresentative of the real populations to be distinguished.

The histogram that records the values of the discriminant function must have narrow enough bins to count the number of objects so that small changes in the cutoff value can be evaluated. As the discriminant function is a dummy variable that ideally varies from a mean value of 1 to 2 for the two categories, it can be expected that the cutoff value will normally lie between these extremes. Instead of having narrow histogram bins over the entire range, it is often satisfactory to have fine bins to count objects in the range of values near 1.5, with much coarser bins at more extreme values. There are not comprehensive guidelines that can be given for establishing these bins, but usually the histograms resulting from the training set of objects will serve to indicate a suitable setup.

The idea behind the storage of the histogram is that as many objects are subsequently measured, their discriminant function will be evaluated, and the value compared to the cutoff value. This will indicate whether the object is in category 1 or 2, and the degree of confidence in the result (the probability that the selection may be in error, as described above). Whenever the error probability is greater than some preset amount, the system will ask the operator to confirm the selection based on his or her visual examination and independent judgement.

This will initially result in the correction of some missed classifications because the training set of objects may not have accurately represented the category populations. As each corrected value is assigned to the proper histogram, the probabilities of error will change and the cutoff values must be re-evaluated periodically. Depending on the rate of measurement, this may occur after every altered classification, or every ten or one hundred such cases. In this way the system continues to learn. The histograms acquire many more data points so that the shape is better defined, the cutoff point is refined, and the error probabilities generally decline or at the very least become more precisely evaluated.

The ability to select a cutoff point based on the area fraction of the tail of the distribution that extends past the decision point allows different error probabilities to be assigned to each population, if this is appropriate. The use of the area fraction (the area of the tail divided by the total area of the histogram for each population) makes the method insensitive to the actual size of the distribution, except for statistical precision. Consequently, it is appropriate when the two populations are of different sizes.

The re-evaluation of the cutoff point and the related probabilities as more data are acquired uses all information available and is a Bayesian statistical approach to the data in that it minimizes the probability of errors of erroneous classification (Russ & Russ, 1987). It allows the system to continue to learn as more data become available (as more objects are measured). It is not at all unusual for the system, after a brief period of training, to become better at object identification than any of the people who trained it. My experience has been that after an initial training with fewer than a hundred objects, the ones that are presented for possible user reclassification are often ones that generate heated discussion as to their proper category amongst the users. In other words, the system is at least as confident as the human users in its assignments.

The ability of the system to derive its own context line by stepwise multivariate regression, and even to discard measurement parameters which do not improve the statistical quality of the fit, produces a knowledge base regarding the objects of interest. The ability to apply that knowledge base to new objects constitutes an expert system. The ability to refine the knowledge base continuously and utilize all new incoming data (including that which simply serves to better define the shape of the histogram) without user input or reclassification, qualifies as a learning system. Yet at each step of the way we have asked very little from the software, and can readily comprehend what calculations, comparisons or other tasks the system is performing.

Watching such a system perform virtually instantaneous classification of objects makes it appear to be very intelligent. The realm of computer artificial intelligence is a very nebulous one, which has at various times been associated with different types of tasks ranging from playing chess to behaving like R2-D2. Some researchers claim that artificial intelligence is whatever no one has managed to make computers do yet, so that once a task has been accomplished using computer methods, it no longer deserves the name. Perhaps this discussion will serve to increase confidence in the results, as well as making the realm of "artificial intelligence" less mysterious.

There are a few refinements that can be added to the procedure described above. The equation used for the context line in the previous discussion produces a straight line. There is no fundamental reason that this could not be a curved line in n-dimensional space, so that the projection of all object points onto the line gives the best discrimination of the features. In practice, it is difficult to use nonlinear functions for several reasons. First, it is not easy to decide *a priori* what other function shape should be used. Just allowing all possible quadratic terms (including the cross products of all measured variables) increases the dimensionality of the fit very rapidly, and would require many more data points for a statistically good fit (reasonable uncertainties in the coefficients that define the context line). This in turn requires many more objects in the training set, and correspondingly more work on the part of the user.

Furthermore, there is no reason to arbitrarily limit the function to polynomial form. Logarithmic, inverse, or other terms are equally plausible. It is rarely practical to evaluate all of these possibilities. If there is some *a priori* reason to expect a nonlinear contribution from a particular measured parameter, it can be converted to a new derived variable (such as the logarithm of the density, the square root of the area, or the inverse of the formfactor) before starting the process of selecting variables to perform the fit.

An additional difficulty with nonlinear context lines is knowing how to project the data points onto them to build a histogram. Some points may lie on more than one plane perpendicular to the curved line.

Other identification methods

It is interesting to compare this method of object recognition to several other commonly used approaches (Derde et al., 1987). The first of these, range analysis, also begins by plotting each point (representing a series of measurements on a feature) in an n-dimensional space, where each axis is a measured parameter (Figure 9-10). The points are then examined to locate clusters, which are considered to be categories or types. Newly measured points (new objects) are classified according to which cluster they are closest to, usually in a vector sense. This method is very difficult to apply in most real situations. Finding clusters is a highly nontrivial operation, especially when there is no *a priori* way to know how many there are or where they may lie. Consider as an example the ongoing debates about clusters of galaxies in the universe. It is made considerably more difficult in this case because the axes (the measurement parameters) are not necessarily orthogonal (some parameters depend on others), and are not equally

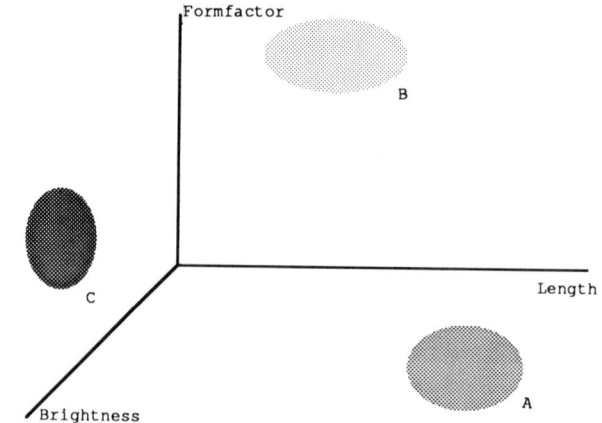

Figure 9-10: Regions suitable for identification with range analysis

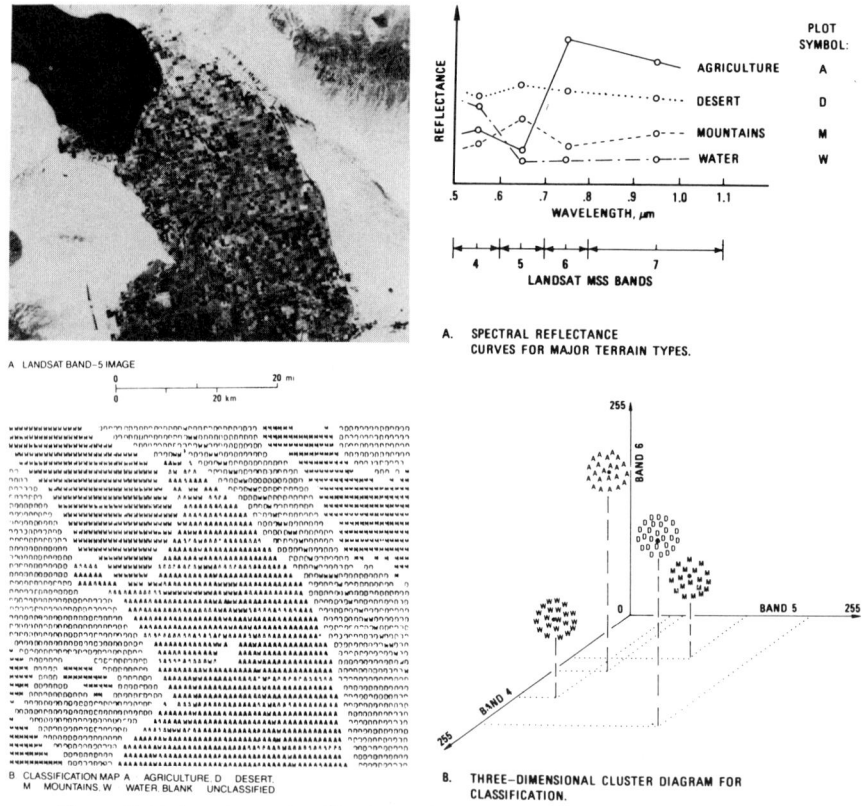

Figure 9-11: Land use classification using TM (Thematic Mapper) images. Left: a) one image showing reflectivity in a selected wavelength range; b) classification of land use; Right: a) reflectance vs. wavelength plots for different terrain types; b) cluster diagram showing terrain identification criteria. Reflectivity in different wavelength ranges provide independent parameter values that define a space in which different clusters identify terrain types (Sabins, 1986).

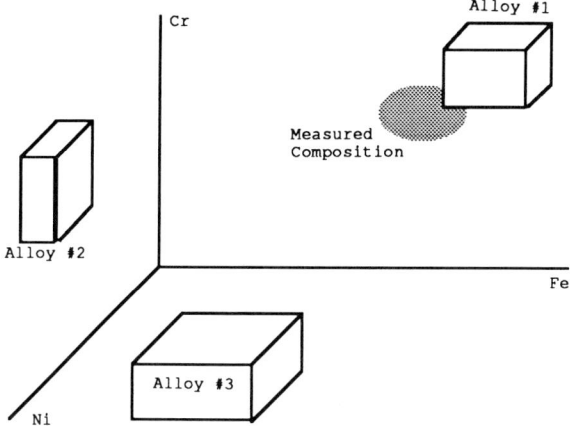

Figure 9-12: *n*-dimensional search for alloy identification.

important in separating the types. This is another way of saying that the clusters need not be spherical (or have any other predetermined shape). Finally, the limits of a clusters are not well defined, so that deciding when a new point lies in an existing cluster is difficult at best.

One field in which *n*-dimensional classification is routinely applied is the identification of types of terrain or land use from satellite or aerial photos (Grasselli, 1969; Sabins, 1986). Most such imaging utilizes several different bands of radiation, including the infrared as well as the visible. Patterns of spectral intensity are associated with particular kinds of vegetation, minerals, and so forth in a supervised training operation in which the operator marks locations and the image analysis system plots a point in terms of the various spectral intensities. These points are then grouped into regions that provide the classification ranges, and the entire image or series of images can then be classified pixel by pixel (Figure 9-11).

Another situation in which this *n*-dimensional search is successfully used is a more limited one. For alloy identification based on measured elemental composition, it is common to plot the new analysis point in an *n*-space where the axes are the concentrations of each of the elements (Figure 9-12). The point may furthermore be enlarged to take into account the uncertainty in the results for each element (producing an ellipsoid since the precision may vary for each). The allowed range of composition for each possible alloy is known beforehand and can be plotted, usually as a rectangular prism. If the ellipsoid representing the analysis intersects any of the prisms, then the alloy is identified. If not, then the nearest alloy (in a vector sense) can be reported (Russ et al., 1980). In practice, the comparison of measured vs. permitted compositions is carried out with a series of "IF" tests in the program, and the order of the tests and conditional branches must be tailored to each specific application. By comparison, the method described in this chapter is far more flexible, able to learn for itself (both initially and continuously) what parameters are important and what distinguishes the categories. It is a parallel rather than sequential method, and does not call for either custom programming for each application nor for exotic methods such as cluster finding.

The general approaches used to locate clusters of points are computer intensive. One is to give each point a mass (perhaps unequal values to weight some more important or "archetypical" objects), and then calculate the gravitational attraction of the points for

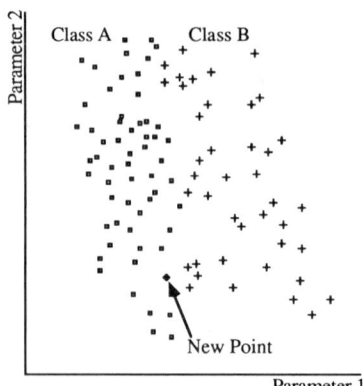

Figure 9-13: Clusters of points with a *kNN* neighbor method. The nearest neighbor to the test point lies in Class B, and so do 3 of the nearest 5 neighbors.

each other (Russ & Hare, 1983). The net force on each point causes it to move. After enough iterations of this procedure, the points cluster tightly so that the centers of the clusters and the number of points in each can be readily determined. In effect, the procedure simulates the collapse of galaxies from scattered stars. Its principal drawbacks are that it takes a very long time to perform the calculations, and they are very sensitive to just how representative the original points were. Adding a few more points from a few more training objects may completely alter the resultant pattern.

Pattern recognition can also proceed by keeping the location of all points in n-space and comparing each new measurement to them to decide which class it belongs to (Figure 9-13). In general, this is done by a *kNN* method in which the k nearest neighbors (typically a small number, from 1 to 5) are found and used to decide the type of the new point (Duewer et al., 1975). Again, the points may not all have the same weight. If not all of the k neighbor points are of the same type, a probability or confidence limit on the identification can be calculated.

The principle drawback of the *kNN* approach is that all data points from the training set must be retained in memory, and searched each time to find the nearest neighbors. The time required to do this increases rapidly with the number of neighbors, the number of dimensions, and the size of the training set, yet quite a large initial set is desirable to produce good definition of the clusters. Additional training points can be easily added to the data base, however. The method is very sensitive to the effect of outliers, since only the few nearest points are considered each time. This makes it somewhat more probable to find a match with a population of points that is numerous, since outliers are apt to be more numerous as well, and relatively difficult to use for the identification of minor types which may have only a few points in the data set.

The method is difficult to adapt to situations in which the parameter axes are not orthogonal, have nonlinear scales, or have different weights for importance. The concept of a neighbor "distance" in such a space is not well defined and may be distorted in different directions or regions of the plot. Unfortunately, this is often the case when shape factors are used for identification since they are often not all independent (so the axes are not orthogonal) and are certainly not linear (a change in aspect ratio from 1.1 to 1.5 is far different from a change from 4.1 to 4.5, for instance).

An example

The most robust method is usually the context line or linear discriminant analysis method as described above. An example of the operation of this method may help to clarify the process (Russ & Rovner, 1987, 1989). In the study of archaeological sites in the American southwest and much of central America, one subject of considerable interest is the process of domestication of corn, *Zea mays*. It is widely believed, although not proven, that corn is a domesticated offshoot of the wild grass teosinte, of which there are many varieties. Little remains of corn in an archaeological context that can be dated, of course, but it happens that corn, like all grasses (and many other plants) produces small silica bodies in and between the cells on the stalk, leaves and other parts of the plant. These are called opal phytoliths, and reflect internal cell structure (Figure 9-14). They are generally a few micrometers in size, and act to stiffen the plant tissue. They are also selectively produced at any site of injury (including cropping by animals).

From an archaeological point of view, phytoliths are of interest for two reasons. First, they survive in the soil for long periods of time and can be recovered from layers that are datable and show other signs of human habitation. Second, they have shapes that are distinctive. Research over several decades has shown that phytoliths can be used taxonomically to identify the species of grasses, including corn and its precursors (Twiss et al., 1969; Rovner, 1971; Pearsall, 1978; Piperno, 1984). Most of this identification has been carried out by humans, who have accumulated photographs (mostly using the scanning electron microscope) and built catalogs from which matching comparison to unknowns can be performed. The success rate for skilled observers in blind tests is generally better than 95%. However, this has been at the cost of a very substantial amount of effort, and the knowledge and experience is not readily transferred to other researchers. Such a situation is ripe for the use of computer image analysis methods.

These were applied to the phytolith case using a training suite of 5 known species of maize and teosinte. Slides were prepared with phytoliths extracted from plant tissue by chemical dissolution, and imaged with the SEM. The images were then subjected to computer image analysis to determine several size and shape parameters. Since the

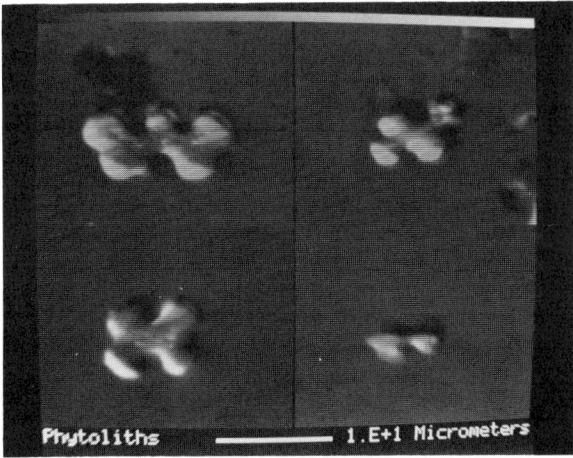

Figure 9-14: Corn phytoliths viewed in the SEM.

Table 9-1: Results for taxonomic identification
(T1 and T2 are teosinte, M1, M2 and M3 are *Zea mays*)

Length data and comparison results

Sample	Bilobate	Cross-Body
T1 & T2	18.1	19.5
M1 & M2 & M3	20.4	20.8
Anova F Value	6.647	4.013
Probability of signif. diff.	98.92	95.58

Anova Probability of significant difference in length

Sample	T2	M1	M2	M3
T1	17.93	95.22	99.00	96.64
T2	-	87.32	96.18	96.80
M1	-	-	36.37	27.87
M2	-	-	-	20.65

phytoliths could settle onto the substrate in any position, and their brightness was determined primarily by SEM operating conditions, no position or brightness parameters were used. The shape parameters used were those described in Chapter 7, which are all insensitive to feature orientation.

Discrimination based on feature brightness produced a binary image which was then subjected to a simple erosion/dilation (or opening) operation to remove small bits of noise and debris present in the original image. Some of the phytoliths settled on their edge so that no diagnostic shape measurement was possible. All such features were rejected from further measurement and analysis, based on their area, roundness and aspect ratio. This also served to eliminate some tubular phytolith shapes which are generally considered to be non-diagnostic for taxonomic work.

One set of published work representing more than a decade of visual examination of phytoliths from corn and other grasses has suggested that the ratio of the number of phytoliths with a "bilobate" (i.e., dog-bone) to those with a "cross-body" (i.e., maltese cross) shape is of taxonomic importance. It has also been suggested that the size of the cross-bodies serves to distinguish species. Both of these propositions were tested within the first few hours of work.

The system was first trained to distinguish and recognize the bilobate and cross-body shapes, using a training set of 30 objects which were manually identified as one or the other (by touching them with a mouse-controlled pointer). Correlation and regression on this data set produced a context line using only four measured shape variables

$$\text{TYPE} = 3.5068 - 3.9526 \cdot \text{Formfactor} + 4.2618 \cdot \text{Convexity} - 1.4158 \cdot \text{Solidity} - 1.2242 \cdot \text{Aspect ratio}$$

This calculates values near 1 for bilobates and near 2 for cross-bodies. It was not apparent to the human users why these particular measures of shape were better for distinguishing the two types of phytoliths than some other parameters such as roundness, extent, etc., but that is because none of these shape factors correlates closely with whatever it is that humans use for recognizing shapes. Few physiological studies have been carried out in this area.

On the original set of objects, this produced a better than 99% separation. This was then tested on a second set of training objects from a different species of corn, and produced a better than 98% separation. After applying the results to a total of 300 objects from 5 different species of plants (3 maize and 2 teosinte), the results had improved to better than 99.6% accuracy in distinguishing the two shapes.

The initial measurement of size was based simply on the single parameter Length (Table 9-1), although many others performed about as well. The original published report (based on visual recognition) that the size of cross bodies was species diagnostic was confirmed, and produced a 95%+ probability of being able to distinguish the cultivated species from the teosintes. However, we also found that using both the bilobates and the cross bodies produced an even better result, being able to distinguish the three maizes from the two teosintes 99%+ of the time. Automatic image measurement with more parameters was even able to distinguish the individual varieties from each other more than 95% of the time.

A second blind test included two sets of 100 phytolith images that were actually taken from the same plants, but selectively from leaf and stem areas. Not only did the analysis routine report correctly that the two populations were indistinguishable as to species, but it also showed that they could be successfully distinguished based on roundness, at a better than 99% confidence level. Recall that in the original context line analysis, roundness was omitted as it did not help to discriminate species. But there are good evidences that it is sensitive to the rate of growth of tissue, and therefore can be used to discriminate between phytoliths from different parts of the same plant.

The fact that the performance of the computer assisted image analysis system for feature recognition and measurement produced results better than those of the skilled researchers who first proposed and developed the basic ideas is not surprising to those familiar with the system. Nor is the speed with which results can be obtained. We can examine hundreds of phytoliths (enough to produce statistically significant results) from more species of corn or other grasses in 1 month than have been done by hand and eye in the last 20 years. Further, more parameters can be examined with more consistency, and the results give greater assurance of success.

We believe that the reason human observers decided that the key to distinguishing the species lay in the size of the cross bodies is that they vary less in size and shape than do bilobates within a species, and thus appear easier to measure. Actually, the size of the bilobates proves to be more useful in making the distinction, in spite of their greater variation within species, but was overlooked by the human researchers, probably because the data were too subtle for the human eye (which does not deal too well with the need to characterize variation).

Comparing multiple populations

In many cases, the classification problem to be solved is not simply to distinguish objects in one class from those in another. It may instead be to distinguish objects in many classes, or to decide whether an object lies within a given class or not (regardless of what else the object might actually be). These present quite different problems, and will be discussed separately.

If there are N classes, then they can in principle be broken down into all possible pairs of classes. Each pair will have a context line (usually these will be skew in n-space), and we will assume that all classes can be distinguished pairwise. However, that does not offer a guide to how to identify an individual object. Its various measurement parameters can of course be used to calculate the values of each discriminant function, so the position along each context line can be determined. But identification is more complicated.

286   Chapter 9

Figure 9-15: A mixed population of letters in different sizes, fonts and orientations.

The usual approach is to have a sequential set of decisions, arranged in the form of a logic tree. Each individual step involves a straightforward comparison along a single context line, which may provide discrimination between groups of several classes of objects. As an illustration of this process, we will consider some familiar features, the first five capital letters (A,B,C,D,E) which we "know" how to recognize (Figure 9-15). Let us see what is required for the computer to accomplish this task. First, to complicate matters slightly (but in a very realistic way), we will require that letters from several different printing fonts (with serifs and without) be used, as well as letters of different sizes and in any possible orientation.

The range of sizes will preclude the use of any size-specific measurement parameter. The free orientation rules out some of the very simple algorithms that are actually used in computer character recognition. This is most often done using a template matching technique, as described in an earlier chapter. Such techniques are fast, and quite reliable, but very specific. This example is intended to demonstrate a more general solution to a more general problem. Various dimensionless shape factors can be used to distinguish individual pairs of letters, but there is no single one that directly separates all five. Furthermore, most of the parameters take on a range of values when real printed characters are digitized, especially when they come in different sizes, orientations and fonts.

A training set of characters was used to establish the required information. It consisted of from 25 to 40 examples of each character, identified to the computer by the human operator. More than a dozen shape factors were recorded, but only three are actually needed to achieve the discrimination, so only these will be described here.

The first one is simply the number of holes within the feature. The letters C and E have no internal holes, while the letters A and D have one and the letter B has two. This partial classification of the features reduces the problem to two pairwise separations, for which our previous understanding of context lines is directly useful. In this case, each separation can be achieved with a single parameter rather than a combination, so the context line is parallel to a single parameter axis.

The difference between the letter A and D can be distinguished by the formfactor of the filled object (i.e., neglecting any internal holes). For an ideal sans-serif character, the D is convex and quite rounded, so this this ratio should be close to 1.0, while the A has projecting legs and produces a smaller value. But when serif characters are included and variations in the detailed shape of the digitized characters are taken into account, there is a somewhat greater variation. Based on the training population, the external formfactor for

27 letter A's and 36 letter D's gave the results shown:

	A	D
Mean ± S.D.	0.3547 ± 0.0223	0.7939 ± 0.0481
Range	0.3057 – 0.3890	0.7257 – 0.8821

Notice that in this case, the two classes are widely separated. The decision point can be placed anywhere within a wide range of values and still produce virtually 100% confidence in the discrimination. We will arbitrarily use 0.56 for the example that follows.

The C and E are more difficult to separate. One useful parameter is the solidity (the ratio of the feature's area to the convex area, or the area within a taut string outline around the feature). It would also be possible for the simple sans-serif letter shape to use the number of branch points in the skeleton (one for the E, none for the C), but when serifs are present this is not reliable. A combination of several parameters performs better than the use of solidity alone, but it is adequate for the example task at hand. Based on the training population, the solidity for 38 letter C's and 40 letter E's gave the results shown:

	C	E
Mean ± S.D.	0.4318 ± 0.0275	0.7781 ± 0.1936
Range	0.3730 – 0.4837	0.4980 – 0.9100

In this case, the mean and standard deviation values are very misleading. The C values form a rather compact cluster of points, with little variation, but the E's are spread out over a very large range with no hint of a normal or Gaussian distribution. This is because the serifs on characters strongly affect the convex area, especially in conjunction with the effect of rotation. Nevertheless, setting a decision point at 0.49 will serve to distinguish the two letters from each other in the training set, and actually works adequately with only rare errors in an even broader set of characters drawn from more different fonts.

Given these individual discrimination operators, how should the overall classification proceed? The most direct and efficient method is to begin with the number of holes (Figure 9-16). This will identify a B if the value is two. If not, then if the value is 1 it is

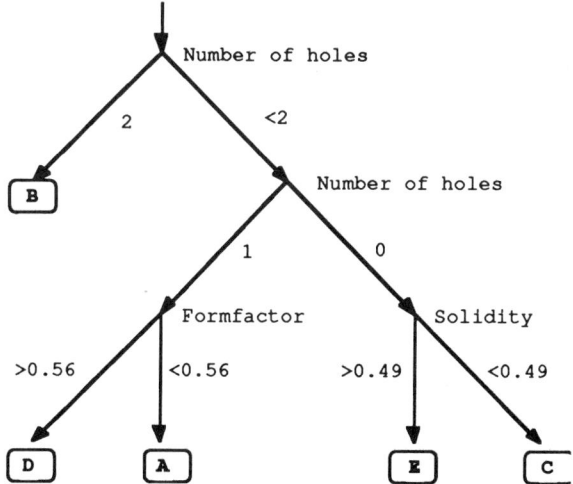

Figure 9-16: Optimal decision tree for feature recognition.

necessary to calculate the external formfactor and compare it to a value of 0.56. This will identify the result as either an A or a D. If the value for the number of holes is 0, then the solidity must be calculated and compared to 0.49, which will distinguish the C from the E.

Even for this very simple example, there are other decision tree sequences that are significantly less efficient. For instance, it would not be desirable to begin with the formfactor, because this parameter does not enter into most of the discriminations. For this simple example, the human user can be expected to produce the optimum order of testing because the possibilities are few and the use of single parameters rather than more complex context lines makes them intuitively clear.

Constructing the proper sequence of tests becomes much more difficult when more parameters and more classes are involved. A particular type of expert system software called a decision tree designer can be used to search for optimized sequences of tests (Quinlan, 1983). This is not a fast operation, but it needs to be performed only once. After that, the education of the system as to the individual decision points can proceed as described earlier, but the order in which the tests are applied does not vary.

The optimum decision tree is not necessarily the one with the fewest branches. Some parameters may be easier to measure than others, and the more costly ones would normally be placed at the ends of the tree branches rather than near the root. When the set of rules used for arriving at decisions also includes information on the relative cost or difficulty of obtaining various observations or inputs, this can be used to assist in determining an optimum search and solution tree (Cockett & Herrera, 1986; Cockett, 1987). This is usually called "knowledge shaping" and may greatly improve the overall system efficiency.

The result of designing a decision tree is a set of comparison steps and conclusions, called production rules or IF...THEN rules. They express the system's knowledge about a particular body of data in a way that can easily be applied by even rather small computers. They usually deal with individual measurement parameters separately.

The use of a decision tree with rule building is a classic one for expert systems. The various IF...THEN tests can be weighted as to probability if several individual discriminant functions are used and no single one gives total separation. Of course the order of the tests is highly significant, but methods from information theory are available that utilize entropy and other considerations to develop the best order. If only fixed decision points are used (a sequential, univariate classification scheme), it is not possible to add more objects to the training set without recalculating the entire decision tree, and the result becomes very sensitive to outliers. The method described here, with incorporation of context lines between populations and a linear discriminant analysis along each line, allows the method to learn from experience and makes a much more robust system.

The problem with this approach is that it is completely specific to a given application. For instance, in terms of the discrimination of the letters A...E shown above, adding the letter F would require adding a different parameter (or more), another rule, and perhaps a different order of applying the rules. Adding other letters, perhaps the entire alphabet or the Japanese character for "Men's Room," would again require starting over from scratch. Little of the previous knowledge base could be utilized, and even the original raw data (measurement parameters from the training set) would have to be consulted to derive new context lines.

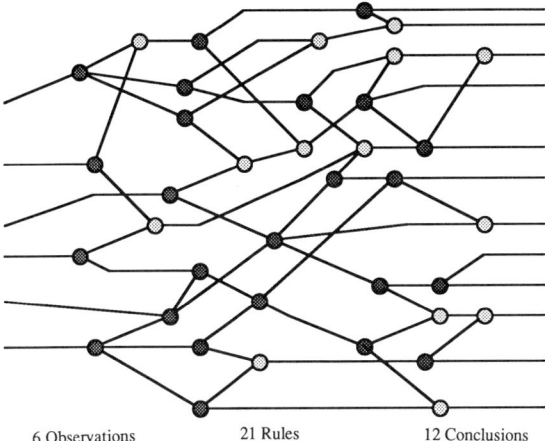

6 Observations   21 Rules   12 Conclusions

Figure 9-17: Schematic diagram for an expert system.

The complexity of production rules increases rapidly. It severely taxes a small computer to recognize the entire Roman alphabet in a limited variety of fonts, even if they are restricted as to orientation, and the diversity of Chinese ideographs would certainly be beyond present capability. Extension to the complexity and variability present in handwriting (Brady & Wielinga, 1978) would require a more flexible approach altogether.

A general-purpose classic expert system would consist of a series of rules that would offer many possible paths from measurement values (observations or facts) to conclusions. These rules are stored separately from the program that sorts through them to find a successful path, known as the inference engine.

In the very schematic diagram for an expert system shown in Figure 9-17, there are 6 initial observations, 21 rules (many of which may call for additional observations or data) and 12 possible conclusions; real systems are several orders of magnitude more complex. Since the number of observations is smaller than the number of possible conclusions, forward chaining would normally be preferred in this example (starting from observations and testing each possible conclusion to see if it is supported by other available facts). In some other situations, a more efficient approach is backward chaining in which each possible outcome is examined to see if the required preliminary conclusions or observations are present. It is possible in this example to get from an observation to a conclusion with as few as two rules, or as many as five. Notice that some of the rules have more than two possible outcomes (branches leading to the right). In some systems, each possible outcome would represent a separate rule.

Normal forward searching for a successful path through this network would start at some point on the left, follow a path to a rule, and follow the outcome until a contradiction was met (a rule could not be satisfied). The system would then backtrack to the preceding node and try a different path, until a conclusion was reached. This approach does not test all possible paths from observations to conclusions. Heuristics to control the order in which possible paths are tested are very important. In cases of realistic complexity, it is not possible to test all consequences of applying one rule, and so simplifications such as depth-first or breadth-first strategies are employed. Pruning, or

Figure 9-18: Images of mixed nuts.

rejecting a path before working through it to exhaustion, is also used. (You may use as a rough analogy the search for a move in a chess program. Some initial moves are rejected immediately, while others are searched to various depths to evaluate the various responses possible and the outcomes). Reverse searching works in the same way, starting from possible conclusions and searching for paths that lead to observations. Some search "engines" combine both forward and reverse searching methods.

When fuzzy logic is used, the various rules contain probabilities. In that case, it is ideally necessary to construct a total probability by combining the values for nodes along each path, to select the most likely path and hence the most probable result. The heuristics that control search order are less important, but the total computation load is much greater. Properly constructed knowledge bases include both the probabilities of belief and disbelief. For instance, {if A can fly then there is a 70% chance that it is a bird; if A cannot fly there is a 5% chance that it is a bird} would be necessary to accommodate butterflies, which can fly but are not birds, and penguins, which are birds but cannot fly. Normally these probabilities are supplied by the "expert" who is external to the program.

Object Recognition

In rare situations they can be obtained by statistically meaningful sampling methods, but usually they are based on human experience and judgement, and may therefore be incomplete or biased.

An example of contextual learning

As a illustration of an application combining many of the different approaches to identification (and showing a realistic level of complexity and some of the typical problems encountered), populations of several kinds of natural nuts were used (Figure 9-18). Some of these (acorns, walnuts, hickory nuts, pecans) were collected from local trees, and represented different species (e.g., red and white oak acorns, both with and without their caps, and several kinds of pecans) so that the populations were bi- or multimodal (e.g., see Figure 9-19), and this was reflected in the measurement data. Others (peanuts, pistachios, filberts, brazil nuts and almonds) were purchased in local stores and also showed a considerable chaotic variation in their appearance and data.

In this study, conventional image acquisition, processing and measurement methods were employed. The images were obtained using standard video technology, digitized into a desktop computer system with 256 pixel width and 8 bit grey-scale. The nuts were scattered on a grey surface, with no attempt to control their orientation. The lighting of the nuts was diffuse, but not perfectly uniform. Furthermore, one side of the nuts was generally darkened by shadows, which sometimes fell on nearby objects. Nuts were never piled on top of one another, so the requirement that objects be fully visible was met without difficulty. The textured cloth surface on which the nuts rested generally served to separate them slightly, and as a matter of convenience, nuts were sometimes moved slightly to minimize touching. However, this had no effect on the results as identical measurement results can be obtained after additional image processing steps even if the objects do touch at their periphery.

In order to define the features in spite of their nonuniform illumination, a Kirsch (gradient) operator was applied to the images. This produced well defined nut outlines which were then thresholded and filled to produce binary images of each nut as shown in Figure 9-20. When adjacent objects touched, a watershed segmentation routine was applied to separate them. All of the image processing and editing techniques used were selected from standard automatic techniques discussed in earlier chapters.

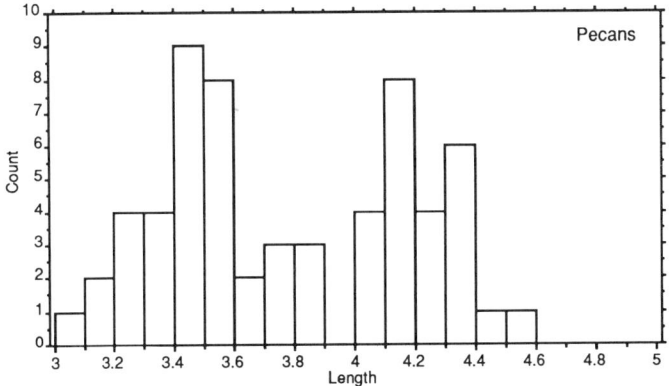

Figure 9-19: Bimodal length histogram for two species of pecans.

Using the binary images, a series of measurement parameters were obtained for each feature in each image. For the training populations (25-30 of each nut type, chosen at random but without any attempt to achieve a statistically representative sample), images typically contained 8-15 objects, all of one type. Subsequent images for continuous learning included random mixtures of many different nut types.

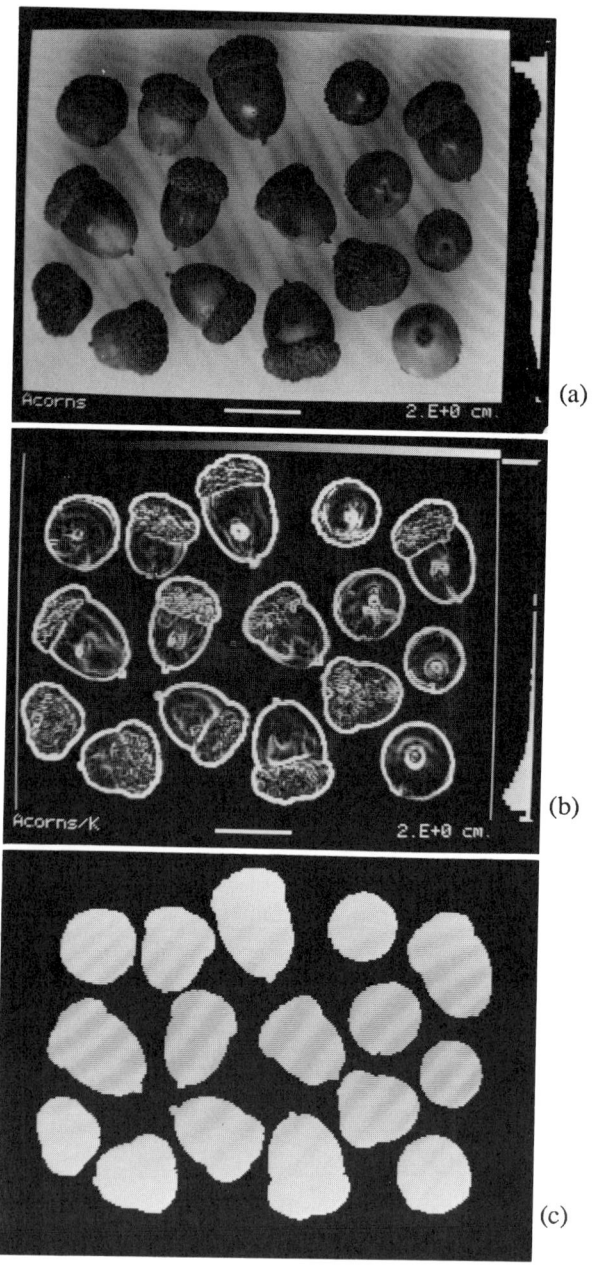

Figure 9-20: Image of acorns (a), application of a Kirsch filter (b), and the filled binary image (c).

For each feature, the types of measurement parameters available in the system fall conveniently into the four classes discussed in Chapter 7: measures of size, shape, position and brightness. For the nuts, because of the way they were handled, the position data (centroid coordinates, orientation angle, neighbor distance, etc.) were not considered useful and were therefore ignored. As only monochrome images were obtained, no color information was available.

A total of 40 measurement parameters were provided by the image analysis system. the regression method discussed below was allowed to select the ones that were important in this application, ending up with 17 parameters actually used in subsequent object identification. Some, such as area or brightness, have an immediately obvious role in this process as judged by a human (for instance, distinguishing a walnut from a pistachio based on size, or a brazil nut from an almond based on lightness. No claim is made, however, that these measurement parameters are in general related to the judgement criteria applied by humans. However, determining which parameters and values are to be used to identify different population types is very difficult, because the populations are chaotic, and may be bi- or multimodal. Humans do not easily characterize such populations, to extract the key "rules" required for the expert system (Duda et al., 1976; Tversky & Kahneman, 1974). This is especially true for measures of shape, which do not even use terminology common to human language and experience. Human shape description often cites a prototypical object as an example of a shape. This is inherently vague since the aspect of shape being identified is unclear. This vagueness is itself very robust in human language (Alston, 1967) but that does not help us here.

The artificial intelligence (AI) methods that are actually used for practical problems at the present time include several that seem at first glance to be applicable to the nut identification problem posed here. However, each possesses serious drawbacks that are best understood by comparison to the method that was actually used. Contextual learning is a two stage process. The first uses the information from the relatively small training population to define the context within which identification will subsequently be performed, and the second learns to make the identifications by using Bayesian statistics on the distributions of values from all further object measurements.

For the 17-parameter measurements obtained on the nuts, it is possible to plot each object as a point in a 17-dimensional space (one axis for each parameter, even though some of the parameters are not orthogonal or independent in a mathematical sense). It would be expected that clusters of points corresponding to each population would form, and indeed, cluster analysis is one of the methods to which comparison will be made below. Lines joining each pair of population clusters can be defined unambiguously by multiple regression as those lines about which the points in the populations have the smallest moment. Along these lines, the clusters are maximally separated as shown earlier in Figure 9-5.

These "context" lines can be defined by their direction cosines. As a practical matter, instead of retaining all 17 direction cosines, it was acceptable to use stepwise multiple regression in which parameters are added to or removed from the equation of the context line based on the F-value (an arbitrary cutoff of 4.0 was used). This was done to eliminate axes essentially perpendicular to the context line, as indicating parameters not useful in distinguishing the particular population groups. In this way, context lines were determined with as few as two or as many as eight terms.

A total of nine populations (the nine nut types listed above) were eventually added to the knowledge base. Initially, five types were measured, giving rise to 10 pairwise context lines (5 · 4/2). The later addition of 4 more nut types brought the total to 36 (9 · 8/2). The regression lines used the 17 size, shape and brightness parameters mentioned above. Some were used frequently (for instance brightness in 30 out of 36 lines, and width in 21 out of 36) but were generally not the most important terms in the regression equation. Others were used only occasionally, but were then quite important (for instance, convex area was used in only two context line definitions, but was the single most important factor for distinguishing pistachios from almonds). On the average, each parameter was used just under 11 times, and each context line contained about 5 parameters.

The projected positions of each measured object along these context lines can be directly calculated as a pseudo-variable using the equations determined by the regression. These values are effectively "type" values, which should ideally have integer values corresponding to the identification of the various populations. In practice, of course, they have real values that only approximate these values, and it is necessary to determine where to establish cutoff values or decision points to make the identification. In order to facilitate this linear discriminant analysis, frequency distribution histograms are obtained along each context line. It is important that the size of the bins be sufficiently fine to permit adjusting the decision point in fine steps, yet coarse enough to require modest storage for the histogram data and to provide a reasonable description of the shape of the distribution. It is not necessary to maintain equal bin sizes; instead, very coarse bins can be used within the clusters, and fine bins in the region between them. This can either be done using a continuous function, such as a parabola, or using discrete values.

The resulting bins are used during the learning process. The individual parameter measurements for each object are used only to calculate the values along the context lines, and those derived values are used only to address the corresponding bin of the distributions, which are incremented for each object in the initial training population as well as each subsequently viewed object. The context lines themselves are fixed and do not move, so that in principle they do not point accurately toward the center of each distribution. However, even with very small and non-representative training populations the lines wander only slightly in the high-dimensionality parameter space in which they

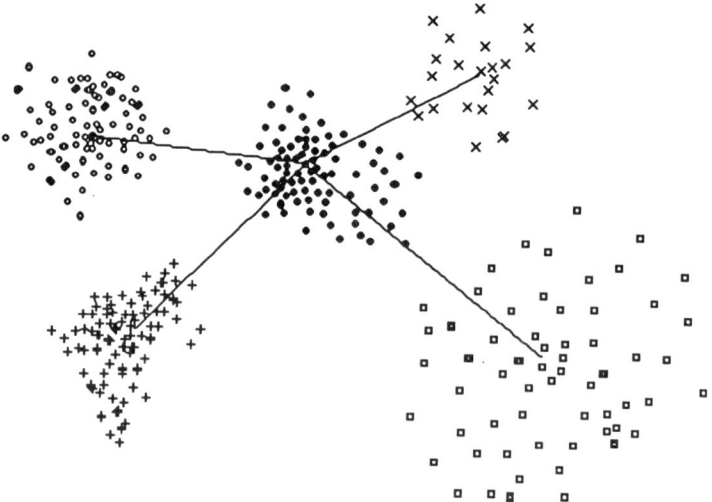

Figure 9-21: Two-dimensional schematic showing context lines between population clusters.

Object Recognition   295

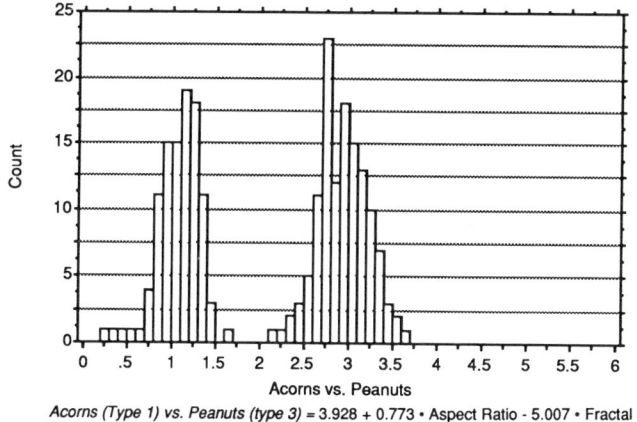

Figure 9-23: Hickory nuts/almonds context line and frequency distribution.

exist, and serve quite adequately to define the lines along which distribution limits can be used to decide identity.

As the population histograms evolve with further learning, the limits of each population become better defined. The situation is schematically as shown in Figure 9-21, reduced to a two-dimensional simplification. Each population has a shape that is not known in detail, but its extent in the particular directions toward other populations is well characterized by the frequency distributions along the context lines (Strat, 1984; Duda et al., 1976).

In some cases, these distributions are entirely distinct. Figures 9-22 and 9-23 show representative examples, in which the shapes of the distributions vary but there are no overlaps. For the nut application used here, 30 of the 36 pairwise context lines had no overlap in the distributions, even after a total of more than a thousand nuts had been added to the data base. In these cases, the decision point can be taken at the midpoint value between the limits of the two distributions.

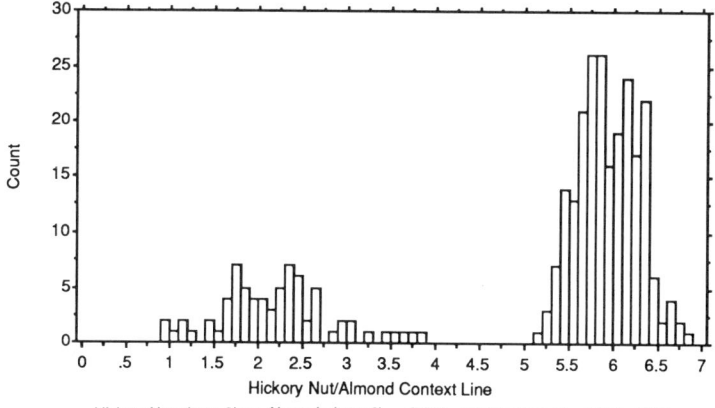

Figure 9-22: Acorns/peanuts context line and frequency distribution.

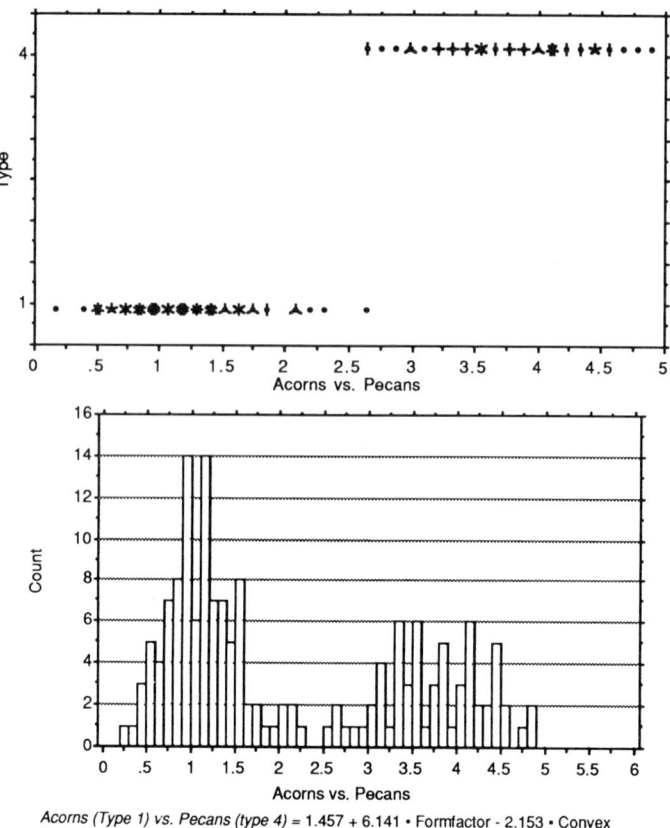

Figure 9-24: Acorns/pecans context line, scatter plot, and frequency distribution.

In the remaining 6 cases, some overlap was present. The probability of identification of an object along that context line can be determined by straightforward Bayesian statistics. The decision point is normally positioned to make the errors of misidentification equal as shown earlier for manually located decision points in Figure 9-7. In other words, the tail of each distribution extending past the decision point, normalized by dividing by the total area of its histogram, represents the probability of misidentification (identifying the object as type A when it is actually type B, or type B when it is actually type A). The system makes these errors equal, within the limits of the discretized frequency distribution, by adjusting the decision point.

In normal use, the system compares each new observation to the decision points along the various context lines. If all values for one population agree, then that type is assigned to the object. If one context line value lies within a specified error probability of the decision point, then the operator is asked to confirm the tentative identification. If more than one value is undecided at the specified error probability, then the operator is told the most probable identification and asked to provide the correct identification (for the case of the nuts, this action could only be triggered by introducing entirely foreign objects). Whenever user input is required, the decision points along the various context lines are updated to reflect the changes due to the added points from all past observations.

Object Recognition 297

Otherwise, updating is carried out automatically after a preset number of objects. The arithmetic involved is trivial and this operation is so fast as to be undetectable in routine operation.

As the identification of objects is shown on the display screen, it is also possible for the user to inquire about any particular object identification. The system reports the single-tailed probability associated with each context line (the area of the histogram past the measured context-line value as a percentage of the entire histogram area), for the object. Thus the system can explain its reasoning to the operator. The operator can also override any decision.

Experience with the nut application described here, and with several others, indicates that objects which the system cannot identify with high probability are often ones that the operator also has difficulty with, at least using the image presented on the computer display. Only by using color information, or by looking at a different view of the object, can a conclusive decision be made. For instance, the overlap along the acorn-pecan context line (Figure 9-24) indicates an error frequency of about 1% for misidentifying these nut types. The error is due to a similarity in appearance between a short variety of pecans and large acorns, viewed from the sides and missing their caps, as shown in Figure 9-25.

The most probable error is that between filberts and acorns (Figure 9-26). The confusion arises principally when either type of nut is sitting on its base, hiding the scar which is a fairly distinctive difference between the two types. The remaining clue is that the acorns tend to be more nearly round when viewed end-on, while the filberts are slightly triangular. This difference shows up in the formfactor, but there is still a roughly 10% probability of error in the identification. It is interesting that given only the same photo of nuts as appears on the computer screen (Figure 9-27), naive human observers make about the same number of errors as the program.

The contextual learning approach to feature recognition is easily implemented in an image analysis computer system, with modest computational and storage capacity. In the application shown, the system quickly learns to identify nine different but grossly similar

Figure 9-25: Image of acorns and pecans.

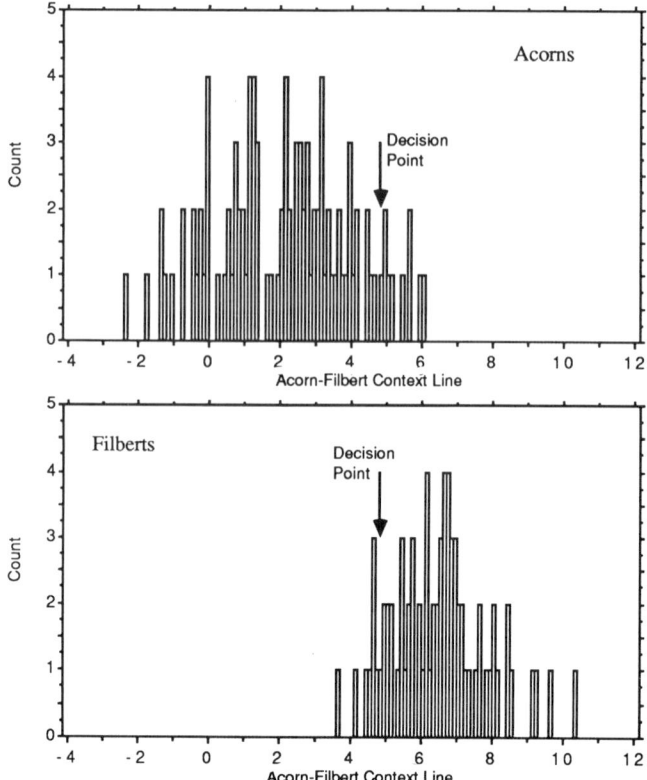

Figure 9-26: Details of distribution histograms along acorn-filbert context line showing magnitude of overlap at decision point.

Figure 9-27: Image of filberts and acorns.

Table 9-2. Confusion matrix showing errors in nut identification

Actual Type		Identified Type							
		1	2	3	4	5	6	7	8	9
Acorns	1	91	0	0	1	0	0	0	10	0
Hickory Nuts	2	0	70	0	1	0	0	1	0	0
Peanuts	3	0	0	126	0	0	1	0	0	0
Pecans	4	2	1	0	54	0	0	3	0	0
Pistachios	5	0	0	0	0	126	0	0	0	0
Almonds	6	0	0	0	0	0	224	0	0	0
Brazil Nuts	7	0	3	0	6	0	0	115	0	0
Filberts	8	8	0	0	0	0	0	0	67	0
Walnuts	9	0	0	0	0	0	0	0	0	107

nut types, with better than 96% total accuracy, and better than 99% for most nut types (Table 9-2). The time required for initially determining the context lines is several minutes, but this is a one-time operation. Subsequent identification of features takes no longer than the time needed to perform the various image processing and measurement operations (a few seconds per image field, in which tens of nuts may be present), and learning is essentially continuous.

Beyond its success in the immediate application, the contextual learning method offers several important advantages over other AI methods. It develops its own recognition criteria, is capable of continuous learning, requires limited storage space since the knowledge base is contained in the structure of the data and program, and uses the operator as a trainer but does not seek to duplicate human recognition methods. While it is capable of being implemented in a serial computer, the method is inherently parallel and can take advantage of increased processing power as it becomes available. In principle, the method should be equally adaptable to recognizing and distinguishing a broad variety of other types of objects, including those encountered in quality control, surveillance, and medical diagnostic situations. It is also adaptable directly to use with other types of data than those provided by image measurements.

Other applications

This one rather simple example of automatic feature recognition can be extended to many other areas of application. Different inclusions and precipitates in metals often appear different in size, shape and brightness (for a given preparation method). To properly characterize the cleanliness of the material, they should be separately identified and measured.

Industrial quality control is an obvious area of use for these techniques. From examining parts for dimensional accuracy to checking packaging for defects, automatic systems are finding increasing use. Logically enough, this is especially true in the field of semiconductor technology, where more and more complex devices require many interconnections. Some of these same devices are ultimately used in the computers that perform the inspection.

The final problem to be considered is deciding whether an object is or is not part of a particular class. This is often a forensic or quality control task. Given a series of production parts, for instance, the purpose might be to decide whether each is sufficiently like the "standard" part in size, shape, etc. to be accepted. Usually in a quality control context it is possible and economically acceptable to use a large training set of objects and for the operator to specify the important parameters (based on the technical specifications of the part, for example).

The forensic problem is more general. Given a series of capsules, which, if any, show signs of tampering? Given several paint flakes from a known site and a suspect's pocket, is it possible to conclude that the paints are different. This can be a difficult test to make with high confidence and more parameters than those provided by image analysis may be required. For the case of the capsules, measuring the length may indicate that tampered capsules have been opened and reclosed. But uncontaminated capsules also show some natural variation in length, and enough known good capsules must be measured to establish the entire shape of the distribution so that the probability of any test specimen exceeding the acceptable range can be known with some confidence. This is of course a less discriminating test than performing chemical analysis, but it is much faster. By combining a color test (perhaps with special illumination) to look for particular compounds (either the suspected contaminant, or simply oils from handling) additional information can be combined in a more discriminating test.

Paint flakes are normally compared as to color, and the number, color and order of any underlying layers. Chemical identification of the inorganic pigment elements (for instance by electron microprobe X-ray analysis) is also helpful, although more time consuming, as are chemical and physical tests. All of this information can be combined, but it is usually impossible to assess the natural range of variation in the original paint. The report of this type of matching is usually heavily weighted with protective phrases such as "cannot be distinguished from" or "could be the same as". Only in the case of a definitely distinct difference, such as a different color or composition pigment, can a firm conclusion be given, and this kind of evidence is often not used. Finding that a paint flake on a suspect's coat does *not* match the paint at the crime scene obviously does not prove that he *wasn't* there!

Cancerous or otherwise diseased cells viewed in tissue preparations, biopsies, smears, etc. in the light microscope are recognized by trained pathologists. The automatic image analysis system can be trained to perform the same search, at least for the initial screening (Figure 9-28). This requires deciding what parameters will reliably distinguish the abnormal cells from healthy ones. This is not simple, as the range of variation in cells is large. Training an image analysis system to make the distinction at least for a restricted class of problems appears to be within the range of capacity of modern systems (Castleman & Melnyk, 1976; Bins & Takens, 1985; Frank et al., 1988; Bauer et al.,

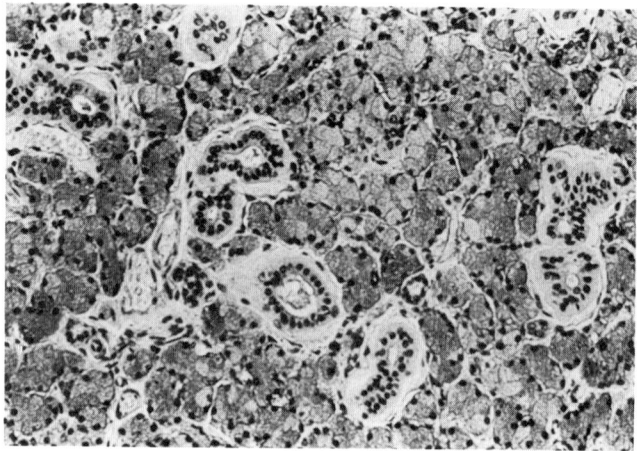

Figure 9-28 : Tissue in the light microscope - how to recognize healthy and cancerous cells?

Figure 9-29: Example of license plate image.

1988, Nomori et al., 1988). This would then save the stage coordinates of any suspicious cells for the human expert to review, effecting a great savings in the time required for the human.

Surveillance photos, ranging from satellite images to video cameras watching cars enter secure areas (Figure 9-29), must deal with a very large number of images, and with a small number of operators for whom fatigue can introduce a serious decline in efficiency. Automated recognition systems can take up much of the slack. Recognizing the signature of a target of military significance, such as a ship or tank in a satellite photograph has been demonstrated many times, and examples are used in some textbooks on image analysis. Dedicated systems that will find the license plate and read it in a video image of the rear of a car passing through a checkpoint have been in routine service for several years.

Most of these application examples are distinguished by an economic justification for using automatic image analysis and recognition. They require the scanning of very large numbers of images, for which human labor would be costly, inefficient or error prone.

However, as the cost of the systems themselves has declined and the flexibility of the software has improved so that training such a system is not prohibitively costly, it seems reasonable to anticipate the same methods being used for many more applications in research as well as more routine situations. The magnitude of these tasks can be better understood by comparison to the simple yet very robust methods by which, for example, the frog's visual apparatus identifies and locates food (Lettvin et al., 1959).

Artificial intelligence

Although very little of the terminology of artificial intelligence research has been used in this chapter, the generalized problem of object recognition from images is one of the classic tasks addressed in AI studies. Indeed, two important threads in the development of "intelligent" computers (which are thereby able to deal with problems that are not fully defined beforehand, and may approach in complexity the tasks we customarily assign to human assistants) are image processing and artificial intelligence.

Many of the potential areas of application have been mentioned: industrial quality control, surveillance including military applications and remote (satellite) sensing, robotics vision, diagnostics in biology and medicine, and so on. Such applications have been the goal of much combined research on computer vision and artificial intelligence, as well as on a better computational model of human visual methods (Grasselli, 1969; Duda & Hart, 1973; Nevatia, 1982; Braddick & Sleigh, 1983; Levine, 1985; Lowe, 1985; Ohta, 1985; Rosenfeld, 1986; Beck et al., 1983, 1986; Browne & Norton-Wayne, 1986; Gonzalez & Wintz, 1987; Zuech & Miller, 1987; Dougherty & Giardina, 1987, 1988 a,b).

Much of the emphasis here has been on images in a somewhat restricted class known as 2-1/2D (images of 3-D objects that are dispersed so as to avoid precedence problems). In many application areas these conditions are met, e.g., QC (viewing individual products or packaging in a controlled situation; satellite imagery (the viewed surface is practically speaking, flat); and computer-assisted microscopy (samples are either surfaces with local irregularities, dispersed powders, etc., or thin slices through materials viewed in transmission).

The image acquisition, processing and analysis steps covered in preceding chapters are a necessary preliminary requirement for most "computer vision" applications. We have defined *processing* as manipulating images to correct defects or emphasize characteristics of interest, and *analysis* as reduction from massive amounts of data to a small set of numbers describing the features or surfaces present). In this chapter we have addressed some of the requirements for applying the methods of AI (e.g., expert systems, cluster analysis, neural nets) to select the important image characteristics as determined by the application, with or without expert human assistance, and to use the data to draw proper conclusions (identification, classification, presence or absence of types, etc.).

The hardware considerations for successful AI applications in terms of computational speed, memory and storage requirements are surprisingly modest for most of the approaches that have been described, and for most applications. For "real time" response, a massive investment in parallel computers with specialized circuitry may be needed, but many important applications can be handled by existing micro and minicomputers in seconds or minutes. Indeed, the hardware requirements of image analysis are quite similar to those for AI (Kowalski, 1988; Uhr, 1987).

Image acquisition using video technology is generally adequate and quite inexpensive. Choice of appropriate image magnification usually permits "conventional" (i.e., inexpensive to acquire, store, process, etc.) images with approx. 512x512 (or 640x480) pixel resolution x 8 bits (grey scale) or 24 bits (when RGB or other color encoding is used). Conventional serial implementation or Von Neumann computers are adequate for the development of algorithms and for many applications, but both the image analysis and the artificial intelligence components of the overall task are well suited to parallel computers, and will be able (indeed, stand ready) to take advantage of developments in this area.

Languages appropriate for these tasks are less straightforward to select. Both AI and Image analysis have been well served by rather specialized developments in high-level languages. In the image analysis case, these typically incorporate low-level tightly constructed and optimized machine language or hardware kernels to conduct specific operations, and a conventional scientific front end language such as Pascal, C, Fortran or even Basic since most of the relevant data are obtained in numeric (real number) form. Much of the important AI development has utilized LISP or Prolog type languages which emphasize list or semantic processing, and are less adaptable to numeric calculations. Specialized "languages" associated with statistical analysis programs, spreadsheets, etc.

are also useful in some of these applications. Few attempts to connect these different fields have been made, but at least one LISP-based image processing system has been reported (Bright, 1987).

Several different AI methods have been mentioned in this chapter. Classic expert systems use predicate calculus and production rules (if..then). They usually separate the data base (rules) from the inference engine used to find a solution (Quinlan, 1983; Winston, 1975; Black, 1986; Forsyth & Rada, 1986; Michalski et al., 1986; Schutzer, 1987; Slatter, 1987; Winstanley, 1987; Rolston, 1988; Walters, 1988). This is performed by an iterative search for a path from observations to conclusions by forward and reverse inference. In problems of real complexity, specialized routines for pruning the search tree and specialized search heuristics are employed. These methods are difficult to combine with either supervised or unsupervised learning. They can explain their reasoning in terms of the rules applied, but are susceptible to noise or confusion due to missing or incorrect information. When this occurs, algorithms for conflict resolution are needed to permit the search to continue.

Fuzzy logic has been applied to expert systems to make them more adaptable to situations in which incomplete information or several possible outcomes are present (Zadeh, 1965; Negoita & Ralescu, 1975, 1987; Goodman, 1985; Sanchez & Zadeh, 1987; Zimmermann, 1987). The rules in this case include probabilities of belief and disbelief for each outcome, which usually must be supplied by an expert. With fuzzy logic, expert systems are even slower to iterate than the classic expert system. This is because they must locate the most likely path (usually defined in a minimum energy sense) rather than just any path from observations to conclusions. In a formal sense, the use of fuzzy logic is equivalent to Bayesian logic with levelled classic sets.

Model-based methods for recognition involve fitting planes, cylinders, boxes to edge lines (Tanimoto & Klinger, 1980; Ikeuchi & Horn, 1981; Rosenthal, 1981; Shapiro & Haralick, 1982; Roberts, 1982; Brooks, 1984; Kuhl et al., 1984; Pentland, 1986; Goad, 1986; Horaud & Bolles, 1986). Chapter 10 discusses methods for interpreting surface images. Once identified, the individual surfaces are ordered into a hierarchy based on presumed connections, and then into simple shapes and finally into more complex arrangements. These can then be compared iteratively to real objects in arbitrary orientations;. Model-based recognition is primarily oriented to 3-D images with precedence in which one object may partially obscure another. Examples of its use include "blockworld," as well as robotics and surveillance. Modelling also uses time sequences of images in which distances between points that remain constant are used to infer rigid body motion (human vision also makes this interpretation). When precedence is present in an image, model building locates the topmost feature, searches for a second feature, and then predicts a third. As each prediction is confirmed, it permits defining an object and its orientation, a very iterative procedure. For instance, identifying an airplane built up from cylinders and other simple shapes using model building with an expert system may require the application of about 6000 individual rules.

Cluster analysis is another classic AI tool (Devijver & Kittler, 1982; Wee, 1986; Derde et al., 1987; Browne & Norton-Wayne, 1986). As pointed out in this chapter, it requires a high dimensionality parameter space and has large storage requirements. The main difficulty lies in finding natural clusters as opposed to imposing them based on independent knowledge. The *kNN* method is straightforward, and can learn from additional measurements, but requires unlimited storage and increasing time. It does not use all available information, and is sensitive to different set sizes and to irregular cluster shapes. Other approaches such as the principal components method (Thurstone, 1935;

Yates, 1987) are sometimes used to reduce the dimensionality of the space by locating simplified axes, but the physical meaning of the results is often questionable.

Neural nets are enjoying a strong resurgence in interest at the present time. The origin of this approach was the "Perceptron" and the McCulloch and Pitts neuron model consisting of a summation device and comparator (McCulloch & Pitts, 1943; Rosenblatt, 1957, 1958, 1962; Nilsson, 1965; Minsky & Papert, 1969; Minsky, 1975; Palin, 1982; Duff & Fountain, 1986; Finkbeiner, 1988; Grossberg, 1988; Anderson & Rosenfeld, 1988). Hebb's (1949) rule for modifying the weights or strengths of various inputs allows this system to learn. Modern multilayer systems hide the decision rules, and employ more complex neurons which have proportional (although not necessarily linear) outputs, rather than simple binary outputs. Learning requires a back propagation method (Rumelhart et al., 1986) to modify the weights of the hidden layer elements, which has been shown to work well although it is clearly not similar to the functioning of neurons in the human brain.

With these systems, it is difficult to find just where the "knowledge" resides in the system (like our own brains, it seems to be distributed widely throughout the network). It is difficult to assign probabilities or confidence limits to the result, but neural net systems usually fail gracefully. Unlike classic expert systems, the solution time for a neural system decreases as more information is made available. Their principal shortcoming is the need for unusual computer architectures for efficient realization. Current systems, whether using custom hardware or emulation in conventional computer, are limited to what are often described as "toy" problems. In order to learn, neural systems must be given a representative sample of the object populations.

The contextual learning method combines some of the attributes of these earlier methods. The role of the system user is as a teacher but not an expert (he or she must be able to recognize that features are of different types, but need not know how this decision is reached). This method establishes fuzzy cluster boundaries along context lines determined using stepwise multiple regression with exact Bayesian probabilities (Draper & Smith, 1981), or other methods such as principal component analysis. It exhibits continuous learning even after non-representative training populations are used, and can report confidence levels for its solutions, and ask for help when confidence levels are low. The knowledge base can evolve without requiring more storage space. Contextual learning offers solution times much faster than classic expert systems, because it is fundamentally a parallel method. It is classified as a knowledge-based system because the models reside in structure of data.

The reason for presenting this somewhat abbreviated discussion of AI methods here in the context of feature recognition is to point out the usefulness of these methods. The focus of much current interest is in the development of new algorithms that are successful because they better mimic human physiology. However, our purpose is NOT to research human reasoning *per se*, but to make more successful computer expert vision systems. And our emphasis is on the software methods rather than the hardware, which may be a conventional serial computer, one of the latest massively parallel systems, or even a custom device emulating the response of the human eye or other senses (Sivilotti et al., 1987), to directly extract "interesting" information.

References

W. Alston (1967) *Vagueness*, in Encyclopedia of Philosophy (P. E. Edwards, ed.) Macmillan, New York NY

J. A. Anderson, E. Rosenfeld, ed.) (1988) Neurocomputing: Foundations of Research, MIT Press, Cambridge MA

D. H. Ballard, C. M. Brown (1982) Computer Vision Prentice-Hall, Englewood Cliffs NJ

D. H. Ballard, C. M. Brown, J. A. Feldman (1984) *An approach to knowledge-directed image analysis* in Computer Vision Systems (A. R. Hanson, E. M. Riseman, ed.) Academic Press, New York, NY, 271-282

H. C. Bauer, A. Kreicbergs, C. Silfversward, B. Trihukait (1988) *DNA Analysis in the Differential Diagnosis of Osteosarcoma* Cancer 61, 2532

J. Beck, B. Hope, A. Rosenfeld, ed. (1983, 1986) Human and Machine Vision vol. I & II, Academic Press, New York NY

M. Bins, F. Takens (1985) *A Method to Estimate the DNA Content of Whole Nuclei from Measurements made on Thin Tissue Sections* Cytometry 6, 234

W. J. Black (1986) Intelligent Knowledge-based Systems: An Introduction Van Nostrand Reinhold, London

O. J. Braddick, A. C. Sleigh, eds. (1983) Physical and Biological Processing of Images Springer Verlag, Berlin

J. M. Brady, B. J. Wielinga (1978) *Reading the Writing on the Wall* in Computer Vision Systems (A. R. Hanson, E. M. Riseman, ed.) Academic Press, New York NY, 283-302

D. S. Bright (1987) A LISP-based image analysis system with applications to microscopy J. Microscopy 148, 51-87

R. A. Brooks (1984) Model-Based Computer Vision UMI Research Press, Ann Arbor MI

A. Browne, L. Norton-Wayne (1986) Vision and Information Processing for Automation Plenum Press, New York NY

K. R. Castleman, J. Melnyk (1976) *An Automated System for Chromosome Analysis: Final Report*, Document 5040-30, Jet Propulsion Lab., Pasadena CA

J. R. B. Cockett (1987) *Decision Expression Optimization* Fundamenta Informaticae X, 93-114

J. R. B. Cockett, J. A. Herrera (1986) *Prime Rule-Based Systems Give Inadequate Control* Proc. ACM SIGART ISMIS, Knoxville TN, p.441-449

M-P. Derde, L. Buydens, C. Guns, D. L. Massart, P. K. Hopke (1987) *Comparison of Rule Building Expert Systems with Pattern Recognition for the Classification of Analytical Data* Anal. Chem. 59 1868-1871

P. Devijver, J. Kittler (1982) Pattern Recognition: A Statistical Approach, Prentice Hall, Englewood Cliffs NJ

W. J. Dixon (1983) *BMDP Statistical Software* Univ of Calif. Press, Los Angeles

E. R. Dougherty, C. R. Giardina (1987) Image Processing: Continuous to Discrete, Prentice Hall, New York NY

E. R. Dougherty, C. R. Giardina (1988a) Mathematical Methods for Artificial Intelligence and Autonomous Systems, Prentice Hall, New York NY

E. R. Dougherty, C. R. Giardina (1988b) Morphological Methods in Image and Signal Processing, Prentice Hall, New York NY

N. Draper, H. Smith (1981) Applied Regression Analysis (second edition) Wiley, New York NY

R. O. Duda, P. E. Hart (1973) Pattern Recognition and Scene Analysis Wiley, New York NY

R. O. Duda, P. E. Hart, N. J. Nilsson (1976) *Subjective Bayesian Methods for Rule-Based Inference Systems* Proc. Nat'l Comput. Conf, AFIPS Proc. 45, 1075-1082

D. L. Duewer, J. R. Koskinen, B. R. Kowalski (1975) *Arthur* Lab. of Chemometrics, Dept. of Chemistry BG10, University of Washington, Seattle

M. J. B. Duff, T. J. Fountain, ed. (1986) Cellular Logic Image Processing Academic Press, London

A. Finkbeiner (1988) *The Brain As Template* Mosaic, National Science Foundation, 19#2, 2-15

R. Forsyth, R. Rada (1986) Machine Learning Applications in Expert Systems and Information Retrieval Halsted Press, New York NY

J. Frank et al. (1988) *Classification of images of biomolecular assemblies: a study of ribosomes and ribosomal subunits of Escheria coli* Journal of microscopy 150 99-115

C. Goad (1986) *Fast 3-D Model-Based Vision* in From Pixels to Predicates (A. P. Pentland, ed.) Ablex, Norwood NJ, 371-391

R. C. Gonzalez, P. Wintz (1987) Digital Image Processing, Addison Wesley, Reading MA

I. R. Goodman (1985) Uncertainty Models for Expert Systems: A Unified Approach to the Measurement of Uncertainty North-Holland, Amsterdam

A. Grasselli (ed.) (1969) Automatic interpretation and classification of images Academic Press

S. Grossberg (ed.) (1988) Neural Computers and Natural Intelligence MIT Press, Cambridge MA

D. O. Hebb (1949) The organization of behaviour John Wiley, New York

P. Horaud, R. C. Bolles (1986) *3DPO: A System for Matching 3-D Objects in Range Data* in From Pixels to Predicates (A. P. Pentland, ed.) Ablex, Norwood NJ, 359-370

K. Ikeuchi, B. K. P. Horn (1981) *Numerical shape from shading and occluding boundaries* in Computer Vision (M. Brady, ed.) North Holland, Amsterdam, p.141-184

J. S. Kowalski, ed. (1988) Parallel Computation and Computers for Artificial Intelligence Kluwer Academic Publ., Boston MA

F. P. Kuhl, O. R. Mitchell, M. E. Glenn, D. J. Charpentier (1984) *Global Shape Recognition of 3-D Objects Using a Differential Library Storage* Computer Vision, Graphics and Image Processing 27 97-114

J. Y. Lettvin, R. R. Maturana, W. S. McCulloch, W. H. Pitts (1959) *What the Frog's Eye Tells the Frog's Brain* Proc. Inst. Rad. Eng. 47#11, 1940-1951

M. D. Levine (1985) Vision in Man and Machine McGraw Hill, New York NY

D. G. Lowe (1985) Perceptual Organization and Visual Recognition Kluwer Academic Publ, Boston MA

W. S. McCulloch, W. Pitts (1943) *A logical calculus of the ideas immanent in nervous activity* Bull. Math. Biophys. 5, 115

R. S. Michalski, J. G. Carbonell, Jr., T. M. Mitchell, eds. (1986) Machine Learning, 2nd Edition, Tioga Press, Palo Alto CA

M. Minsky (1975) *A Framework for Representing Knowledge* in Psychology of Computer Vision (P. Winston, ed.) McGraw-Hill, New York NY

M. Minsky, S. Papert (1969) Perceptrons: An Introduction to Computational Geometry MIT Press, Cambridge MA

C. V. Negoita, D. A. Ralescu (1975) Applications of Fuzzy Sets to Systems Analysis, Halsted Press, New York NY

C. V. Negoita, D. A. Ralescu (1987) Simulation, Knowledge-based Computing, and Fuzzy Statistics Van Nostrand Reinhold, New York NY

R. Nevatia (1982) <u>Machine Perception</u>, Prentice Hall, Englewood Cliffs NJ

N. J. Nilsson (1965) <u>Learning Machines</u> McGraw Hill, New York NY

H. Nomori, H. Horinouchi, S. Kaseda, T. Ishihara, C. Torikata (1988) *Evaluation of the Malignant Grade of Thymoma by Morphometric Analysis* Cancer <u>61</u>, 982

Y. Ohta (1985) <u>Knowledge-based representation of outdoor natural scenes</u> Pitman, Boston MA

G. Palin (1982) <u>Neural Assemblies, an Alternative Approach to Artificial Intelligence</u> Springer Verlag, Berlin

D. M. Pearsall (1978) *Phytolith analysis of archaeological soils: evidence for maize cultivation in formative Ecuador* Science, <u>199</u>(4325), 177-178.

A. P. Pentland, ed. (1986) <u>From Pixels to Predicates</u>, Ablex Publ., Norwood NJ

D. R. Piperno (1984) *A comparison and differentiation of phytoliths from maize (Zea mays L.) and wild grasses: Use of morphological criteria* American Antiquity, 49(2):361-383.

J. R. Quinlan (1983) *Machine Learning: An Artificial Intelligence Approach* (R. S. Michalski et al., ed.) Tioga Publ. Co. Palo Alto, CA p.463

L. Roberts (1982) *Recognition of three-dimensional objects* in <u>The Handbook of Artificial Intelligence vol III</u> (P. Cohen, E. Figenbaum, ed.) Kaufmann, Los Gatos CA

D. W. Rolston (1988) <u>Principles of Artificial Intelligence and Expert System Development</u> McGraw Hill, New York NY

F. Rosenblatt (1957) *The Perceptron: A Perceiving and Recognizing Automaton* Project PARA, Cornell Aeronautical Laboratory Rept. 85-460-1

F. Rosenblatt (1958) *The Perceptron: A Probabilistic Model for Information Organization and Storage in the Brain* Psych. Rev. <u>65</u>, 358-408

F. Rosenblatt (1962) <u>Principles of Neurodynamics: Perceptrons and the Theory of Brain Mechanisms</u> Spartan Books, Washington DC

A. Rosenfeld, ed. (1986) <u>Techniques for 3-D Machine Perception</u> North Holland, Amsterdam

D. A. Rosenthal (1981) <u>An Inquiry Driven Vision System based on Visual and Conceptual Hierarchies</u> UMI Research Press, Ann Arbor MI

I. Rovner (1971) *Potential of opal phytoliths for use in paleoecological reconstruction* Quaternary Research, <u>1</u>#3, 345-59.

D. E. Rumelhart, G. E. Hinton, R. J. Williams (1986) *Learning representations by back-propagating errors* Nature <u>323</u>, 533-536

J. C. Russ, T. M. Hare, F. U. Luehrs (1980) *A Novel, Efficient Alloy Sorting Algorithm*, Proc. Ann. Denver X-ray Conf., Denver Univ., Denver CO, 130-131

J. C. Russ, T. M. Hare (1983) *A Self-Educating Classification Scheme for Particle and Phase Identification*, Microbeam Analysis 83, San Francisco Press, 111

J. C. Russ, J. C. Russ (1987) *Teaching computers to see*, <u>Microbeam Analysis 87</u>, San Francisco Press

J. C. Russ, I. Rovner (1987) *Stereological verification of Zea Phytolith taxonomy*, Phytolitharien Newsletter, <u>4</u>#3, 10

J. C. Russ, I. Rovner (1989) *Stereological identification of opal phytolith populations from wild and cultivated Zea Mays* American Antiquities (in press)

F. F. Sabins, Jr. (1986) <u>Remote Sensing: Principles and Interpretation</u> second edition, Freeman, New York

E. Sanchez, L. A. Zadeh, ed. (1987) <u>Approximate Reasoning in Intelligent System Decision and Control</u> Oxford Press, New York NY

D. Schutzer (1987) <u>Artificial Intelligence: An Applications-Oriented Approach</u> Van Nostrand Reinhold, New York NY

L. G. Shapiro, R. M. Haralick (1982) *Organization of relational models for scene analysis* IEEE Trans. Patt. Recog. Mach. Intell. <u>PAMI-4</u>#6 595-602

M. A. Sivilotti, M. A. Mahowald, C. A. Mead (1987*) Real-time visual computations using analog CMOS processing arrays* <u>Advanced Research in VLSI</u>: Proc. 1987 Stanford Conf. (P. Losleben, ed.) MIT Press, Cambridge, MA, 295-312

P. E. Slatter (1987) <u>Building Expert Systems: Cognitive Emulation</u> Halsted Press, New York, NY

T. M. Strat (1984) *Continuous Belief Functions for Evidential Reasoning* Proc. of the Conf. of AAAI, 308-313

S. Tanimoto, A. Klinger, ed. (1980) <u>Structured Computer Vision: Machine Perception through Hierarchical Computation Structures</u> Academic Press, New York NY

L. L. Thurstone (1935) <u>The Vectors of Mind</u> Univ. of Chicago Press, Chicago IL

A. Tversky, D. Kahneman (1974) *Judgement under Uncertainty: Heuristics and Biases* Science <u>185</u>, 1124-1131

P. C. Twiss, P.C., E. Suess and R.M. Smith (1969) *Morphological classification of grass phytoliths* Soil Science Society of America Proceedings, 33(1), 109-115.

L. Uhr (1987) <u>Multicomputer architectures for artificial intelligence: toward fast, robust, parallel systems</u> Wiley, New York NY

J. R. Walters (1988) <u>Crafting Knowledge-based Systems: Expert Systems Made Easy</u> Wiley, New York NY

W. J. Wee (1968) *Generalized inverse approach to adaptive multiclass pattern classification* IEEE Trans Comp <u>C-17</u>, 1157-1164

G. Winstanley (1987) <u>Program Design for Knowledge-based Systems</u> Halsted Press, New York NY

P. H. Winston (1975) *Learning Structural Descriptions from Examples* in <u>The Psychology of Computer Vision</u> (P. H. Winston, ed.) McGraw Hill, New York NY

A. Yates (1987) <u>Multivariate Exploratory Data Analysis</u>, State Univ. of New York Press, Albany NY

L. A. Zadeh (1965) *Fuzzy Sets* Information and Control <u>8</u>, 338-353

H.-J. Zimmermann (1987) <u>Fuzzy Sets, Decision Making and Expert Systems</u> Kluwer Academic Publ., Boston MA

N. Zuech, R. K. Miller (1987) <u>Machine Vision</u>, Fairmont Press, Lilburn GA

Chapter 10

Surface Image Measurements

In the preceding chapters, the assumption has been made in most cases that images are planar. Either they result from viewing a cut section through some opaque body, or they are the projection of objects onto a plane. Most of our experience in the real world is, of course, with neither of these restricted types of images. While still stopping well short of the complexity of real-world images, which contain a variety of surfaces and objects with great depth, it can be important and useful to relax the criterion of planar surfaces somewhat. There are many interesting surfaces that are approximately flat on a large scale, but locally quite rough. Examples range from the surface of integrated circuits, where strips of metallization or photoresist produce surface relief, to skin with its folds and wrinkles, to the surfaces of planets, with such relatively minor irregularities as mountains and oceans.

Depth cues

For the present, we will continue to deal with images that represent a single point of view. Stereoscopic images in which two viewpoints are used to obtain relief information will be discussed in the next chapter. Since humans and all "higher" animals are two-eyed, it is easy to dismiss the subject of viewing rough surfaces as requiring this stereoscopic approach. However, this is far from bring the case. In fact, the actual human uses of stereoscopic vision are comparatively few. We judge distance, as we judge most other things, relatively. The sine and cosine terms needed to calculate exact distances seem not to be computed in our brains. If we are trying to touch something, the images of the object and our hand may be compared in parallax to decide which way to move it, but we don't work out the distance in cm. along the way.

There are additional cues to relative depth of objects in an image besides the parallax they produce in two different eye views (Figure 10-1). These include precedence (one object partially obscuring the other, indicating it is closer), relative size (of two similar objects, the larger one is closer), haze (more distant objects generally appear dimmer and may be tinged with blue from light scattering by the intervening air), and position. The latter criterion results from our learned experience with linear perspective. The "vanishing point" in a perspective drawing lies on the horizon. Objects with higher positions on the ground are therefore farther away (and vice versa for things on the ceiling).

Most of the uses of these criteria for judging distance or depth of objects are not necessary for the somewhat limited class of images we will be dealing with here. However, they illustrate the fact that our reliance on stereo vision is not that heavy. Many people suffer from amblyopia ("lazy eye") in which only one eye really contributes any information to the brain. Others may have lost an eye. These people generally deal quite well with the world around them, including driving cars and other activities that would seem to require good depth perception.

Figure 10-1: Cues to depth as used by Renaissance painters (see text).

It is true that two sequential images obtained by moving the head can be used to generate depth information from parallax in almost the same way as two simultaneous images. Some animals whose eyes are placed to have only a minimal field of overlap where true stereo would be possible may use such a method (one theory holds that snakes "weave" and move their heads from side to side to better triangulate the distance to strike). But the chief reason that humans and other animals have evolved with two eyes, beyond a general argument of bilateral symmetry, is that one eye is a backup. If one is lost, we have and need the other. Our reliance on visual input is such that even with modern aids and concern for the sightless, functioning in the world is certainly impaired without vision (and not too many generations ago it simply meant death). Yet one-eyed individuals cope fairly well with depth perception using these other cues.

For the limited class of images we will discuss here and try to measure, the most important cue is texture and shading. Precedence, in which one feature or portion of the surface lies in front of and partially obscures another, is a much more powerful cue to organizing a three dimensional scene in our minds, but it makes difficult or impossible the task of mapping an entire surface in the computer. Most microscope images are intentionally restricted to surfaces simple enough to avoid this, or use viewpoints which eliminate it. From the appearance of surfaces we judge how far away they are, how locally rough they are, and how they are oriented with respect to our viewing direction. To examine each of these and find ways to measure them, we will first have to discuss the factors that produce images from the scattering of light and electrons from surfaces.

Image contrast

When light photons strike a surface, scattering occurs. In the simplest case, that of a perfect plane surface, the scattering is mirror-like or specular. The fraction of the incoming light that is reflected is proportional to the square of the ratios of the indices of refraction of the materials on either side of the boundary (one is usually air). This means that surfaces made up of different materials would reflect different amounts of light in different regions, producing a brightness difference.

Few surfaces are sufficiently flat on a fine scale to function as good mirrors. If the surface is irregular, with facets at various orientations, then the light is reflected in many different directions, and the light that reaches the viewpoint (represented by the eye or a

Surface Image Measurements 311

camera) is reduced. For a perfectly diffuse surface (or a surface viewed under perfectly diffuse lighting), the intensity of the light reaching the viewpoint varies as the cosine of the angle between the surface and the viewer. As shown in Figure 10-2, most real surfaces scatter light in a combination of diffuse and specular ways, resulting in an overall surface brightness from diffuse scattering plus a specular "hot spot" from locations at the correct orientation to act as a mirror between the light source and the eye (equal angles for the incident and reflected beam).

More complicated situations are produced by colored surfaces, which do not scatter all wavelengths of light equally. Like transmitted light, for instance through a colored filter, light scattering produces color by selectively absorbing some wavelengths. This absorption is usually due to the individual photons having enough energy to excite some mode of vibration of molecules present in the material, although other kinds of mechanisms also operate. Metals appear colorless, because no selective absorption of particular wavelengths occur. In many materials, very small changes in the concentration of minor impurities (either accidental or intentional) produce profound changes in color. However, in most cases the orientation of the surface with respect to the light source affects only the brightness of the scattered light and not the color. This means that monochrome grey scale images will usually be sufficient to judge surface orientation, and color will not be needed.

When surfaces vary both in orientation and composition, so that color and brightness may both change due to composition or structure, the human visual system uses the dependence of brightness or intensity solely on orientation to separate the two effects. This is what enables us to recognize objects with complex shapes as being one single thing, even though the different surfaces have different brightnesses. The same mechanism works to correct for differences in illumination intensity. We automatically compensate for shadows falling across objects, and do not consider the edge of the shadow to represent a boundary, because of color constancy on both sides of the shadow.

Fortunately, in examining surfaces in the context of this chapter, we will be able to ignore this kind of effect, by controlling the light source and assuming that relief is low enough to prevent one object from casting its shadow onto others. We will also ignore any effects of polarization of light by scattering, or scattering of polarized light.

The cosine dependence of brightness on the angle between the surface and the viewer is termed a "Lambertian" relationship. With this simple model we can construct a map of surface orientation from measured brightness, assuming we know the geometry of the source of illumination. It is then possible to begin at any point, usually an edge, and integrate slope times distance to determine the height across the image. This operation is generally known as "shape-from-shading". The general form of Lambert's law, as used

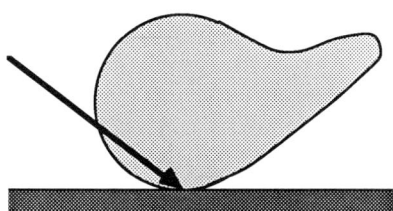

Figure 10-2: Combined diffuse and specular scattering of light from a surface.

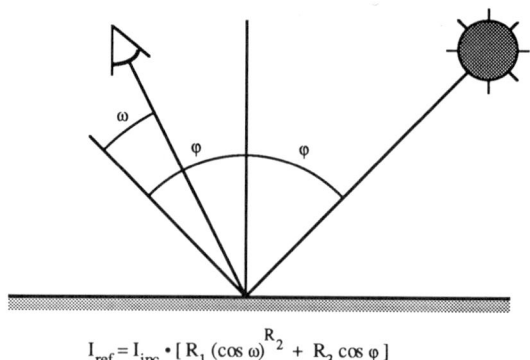

$$I_{ref} = I_{inc} \cdot [R_1 (\cos \omega)^{R_2} + R_3 \cos \varphi]$$

Figure 10-3: Relationship between Incident and reflected or scattered intensity.

in computer modelling of surfaces, can be written as shown in Figure 10-3, to include both diffuse and specular reflection.

The three R values can be varied between 0 and 1 to match real material reflective properties, and may furthermore be a function of wavelength if the surface has color. Many modelling programs divide the surface into facets, much as the stereo pair and serial section surface modelling routine discussed in the next chapters. The angles the surface makes with respect to the viewer and the source of illumination are calculated, and then the brightness of the surface is calculated from the expression in Figure 10-3.

This modelling approach ignores secondary effects such as shadowing of one surface by another, or fill illumination from one surface on another. A complete solution to this problem requires a much more intensive computer approach called "ray-tracing" in which each line of sight from the viewpoint to the scene is traced through all of its reflections and refractions to find its source of light, and keep track of modifications in intensity and color. For diffuse surfaces this requires splitting each ray into many parts, and the complexity increases very rapidly. These methods are practically never required in conjunction with the surface representation methods of interest for image analysis, because the source(s) of illumination and the complexity of the surface are not allowed to become that "interesting."

The faceted appearance of surfaces modelled in the discrete way just discussed can be improved by Gouraud shading. This averages the brightness (or color) across nodes and lines where the facets join to produce a smooth variation instead of an abrupt change. It creates a problem if the surface contains some edges that really are sharp. This is often handled by setting an upper limit on the brightness difference (which is easier to handle than an upper limit of the angular orientation difference). If the brightness change exceeds the threshold, then it is assumed that a real edge is present and no shading correction is applied.

Modelling of real scenes involves precedence calculations (one surface or object can partially hide another), intersection of surfaces, and other complications. These are often unimportant in dealing with the more restricted class of images of surfaces and objects suitable for measurement. While surface modelling can be important in presenting the results of image measurement, our present concern is how to measure surfaces from the image brightness. Usually a somewhat simplified version of the Lambert law is used for integration. The result is a series of elevation profiles as shown in Figure 10-4.

With a single light source, the shape-from-shading integration can only function in one direction. For instance, if the light source is to the right (along the positive X axis, using normal notation), then surface inclinations up or down (in the Y direction) are not distinguished. If two light sources, first one in the X direction and then another in the Y direction, are used this permits collecting two separate images (one with each light source), and performing the integration in all directions to construct a complete height map. It is not necessary that they be at right angles or exactly in line with the rows of pixels in the image, of course, but that is the simplest arrangement.

The main difficulty in applying shape-from-shading is that it assumes the surface to be quite uniform in composition and local structure, and the images to be quite noise free. Otherwise the process of integration will diverge, producing poor or even nonsensical results. Also, any variation from the Lambertian cosine law will produce a large error, so specular reflection must be avoided. This is usually possible for surfaces whose local orientation varies over a small range of angles, but when all possible angles are present, the occurrence of bright spots due to local specular reflection is hard to avoid. Sometimes, these can be edited out of the image because they are brighter than anything else, and can be replaced with a substitute value based on knowing the geometry of the light source.

The Lambertian relationship for light intensity from a perfectly diffuse surface is $dI/d\Omega = \sigma \cos \delta$, where I is the intensity of light coming from the object, and $dI/d\Omega$ is the fraction of that light in a solid angle $d\Omega$, σ is the surface reflectivity, and δ is the angle between the viewing direction and the local surface normal. This applies to diffuse lighting. If the light source is a point and the surface is a perfectly diffuse reflector, this relationship must be altered. If the angle of incidence of the illumination with respect to the surface normal is φ, then the intensity of light falling on the surface is proportional to $\cos \varphi$ and the overall relationship becomes

$$dI/d\Omega = I_0 \cos \varphi \, \sigma \cos \delta$$

where I_0 is the source brightness, as shown in Figure 10-5.

Shape from texture

The "classical" approach to shape-from-shading just described utilizes the Lambertian intensity relationship for diffusely scattering surfaces, which simply states that the observed intensity of light scattered from a surface is

$$I = \rho \lambda (N \cdot L)$$

where ρ is the surface reflectivity or albedo, λ is the intensity of the illumination, and N

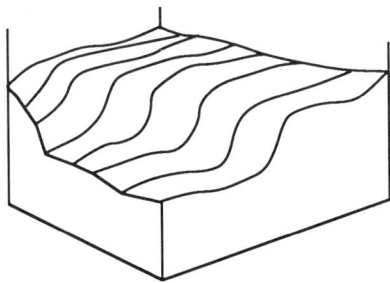

Figure 10-4: Example of shape-from-shading reconstruction.

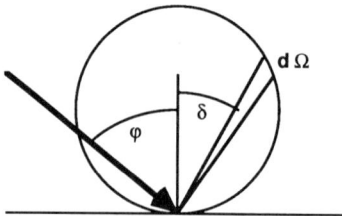

Figure 10-5: Lambertian relationship for light intensity from a diffuse surface.

and L are unit vectors normal to the surface and pointing to the light source, respectively. By measuring the brightness at many points in the image of a surface, and calculating from those values the derivatives and second derivatives of brightess, it is possible to construct equations that can be integrated to give the surface slope and elevation everywhere (Horn, 1970; Ikeuchi & Horn, 1981; Horn & Brooks, 1986). Brooks (1981, 1983) has shown examples of model-based image interpretation in which shape-from-shading reconstruction is used to fit simple geometrical shapes such as cylinders to parts of images, and then assemble those into an understanding or recognition of objects in surveillance images.

However, this type of integration requires information on the direction of the illumination and boundary conditions that is rarely available, with the result that the technique cannot be applied in many real cases. Furthermore, the deviations from diffuse Lambertian surface behavior that may be present, or variations in local surface albedo because of differing material characteristics, introduce nonrecoverable errors.

Pentland has shown that there is a much simpler way to obtain information on surface orientation and curvature from analysis of the local brightness information (Pentland, 1984). The local surface tilt and slant angles are defined respectively as the angle that an iso-elevation contour on the surface makes in the plane of image projection, and the angle at which the surface intersects the plane of projection. In terms of the components x, y, z of the surface normal vector to the plane N shown in Figure 10-6, these angles are arc tan (y/x) and cos z respectively, where the (x,y) plane is taken parallel to the projected image plane and z is the direction of view.

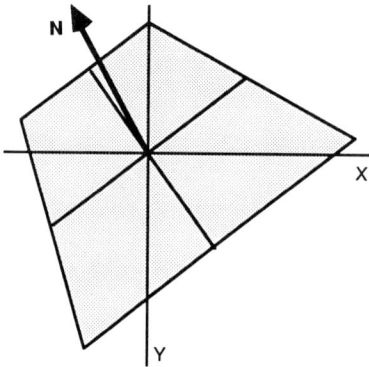

Figure 10-6: Orientation of a surface with respect to the viewer.

By restricting the measurements in the image to brightness and to the first and second derivatives of brightness, there is not enough information to determine the entire surface orientation and curvature. However, by making some simplifying approximations that are not locally restrictive, much becomes possible. If the surface is modelled as one with the two principal curvatures either zero or equal in magnitude, then orientations and curvatures can be determined. The five types of surfaces that are possible are listed below, and several are shown in Figure 10-7.

1. A plane, for which the curvature values are both zero
2. A cylinder, for which one curvature is zero
3. A convex spherical element, for which both curvatures are positive
4. A concave spherical element, for which both curvatures are negative
5. A saddle surface, for which one is positive and one negative

One intriguing result of this is that lines for which the curvature is zero in some direction will mark the cylindrical regions that are the transitions where surfaces change type, or where different surfaces intersect. The location of these lines permits segmenting the image into regions corresponding to different surfaces. This kind of segmentation using zero-crossings has often been utilized based on its empirical success. This insight indicates why that method works.

For Lambertian conditions, these surfaces give rise to images with brightnesses that behave as follows, using the notation that dI is the derivative of I in some direction in the x, y plane, d^2I is the second derivative, and I_x, I_y, I_{xx}, I_{yy} and I_{xy} are partial derivatives of the form $\partial I/\partial x$, $\partial^2 I/\partial x^2$, and $\partial^2 I/\partial x \partial y$, as appropriate:

1. If $d^2I = 0$ in all directions, then the surface is a plane
2. If $d^2I = 0$ in some direction, then that is the axial orientation of a cylinder
3. If $I_{xx}/I_{yy} < 0$ in some direction, then the surface is a saddle
4. If $I_{xx}/I_{yy} \geq 0$ in every direction, then the surface may be concave, convex or a saddle

In addition, the local surface orientation (the tilt and slant mentioned before) can be determined for these surfaces using the values of the second derivatives I_{xx}, I_{yy} and I_{xy}. The direction on the image for which d^2I is greatest is the direction of tilt. This is illustrated in Figure 10-8 for a spherical surface which (depending on where you assume the source of illumination to lie) is either a bump or a depression. At various locations on the surface, vectors are drawn in the direction of maximum and minimum d^2I. The line of maximum d^2I gives the tilt direction. The magnitude of the cross derivative I_{xy} gives the surface slant (angle of dip toward or away from the viewer). It is particularly useful that the first derivatives of brightness and the absolute values of brightness do not enter into

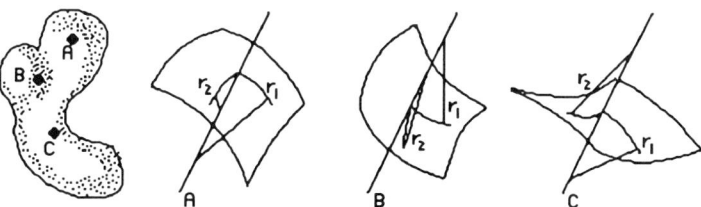

Figure 10-7: Surface of a complex shape, showing convex, concave and saddle points in terms of the principal radii.

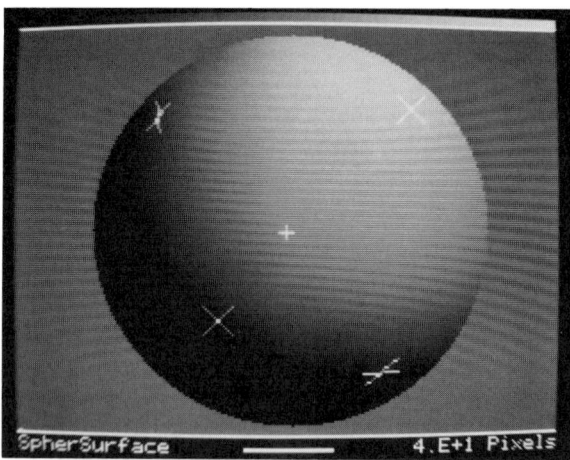

Figure 10-8: Directions of maximum and minimum brightness gradient indicate surface orientation on a spherical surface patch.

this classification scheme, since they often vary depending on changes in the surface albedo or in the location of the light source(s).

Pentland has demonstrated that these rules permit determining the local orientation, and from those data building up a map of surface elevation, for images as diverse as manmade machined objects, natural rocks and terrain, and even a human face. He was furthermore able to show subsequently (Pentland, 1986) that similar techniques could be applied to fractal rough surfaces. The mean value of d^2I/I_0 is maximum in the direction of tilt, and the slant is given by the spread between the angles at which the maximum and minimum values are obtained. Others (Kender, 1979; Witkin, 1981) had previously introduced a similar relationship using the maximum and minimum scale of texture present as surface markings on smooth surfaces. Figure 10-9 shows an example of an image in which local ellipses have been drawn whose axes are the orthogonal derivatives of brightness; the ellipses show the local orientation and its spatial variation quite well. However, this approach makes systematic errors when non-isotropic surfaces are encountered (for instance, surfaces such as textiles that have a natural preferred orientation of texture), and will not work on natural fractal surfaces because, by definition, the roughness is insensitive to scale.

Another approach to surface measurement is, of course, stereoscopy. The classical stereoscopic technique requires two different images of the same region from slightly different points of view. Matching of points present in both images then permits calculating their distance from the parallax, or displacement of the point between the images, using straightforward trigonometry. One of the most important drawbacks of this method is the computational difficulty in matching points between the two images. Several different approaches have been used; they are discussed in more detail in the next chapter.

A quite different method, more related to shape-from-shading, is known as "photometric stereo" (Woodham, 1980, 1981). In this case, two images are obtained from the same point of view, but with two different sources of illumination. From the variation in brightness of points in the image, the surface orientation based on a matrix of partial derivatives can be determined (Carrihill & Hummel, 1985). In a few cases

Surface Image Measurements

multiple solutions are possible, but these can be resolved by the addition of a third image with a different light source. Photometric stereo is in many respects complementary in its capabilities to traditional stereo. It determines local surface gradients rather than the range or elevation of points, and it works best on smooth surfaces with uniform properties and few discontinuities, whereas traditional stereo works best with textured surfaces and sharp discontinuities (which provide the interesting points or variation in cross correlation that enable matching of points).

Photometric stereo does not require the matching of points between the two images, which is the major difficulty with traditional stereo, because the two images are spatially identical. However, the amount of computation is formidable, requiring extensive real-valued mathematics at each point, and the results are effectively limited to rather simple and idealized surfaces such as man-made objects with regular shapes and uniform properties. The method has found several applications in quality control ("QC") inspection, where these criteria are met.

The scanning electron microscope

One of the most interesting applications of shape-from-shading has not used light at all, but instead relies on the fact that electron backscattering can also be approximated by a Lambertian law (Reimer et al., 1984, 1986; Carlsen, 1985). The application has been to map surface elevations on integrated circuits, viewed in the scanning electron microscope. Usually the point of view will be restricted to a normal view of the surface. The use of shading correction and rubber-sheeting to deal with oblique views can be added if necessary. The key to this method is a fundamental reciprocity between light and electrons. The SEM can be considered as an analog to light illumination if the point of view is taken to be the electron gun, and the light source is considered to be placed at the electron detector.

Electrons in the SEM strike the specimen surface at one point at a time, and the beam is scanned, usually slowly, in a raster pattern to acquire the entire image. At each point, the electrons penetrate into the sample and lose energy by interacting with the atoms that constitute it. A typical incident electron has an initial energy in the range of keV (from 1 to 30 keV is common). Individual interactions of the electron with the atoms in the sample are of two types: elastic, in which the electron changes direction but does not lose energy, and inelastic, in which a energy is lost. If the electron gives up energy by

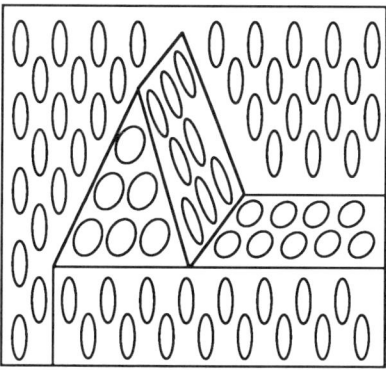

Figure 10-9: Orientation of each surface patch is shown by ellipses based on texture scale or gradient.

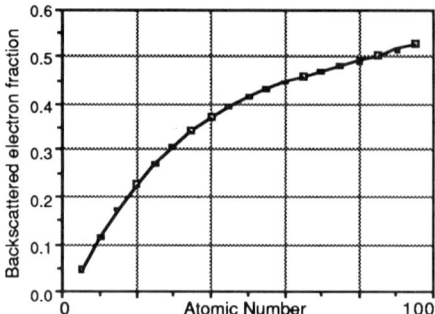

Figure 10-10: Variation of fraction of electrons that backscatter from the sample as a function of its atomic number.

creating ionizations (knocking bound electrons out of their shells), it can produce characteristic X-rays and heat in the sample. A large fraction, usually more than half of the incident electrons, lose all of their energy in this way. The total depth of penetration is of the order of a micrometer before the electron stops.

Inelastic scattering causes the electron paths to zig-zag through the sample, while each is slowing down. The scattering angles depend on the atomic number of the atoms which the electron encounters, since higher atomic number atoms have more electrons and more protons, and so produce larger electrostatic fields around the nucleus to deflect the moving incident electron. If the zig-zag path followed by the moving electron re-emerges from the sample surface before the electron has lost all of its energy, it is called a backscattered electron.

The fraction of the incident electrons that backscatter depends on the mean atomic number of the sample and on the orientation of the surface (Figures 10-10 and 10-11).

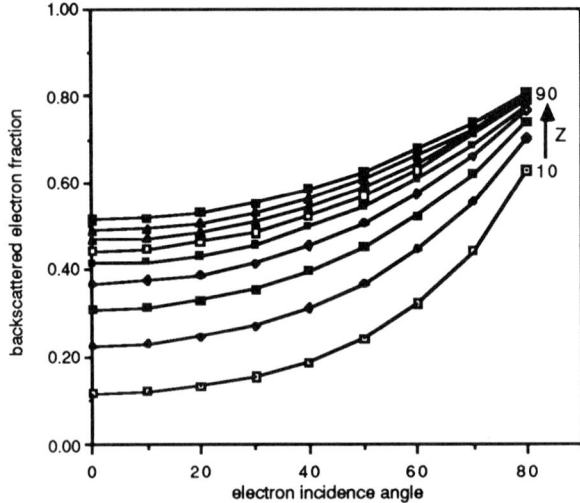

Figure 10-11: Variation of backscattered electron fraction with incidence angle for different atomic number targets.

Surface Image Measurements

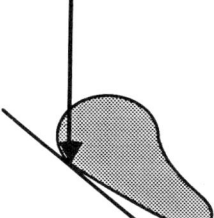

Figure 10-12: Pronounced forward scatter of electrons from tilted specimen.

The backscattered fraction is typically from 10 to as much as 50%. An empirical equation for the total backscattered intensity η as a function of the beam incidence angle φ and atomic number Z is

$$\eta(Z,\varphi) = B\, (\eta_0/B)^{\cos \varphi}$$

where $B = 0.89$ and $\eta_0 = -0.0254 + 0.016\, Z - 1.86 \cdot 10^{-4}\, Z^2 + 8.3 \cdot 10^{-7}\, Z^3$. This breaks down for electron energies below about 5 keV. Also, it says nothing about the energy distribution or the spatial distribution of the electrons. The Lambertian cosine approximation ($I_{BSE} = \eta \cos \delta$) is acceptable for incidence angles close to normal, but for large tilt angles ($\varphi > 45°$) there is a pronounced forward scatter (Figure 10-12), broader than a specular reflection, that is not well modelled or accounted for in present surface measurements.

There are other complications that hamper the application of this relationship to surface reconstruction. The energy distribution changes significantly with atomic number and tilt angle, and since most electron detectors have an output signal that varies with energy, this also changes the measured brightness value.

Backscattered electrons (BSE) can be detected in a variety of ways, one of the simplest being a simple Schottky diode. If two such detectors are positioned on opposite sides of the sample, then the signal strength from each will vary with the composition and orientation of the local sample surface as the incident beam moves across the specimen (Figure 10-13). However, by combining the signals from the two detectors these effects can be separated. Since each detector's individual signal is proportional to the quantity $(1 \pm \sin\varphi \cdot \cos\chi)$, where φ is the horizontal tilt and χ is the azimuthal angle of the local surface orientation, the sum of the two detectors $(A + B)$ is proportional to

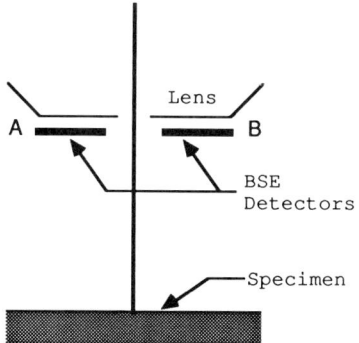

Figure 10-13: Pair of backscatter detectors mounted in the SEM.

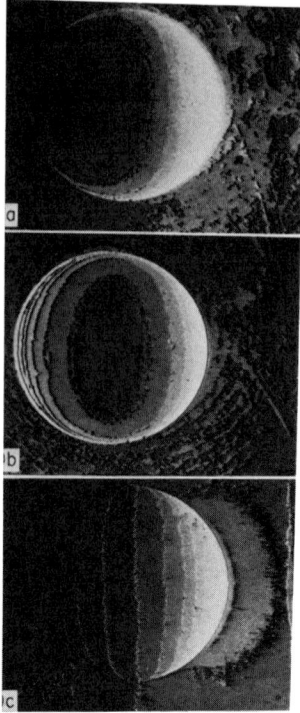

Figure 10-14: Example images from a spherical sample: a) *A* detector only; b) *A* + *B*; c) *A* − *B* (Reimer, 1984; Reimer et al., 1986).

η_0, the total backscattered electron fraction. For low tilt angles, this is primarily dependent on the atomic number and largely independent of the local surface orientation, while the difference is a function of the surface angle. On the other hand, the difference between the two signals depends on the surface angles as well as the total backscattering. The ratio $(A - B)/(A + B)$, where *A* and *B* are the two detector outputs, produces a signal that is essentially independent of sample composition, and furthermore depends on local surface inclination with a functional relationship that is closely similar to the Lambertian cosine law for light scattering. Figure 10-14 and 10-15 show the use of two detector systems; the sample in Figure 10-14 is a spherical ball viewed in the SEM. The images show a single detector, the sum of two detectors, and the difference between the two detectors.

This means that shape-from-shading can be performed with backscattered electron images of surfaces in the SEM. To obtain two separate images with shading information in two orthogonal directions, two pairs of backscattered electron detectors can be used. Indeed, many SEM's are available with a set of four detectors symmetrically arranged above the sample, whose outputs can be directly used for this purpose.

For small structures such as photoresist or metallization stripes on semiconductors, there are two problems with using this approach, in addition to the usual one of integrating noisy images. The first is that the range of the electrons is not small, and this limits the image resolution. Backscattered electrons sample a region that is from 0.1 to 1 micrometer in size, which is not much smaller than the dimension of some of the devices now being fabricated (Figure 10-16). The image resolution should ideally be an order of

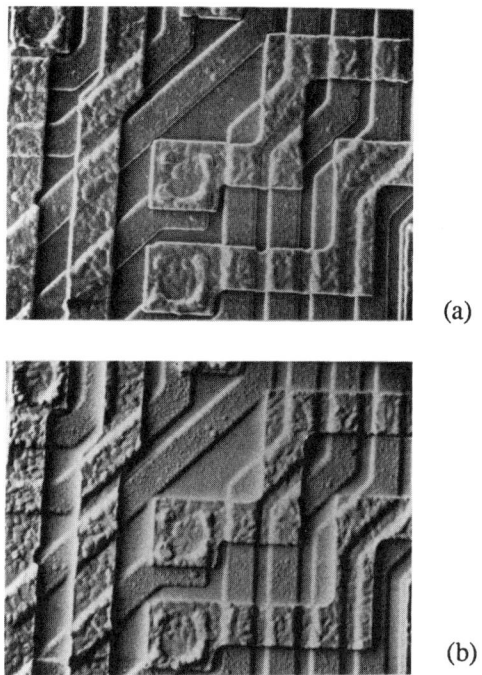

Figure 10-15: Images on a microelectronics specimen: a) $A + B$; b) $A - B$ (Reimer, 1984; Reimer et al., 1986).

magnitude smaller than any structure to be measured. This range can be reduced somewhat by dropping the incident electron energy. In fact, many SEM's used to examine integrated circuits are now operating at quite low voltages, from a few keV down to a fraction of a keV. This was done not so much to reduce the range of the incident electrons for resolution purposes as to reduce damage to devices and exposure of photoresist, and to reduce charging of the samples because the electrons which do not backscatter must be conducted to ground, and this is easier to do when they stay near the surface than when they are deep inside the material.

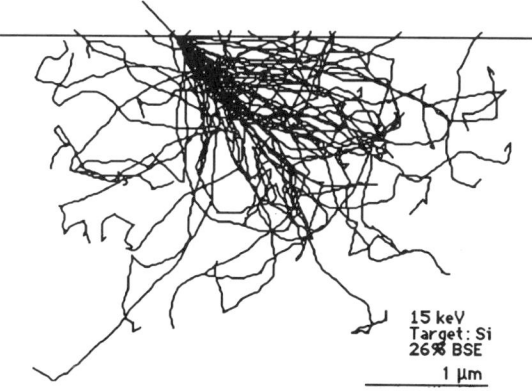

Figure 10-16: Size of electron capture volume (shown by a Monte-Carlo simulation) is comparable to small size of modern device dimensions.

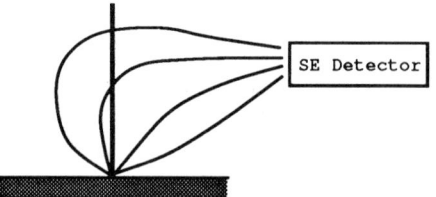

Figure 10-17: Secondary electron detector collects electrons that leave the sample in many directions.

But the byproduct of dropping the incident electron voltage is that the energy of the backscattered electrons is also reduced. This makes them harder to detect, indeed the surface barrier detector is hardly sensitive at all to electrons under 1 keV. and its output rises proportional to electron energy. Other types of detectors, such as the secondary conversion detector or grids to accelerate the backscattered electrons before they strike the detector, may be able to fill this gap.

Another approach to dealing with the problem is to use secondary rather than backscattered electrons. Secondary electrons (SE) are ones that were initially within the sample, occupying bound orbital positions around the atoms there. Struck by the incident electron, they acquire enough kinetic energy to escape the atom. Those near the surface may leave before being recaptured, with energies of a few tens of electron volts. They carry information about a much smaller region of the sample, both laterally (coming primarily from very near the focussed incident electron beam) and in depth (representing only a few nanometers). This produces images with much higher resolution, and also means that satisfactory secondary electron images can be obtained with quite low accelerating voltages for the incident electron beam.

The interpretation of the secondary electron signal in terms of the surface structure is much more complicated than for backscattered electrons, however. This is partly due to the nature of the electron detector, which uses an electrostatic field to collect electrons that leave the sample with a considerable range of initial directions (Figure 10-17). This is possible because the secondary electrons have very low energies. Also, edges or other very local protrusions that expose more surface area appear bright in the secondary electron image. This kind of "cartoon-like" outlining does not disturb the human visual system, indeed we have seen that such edge outlining is used by the early vision processing itself, and secondary electron images appear very sharp and easy to view.

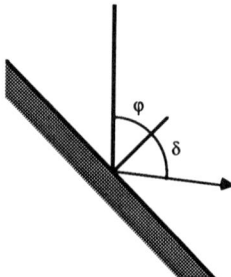

Figure 10-18: Electron incidence and emission angles.

Surface Image Measurements

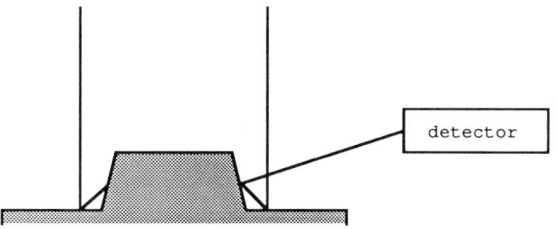

Figure 10-19: Shadowing and fill illumination.

An approximate and empirical model relating the intensity of the SE signal to beam incidence angle φ and the viewing angle δ (Figure 10-18) is

$$dI/d\Omega = I_0 \sec^\kappa \varphi \cos \delta$$

where κ is an empirical value observed to vary in the range $0.65 < \kappa < 1.3$ for various surfaces, and I_0 is a function of electron voltage and material composition.

Integration of either the SE or BSE brightness to model actual surface shape is further complicated by shadowing and secondary illumination (Figure 10-19). Shadowing occurs when one portion of the irregular surface partially blocks the detector, lowering collection efficiency and producing a darker region in the image. Fill illumination results when electrons backscattered from one portion of the surface strike somewhere else, giving rise to an increase in signal. Both effects alter the calculated profiles. In principle, they can be corrected by an iterative calculation, but if the effect is large, the slopes are steep, or there is some additional effect such as a composition variation, this may not be possible.

The quantitative relationship between brightness and surface structure is quite complicated for secondary electrons. It must take into account surface contamination and electronic states that would not affect higher energy backscattered electrons. Modelling of secondary electron contrast from surface structures is still an infant science. It is carried out with Monte-Carlo simulation programs that use detailed knowledge about the physics of electron scattering, and computerized random number generators to sample the distributions of possible events, to predict what will be measured from particular surface configurations. There is not yet any comprehensive model that would make a complete shape-from-shading integration possible.

Furthermore, the entire shape-from-shading method is really only applicable to structures that are large compared with the dimensions of electron penetration. Most of the interesting problems requiring SEM measurement are not in that class, but rather involve quite small structures for which a much more exact quantitative model is required. An example of this is discussed below in conjunction with line width measurements, which use the result of Monte-Carlo simulations directly.

A modification of the shape-from-shading approach in the SEM has been suggested (Beil et al., 1989) which uses stereoscopy to locate abrupt changes in orientation. As discussed in more detail in Chapter 11, stereoscopy is best suited to sharp discontinuities since points can be more easily identified and fused in the two images than can points on gradually varying surfaces, while the shading integration tends to break down at those points. Then the shape-from-shading calculation is carried out between those points to fill in the more smoothly varying regions for which it is best suited.

324   *Chapter 10*

Line width measurement

The parameter of great interest in characterizing the surface of integrated circuits is the width of "lines", or traces of material left after etching, or of metallization that has been laid down for interconnections or (more often) photoresist that has been exposed, developed and etched to leave material that will block the diffusion or implanting of ions that act as donors or acceptors in the electronic structure of the material. Measurement of these dimensions is known as critical dimension metrology (CDM). Figure 10-20 shows a top view and cross section of photoresist lines about 1 µm wide on silicon.

As device dimensions have shrunk from many micrometers to one micrometer and to submicrometer sizes, these measurements have become more critical and more difficult. The light microscope is inadequate both in image resolution and depth of field for these small sizes, so the scanning electron microscope must be used (Nyyssonen & Larrabee, 1987; Postek & Joy, 1987; Atwood & Joy, 1987; Postek, 1987). Both backscattered and secondary electron signals are used, each having the limitations and problems explained above.

In most cases, an entire image is not required for line width measurements. Instead of scanning the electron beam over a full raster to store a complete image, it is swept over a line that crosses the feature to be measured. The restriction is usually made that in addition to the specimen surface being horizontal with respect to the beam, the stripe to be measured must be oriented at right angles to the raster direction in which the beam moves. We will see later how this restriction can be relaxed.

When the beam scans across a single line traversing the feature to measured, the result is not an entire image but a simple waveform or signal profile across the feature. Even if the feature has a simple shape, such as a rectangular cross section with flat top and vertical sides sitting on a flat surface, the waveform may not be simple, or even symmetrical. This depends on the position of the detector(s) and the signal being

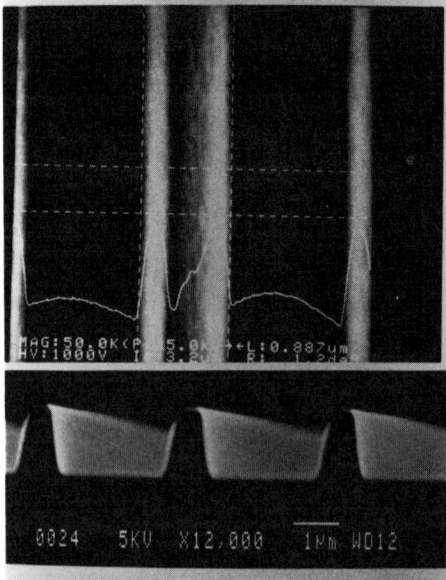

Figure 10-20: Top and cross section views of photoresist lines.

Surface Image Measurements 325

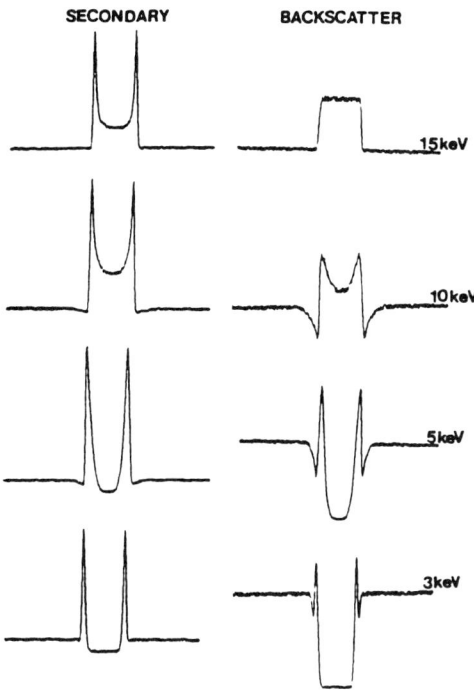

Figure 10-21: Secondary and backscattered electron signal profiles across a trace, using different accelerating voltages (Joy, 1987b).

acquired, and perhaps on the speed with which the beam is scanned and the frequency response of the video amplifier. The signal also varies with the composition of the materials, the accelerating voltage, and the shape of the specimen in a complicated way (Figure 10-21).

Generally, the backscattered electron signal, coming from multiple detectors above the sample, produces a symmetrical and fairly simple result, while the secondary electron signal is much more complex. Complexity itself is not necessarily a handicap to measurement, of course. In many cases what is needed is simply a reproducible profile shape from stripes of fairly consistent composition and cross-sectional shape (Monaghan et al., 1984). In these cases, prior calibration with independently measured standards, or even production samples that are considered "satisfactory" but are not measured against any standard scale, can be used to establish quality control requirements. Then successive measurements on production samples are used to control variation without regard to exact dimension.

In the more general case, however, constant variation in resist composition, size, shape of the stripes, and other factors are encountered. The shape is never the simple rectangle suggested above. Instead, the sides of the stripes are angled, and often irregular, and the "flat" surfaces of both the stripe and the substrate may be rough or even undercut. The corners of the stripe at both the top and bottom (where it meets the substrate) may be either cusped or rounded (Figure 10-22). These irregularities affect the signal profile, of course. They also raise the very fundamental question of just what "width" is to be measured.

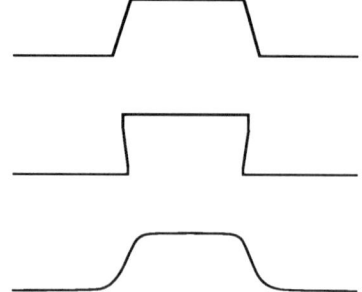

Figure 10-22: A few different physical profiles encountered in linewidth measurement.

Two problems begin to interact here. First is the definition of the "edge" of the stripe or line in terms of the signal profile, and second in terms of the physical profile. The two are not necessarily the same, or even consistently related.

For instance, with even a very simplified shape, the profile may show either a drop or rise in brightness while the beam is on the flat region near the raised portion of the trace (Figure 10-23), depending on whether shadowing or fill illumination is predominant (or in other words whether the electrons leaving the original point of impact and striking the trace are absorbed there or whether in the process they generate even more electrons). This depends on the specific geometry of the sample and detector, the composition of the material (both the substrate and the trace), and the electron voltage. Likewise, the edge will usually appear bright due to increased electron production from the side of the trace, and this may or may not depend on the edge orientation and shape, including local roughness not resolved in the image.

Simulation methods can be used to generate the signal profiles that would be obtained from various types of physical lines, taking into account the material (of both the strip and the substrate), shape (thickness, surface roughness, angle of sides, etc.), and the instrument. The important instrument parameters are the accelerating voltage, the orientation of the sample surface, and the type and location of the detectors used (Miyoshi & Yamazaki, 1986; Joy , 1987a, b).

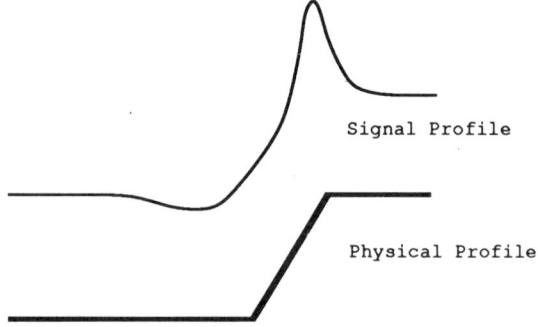

Figure 10-23: one typical relationship between a physical profile and the measured signal profile. The signal maximum and minimum do not necessarily occur at the discontinuities in the surface.

Surface Image Measurements

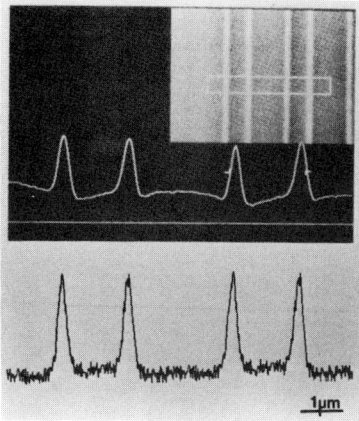

Figure 10-24: Agreement between measured secondary electron signal profile and Monte-Carlo simulation (Joy 1987b).

Accelerating voltage is a critically important variable. In most cases it is now considered desirable to keep it quite low (1-5 keV), to minimize radiation damage to the specimen (Rosenfield, 1987). This also produces a signal trace for the secondary electron image that emphasizes the edges of the trace being measured, which is sometimes helpful. On the other hand, low beam voltages make it difficult to generate or focus an intense electron beam, so the image is noisy. Also, surface contamination is somewhat more rapid with low voltages and has a greater effect on the image. Finally, the simulation methods that are commonly used employ Monte-Carlo methods that sample known probability distribution functions for the scattering angles and interactions probabilities ("cross sections") for the production of secondary electrons or other signals (Newbury, 1987). The formulae for the angles, mean free paths, and rates of energy loss are based on data measured at much higher voltages (10-50 keV) and it is believed that they do not accurately predict quantitative results at these lower voltages. Nevertheless, the simulations qualitatively match some measured results (Figure 10-24), and further work to refine the methods is underway.

The use of secondary electrons offers potentially better resolution than backscattered electrons (particularly at these low beam voltages where the backscattered electrons are poorly detected by most conventional detector systems). However, conventional secondary electron detectors produce lateral shading of images from rough surfaces, which severely complicates the measurement. Collection of the electrons through the final lens seems to overcome this problem, but is used on only a few instruments.

This places the burden of measurement back on the processing of the signal. Depending on the type of detector, its placement, the physical shape of the line to be measured and the material of the stripe and substrate, and the voltage, the shape of the waveform may be quite variable. In most cases it is necessary and hopefully possible to assume that it is symmetric, namely that the two sides of the stripe or line are shaped the same and produce the same variation in signal. If this is not the case, much more complicated processing (and simulation, and calibration) will be required.

Problems with the measured signal such as noise, finite bandwidth of the amplifiers (which cause loss of high frequency information in the signal profile, as well as a slew rate distortion that can make the profile nonsymmetric), and so forth must be dealt with

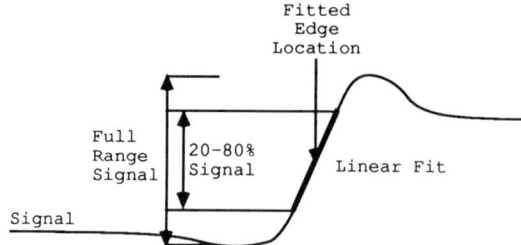

Figure 10-25: defining an arbitrary edge location from the 20-80% points on the signal profile.

beforehand. Noise is generally handled by signal averaging, especially when only a single line is being scanned across the feature. In this case, adding data from a large number of scans and normalizing it will give a relatively noise free result even when low beam currents produce a noisy image or signal on each scan. When an entire image has been recorded, it is sometimes practical to average together the signal from a number of raster lines. If the stripe being measured is not perfectly orthogonal to the raster, this can introduce an additional problem. But information from several scan lines across the feature can be combined to reduce shot noise in the signal profile to be measured.

Finding the edges from the signal profile is not trivial (Frosien, 1986). The shape of the signal profile can vary over wide extremes, with the maximum signal strength occurring when the beam position is anywhere from the sides of the stripe to inside the top edges. The first problem is usually to reproducibly measure the signal profile, and then the second step is to calibrate this against known standards or simulation results.

To measure the profile, it is possible to work with the profile itself, or with some derivative which will enhance the edges. The use of a first derivative is likely to produce problems because the result is different depending on whether the signal rises or falls. This problem can be overcome by squaring the first derivative.

The second derivative does not have this problem, and also usually marks the center of an edge rather than one or the other corners where the original slope changed, which may sometimes be the preferred location of measurement. In general, it is possible to use any even derivative, or any even power of an odd derivative, to obtain a signal that retains its symmetry and marks transition points of one kind or another across the profile. In practice, any higher derivatives than the second are almost never used because they respond strongly to any noise in the signal, and this tends to swamp the real information.

When the original signal profile is used, there are several definitions commonly employed to locate the edge for measurement. For instance, arbitrary criteria are often used to decide when the signal has deviated from the constant level of the substrate or stripe. Points representing 20% and 80% of the variation between the two constant levels are often used. These may either be used themselves as edge points, or may be used to interpolate a midpoint or some other arbitrary location.

In a somewhat more elaborate version of this approach, all points between the 20% and 80% limits are used to fit a linear least squares line, whose intercept with a similarly fitted baseline is then used to establish the width (Figure 10-25). This is a purely empirical approach that does not produce consistent results with specimens of varying geometry or composition (Miyoshi et al., 1986).

Surface Image Measurements

Another *ad hoc* approach to finding consistent measurement points on the signal profile is to first have the user mark an approximate location. Within a finite search width of this region the maximum value of the second derivative is located. This usually marks the point at which the curve begins to bend sharply upwards. Then the nearest extreme value of the first derivative on the side of that point toward the center of the stripe is taken as the point to be measured. This is the steepest part of the signal, rising along the side of the stripe. Notice that it may not be the steepest part of the physical stripe, and even if it were, that might not be a representative spot on the line, because standing waves, roughness and other phenomena can alter the side slopes without changing the width of the top or bottom.

A major problem with these approaches and their variations is that the physical profile is rarely a simple step, and the signal profile is even more complicated. The background level often varies near the stripe, depending on the geometry of stripe and detector. If the secondary or backscattered electrons from the substrate strike the stripe and are absorbed, the signal will drop. If they produce additional electrons from the stripe, the signal will rise. Likewise, the signal may not level off until the beam has moved well onto the stripe, away from the edge. It may be either rising to that level or dropping to it, depending on the signal used and the accelerating voltage. This makes arbitrary definitions of locations to measure somewhat dubious. They do not correspond to any physical basis in the sample or the electron interactions taking place. Only detailed simulations of the particular situation being measured can predict the proper measurement location.

There remains the problem of calibration. The most common approach is to use real devices and measure the distance between repetition of the same kind of stripe (Figure 10-26). Generally the spacing of stripes is rather well known from the device design, and this repeat distance must therefore give a suitable calibration value. It is possible to use template pattern matching in the image to automatically locate features. This is usually done to verify the accuracy of device placement (Kayaalp & Jain, 1987a) but can also be used to automatically locate features whose distance provides calibration information. In some cases, device spacing cannot be used because there is no suitable repetition. In others, if the image has shading or other asymmetry problems, it is not valid to make this assumption because the repeat is measured between two points on the same side of two stripes, whereas line widths will be measured between opposite sides.

Calibration of measurements on specific devices is often accomplished by destructive cross-sectioning of wafers, so the actual dimensions can be determined on the cross section. This is time consuming, costly and difficult. It is sometimes practical to apply stereo pair matching techniques to this task (Kayaalp & Jain, 1987b), taking two SEM images at different viewing angles and determining the profile shape to determine a calibration dimension. This is not practical for on-line critical dimension metrology, but produces good calibration data when different line shapes and compositions of substrate and trace material are encountered from time to time. Stereoscopy is discussed in detail in the next chapter.

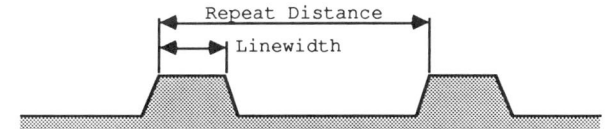

Figure 10-26: Using spacing to calibration line width measurements.

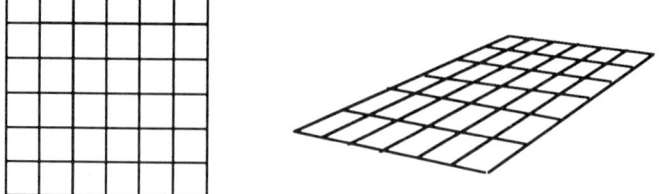

Figure 10-27: Foreshortening due to specimen tilt produces trapezoidal distortion which can be corrected by image processing.

To date there are very few independently known standards for line width measurement. For instance, there is a silicon on silicon standard made at the U.S. National Bureau of Standards, but no photoresist standards (for the rather obvious reason that they are less durable). In consequence, more reliance must be placed on good simulation routines, and as explained before these are still very much in a stage of development. Since the simulations produce signal profiles that should be obtained from various physical line profiles, it is necessary to do the simulation beforehand, setting up the parameters to correspond to the desired device characteristics and the imaging conditions, and then generating the expected signal. It is not so easy to begin with a measured signal and iterate the simulation program to match it.

The discussion so far has assumed that the sample surface is viewed perpendicularly, and the stripes to be measured are oriented at right angles to the raster scan direction. The current generation of instruments that attempt to perform this type of metrology automatically and repeatedly use this geometry, as much as possible. However, for occasional measurements in a standard SEM, it may be difficult to achieve this alignment. Furthermore, there are some good reasons to consider deviating from them.

Specimen charging due to the incident electron beam depositing more charge into the sample than is emitted by combined secondary electron emission and primary electron backscattering produces serious image defects for the secondary electron signal that can make measurement impossible. Indeed, one reason to use very low accelerating voltages is to reduce this charging. If the sample surface is tilted with respect to perpendicular to the incident beam, the effect is reduced. The electron voltage at which the combined secondary emission and backscattering coefficients exceed 1.0 (meaning that no surface charging takes place) is increased by more than a factor of two when the sample surface is inclined, for example. For some photoresists, this can be a critical difference (for instance, for HPM resist the critical energy is raised from 0.55 to 1.5 keV by tilting the surface to 45°). Tilting the sample may also be useful to increase the detector signal strength, and of course not all devices have lines that conveniently run at right angles to each other and to the scan direction.

For whatever reason the sample may be inclined or rotated with respect to the beam, it is necessary to rectify it before measurements are performed (Figure 10-27). This can be done by "rubber sheeting", discussed in an Chapter 3. This image processing operation produces a derived image in which each pixel has an address calculated with respect to the original image. The brightness information from that location (perhaps interpolated from the nearest neighboring pixels to the mathematical address point) is mapped to the new image, to produce an image that is aligned in the required way. For most SEM images, tilt involves not only a difference in X and Y dimensions, but also a foreshortening that makes squares appear as trapezoids. Rubber sheeting corrects this. It can also rotate the image to align traces perpendicular to the scan lines.

Of course, this rubber sheeting operation requires that the amount of rotation and tilt be known. Sometimes this information can be obtained from an independent source, such as the alignment of the specimen stage, but in the most general case it is not known beforehand and must be extracted from the image itself.

Fortunately, this is not hard to do. The linear Hough transform described in an earlier chapter is ideally suited to the purpose. Bright lines in the original (tilted, rotated) image are used. If the important edges are not bright, but are step changes from one brightness level to another, then a prior image processing step to obtain a gradient image (using a derivative operator such as the Sobel or Kirsch, or even a simple Laplacian) will produce an image in which the feature edges are outlined.

The linear Hough transform then maps each point in the original image, weighted by it brightness, into a space whose axes represent the slope and intercept of all possible lines through that point. Each point becomes a line, and the superposition of all the possible lines produces point in Hough space. The result is a family of bright points that can be discriminated in the Hough transform image. Each point represents one of the straight line segments corresponding to a feature edge in the original image. The orientation of these lines is well characterized, and can be used to derive the appropriate rubber sheeting correction equation that will generate the addresses used for the geometric mapping operation. For a given class of images, this process can be automated so that images are acquired from samples with a general orientation and then are justified so that the displayed and measured image is rectified.

Depending on the sample's surface relief and the type and position of the signal detector used, there may also be a shading asymmetry when this type of rectification is carried out. The most simple such case results when the detector is at one side of the specimen, so that edges facing it are brighter than ones facing away. To a limited degree, this can be corrected using a spatial derivative operator in the appropriate direction. However, if the relief is sufficient to cause some sides of features to disappear then it is obviously not possible to reconstruct them. Depending on the direction of tilt with respect to the direction of the lines whose widths are to be measured, this may or may not represent a problem.

In the discussion of metrology of line widths above, it has been assumed that it is the width alone that is of interest. Of course, height or other dimensions may also be important, but in general these require either very specific calibration of some signal characteristic against simulation or standards, or more usually stereoscopic measurement. The latter method will be dealt with in the next chapter, and is the principal tool available for measuring height dimensions on surfaces.

Roughness and fractal dimensions

There remains another kind of surface irregularity that can be measured from single images, whose importance is now being increasingly recognized. This is the fine-scale surface roughness, and it is revealed in various textures on the grey scale surface images. Quantitatively describing and comparing the roughness of surfaces is important in many fields. Adhesion of bonding and coatings is an outstanding example, in which the preparation of a rough surface is necessary for good bond strength. Preparation methods (electrolytic, chemical, mechanical, etc.) vary widely, and are generally controlled by some variable such as length of time which is only indirectly related to roughness.

Surface roughness, in turn, may be evaluated by measuring bond strength. This is poorly suited to real control of the preparation process, or to an understanding of the underlying parameters related to the roughness of the surface. Indeed, it is often found that optimum strength bonds to rough surfaces require an intermediate degree of roughness, since very flat surfaces provide too little surface area or adhesion, while very rough surfaces may cause a reduction in contact area because not all of the surface is wetted.

Measurement of roughness can be accomplished in several traditional ways, including profilometry (physically tracing a stylus across the surface), light scattering, and stereoscopy. Each of these has some important limitations, for instance profilometry cannot follow very deep or intricate surface configurations, light scattering is suitable only for nearly flat surfaces, and stereoscopy is notoriously time consuming and hence expensive. Transverse sectioning of prepared surfaces is a relatively straightforward method, but is destructive to the sample so that further tests cannot be performed. There also remains the problem of how to numerically represent the roughness.

One parameter sometimes used to describe surface roughness is simply the ratio of actual surface area to projected area. Sometimes called a "wrinkle factor," this can be calculated directly from the length of a line drawn along the surface divided by the distance between the end points (assuming that the line is randomly oriented with respect to the surface, of course). But this neglects the important consideration that most natural surfaces do not have a simple Euclidean dimension. Instead, the surface area generally increases as finer tools (or higher magnifications) are used to measure it. This gives rise to a "fractal" dimension which describes this kind of roughness, and indeed the fractal dimension itself is a valuable parameter for the characterization of roughness of boundaries and surfaces.

In Chapter 6, a fractal dimension for a feature outline was described. The formal definition of the dimension comes from the slope of a plot of the length of the perimeter measured as a function of the length of the measuring tool, or image resolution. For surfaces, the same type of definition is possible (Paumgartner et al., 1981). Fractal surfaces can be generated in the computer by continuously elaborating the surface with structures of smaller and smaller dimension (Figure 10-28). It is immediately apparent when looking at computer-generated random fractals, that they somehow look "natural." Indeed, one of the first "important" uses of fractal concepts was the computer generation of landscapes with believable appearances for movies such as the landing scene in "Alien." It has been shown that such diverse real-world objects as the length of rivers and the bending and branching patterns of trees, the surface irregularities of mammalian brains and the internal structure of the lung, the formation of cracks in materials and of electrical discharges, and the agglomeration of soot particles, can all be described with this simple parameter (Mandelbrot, 1983; Feder, 1988).

Some of these applications include materials which can be examined with the light or scanning electron microscope (Laibowitz et al., 1985). Examples include: a) the characterization of the roughness of fracture surfaces in materials (Underwood & Banerji, 1986; Underwood, 1987; Mecholsky & Passoja, 1985); b) the roughness of metal and ceramic surfaces etched by chemicals to improve the bonding of dental materials, or prepared by high precision grinding or machining operations (Russ & Russ, 1987b); c) the porosity of sintered compacts (Coster & Chermant, 1985); d) the size and shape distribution of particulates prepared by milling or by natural processes of fracture (Kaye, 1986); e) the growth of particle agglomerates by diffusion limited aggregation (DLA) (Sander, 1986; Stanley & Ostrowsky, 1986); and f) the increasing roughness of skin with aging and the effect of cosmetic treatment (Corcuff et al., 1983).

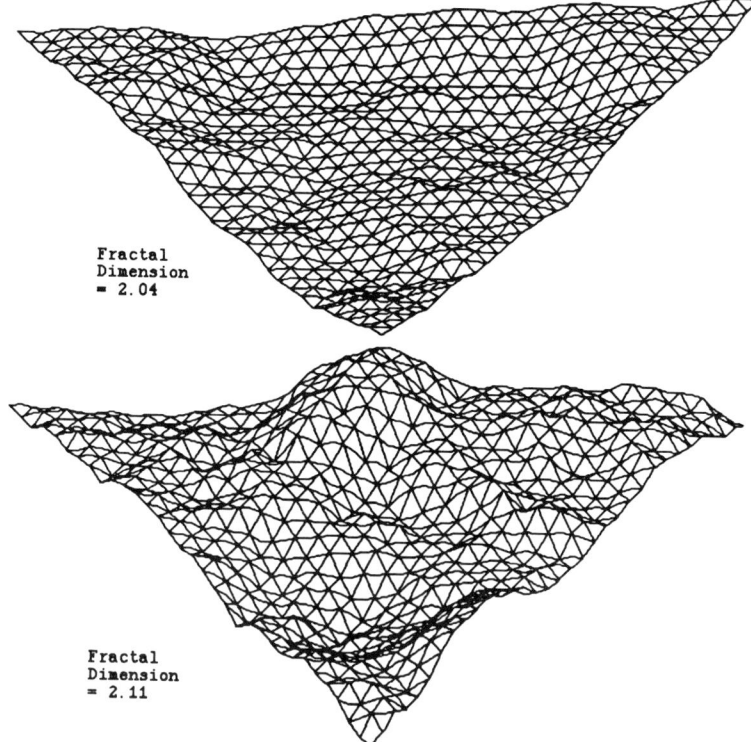

Figure 10-28: Computer-generated fractal surfaces.

Most of these applications involve the fractal dimension of a surface rather than that of a line. For example, Mecholsky et al. (1986, 1989) have shown that the surface roughness of brittle fractures as measured by their fractal dimension is related directly to mechanical properties. The plane-strain fracture toughness $K_{IC} = E \cdot D^{1/2} \cdot a_0^{1/2}$, where D is the fractional part of the fractal dimension, E is the modulus of elasticity, and a_0 is a dimension characteristic of the material, varying from about 0.1 nm for single crystals to 0.3-0.4 nm for polycrystalline oxides, and to about 5 nm for glass ceramics. These dimensions are similar to the size of structural units in the materials (atomic dimensions for the single and polycrystals, and the size of molecular clusters for the glass ceramics).

If the surface area of an object is measured with varying tool sizes, or at different resolutions, a Richardson plot is obtained. The fractal dimension is then given as a real number greater than 2.0 (which is the limit for a Euclidean surface). The greater the number, the "rougher" the surface and the greater its tendency to spread into the space near the nominal surface. Further, the character of a fractal line or surface is one of self-similarity. That is, the degree of roughness and hence the general appearance of the surface is the same at any magnification (shown by the linearity of the Richardson plot). Most real objects only obey this ideal fractal relationship over a finite range of dimensions, of course (Kaye, 1984). Underwood (1987) has suggested modifications to the construction of the Richardson plot to deal with this.

For surfaces, it is theoretically possible to use some minimum structuring element such as a triangle to repeatedly lay on the surface to measure its area (just as a line is used

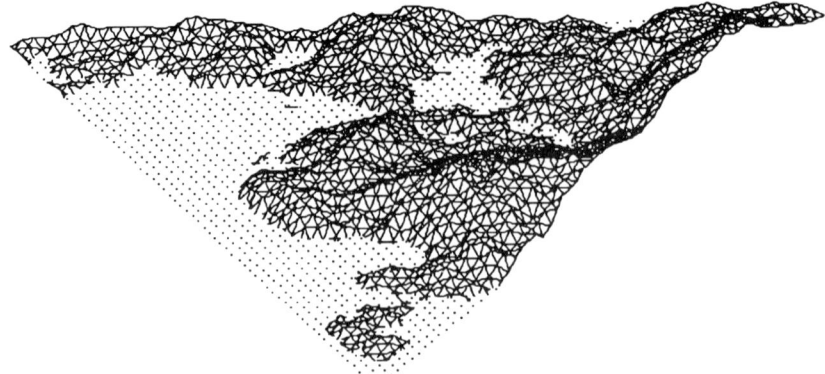

Figure 10-29: A fractal surface cut by a plane. The outline is also fractal, with a dimension 1 less than the surface.

to measure the perimeter of a feature). In practice, this is usually prohibitively difficult (Thompson, 1987; Flemmer & Clark, 1987). However, it was shown by Mandelbrot that passing a plane through a fractal surface would generate an intersection whose boundary length is also fractal, and whose dimension, of the form 1.D, is exactly one less than the dimension 2.D of the surface. Since measuring the fractal dimension of the boundary line is comparatively straightforward, this provides a useful tool to measure surface roughness.

Two different approaches have been used to create the plane of intersection. One obvious method is of course to mechanically section the surface by cutting and polishing a plane perpendicular to its nominal direction, and then measure the fractal dimension of this line either manually (tracing the boundary with a tablet) or automatically, by imaging the cross section. This method has been used to study metal fractures, etched metal surfaces, and so forth (Underwood & Banerji, 1986; Underwood, 1987).

The second approach is to plate the original surface with some contrasting material, and then polish down into the surface so that islands of the original material appear surrounded by a sea of the plating material (Figure 10-29). The boundaries of the islands can also be measured to obtain a fractal dimension, and this has been used to study fractures in ceramics, for instance by applying a nickel plating to fracture surfaces in alumina before polishing (Mecholsky & Passoja, 1985).

Both methods require a significant amount of specimen preparation, and are destructive of the original surface. However, it happens that there is another way to get the roughness information from images of the surface itself. It was first shown that the image of an ideally fractal surface illuminated by diffuse light would have brightness values that were also "fractal" in a formal sense. Pentland (1983) has shown that for an ideally fractal surface with diffuse reflectance, illumination by visible light will produce an image in which the brightness varies in a way that can also be described as "fractal" (in the sense that if brightness is interpreted as a third dimension, the resulting "surface" is fractal). He showed that fractal dimensions of these surfaces could be determined from the power spectrum of the Fourier transform of the image, and also that human observers ranked the expected physical roughness of surfaces with high correlation to the fractal dimension of the grey scale image (Larking & Burt, 1983; Peleg et al., 1984). The dimension of this brightness surface was shown to be correlated with the physical fractal dimension of the original surface, and the same correlation was extended to secondary

Surface Image Measurements

electron images from the SEM (Russ & Russ, 1987a). The method presented below for calculating a slope for the brightness difference vs. distance curve approximates the same measure of roughness.

This is true to a surprising degree for both light and scanning electron microscope pictures, which are the two predominant methods used to examine rough surfaces. As was discussed earlier in this chapter, under the topic of shape-from-shading, both the secondary and backscattered electron images in the SEM have an approximately Lambertian relationship between brightness and slope at low magnification, where the details of the electron/specimen interaction do not come into play. The result is that the derivative of brightness with respect to slope is also quite similar to that for surfaces illuminated by diffuse light. Texture operators employ that relationship to measure the local roughness.

Haralick et al. (1973) first developed a series of texture parameters for use with satellite photographs, which were introduced in Chapter 5. By combining the brightnesses of neighbor pixels in various combinations, he obtained quantities that could be used to distinguish regions of water from grasslands, forests, etc. Subsequent refinement of these methods (Haralick, 1978; Davis et al., 1979; Tomita et al., 1982; Sun & Wee, 1983; Jernigan & D'Astous, 1984; Bovik et al., 1987), particularly with color or false-color (e.g. infrared) images now enables very specific mapping of the acreage devoted to particular crops and their maturity, as well as the spread of diseases and acid rain damage to forests.

For rough surfaces, it is necessary to examine more than the immediate neighbors to determine a texture parameter that correlates directly with the surface configuration. Comparison of pixel brightness is only made between pixels within arbitrary outlines (corresponding to feature outlines, as determined by prior segmentation of the image). Figure 10-30 shows an example in which the trend of mean brightness difference between pairs of pixels within features are determined. Tables of the number of pixels with various brightness differences, as a function of distance out to a limit of (typically) 6 to 10 pixels are used. These tables are quickly obtained from the stored image in the microcomputer, as they require only a subtraction and counting operation.

The comparison of pixel brightness is normally carried out only in the 90 degree orthogonal directions (along and perpendicular to the scan lines). It is sometimes desirable to further restrict this to only one direction. If scan speed is high relative to bandwidth of video amplifier, then we should ignore variations along scan lines (which may be blurred). Likewise, at high image magnification relative to the number of scan

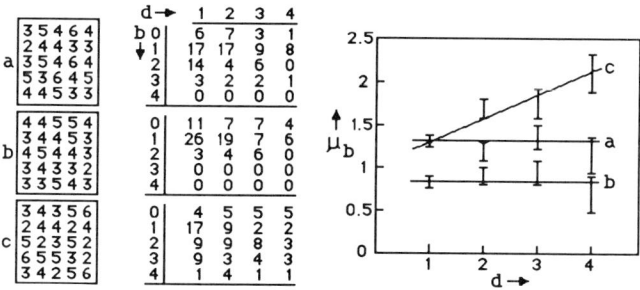

Figure 10-30: Schematic diagram showing three image fragments with different intercept and slope values for the plot of mean brightness difference vs. distance.

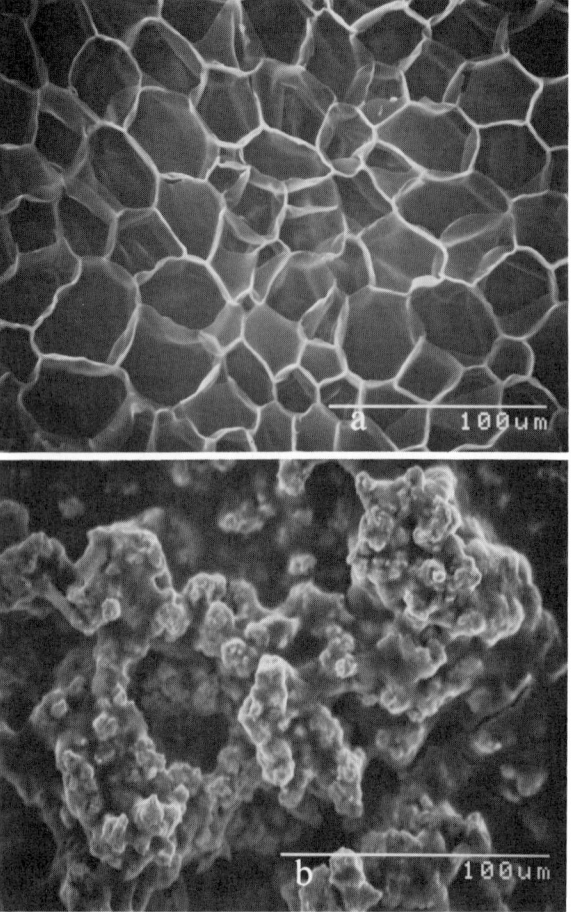

Figure 10-31: SEM images of surfaces with different roughnesses:
a) cork; b) calcined phosphate; c) niobium powder; d) graphite.

lines (line spacing less than or comparable to the microscope resolution), overlap between adjacent lines can produce correlation, so we should not compare neighbors in this direction. In either case, loss of resolution is evident in change of slope (flattening) of delta-brightness vs. distance curve at low spacings. The elimination of the 45 degree directions also eliminates the problem of different spacing (by a factor of 1.414) of pixels in these directions.

As a real example, Figure 10-31 shows a series of surfaces viewed in the SEM, which have a visual difference in apparent roughness that we wish to characterize. The cork image is actually a cut section through cork, in which the individual cells are quite uniform and "flat" in brightness. The remaining images are all of particulates with varying degrees of roughness.

Table 10-1 shows the mean difference in pixel brightness as a function of distance. Plots of these data (Figure 10-32) show a monotonic and nearly linear variation of brightness difference with the log of distance. The intercept and slope of the curve are

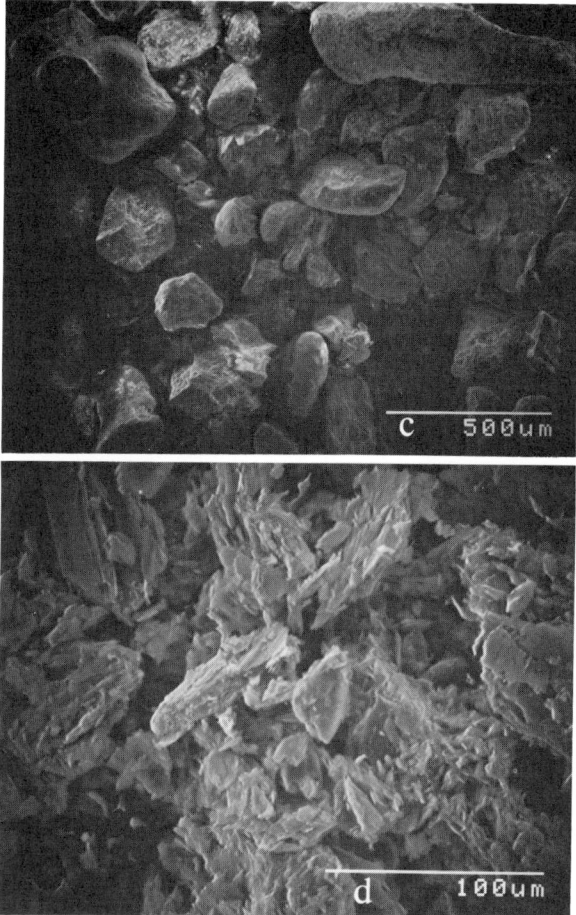

obtained for each type of feature (data are shown in Table 10-2); for convenience, these are referred to as *contrast* (the intercept, which gives the expected variation in brightness between each pixel and its immediate neighbors), and *texture* (the slope, which describes the increase in the brightness difference with distance). A logarithmic scale for distance was used to improve the linearity of the plots. It is also consistent with the fractal interpretation of the data, as described below, by analogy to Richardson plots which also use a logarithmic distance scale.

Table 10-1. Mean brightness differences as a function of distance (in pixels) for the images shown in Figure 10-31

Material	Distance: 1	2	3	4	5	6
Cork	1.390	2.054	2.522	2.852	3.094	3.285
Calc.Phosph.	1.974	3.027	3.679	4.134	4.488	4.786
Nb Powder	2.091	3.402	4.218	4.691	5.003	5.221
Graphite	2.477	3.911	4.651	5.042	5.304	5.502

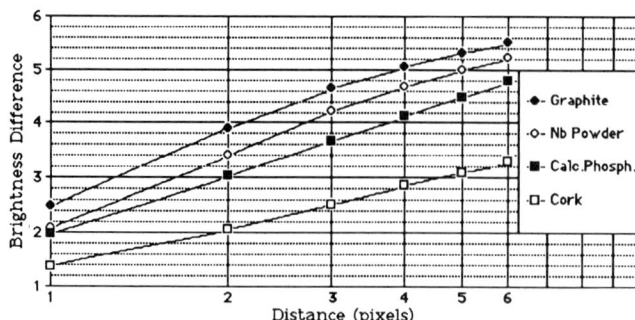

Figure 10-32: Plot of the mean difference in brightness between pixels within feature outlines as a function of the distance between them, for objects in the Figure 10-31.

Applying the method just described produces the data of Table 10-2 and the plots shown in Figure 10-33. The slope and intercept values were obtained by linear least-squares fit of straight lines to the data points. As expected, the slope (texture) and intercept (contrast) are both lowest for the cork. The graphite has the highest contrast, but the texture is less than for the niobium powder. Notice that the intercept values for the oxidized niobium powder and the calcined phosphate are nearly identical, but the slopes are not. These numerical measurements can all be qualitatively confirmed by visual inspection and comparison of the images.

Obviously, the automatic determination of these roughness parameters on individual features within SEM images permits classification of the objects. By measuring and saving the contrast and texture, along with the area, perimeter, length, breadth, location, orientation, and other descriptions of particle size and shape, the measurement system can subsequently produce distribution plots for only a specified class of objects, carry out correlation or Anova tests between such classes to compare size or shape, or apply different transform equations to estimate volume or surface area for particles in different classes based on surface roughness. It is also possible to form a derived image in which the pixel value is a function of texture, and to discriminate this image as discussed in chapter 5 to achieve segmentation of the original image based on feature texture.

In addition to this principal use of the new parameters describing surface roughness, it is desirable to relate the values more directly to the physical roughness of the sample surface. It is well known that the relationship between surface orientation (and roughness) and the brightness of the secondary electron image (or, for that matter, the backscattered electron image) is not straightforward. It would be surprising, therefore, to find a quantitative relationship between the parameters described above and the physical characteristics of the sample surface. However, by looking at brightness differences

Table 10-2. Best-fit (linear regression) values for slope and intercept in the plots of Figure 10-32

Mean Brightness Difference = Intercept + Slope · $\log_e$ (Distance)

Material	Intercept	Slope
Cork	1.359	1.070
Calc.Phosph.	1.959	1.571
Nb Oxide	2.157	1.776
Graphite	2.622	1.695

Surface Image Measurements

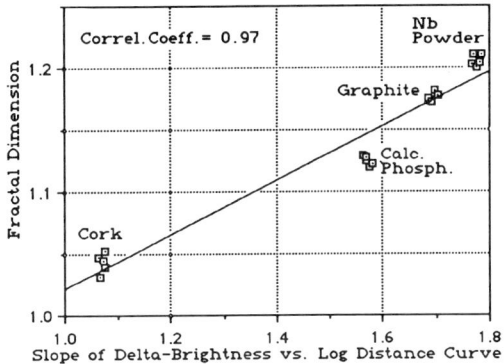

Figure 10-33: Plot of the manually measured fractal dimension of profiles of features versus the slope of the curve of brightness difference plotted against the logarithm of distance between neighboring pixels.

some of the factors causing variation in brightness due to sample composition, location of the electron detector, overall height or tilt of the surface, etc. are relaxed.

Fractal dimensions of the perimeters (profiles) of the particles were obtained for each of the objects in the images shown before, by using the manually digitized outlines of objects as described by Schwarz and Exner (1980) with 250-500 points around each of 5 separate particles, viewed at an appropriate magnification. The perimeter was determined using every nth point on the periphery, and a Richardson plot of log perimeter vs. log step length constructed, as discussed in Chapter 6. The fractal dimension was taken as the slope of the curve. For projected particle images, the profile fractal dimension gives only a lower bound to the actual surface dimension. As expected, the cork gave the lowest fractal dimension because the cells are practically Euclidean in outline. The other objects have fractal dimensions which increase as the surface roughness increases. As shown in Figure 10-33, there is a significant correlation (0.97) between the fractal dimension of the outline and the texture as defined above. The scatter amongst the points for each separate object of the same type indicates the numerical significance of the measured values.

Images obtained over a range of magnifications give virtually identical parameter values as long as the magnification stays in the range of self-similarity of the objects. As described by Kaye (1984), many objects exhibit a fractal behavior (and hence have a descriptively useful fractal dimension) in a size range corresponding to structural or other dimensions. Just as the measured fractal dimension stays constant in images which resolve surface detail in this size range, so do the measured contrast and texture (for the same reason, namely that the images are identical in appearance).

To demonstrate the method with both light and electron microscopy, several sheets of electro-deposited copper with surfaces roughened by a proprietary process intended to facilitate bonding of electronic components were subjected to measurement. Each was sectioned transversely and examined in the light microscope at a magnification of 750x (Figure 10-34a) to provide a direct measure of roughness, and a fractal dimension determined. This was done by digitizing the image and determining the boundary length (perimeter) while coarsening the binary image (mosaic amalgamation as discussed in Chapter 6), to form a Richardson plot (Russ & Russ, 1986, 1987b). The surface images (Figure 10-34b,c) obtained with a scanning electron microscope (at a magnification of

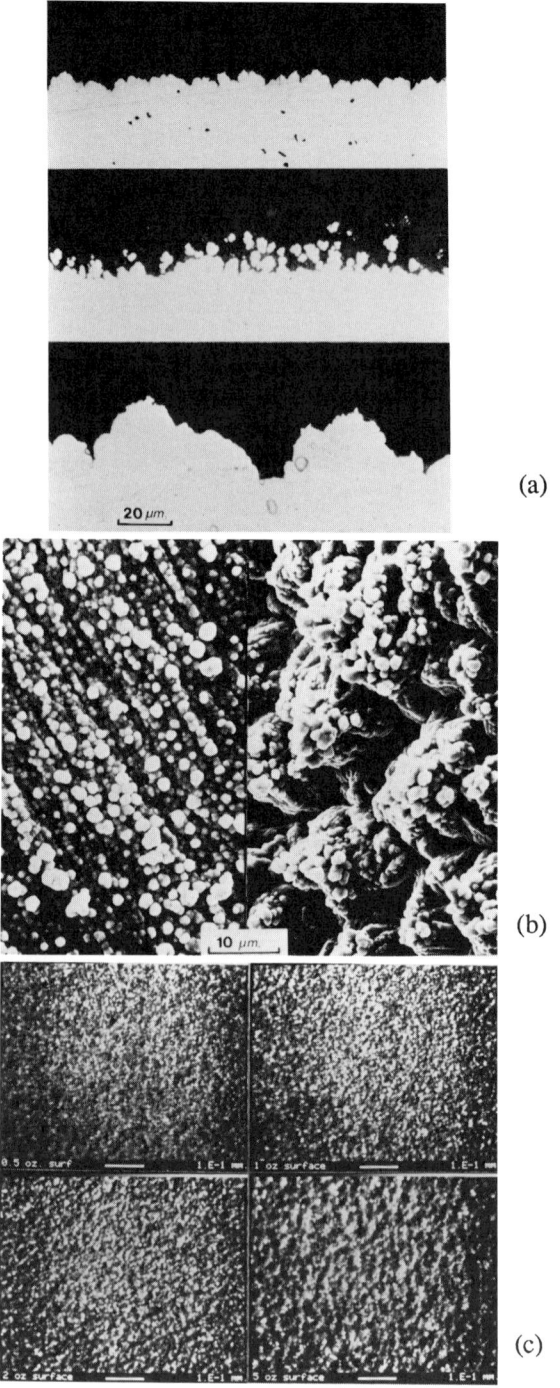

Figure 10-34: a) Transverse sections of 0.5, 2 and 5 oz coating specimens; b) SEM surface photos of 1 and 5 oz specimens; c) Light microscope photos of 0.5, 1, 2 and 5 oz specimens.

Table 10-3. Measured fractal dimension and texture data

Sample Designation	Transverse Section Fractal Dimension	SEM Image Texture	Light Image Texture
0.5 oz	1.17	1.536±.035	1.244±.022
1 oz	1.24	1.598±.039	1.262±.020
2 oz	1.31	1.641±.039	1.323±.019
5 oz	1.34	1.667±.032	1.350±.016

1000x) were subjected to a texture measurement, and visible light surface images were also obtained and similarly measured at a magnification of 160x.

The raw data are summarized in Table 10-3. Standard deviations reflect the variance between measurements on eight different regions of each image. The absolute magnitudes of the texture measurements depend on the imaging and illumination conditions used, but the values correlate closely with the physical roughness as measured by the transverse fractal dimension. The very good correlation demonstrates the value of the surface image texture method for quickly and nondestructively characterizing roughness.

The slope of the plot of the trend of mean brightness difference as a function of (log) distance correlates directly with the surface fractal dimension (Figure 10-35). However, it is not numerically equal to it. The calibration constant depends upon the imaging conditions used, the gain on the video or SEM signal, the type of material and strength of illumination, accelerating voltage, etc. For any given setup, this is most easily calibrated by measuring this texture parameter for a sample of known surface roughness, and then comparing other unknown surfaces to it.

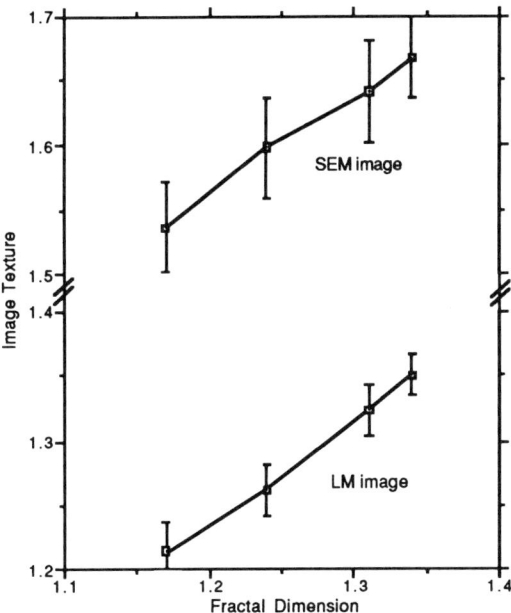

Figure 10-35: Correlation between the fractal dimension measured on transverse section, and texture in SEM and light microscope images.

Table 10-4 Bond tensile test and surface roughness data

Sample	Tensile Strength	Surface Texture	Fractal Dimension
1	19.02 ± 6.85	1.29 ± 0.058	1.11 ± 0.015
2	2.46 ± 2.41	1.15 ± 0.079	1.02 ± 0.005
3	29.25 ± 5.77	1.53 ± 0.153	1.19 ± 0.030
4	22.65 ± 8.94	1.35 ± 0.067	1.14 ± 0.015
5	22.59 ± 6.52	1.33 ± 0.118	1.12 ± 0.020

The surface roughness may also be related to other surface properties. In one study of the bonding of dental materials, we prepared surfaces by different chemical and electrochemical etching techniques. After the bonds were made, they were tensile tested to determine the adhesive strength. Then the surfaces were heated to remove the adhesive, and the test cycle repeated. The strength of surfaces prepared by different methods was compared to the fractal surface dimension obtained in two different ways. First, the texture of SEM images of the surface was measured in the middle of the series of repeated bonding and strength tests. Then, when the tests were complete, the surfaces were examined in cross section in the SEM to directly measure the fractal dimension. Table 10-4 shows the results.

The data show a very large scatter, because the surfaces are not really uniform. The microstructure of the cast metal contains a eutectic mixture of two phases, which etches quite differently than the single-phase regions. It was also noteworthy that the deepest or most severe etch did not produce either the roughest surface or the greatest strength. The etching mechanism of the metal is more complex than that, and overetching causes a drop in strength as the surfaces become smoother. The scatter in the mechanical test results is also considerable. However, plots of the values (Figure 10-36) show a definite trend, and are being used to evaluate improved surface preparation methods that will produce optimum strength values.

Roughness can also be used as a parameter to identify or select features, without necessarily obtaining a value that has been calibrated to yield an absolute meaning. Chapter 5 discussed methods for discriminating features within images based on brightness, edge contrast, etc., and showed that it is also possible to use the image texture or roughness to accomplish this.

Usually, the presence of variation in the brightness of pixels within features or across surfaces is ignored and considered to represent noise in the imaging process. In many

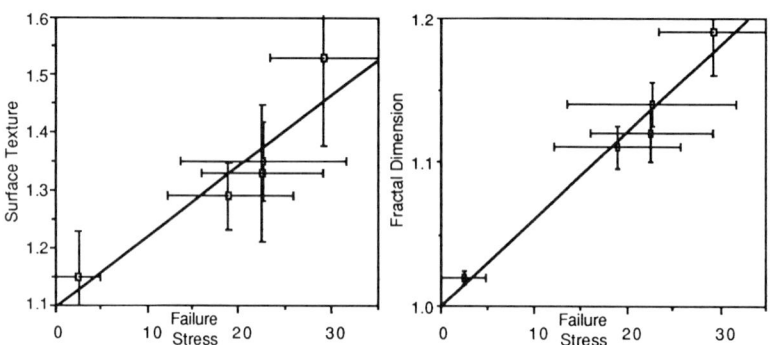

Figure 10-36: Correlation of surface roughness and bond strength.

Surface Image Measurements

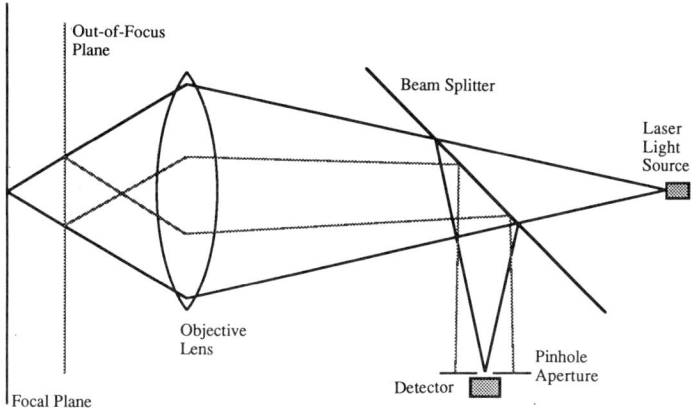

Figure 10-37: Principle of the Confocal Scanning Light Microscope - light reflected from out-of-focus planes or points does not reach the detector.

cases it actually carries information about the surface roughness of the objects being examined. For the case of line width metrology discussed above, it is well known that the edges of the traces at least are not planar, and their surfaces may not be either. The characterization of the local roughness from the texture of the acquired images may make it practical to incorporate this roughness into the simulation programs, to improve the accuracy of the simulation and produce a better match between the generated signal profiles and the measured ones.

Other surface measurement methods

Several other methods are also used to measure surface topography. The interest in developing these methods has been largely driven by the need to characterize microelectronic devices, but the same methods are also used for surface wear and similar applications. One interesting method uses the confocal scanning light microscope. This uses a scanning light beam, usually from a laser, to build up the image point by point. An aperture at the detector is optically superimposed on the source by a beam splitter or partially silvered mirror. This rejects light scattered from locations in the sample that lie at

Figure 10-38: Reconstructed in-focus surface image of a ceramic fracture.

different focal depths or adjacent to the point of focus on the sample surface (Figure 10-37). The result is normally to produce images with a very shallow depth of focus, even within samples that are transparent or translucent, without loss of contrast or sharpness due to light scattering. In this sense the confocal light microscope allows serial optical sectioning, as discussed in Chapter 12.

However, several commercial confocal scanning light microscopes can also be used for surface mapping. This is possible because the strength of the reflected light signal from a surface is sharply maximized at the correct focus. By varying the focus at each point, it is possible to record the surface height at many points to build up two images simultaneously, one of the surface elevation and the other of the surface image, everywhere in focus (even for surface which would be much too rough for conventional light microscope imaging). The data from these images can then be displayed and measured in various ways.

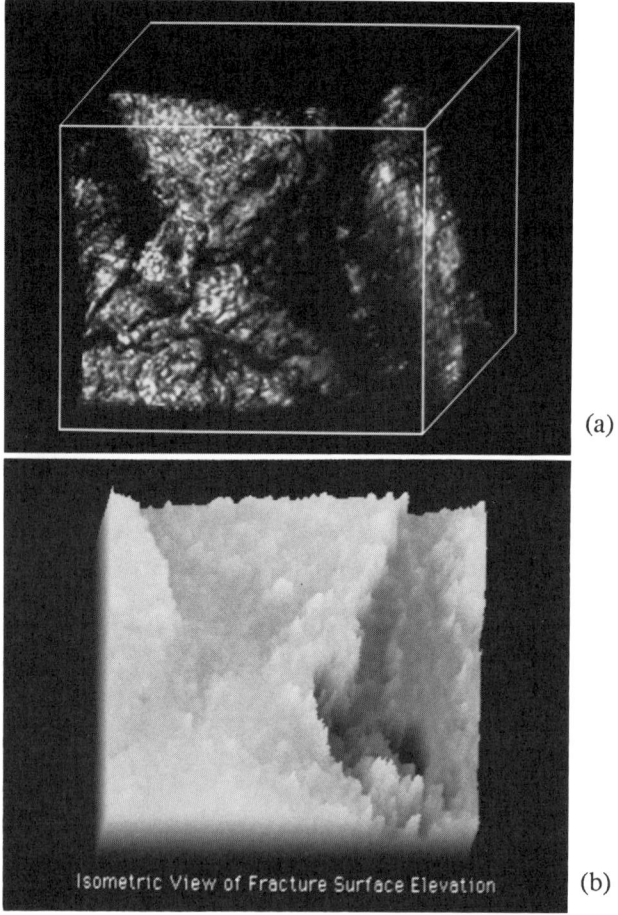

Figure 10-39: Data from Figure 10-38 reconstructed from a viewing angle at 45° to the normal (a) and as an isometric view with brightness coding the elevation (b).

Surface Image Measurements

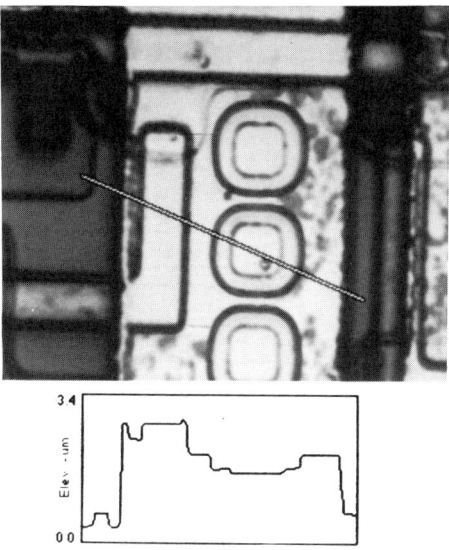

Figure 10-40: Elevation profile across an integrated circuit, with the extended focus image.

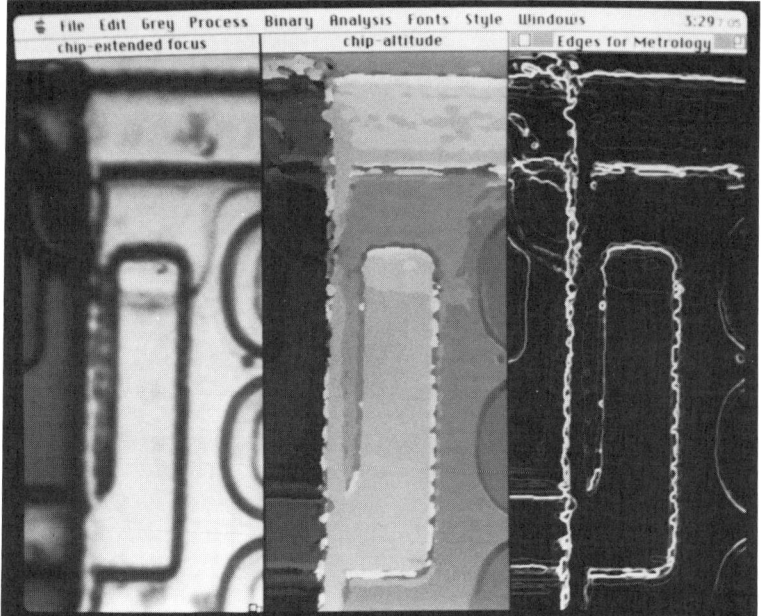

Figure 10-41: Portion of the integrated circuit from Figure 10-40, shown in different modes: a) extended focus in which the in-focus (maximum) brightness is saved for each point; b) range image in which brightness is proportional to height at each point; c) contour lines showing surface profiles, used for device measurement.

Figure 10-38 shows the surface of a fracture in a ceramic reconstructed from a series of 26 confocal images, taken by moving the specimen stage in the z direction in 1 μm steps. At each point the brightest pixel is kept as a representation of the image. The surface can also be viewed in an isometric or perspective view by shifting each plane of pixels as the reconstruction is performed (Figure 10-39). By recording the stage elevation at which the brightest signal is recorded, an elevation profile can be constructed for each pixel position, as shown in Figure 10-40 for the surface of an integrated circuit. Figure 10-41 compares the image of this same device presented as an "extended focus" image, one in which brightness is proportional to elevation, and one with contour lines marking the edges, used for precise measurement of dimensions.

Another approach to measurement of rough surfaces uses "structured light." A simple example of this method is to allow sunlight coming through a window equipped with venetian blinds to strike an irregular object. The narrow stripes of light are distorted from a straight line pattern by the curvature of the object. Measuring the lateral displacement of these stripes in a video image is straightforward, and if the geometry of the light source is known, can be easily converted to surface elevation information (Lee & Fu, 1983). The method is used routinely in microscopes for measuring machining accuracy in tool- and die-making, as well as measuring the curvature of the human eye when fitting some types of contact lenses or before corrective eye surgery, and at larger scales to measure the curvature of the spine and other medical applications. It is limited in two ways for microscopic use: the need to have the individual stripes be wide enough for accurate measurement, rarely less than 5 μm, and the need for the surface irregularities to be simple enough that the continuity of lines is not broken.

References

D. K. Atwood, D. C. Joy (1987) *Improved Accuracy for SEM Linewidth Measurements* Proc. SPIE 775 159-165

W. Beil, I. C. Carlsen, V. Desai, L. Reimer (1989) *Three-Dimensional Digital Surface Reconstruction in Scanning Electron Microscopy using Signals of a Multiple Detector System*, European Journal of Cell Biology 48 Supplement 25, 13-16

A. C. Bovik, M. Clark, W. S. Geisler (1987) *Computational Texture Analysis using Localized Spatial Filtering* IEEE, Proc. of Workshop on Computer Vision, Dec 1987, Miami FL, 201-206

R. A. Brooks (1981) *Model-based Three Dimensional Interpretations of Two Dimensional Images* Proc. 7th IJCAI, Vancouver BC, 619-624

R. A. Brooks (1983) *Model-Based 3-D Interpretation of 2-D Images* IEEE Trans. Patt. Recog. Mach. Intell. PAMI-5#2 140-150

I. C. Carlsen (1985) *Reconstruction of True Surface Topographies in Scanning Electron Microscopes Using Backscattered Electrons* Scanning 7 169-177

B. Carrihill, R. Hummel (1985) *Experiments with the Intensity Ratio Depth Sensor* Computer Vision, Graphics and Image Processing 32 337-358

P. Corcuff, J. deRigal, J. L. Leveque, S. Makki, P. Agache (1983) *Skin Relief and Aging* J. Soc. Cosmet. Chem. 34 177-190

M. Coster, J-L. Chermant (1985) *Precis D'Analyse D'Images* CNRS, Paris

L. S. Davis, S. A. Johns, J. K. Aggarwal (1979) *Texture Analysis using Generalized Co-Occurrence Matrices* IEEE Trans Patt. Recog. and Mach. Intell. PAMI-1 #3 251

J. Feder (1988) Fractals Plenum Press, New York NY

R. L. C. Flemmer, N. N. Clark (1987) *Computer-Based Algorithms for Fractal Analysis of Surfaces* {citation to verbal presentation provided by B. H. Kaye}

J. Frosien (1986) *Digital image processing for micrometrology*, J. Vac. Sci. Tech. B, 4 #1, 261-264
R. M. Haralick (1978) *Statistical and Structural Approaches to Texture* Proc 4th Intl Joint Conf Patt Recog, Kyoto, p. 45-69
R. M. Haralick, K. Shanmugam, I. Dinstein (1973)*Textural Features for Image Classification* IEEE Trans. Syst. Man. Cybern., SMC-3 610-621
B. K. P. Horn (1970) *Shape from Shading: A method for obntaining the shape of a smooth opaque object from one view* (AI Tech Report 79) Mass. Inst. Tech., Project MAC, Cambridge, MA
B. K. P. Horn, M. J. Brooks (1986) *The Variational Approach to Shape from Shading* Computer Vision, Graphics and Image Processing 33 174-208
K. Ikeuchi, B. K. P. Horn (1981) *Numerical shape from shading and occluding boundaries* Artificial Intelligence (special issue on computer vision) 15, 141-184
M. E. Jernigan, F. D'Astous (1984) *Entropy Based Texture Analysis in the Spatial Frequency Domain* IEEE Trans. Patt. Anal. Mach. Intell. PAMI-6, 237
D. C. Joy (1987a) *A model for secondary and backscattered electron yields* J. Microsc. 147, 51
D. C. Joy (1987b) *Image simulation for the SEM* Microbeam Analysis 1987 (R. Geiss, ed.) San Francisco Press, 105-109
A. E. Kayaalp, R. C. Jain (1987a) *Model based inspection of integrated circuit patterns using the scanning electron microscope (SEM)* Proc. SPIE 775 172-179
A. E. Kayaalp, R. C. Jain (1987b) *Using SEM stereo to extract semiconductor wafer pattern topography* Proc. SPIE 775, 18-26
B. H. Kaye (1984) *Multifractal description of a rugged fineparticle profile* Particle Characterization 1 14-21
B. H. Kaye (1986) Proc. Particle Technology Conf., Nürnberg
J. R. Kender (1979) *Shape from Texture: An aggregation transform that maps a class of textures into surface orientation* Proceedings of the Sixth International Joint Conference on Artificial Intelligence, Tokyo, Japan
R. B. Laibowitz, B. B. Mandelbrot, D. E. Passoja (ed.) (1985) Fractal Aspects of Materials Mat. Res. Soc., Pittsburgh
L. I. Larking, P. J. Burt (1983) *Multi-Resolution Texture Energy Measures* Proc IEEE Comput. Soc. Conf. Vision Patt Recog Wash DC, p. 519
H. C. Lee, K. S. Fu (1983) *Three-Dimensional Shape from Contour and Selective Confirmation* Computer Vision, Graphics and Image Processing 22 177-193
B. B. Mandelbrot (1983) *The Fractal Geometry of Nature* Freeman, New York
J. J. Mecholsky, D. E. Passoja (1985) *Fractals and Brittle Fracture* in Fractal Aspects of Materials, Materials Research Society, Pittsburgh
J. J. Mecholsky, T. J. Mackin, D. E. Passoja (1986) Crack Propagation in Brittle Materials as a Fractal Process in Fractal Aspects of Materials II (D. W. Schaefer et al., ed.) Materials Research Society, Pittsburg, PA)
J. J. Mecholsky, D. E. Passoja, K. S. Feinberg-Ringel (1989) *Quantitative Analysis of Brittle Fracture Surfaces Using Fractal Geometry* J. Am. Ceram. Soc. 72 60-65
M. Miyoshi, Y. Yamazaki (1986) *Topographic Contrast in Linewidth Measurement with SEM* Japanese J. Electr. Microscopy 35, 149
M. Miyoshi et al. (1986) *A precise and automatic very large scale integrated circuit pattern linewidth measurement method using a scanning electron microscope* J. Vac. Sci. Technol. B4, 493-499
K. Monaghan, D. Gates, W. Mah, B. Richardson, J. Wilcox (1984) *Effects due to topography and composition*, Proc. SPIE 480 (Integrated Circuit Metrology II), 94-100

D. E. Newbury (1987) *Monte Carlo Electron Trajectory Simulations for Scanning Electron Microscopy and Microanalysis: An Overview* Microbeam Analysis 1987 (R. H. Geiss, ed.) San Francisco Press, 110-114

D. Nyyssonen, R. A. Larrabee (1987) *Submicron Linewidth Metrology in the Optical Microscope* J. Res. Nat. Bur. Stds. 92, 187

D. Paumgartner, G. Losa, E. R. Weibel (1981) *Resolution effect on the stereological estimation of surface and volume and its interpretation in terms of fractal dimension* J. Microscopy 121 51-63

S. Peleg, J. Naor, R. Hartley, D. Avnir (1984) *Multiple resolution texture analysis and classification*, IEEE Trans Pat Anal Mach Intell PAMI-6 #4, 518

A. P. Pentland (1983) *Fractal-based description of natural scenes*, Proc. IEEE Comput. Soc. Conf. Comput. Vision Pat Recog., Wash DC, 201-209; IEEE Trans Patt. Anal. Mach. Intell. PAMI-6 661

A. P. Pentland (1984) *Local Shading Analysis* IEEE Trans, Patt ,Anal Mach. Intell. PAMI-6 #2, reprinted in From Pixels to Predicates (A. P. Pentland, ed.) 1986, Ablex, Norwood NJ, 40-77

A. P. Pentland (1986) *Shading into Texture* in From Pixels to Predicates (A. P. Pentland, ed.) Ablex, Norwood NJ, 253-267

M. T. Postek (1987) *Submicrometer dimensional metrology in the scanning electron microscope* Proc. SPIE 775 166-171

M. Postek, D. C. Joy (1987) *Submicron Linewidth Dimensional Metrology in the SEM* J. Res. Nat. Bur. Stds. 92, 205

L. Reimer (1984) Journal of Microscopy 134#1, p. 1

L. Reimer, M. Riepenhausen, M. Schierjott (1986) *Signal of Backscattered Electrons at Edges and Surface Steps in Dependence on Surface Tilt and Takeoff Direction* Scanning 8 164-175

M. G. Rosenfield (1987) *Analysis of linewidth measurement techniques using the low voltage SEM*, Proc. SPIE 775 (Integrated Circuit Metrology, Inspection, and Process Control), 70-79

J. C. Russ, J. C. Russ (1986) *Shape and surface roughness characterization for particles and surfaces viewed in the SEM* in Microbeam Analysis 1986 (A. D. Romig, ed.), San Francisco Press, 509

J. C. Russ, J. C. Russ (1987a) *Feature-specific measurement of surface roughness in SEM images*, Particle Characterization, 4 (1987), 22-25

J. C. Russ, J. C. Russ (1987b) *The SEM Interpretation of Fractal Surfaces*, Proc. EMSA, San Francisco Press, 540-543

L. M. Sander (1986) *Fractal Growth Processes*, Nature 322, 789-793

H. Schwarz, H. E. Exner (1980)*The implementation of the concept of fractal dimensions on a semi-automatic image analyzer* Powder Tech. 27 207

H. E. Stanley, N. Ostrowsky (1986) *On Growth and Form* Nijhoff, Dordrecht, Holland

C. Sun, W. G. Wee (1983) *Neighboring Gray Level Dependence Matrix for Texture Classification* CVGIP 23 341-352 {more models}

K. A. Thompson (1987) *Surface characterization by use of automated stereo analysis and fractals* Microbeam Analysis 1987 (R. H. Geiss, ed.) San Francisco Press, 115-118

F. Tomita, Y. Shirai, S. Tsuji (1982) *Description of Textures by a Structural Analysis* IEEE Trans Patt Anal Mach Intell PAMI-4 #2, 183

E. E. Underwood (1987) *Stereological Analysis of Fracture Roughness Parameters* Acta Stereologica 6 Suppl II, 169-178

E. E. Underwood, K. Banerji (1986) *Fractals in Fractography* Materials Science and Engineering 80 (1986) 1-14

A. P. Witkin (1981) *Recovering surface shape and orientation from texture* Artificial Intelligence 17, 17-47
R. J. Woodham (1980) *Photometric method for determining surface orientation from multiple images* Optical Engineering 19 139-144
R. J. Woodham (1981) *Analysing Images of Curved Surfaces* in Computer Vision (J. M. Brady, ed.) North Holland, Amsterdam, 117-140

Chapter 11

Stereoscopy

Complete three-dimensional measurements on images require at least two separate viewpoints. In normal human vision, this is provided by the separation of our two eyes. The lateral displacement of objects in these two views is different, depending on their distance. From these parallax displacements, the distance can be computed.

Principles from human vision

In human vision, the distance is not actually calculated. Like most aspects of human vision, we rely on comparative rather than absolute measurements. It is usually sufficient to observe that one object is closer than another, and move one object or the other (or ourselves) to bring them into the desired relationship. Even within the field of view, our eyes fixate on only one or a few features at a time, and only deal with features whose distances lie within a narrow range. Many of the details and features within a typical field of view are never fixated (the process in which the eye briefly pauses with the feature at the fovea, the center of the visual field where the most processing power can be applied). Yet we have an overall impression of the three-dimensional organization of the entire scene, including these "ignored" regions.

Furthermore, the human visual system does not make extensive use of stereoscopy to judge either the absolute or relative distance of objects, as discussed in the previous chapter. The cues to depth in most scenes include relative size of familiar objects (the closer they are, the larger they appear), precedence (if one object partially obscures another, it is closer), and relative position (according to simple perspective, with a vanishing point at the horizon, any object that is on the ground and appears higher than another object is farther away; for objects on a ceiling, the reverse is true). There are also cues in relative brightness and color, since light scattering in the atmosphere makes things far away less distinct and tinged with blue. Finally, the principles of shape-from-shading mentioned in the previous chapter are used extensively by the visual system, and the relative size, shape, distance and orientation of surfaces is obtained by this means as well.

When stereoscopy is used, it is accomplished in humans by vergence. That is, the eyes are rotated inwards by muscles attached to the sockets until corresponding images of a feature both fall on the fovea, the highest resolution array of sensors at the central part of our field of view. This is called "fusion". Notice that this implies that stereoscopy is only applied to one feature in the field of view at a time, and not to the entire scene, whereas the other distance cues mentioned above are available for the entire field of view at once (although humans concentrate on one location or feature at a time). From the amount of vergence motion needed, the relative distance of the object is ascertained. Since only comparative measurements are made, only the relative amount of motion required to fuse the images of each object is required. The brain does not seem to calculate the distance between objects in inches or centimeters. Nor does it perform the comparison very well between objects in a current field of view with those in a previous one, except in the rather special case of tracking the continuous motion of an object and noting the important component of motion toward the viewer.

Not all animals use stereo vision, and those that do so, do not all accomplish it in the same way. Many animals including birds and fish have their two eyes placed on opposite sides of the head for maximum coverage in the field of view, with only a minimum field of overlap in the forward direction which does not include the fovea of the eyes. The relative position of the animal's nose in the field may be more important in judging distance to food than stereoscopy. It is often suggested that the brachiating behavior of monkeys and other animals ancestral to man was aided significantly by the positioning of the two eyes side-by-side where they could be used to judge the distance of a branch, but the evidence for this notion is at best indirect. Many other leaping animals make do without such forward facing eyes.

Nature has evolved the capability for stereoscopy more than once. Few birds have forward facing eyes, but owls do, and appear to use stereoscopy extensively to judge the distance to prey. But owls do not have muscles to move the eyeballs precisely in their sockets as man does, and cannot achieve fusion by vergence. Instead, the owl's fovea is not a spot but a line of cells, which form an axis at the back of the eye that is not quite vertical. The axes of the two eyes incline inwards, along lines that would meet at about the owl's feet. The owl achieves fusion of two images of an object by nodding its head until the images converge on the axes. This nodding motion of the head produces muscle signals that the brain interprets as relative distance to the object. Of course, the nodding motion also makes the owl appear very wise to a human observer, but that is not the owl's problem.

Measurement of elevation from parallax

The utility of a stereoscopic view of the world to communicate depth information resulted in the use of stereo cameras to produce "stereopticon" slides for viewing, which

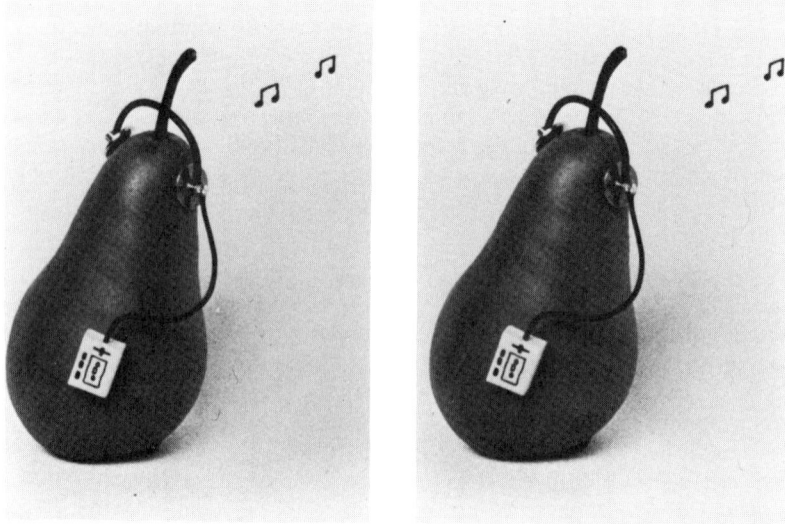

Figure 11-1: A stereo pear (Proceedings of the Royal Microscopical Society).

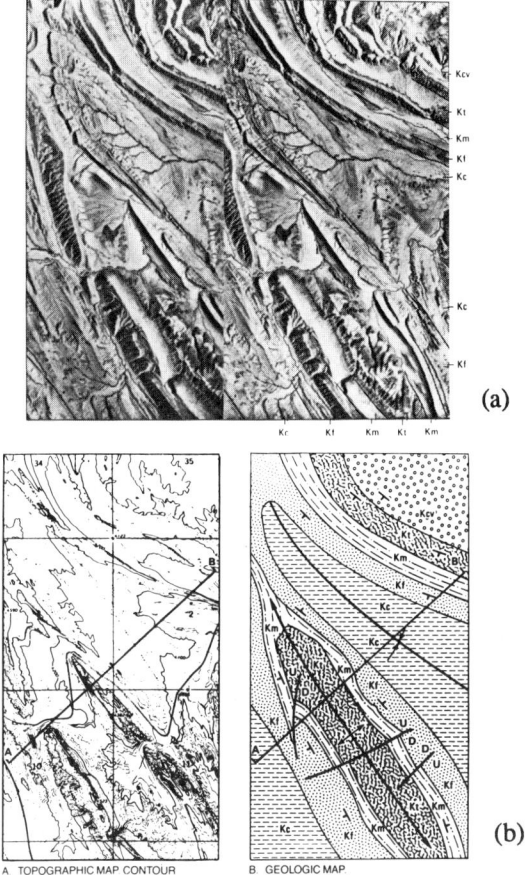

Figure 11-2: Stereo view of the Bighorn Basin in Wyoming: a) aerial photos; b) derived topographic and classification maps (Sabins, 1986).

were very popular more than 50 years ago. Stereo movies (requiring the viewer to wear polarized glasses) have enjoyed brief vogues from time to time. Publication of stereo pair views to illustrate scientific papers is now relatively common, and has given rise to some subtle jokes (Figure 11-1).

Although the human visual system makes only comparative use of the parallax or vergence of images of an object, it is rather straightforward to use the relative displacement of two objects in the field of view to calculate their relative distance, or of the angle of vergence of one object to calculate its distance from the viewer. This is routinely done at scales ranging from aerial photography and map-making to scanning electron microscopy.

There are several different geometries that are encountered (Boyde, 1973). In all cases the same scene is viewed from two different locations, and the distance between those locations is precisely known. Sometimes this is accomplished by moving the viewpoint, for instance the airplane carrying the camera. In aerial photography the plane's speed and direction are known, and the time of each picture is recorded with it to give the distance travelled (Figure 11-2).

Sometimes it is more convenient to move the specimen. In the SEM, the sample and stage can be translated a known distance to produce the same kind of parallax. In both cases, the field of view remains fixed vertically, so some of the original field is moved out of the image, and some new area is included. It is only objects which appear in both fields that can be measured, so this places a limit on how far the airplane or sample stage can be traversed between images (depending on the angle of the field of view, and the distance from the eyepoint to the approximate plane of objects). Usually a series of pictures is acquired in this mode of operation, and each one overlaps the preceding and subsequent images by about 50%.

In Figure 11-3, S is the shift distance (either the distance the plane has travelled or the distance the SEM stage was translated), and WD is the working distance or altitude. The parallax $(d_1 - d_2)$ from the distances between two points as they appear in the two different images, measured in a direction parallel to the shift, is proportional to the elevation difference between the two points.

$$h = WD \cdot (d_1 - d_2) / S$$

If the vertical relief of the surface being measured is a significant fraction of WD, then foreshortening of lateral distances as a function of elevation will also be present in the images. The X and Y coordinates of points in the images can be corrected with the following equations. This means that rubber-sheeting to correct the foreshortening is needed to allow fitting or "tiling" together a mosaic of pictures into a seamless whole.

$$X' = X \cdot (WD - h) / WD$$
$$Y' = Y \cdot (WD - h) / WD$$

Much greater displacement between the two eyepoints can be achieved if the two views are not in parallel directions, but are directed inwards toward the same central point in each scene. This is rarely done in aerial photography because it is impractical when

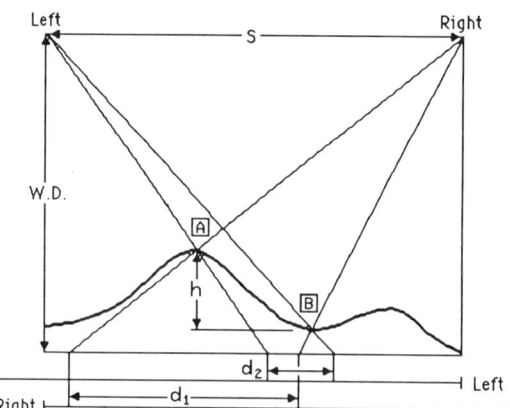

Figure 11-3: Geometry used to measure the vertical height difference between objects viewed in two different images obtained by shifting the sample or viewpoint.

Stereoscopy

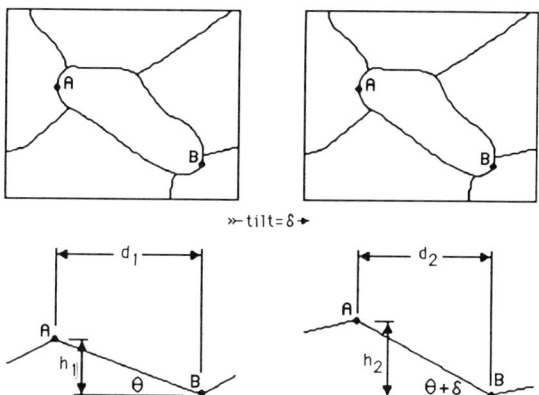

Figure 11-4: Stereo pair images used to measure the vertical height difference between points viewed in two different images obtained by tilting the sample.

trying to cover a large region with a mosaic of pictures, and is not usually necessary to obtain sufficient parallax for measurement. However, when examining samples in the SEM it is very easy to accomplish this by tilting the sample between two views. The same thing would work in the light microscope, but because of the limited depth of field of the light microscope it is rarely used for examining rough objects. Stereo light microscopes utilize two optical paths which are angled toward the same point, to produce low magnification images to the two eyes which magnify the vertical displacements on the objects in proportion to the lateral image magnification, but measurements are rarely made with these microscopes.

In Figure 11-4, the two images represent two views obtained by tilting the specimen about a vertical axis. The points A and B are separated by a horizontal distance d_1 or d_2 that is different in the two images. From this parallax value and the known tilt angle δ applied between the two images, the height difference h and the angle Θ of a line joining the points (usually a surface defined by the points) can be calculated (Boyde, 1973) as

$$\Theta = \tan^{-1} \{ (\cos \delta - d_2/d_1) / \sin \delta \}$$

$$h_1 = (d_1 \cdot \cos \delta - d_2) / \sin \delta$$

Notice that the angle Θ is independent of the magnification, since the distances enter as a ratio.

When two angled views of the same region of the surface are available, the relative displacement or parallax of features can be made quite large relative to their lateral magnification. This makes it possible to measure relatively small amounts of surface relief. Angles of 5-10 degrees are commonly used, but for quite flat surfaces tilt angles in excess of 20 degrees can sometimes be useful. When large angles are used with rough surfaces, the images contain shadow areas where features are not visible in both images, and fusion becomes very difficult. Also, when the parallax becomes too great in a pair of images it can be difficult for the human observer to visually fuse the two images, if this step is used in the measurement operation.

Indeed, most of the measurements made with stereo pair photographs from both SEM and aerial photography are made using extensive human interaction. This is partly because the algorithms which will be described for automatic fusion are relatively new in development and do not always function robustly, so that human verification may be needed in any case. Also, the resolution of photographs used to record images (especially aerial photos used for map-making) is extremely high, while computer image acquisition is much more limited in the number of pixels that can be used. Consequently, the ability to identify small features in both images and to fuse them for measurement is much greater when the human eye is involved, and the parallax distances which can be measured are much smaller. For automatic image analysis the parallax must be at least several pixels to give useful accuracy.

There are several procedures that can be employed with a human observer providing the feature recognition and fusion to enable a simple computer system to perform the calculation of distance or elevation. In the simplest case, two points whose relative elevation is desired are located in both images. The distance between them is measured, either with some kind of marking device such as a drawing tablet connected to the computer, or even with a vernier caliper whose reading is recorded by hand and then typed into the system. From these two measurements, and a knowledge of the viewing geometry (the angle between the two eye points or the lateral shift distance), the vertical distance between the features is immediately calculated as shown above.

Notice that the "distance" between the two points is not the conventional distance, but only the vector distance normal to the tilt axis of the images. This is equivalent to the parallax along a direction parallel to the line connecting the two eyepoints. Displacement in the "vertical" direction at right angles to this direction is basically ignored, although

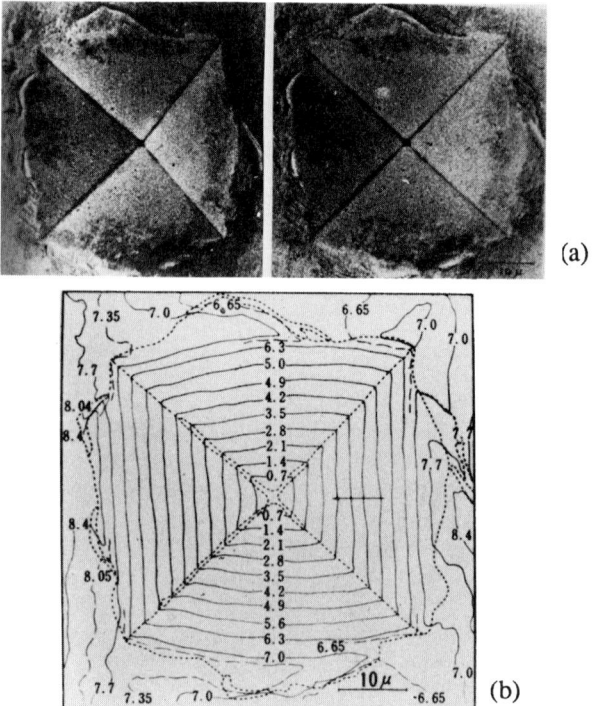

Figure 11-5: SEM stereo pair of hardness indentation (a), and the contour map drawn by a photogrammetry system (b).

when viewing objects from very close distances, it may give some additional information about foreshortening and the distance from the eye to the feature.

In principle, it would be possible to continue this process for many different pairs of points and so construct a complete map of the surface. We will shortly see how this can be accomplished with the computer. But for aerial photos it is usually simpler to work in a different way. Optical aids through which the operator views the two images enable them to be displaced so that points with a specific amount of parallax (corresponding to a given elevation) are aligned. Then the operator manipulates a pointer (usually a spot of light superimposed on the images) to trace lines that appear to be fused at that image offset. The result is a contour line of iso-elevation for the surface represented by the two images. This operation is repeated with different image offsets to produce an entire contour map of the surface. This is essentially the technique by which most of the current generation of contour maps have been generated, supplemented by some on-ground surveying measurements to provide absolute elevations of reference points ("benchmarks" within each map square) to which the relative measurements can be referred.

The prohibitive cost and complexity of such specialized optical devices has restricted their application to other uses of stereo pair photos, such as those obtained with the SEM (Figure 11-5). Instead, most of the measurements of these images has been limited to elevation profiles along lines traversing particular features of interest. This requires only a small number of points whose locations must be be measured in both images. This can be done by hand, but is also easily accomplished with a tablet or other pointing device. Because placing photographs onto a tablet offers greater spatial resolution than marking points on a video raster, and because acquiring the two stereo images usually requires some time and specimen realignment, so that photographic image recording is generally required, the tablet is the most often used method of digitizing point locations.

When two images are to be measured using this procedure, they are first affixed to the tablet surface (for instance with adhesive tape so they cannot move during the measurement). If the tilt axis is known, it is most convenient to either align the images so that this is vertical, or at least in the same direction for each image, and to mark it for the computer. Then only distances at right angles to this direction need to be measured by the program to compute elevations. If the tilt direction is not known, then the program can calculate it by recording the X and Y locations of many marked points in the two images and fitting the tilt axis perpendicular to the mean direction of displacement between the various sets of points.

The measurement procedure is a relative one. Usually some arbitrary location on the images is selected to have a reference elevation of zero or some fixed value. This point is marked on both images. Then each point of interest is also marked, which consists of touching the stylus to the tablet at the corresponding point on the left and right images. From the difference in horizontal distance (i.e. at right angles to the tilt axis) of these points from the reference points, the parallax is obtained, and that together with the known angle of the two views permits calculating the elevation of the point with respect to the (arbitrary) reference location.

Presentation of the data

If points along a linear traverse are measured in this way, the elevations can be plotted as a function of the distance along the line (Roberts & Page, 1981; Nishimoto et al., 1988). These points can then be connected by a polygonal line, or if appropriate

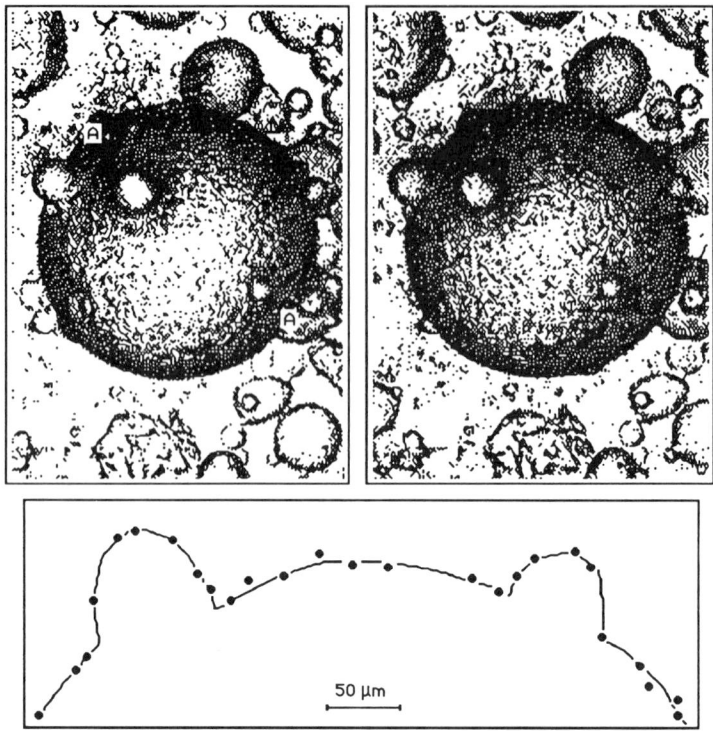

Figure 11-6: Stereo pair images of spheroidal nickel alloy powder, and an elevation profile along the line A-A. The digitized SEM photos are tilted 10 degrees (Roberts & Page, 1981).

mathematical smoothing and fitting of some polynomial or other function can be carried out. The resulting line is then interpreted as representing the physical elevation of the surface along the line. Of course, it is necessary to mark enough points with a close enough spacing to follow the finest irregularities of the surface which are of interest (Figure 11-6). When possible (depending on the availability of surface detail in the images) it is helpful to mark points that are equally spaced along the traverse line. This profile can be used for fractal dimension calculations of the surface as was discussed in the previous chapter, by calculating the length of the line using all points, and then by skipping every second point, every third point, and so on as discussed in Chapter 6. Notice that in marking these points in the two images, it is not necessary to view the images in stereo (with a viewer that permits each eye to focus on the corresponding image), but only to be able to recognize the same feature in both the "left" and "right" eye images on the tablet.

Extending the same procedure to a complete two-dimensional region can be used to construct a surface elevation representation (Figure 11-7). In the simplest case, when fine surface detail is abundant in the images, only one image need actually be mounted on the tablet surface. A grid of regularly spaced points is overlaid on the other image (for instance with transparent plastic). Each of these points is then found on the second image, mounted on the tablet. Because of parallax resulting from varying elevations, these points are not arranged in a perfectly spaced grid. As each is marked with the stylus, the computer can calculate its elevation relative to some arbitrary reference point

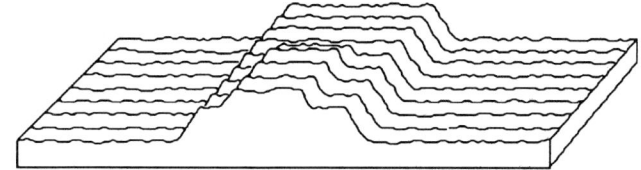

Figure 11-7: A series of elevation profiles across a trace on a semiconductor.

(usually the first point in the grid). This produces a regular matrix of surface elevations which can be stored in the computer memory. When it is complete, a contour elevation map can be easily plotted. Each square of four neighboring points with known positions and elevations is subdivided into two triangles. Within each triangle, any portion of a contour line will be a straight line segment. The ends of the line segment can be directly interpolated between the elevation of the corners of the triangles. In this way, each square in the matrix is examined and all of the required iso-elevation contours can be plotted as shown in Figure 11-8.

The elevation data can also be readily presented in other formats, including as an array of pixels in which brightness or color is used to encode elevation. Actually, this kind of presentation is difficult to view directly because it is not common to our everyday experience, but given such an image, displaying regions based on their elevation (by brightness thresholding) or elevation contours (by brightness contouring) is particularly easy to accomplish. Since the X, Y locations of the points correspond to their locations in one of the two images (the one with the superimposed grid), the contours or other derived information can be directly overlaid on the original image to produce nice displays that communicate information well to the user. Also, brightness or color coding of the contour lines circumvents the usual problem with contour maps of needing to mark the elevation of the contour lines to prevent confusion about what is up and down.

Other displays can also be generated from the matrix of elevation values. These include isometric displays in which a series of elevation profiles is drawn, usually with

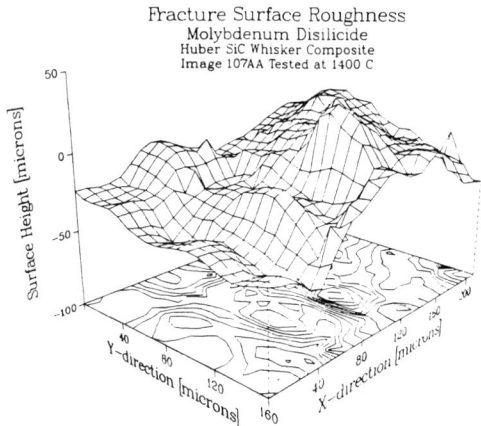

Figure 11-8: Surface elevation on a fracture surface drawn from square measurement grid, with the data in contour-mode (isoelevation lines) format and shown as an isometric display (Los Alamos National Laboratories, Los Alamos, NM).

hidden line removal, to represent the surface as shown in Figure 11-8, or full perspective drawing using CAD/CAM techniques. These will be discussed more later, but it is worth noting that the same computers that are used to make the measurements can often be used to generate these displays, even including surface tiling and shading. When these displays are made, it is possible to magnify the elevation values as compared to the lateral dimensions on the surface, to increase the perception of roughness and the ability to see small details.

The major limitation of the quality of these displays is the number of points that are used. Since each is marked by hand, it can become quite time consuming to mark hundreds of points. Arrays of a few thousand points produce very realistic and detailed surface elevation maps, but may take several hours of operator time to produce.

The data are also in suitable form for the direct measurement of the surface area or the volume above or below the surface. Indeed, this method is often used with before and after pairs of stereo images to measure the volume difference that corresponds to the removal of material in earthmoving operations or volcano eruptions (e.g., the recent Mount St. Helens event in which a large fraction of the side of the volcano's cone was blown away), at one extreme, and such smaller dimensions as the wear on tooth surfaces during cleaning or the wear on dental filling materials as a function of time, at another extreme.

In both cases, it is necessary to locate reference points in the before and after sets of pictures that do not change. With at least three points, a reference plane can be fit which serves as "sea level" for the subsequent volume integration. Each set of three points in the matrix form a triangle. The elevations of the corner points with respect to the reference plane are then used to calculate the volume of the triangular prism between the plane and the triangle. This volume is then summed for all of the triangles to obtain the complete volume. Similarly, the area of each triangle can be calculated and summed to obtain the total surface area. In the expression below, the coordinates X_i, Y_i, Z_i of each corner point for one of the triangular facets on the surface are used to calculate the volume of the triangular prism. The Z values are the elevation above the reference plane. For the present case of points arranged in a regular grid this can be simplified, but we shall want the general version of the expression shortly.

V = (Projected area of triangle) • (Altitude of triangle centroid)

$V = 1/2 \cdot |(X_1 \cdot Y_2 + X_2 \cdot Y_3 + X_3 \cdot Y_1 - X_1 \cdot Y_3 - X_2 \cdot Y_1 - X_3 \cdot Y_2)| \cdot (Z_1 + Z_2 + Z_3) / 3$

The greatest difficulty with this method for obtaining surface maps is that most stereo pairs of images do not have enough detail to permit marking the parallax of points on a regularly spaced grid. An alternative method is to select corresponding points on the left and right images in a random pattern (wherever there is detail that permits fusion to take place). When this is done, it is important to keep some track of where marking has occurred so that good uniform coverage is obtained. This is usually done by employing a stylus that leaves a mark on the image. The procedure requires both images to be placed on the tablet surface, so that the user can sequentially mark the location of a point on the left and right images. The X and Y locations of the point, along with the calculated Z coordinate from the parallax with respect to some arbitrary reference point, are then stored in the computer.

These points do not form a regular grid, so it is somewhat more computationally intensive to plot the contour map. The method is the same (Watson & Philip, 1984; Russ et al., 1986; Russ & Hare, 1986). Each triangle of points has three sides, and along each side the intersection of any contour line segment can be obtained by linear interpolation

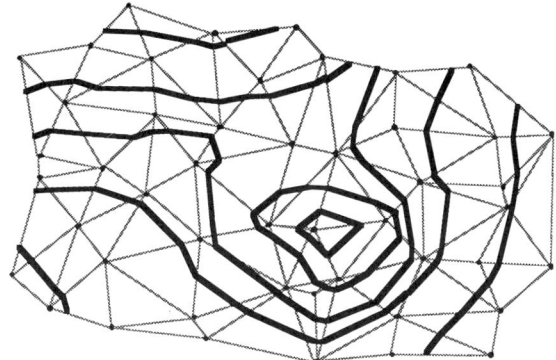

Figure 11-9: Drawing contour lines through triangles between points with measured elevations.

between the corner elevations. The difficulty is in determining how to allocate the randomly arranged points into a unique set of triangles. The proper triangulation of this data set (technically a Voronoi tesselation) covers the entire data space (which may itself be of arbitrary shape) without overlapping. Once this is done, each triangle of points defines a plane which is a facet of the surface. The volume in the prism under the triangle (and above the global reference plane defined) is straightforwardly calculated as shown above. The volumes of all such prisms are summed to determine the total volume under the surface.

A contour map of surface elevation is constructed by determining whether any contour lines should be drawn in each individual triangle. For each triangle, each contour line segment is drawn between points along the triangle edges which are linearly interpolated between the corner or vertex elevations. When the triangulation is complete and the contour line segments have been drawn within each triangle, the result is a complete contour map (Figure 11-9) which can either be plotted as hardcopy, or viewed on the screen in relation to the original image.

As an example of this technique, Figure 11-10 shows a contour map plotted using about 1000 points on the surface of a tooth. Casts of teeth were examined in the SEM before and after dental polishing, to measure the volume of material removed. These values compared well with chemical assay of the fluid aspirated from the patients' mouths during the cleaning operation. The technique has also been used to measure the wear on dental fillings, over periods of time up to 3 years. In these examples, a mosaic of images was needed to cover the surface of the tooth, and the horizontal displacement method (Figure 11-3) was used for measurement. Of course, the technique is equally applicable to the specimen tilting method shown in Figure 11-4.

The array of points specified by their three-dimensional coordinates can also be used for CAD/CAM programs to plot the surfaces as seen from any viewpoint, including hidden surface removal and shading, if this is required. The principles behind this kind of display are discussed in the next chapter, as they are more often used with serial section data.

On a much simpler level, plotting of elevation profiles along any traverse across the surface is easy to accomplish. The traverse line crosses the edges of the triangles, and at each crossing its elevation can be directly interpolated from the corner points. These elevations are plotted against the distance along the line to form the desired profile.

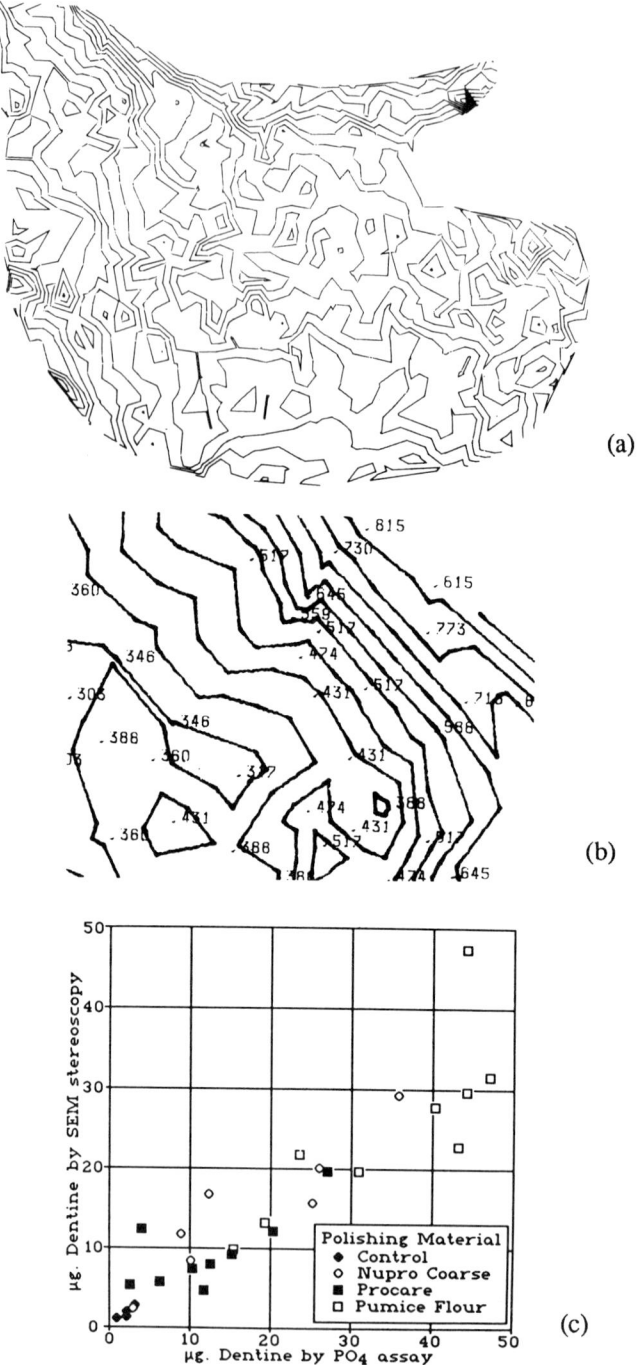

Figure 11-10: Contour map of a tooth surface (a), and enlarged detail (b) showing a few of the measured points and interpolated contour lines, plot of volume (c) determined from the difference between two maps and chemical assay data.

Stereoscopy

Similarly, the integration of volume under (or above) the surface, and the surface area, is accomplished just as was described previously for the case of a regular grid of points.

Note that in all of these manual or interactive measurement methods, we have referred to a single pair of photos. In actual practice, it is not unusual to encounter a mosaic array. If separate mosaics for the left and right eye views are available, but are too large for the tablet surface area, it is a trivial matter to have the measurement program accept commands to shift the X, Y reference point as one pair of images is removed and the next affixed. For the more common type of mosaic obtained by shifting the stage or eyepoint (so that each image has about 50% overlap of the same region of the surface) a more elaborate routine is required. It is usually necessary to mark the same point on subsequent pairs of photos (both laterally and vertically when starting another row of photos) to carry over the relative parallax and tie together the elevation data. This makes for some complicated book-keeping to force the user to mark the required points and to back up to edit mistakes, but does not change the basic calculations or method. Again, arrays of more than a thousand points are generally sufficient to define a surface configuration.

Although it requires a significant amount of computation to tessellate the surface and develop the information required to construct a surface contour map, and even more to present the surface information in other forms such as shaded surfaces, the major difficulty with this operation is not in the required computer resources but the need for several hours of operator time to recognize the corresponding points in the left and right eye images and mark them for the computer. Consequently, it is considered very desirable to find some automatic way to achieve stereo fusion.

Automatic fusion

This has been one of the most computationally intensive tasks attempted with computer image analysis equipment. Several different approaches have been tried. Generally, the method that most nearly approximates the way that the human visual system accomplishes this task is the one that works the best (Marr, 1982; Grimson, 1981; Mayhew & Frisby, 1981; Marr & Poggio, 1976, 1979; Kass, 1986), and methods that apply brute force methods and do not take advantage of what is known about vision from physiological studies have been at best partially useful and more often totally ineffective.

An important insight into how humans achieve fusion came from the work of Bela Julesz at Bell Laboratories (Julesz, 1964). He used random dot stereograms (Figure 11-11), made with a small number of points at the same random locations in a pair of

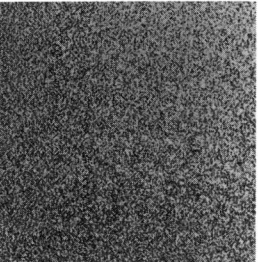

Figure 11-11: Side-by-side random dot stereogram.

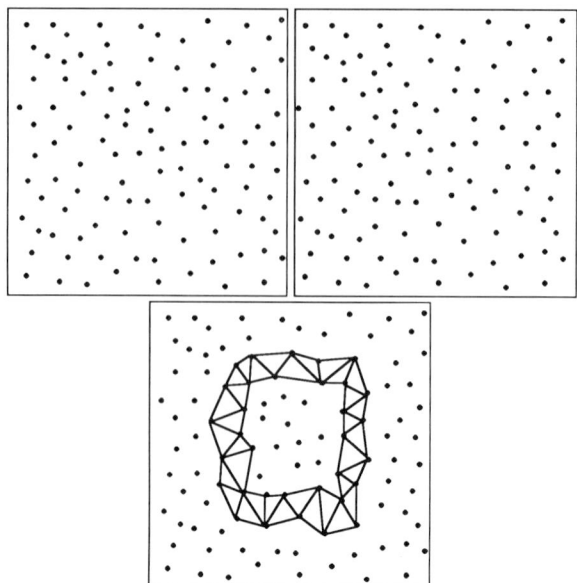

Figure 11-12: A sparse random dot stereogram with a raised central plateau. The cliff edge marked by the triangular tesselation is seen as a square with vertical sides.

pictures but with the dots displaced laterally to produce parallax corresponding to an imaginary underlying surface. When these are viewed, the observer sees the surface, and in fact sees sharply defined outlines of regions even where there are no dots to define such straight or smooth boundary lines.

Figure 11-12 shows an extremely sparse random dot stereogram that is is perceived as a plane with a square raised plateau, even though the data show an irregular shape for the central region. We prefer lines and surfaces in orientations that are multiples of 90, 45 and 30 degrees, planes that meet at these same simple angles, and lines that are straight or smoothly curved. This may be an incorrect bias, especially when looking at microscopic structures, but its general applicability to understanding the macroscopic world is demonstrated by millions of years of evolution. Incidentally, it has been shown that other animals (such as cats and frogs) also have preferences for seeing certain line orientations. One of the problems with an image like this is that any of the dots is a match for any of the others. How does the eye decide which dots to fuse? Figure 11-13 shows that since the dots are identical, there are many possible combinations that could be seen. The key is that we look for the simplest interpretation of the data, in which the dots produce a single surface with minimum distortion and the smoothest or flattest surface.

Furthermore, the visual system is very good at interpolating between dots, or any other structure that it can fuse. Figure 11-14 shows an example of a surface that has no parallax or disparity cues to its depth at all, but is aligned with and has a similar brightness to two other surfaces that do. The visual system automatically connects them so we see a single surface.

The most naive approach to achieving fusion has a rough parallel to the marking of a regular grid of points on one image and then locating them in the second image. It is accomplished by starting with each point in one image, which is just an array of pixels

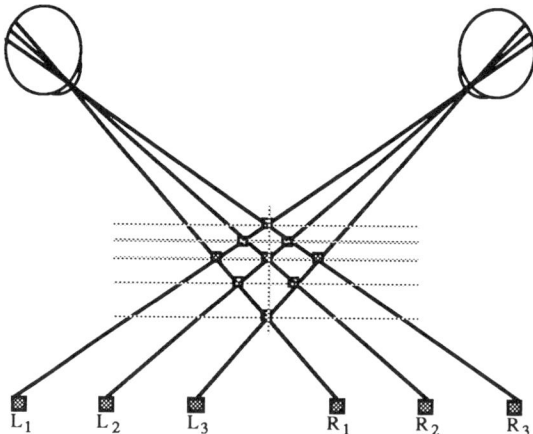

Figure 11-13: Matching up multiple dots in a left and right eye view. The set of possible matches is shown, along with the simplest interpretation that the visual system selects.

with varying brightness values, and searching for the point in the second image that is most "similar" to it.

Of course, it is inadequate to simply look for the same brightness value. Instead, a pattern of brightness values for the pixel and its immediate neighbors is used. Further, because overall brightness levels may change (especially if surface tilting is used to obtain the two images), it is most common to use the difference in brightness between the central pixel and its neighbors. The method used to decide which point in the second image is most similar to the chosen point in the original image is cross-correlation, in which the sum of products of the brightness differences for each of the neighboring pixels is computed, and then normalized by dividing by the product of the mean differences for each image.

If we use $f(j,k)$ to represent the brightness values of one image fragment at pixel address j,k, and $g(j,k)$ to represent the second image fragment, then the normalized cross

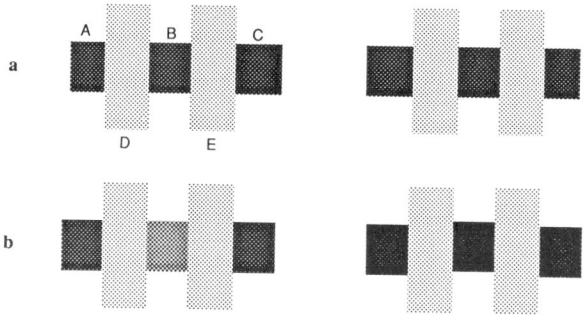

Figure 11-14: Connecting pieces of a surface: a) Stereo pair in which there are no disparity clues for segment B, but it is still perceived as the same depth and contiguous with A and C; b) Small changes in the brightness or vertical alignment of segment B disconnect it from A and C.

correlation function gives a value for each relative position m,n in which they can be lined up and tested.

$$R(m,n) = \frac{\sum_j \sum_k f(j,k) \cdot g(j+m, k+n)}{\left[\sum_j \sum_k f(j,k)^2 \cdot \sum_j \sum_k g(j+m, k+n)^2\right]^{1/2}}$$

To compensate for the different average illumination that may be present for the two images, the $f(j,k)$ and $g(j,k)$ values may be normalized by subtracting their average grey values f_m and g_m, which gives the following expression, in which σ_f is the standard deviation of the grey values in the first image fragment, and $\sigma_{g(m,n)}$ and g_m are the standard deviation and mean for the fragment of the second image in the region being used for the match.

$$R(m,n) = \frac{\sum_j \sum_k f(j,k) \cdot g(j+m, k+n) - f_m \cdot g_m}{\sigma_f \cdot \sigma_{g(m,n)}}$$

Notice that this normalized cross correlation technique is very similar to the pattern matching algorithms described before for feature identification, and indeed it can often be performed by the same hardware or subroutine. A "score" for each pixel position in the second image is obtained. The value of R takes on a maximum for some displacement m^*, n^* which is then assumed to represent the matching point. Rather than searching the entire second image, it is usual to limit the search to that region in which it is considered reasonable for the point to be found. Usually this is a narrow horizontal rectangle, whose height depends on the possible vertical shift of the point due to foreshortening during tilting, or to possible misalignment of the tilt axis of the two images. The width of the rectangle covers the possible parallax values, which in turn reflect the extreme variation in relative height that the surface may have. Depending on how much is independently known about the surface being measured, these limitations can be made tighter to significantly speed up the matching operation.

Within the search rectangle, there will be some location that gives the highest score (Figure 11-15). One approach is to use that location as the match, and from the difference

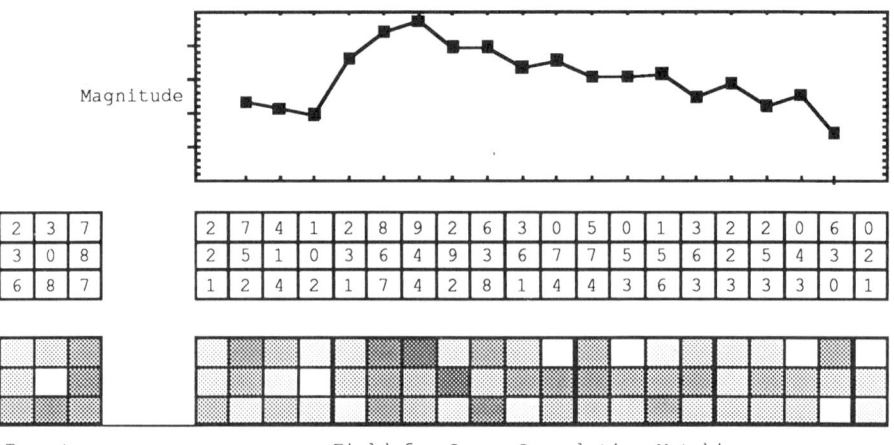

Figure 11-15: Example of cross-correlation matching.

Stereoscopy

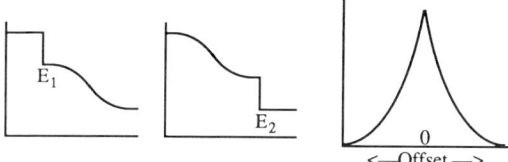

Figure 11-16: An example of two brightness profiles representing the displacement of an edge. The cross correlation function is maximum at zero offset and does not find the true alignment.

in horizontal position to calculate the parallax and hence the elevation of the original point, which can be coded as a brightness and stored in another image memory. This restricts the possible parallax to quantized steps with the dimensions of individual pixels. Since the values of cross-correlation score as a function of pixel location in the search rectangle of the second image usually vary smoothly, it is possible to refine the result by interpolation. Fitting a polynomial in the vicinity of the highest score permits locating the match point to less than one pixel width, and this allows a better estimate of the original point's elevation. The calculations are repeated for each pixel in the original image to build up an entire height map of the surface (Mehlo, 1983; Hood & Howitt, 1986).

Problems with the direct application of cross correlation methods are notorious (Thompson, 1987; Nevatia, 1982). For any but the simplest images under the most idealized conditions, this method simply does not work because too many possible points in the search region have scores that are very similar, and perhaps greater than the "correct" point (Figure 11-16). Accordingly, there have been some attempts to "patch" the method. These generally involve taking the raw score from the cross correlation and either adding or multiplying it by a factor that depends on the parallax it would generate (producing a maximum score from minimum parallax, or in other words favoring a flat surface match of points). An even more complex method uses a running weight that depends on the average parallax of other nearby points that have already been matched. This approach seeks to have elevations vary gradually across the sample surface.

The problem with these and other attempts to modify the cross correlation method is that they presuppose some rather simplified surface structure, such as nearly flat or smoothly varying. They sometimes work for very simple surfaces but fail in the general case. The major reason for the failure is that the cross correlation method does not match points in the more robust way that the human visual system performs, and also that it ignores a fundamental fact about the nature of the surface being viewed. Points that are to the left or right of each other on one image appear in the same order in the second. Only the distances are changed. This is equivalent to saying that the surface is continuous, and not folded or transparent. This is a very strong statement about the surface, and should not be ignored in trying to construct a map of its surface elevation.

Marr and other researchers into the physiology of human vision have shown that the eye is not particularly interested in points. The feature-finding that goes on at low levels in the image processing finds edges, corners and other similar patterns. These are connected into lines by grouping operations, to produce what Marr (1982) calls a "primal sketch". This is a set of short, usually disconnected lines that mark the limits of regions, such as surface facets, edges of objects, and so forth. It is these larger features, the lines in the primal sketch, that are matched between the left and right images to achieve fusion of a stereo pair image. Human vision is particularly responsive to edges (Figure 11-17), which is why cartoon sketches work.

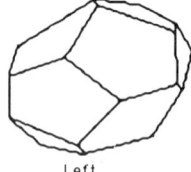 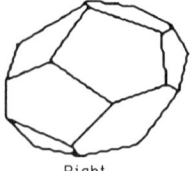

Left   Right

Figure 11-17: Stereo pair views of a simple polyhedron, defined by its edge lines.

Furthermore, the matching of all of the available lines, corners and edges is conducted simultaneously so that no crossing over of points between the images is allowed. This is sometimes referred to as a relaxation method in which the network of features is connected together and allowed to shift relative to each other but not change its topology. Since the shifting of the points develops the parallax values that are subsequently used to calculate elevation, the process is very much like placing a network of features on the image of the surface and then shifting them to represent elevations.

Implementing this approach in a computer requires some compromises. Marr's method for locating the lines in the primal sketch uses a series of operations which will be further discussed in the final chapter, and a hierarchical series of grouping operations. A somewhat simpler method has proved effective for the somewhat simpler types of surface images that we are trying to deal with here. We will generally be able to assume that we are looking at single surfaces, not natural scenes in which different objects may lie in front of or behind others, and so forth. Quasi-real-time fusion of stereo pair images of real scenes has in fact been achieved, using a significant amount of computational power. One demonstration of this capability has been using robotic vision to drive a car down a road (free of any other traffic) while negotiating ways around obstacles.

For our simpler problem we will not derive the full primal sketch. Instead, we locate those points in the images (both left and right eye views) that are "interesting" in the visual sense. These will generally be edges and corners. One operator that finds such points is the Moravec (1977) interest operator. This starts with a value for the rms difference calculated for each pixel and its neighbors in a small region in the image as

$$M = \left\{ \sum_{x,y} [f(x,y) - f(x+i, y+j)]^2 \right\}^{1/2}$$

Another way to find interesting points uses the local entropy calculated for each point based on the brightness values in a local $N \times N$ neighborhood. This is

$$H = \sum -n(g_i) \cdot \log_2 (n(g_i))$$

where the summation is over the g_i grey values, and $n(g_i)$ is the fraction of pixels with each value (Barba et al., 1988). H has units of bits, and represents the amount of different information present in the region. This method has the advantage that it is insensitive to the contrast and brightness values of the image.

The resulting values from either of these operations are stored at the pixel address as a pseudo-brightness values. Each pixel's value is then compared to its neighbors so that only local maxima survive and the other points are set equal to zero. These points represent places where the brightness changes abruptly in the image (Figure 11-18). A list of the thousand or so such points generally lies along the edges of features or facets, and nicely delineates regions that are themselves rather devoid of interest and can be

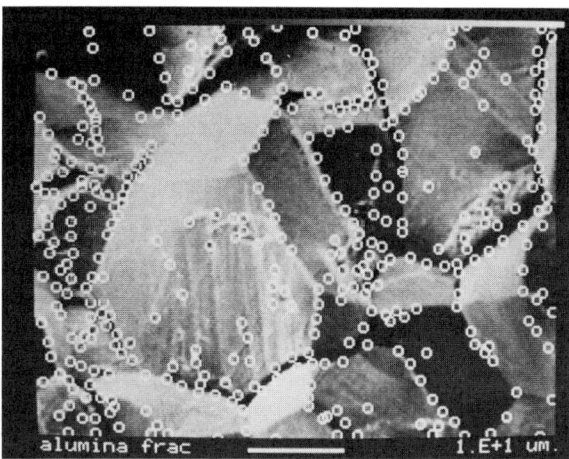
Figure 11-18: "Interesting" points on a fracture surface.

modelled in the resulting elevation map as planes or surfaces linking the interesting points.

This list of points is then matched between the two images (Quam & Hannah, 1974; Medioni & Nevatia, 1985; Kayaalp & Jain, 1987; Lee & Russ, 1989). This involves far fewer matches than in the original brute-force cross correlation approach (a few thousand points versus the original 250,000 pixels, for instance). The matching is usually done with a combination of normalized cross correlation scores for the points, plus relaxation of the positional network to prevent non-physical surface configurations. This means that the points must remain in order, and the height values are bounded. The parallax disparity gives the elevation of each point. The process is relatively fast (the most time consuming operations are the application of the interest operator, and the subsequent tesselation of the points to construct the contour map). Usually more than 90% of the points can be matched, while some of the points that are "interesting" in one image do not match either because they are lost off an edge or otherwise hidden in the other image, or simply do not have interest scores that place them in the top 1000 or so points used for matching. Figure 11-19 shows an example of the method, in which interesting points are located automatically on stereo pair images of wear pits on a metal surface, and used to draw a contour map of surface elevation.

It was mentioned before that the elevation of the points can be stored as brightnesses in an image memory. If this is done for discrete points, as for example the interesting points, then there is a short-cut that can be used to construct the complete height map. The map initially consists of separated pixels whose brightnesses represent elevation, with gaps between them. These gaps can be filled by growing regions around each pixel (by a dilation operation) at the same brightness (elevation). Each growing region has initially the same brightness as its central pixel, forming plateaus that eventually grow until the touch each other. This image can then be smoothed by an operator that compares each pixel to its neighbors. Except for the initial central pixels, which are not allowed to change, all other pixels are smoothed by averaging to produce a faceted surface in which the brightness varies linearly between fixed points. This approach eliminates the need to perform the Voronoi tesselation described earlier, and is essentially parallel (dealing with all points at once) rather than serial. (This is true even if it is performed by a conventional

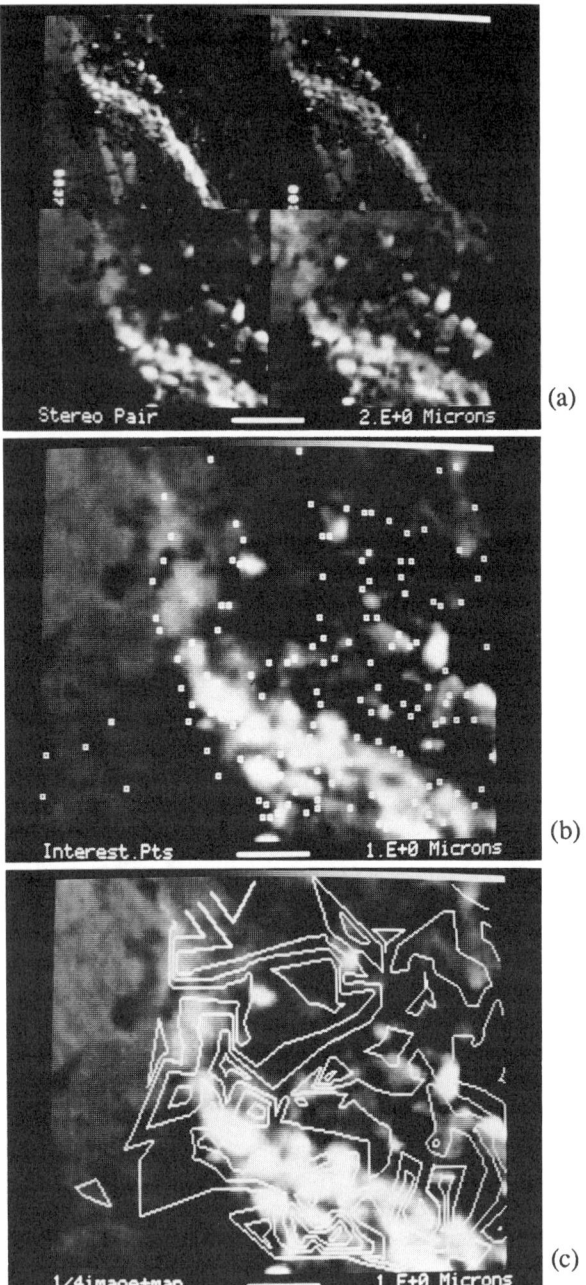

Figure 11-19: Stereo pair of surface with a pit produced by particle impact (a), the interesting points (b), and contour map (c).

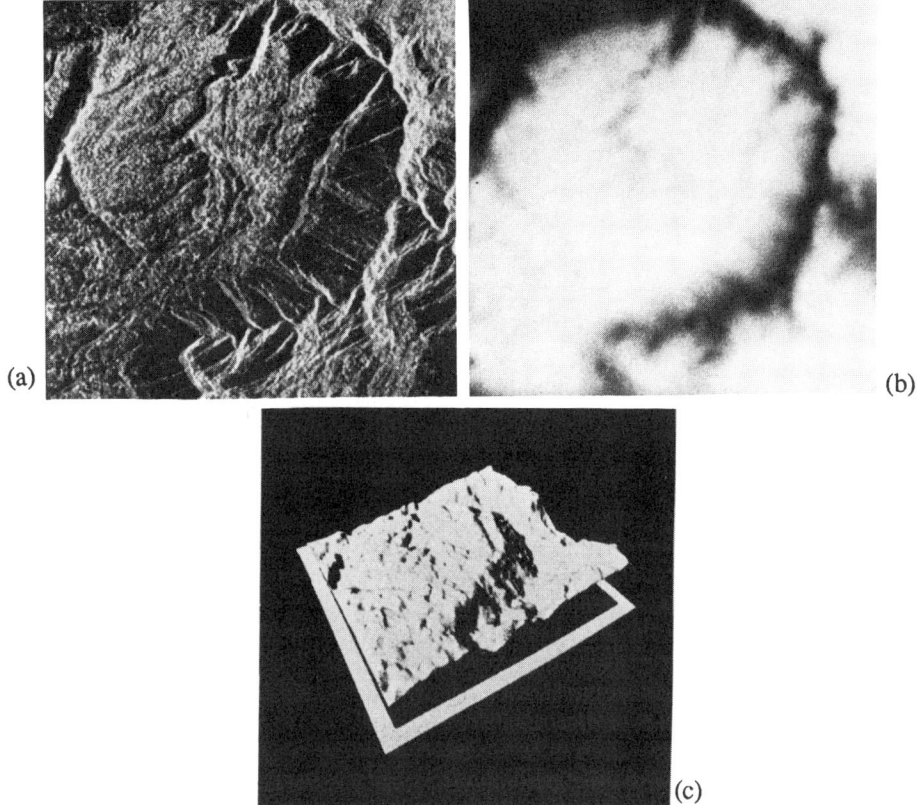

Figure 11-20: Automatically fused images from synthetic aperture radar (area in Northeast India), showing presentation modes: a) original stereo pair shown in re/green; b) disparity (proportional to elevation) coded as brightness; c) shaded perspective view of surface relief.

serial computer.) Once the brightness map has been constructed in this manner, it can be thresholded, color coded, or contoured directly to produce elevation maps, and these can be superimposed on the original image for reference.

The brightness coded elevation image is usually not too easily interpreted by itself. However, because the gradient of brightness (for instance as obtained with a Sobel operator) gives the direction and magnitude of the physical surface gradient, it can be used to quickly construct a surface representation showing the surface as it would appear if illuminated from any direction (Figure 11-20). For each point, the magnitude and direction of the Sobel (or other gradient) operator are used to calculate, or more commonly to look up in a table, a brightness corresponding to the Lambertian cosine law for the surface orientation and the location of the illumination source. It is even possible to include specular reflections (which provide surface highlights) in this kind of a calculation or table. This is a point operation that is very fast. Physical modelling of surfaces by CAD/CAM programs is much slower by comparison, because each facet must be individually drawn, filled and shaded. It is also possible to modify such images by adding varying amounts of random texture to the smoothly modelled surfaces, to simulate local surface roughness. This is just the reverse of the process described above for measuring the fractal dimension of rough surfaces.

Stereoscopy in transparent volumes

The use of stereoscopy is not restricted to surface images, of course. Transmission images of dispersed objects in a transparent medium are routinely obtained by both transmission light and electron microscopy. (They are also obtained from astronomical telescopes, but we have at least for the present no way to obtain a second image from a different point of view. If we could, it would immensely simplify some studies of stellar and galactic distribution.) The method has been applied with notable success to the study of the Aurora Borealis, however. In fact, it was stereo pair photographs taken simultaneously at widely separated locations (linked by the then-new telephone) that first established the true altitude of the aurora as about 100 miles, correcting previous speculative estimates that varied from outer space to just above ground level.

Of course, there are some situations in which the three-dimensional location of features can be determined directly without using stereo methods. One example utilizes the shallow depth of field of the confocal light microscope to view pores in a translucent ceramic, so that the z coordinate is directly available (Russ et al., 1989). However, our interest here is in stereo techniques, The mapping of the distribution of the objects in space usually required manual marking of their locations in the two images. Automatic methods work poorly because the constraint of a surface is not available. Only in the case that the objects are unique in shape, color or some other characteristic that permits matching can automatic fusion be achieved. Of course, it should be noted that the human visual system is not comfortable with this type of image either, and that viewing such images is difficult for most people, so that illusory fusions often occur (matching incorrect sets of points). The subject of illusions will be discussed further in the final chapter. Stereoscopic illusions, when they occur, are among the most persistent and convincing. This testifies to the robust nature of the matching process which the human visual system employs.

When the three-dimensional locations of objects in a transparent volume are determined by stereoscopic measurements, they are not used to draw any kind of representation. Instead, they may be used to measure the number of objects per unit volume or the mean distance between objects. A common example is the study of precipitate particles in materials, as viewed in the transmission electron microscope (TEM). The particles appear as essentially point-like spots. The principles of stereology described in Chapter 8 provide a simple way to obtain the number per unit volume if the thickness of the section being examined is known. It is only necessary to count the number per unit area and divide by the thickness.

However, in many cases the specimen thickness is not known, nor even uniform. The process of thinning TEM specimens often produces wedge-shaped foils. The local thickness can sometimes be measured by using surface markings (either ones introduced by the preparation process, or contamination marks left by the beam). Tilting the specimen produces a parallax shift in the location in the image of points representing the top and bottom surfaces, and from this displacement the local thickness can be calculated.

In the absence of such surface marks, and any other independent knowledge about the thickness (which for instance can sometimes be obtained from convergent beam electron diffraction measurements on the specimen), a direct estimate can be made from the objects themselves (Hare et al., 1988). Since all of the points reside within the thickness of the specimen, some of them will be close to but not outside the upper and

Stereoscopy

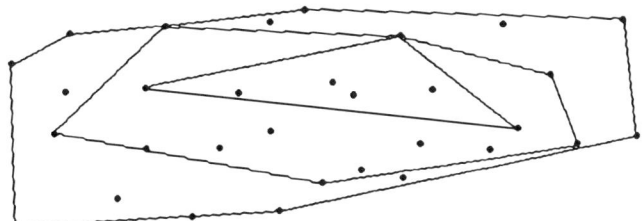

Figure 11-21: Enclosing polygon method in 2-D. Initial triangle formed by randomly selected points encloses non-edge points. Subsequent search through remaining points expands the enclosing polygon to external convex bound, whose lines are tested to find those giving minimum section area.

lower surfaces. These points can be used to fit bounding planes to estimate these surfaces. If enough points (e.g., more than 50-100) are used in the fit, then for the case of randomly distributed points the planes will encompass essentially the full sample volume. In fact, the fraction of the volume enclosed between the fitted planes can be estimated probabilistically as a function of the number of points used in the fitting operation.

A method to find the upper and lower bounding planes that enclose the points operates by eliminating points which lie in the interior of the foil and therefore are not useful to determine the exterior bounding planes. Figure 11-21 shows the method in two dimensions; it is readily extended to the 3-D case. Select any three random points, which form a triangle (in three dimensions, four points are used to form a tetrahedron). Any points within the figure are clearly interior points and can henceforth be ignored. By continuing to select random points from those which remain outside the polygon, and testing them to see if they lie above or below the bounding lines (planes), new boundary points are found. These expand the polygon (polyhedron) surrounding internal points, which can subsequently be ignored. Some points previously on the surface are eliminated as the program goes through the list of new points. The list of surface planes and points grows slowly as the entire data set is scanned. The final external bounding polygon (or polyhedron in 3-D) is efficiently and quickly obtained. The boundary lines (planes) of this figure are then checked to determine which two enclose the minimum area (volume), and they are taken as the boundaries of the sample.

As can be expected, with small numbers of points the error in volume is large because few points actually happen to lie very close to the boundaries. In addition, if all the surface points are on one side of the foil, it becomes possible to obtain a negative value for the volume. By the time 25 random points are used, this possibility becomes extremely unlikely. The standard deviation for the measurement ranges from somewhat more that 10% of the volume for 25 points to under 1% for 400 points, if the points are randomly distributed in the volume.

Two important types of error should be considered for this method. The systematic underestimate of volume for errorless coordinates (the volume correction factor) is a function of the number of data points measured for each of the four shapes. The numerical value of this correction can be determined by simulation, along with the expected mean error for a single measurement. Figure 11-22 shows the effect of the number of points on the underestimate of volume. The error bars show the 95% confidence limits on the values of the volume fraction. A simple volume correction factor

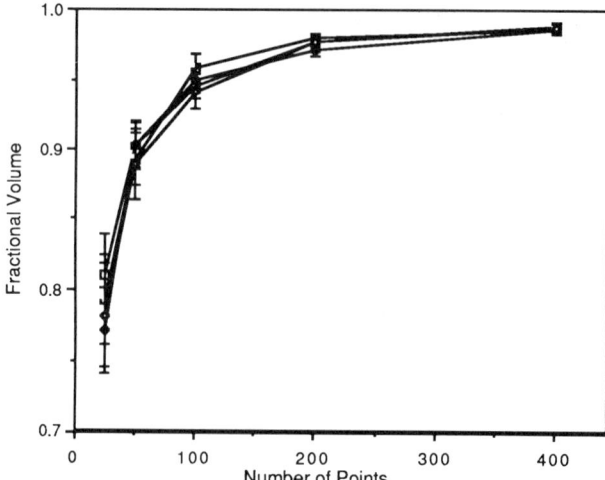

Figure 11-22: Estimated volume fraction using enclosing polyhedron as described in text, fit to varying number of random points (error bars indicate ±2σ for multiple simulations with random points enclosed between boundary planes in various geometries).

can be determined from these results by plotting the calculated volume as a function of the reciprocal of the number of points; this is nearly linear as shown in Figure 11-23. A useful formula for the correction factor is *volume fraction* = $1 - 5/N$, where N is the number of points used in the determination. Multiplying this factor by the estimated volume gives an estimate of the true volume of the foil.

When errors are present into the z coordinates, the value of the correction factor (estimated volume of the foil) and the standard deviation increase, particularly for larger

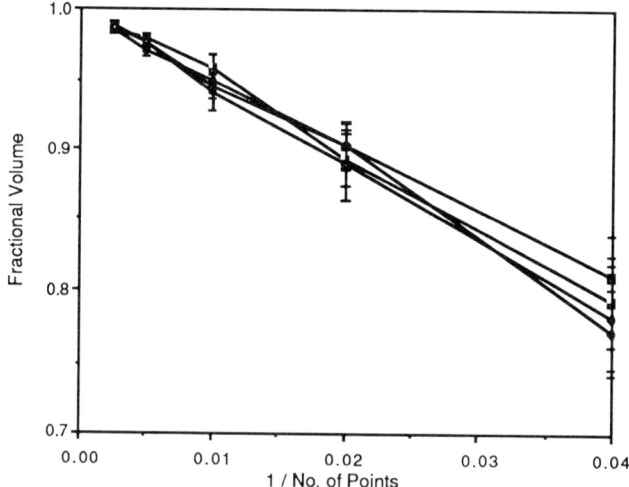

Figure 11-23: Data from Figure 11-22 plotted against reciprocal of number of points. The line Volume Fraction = $1 - 5/N$ provides a reasonable fit to the data.

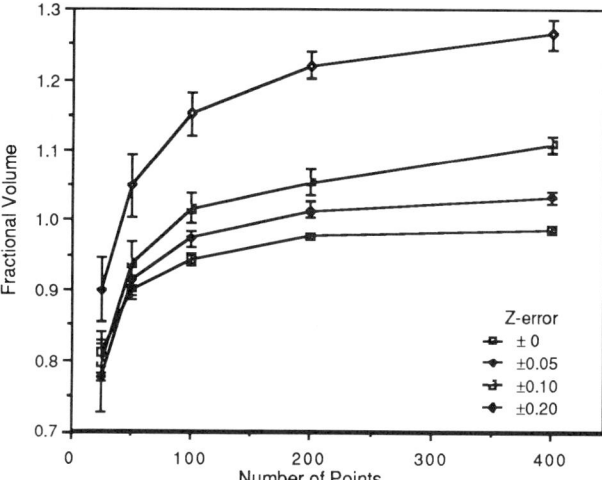

Figure 11-24: Increase in the volume correction factor with relative error in the z-coordinate. Note that volume estimates greater than 1.0 are possible, and increase in probability with the number of points.

numbers of points. This is because the probability of a data point being accidently located above or below the true surfaces becomes larger. Figure 11-24 shows the increase in the correction factor as the uncertainty in z increases. The correction factor exceeds 1.0 as the probability of erroneously finding points above the surface increases. The scatter or the uncertainty in the individual determinations increases as the error in z rises, and for large N becomes the major contributing source of error. For this reason it is important to minimize the uncertainty in the z coordinate.

In summary, stereoscopic methods are useful in a variety of special circumstances which do not require mapping of entire surfaces. These usually employ manual recognition and marking of points where they appear in the two images, and have the computer perform the calculation of height from the parallax. The purpose of these measurements is generally for measurement of object density or spacing rather than the creation of representational displays of the specimen. On the other hand, when surfaces are measured, a variety of manual, semi-automatic and fully automatic techniques for identifying common points ("fusion") are available. The result permits displays of elevation profiles and full representations of the surface, in addition to calculations of the dimensions of the surface and the underlying volume.

References

J. Barba, H. Jeanty, P. Fenster, J. Gil (1988) *The use of local entropy measures in edge detection for cytological image analysis* J Microscopy (in press)

A. Boyde (1973) *Quantitative photogrammetric analysis and quantitative stereoscopic analysis of SEM images*, J. Microscopy 98, 452

W. E. Grimson (1981) From images to surfaces MIT Press

T. M. Hare, J. C. Russ. J. E. Lane (1988) *Volume determination of TEM specimens containing particles or precipitates*, J. Electron Microscopy Technique, 10#1, 1-6

P. J. Hood, D. G. Howitt (1986) *A technique for the 3-D representation of carbon black* Microbeam Analysis 1986 (A. D. Romig, ed.) San Francisco Press, 487-490

B. Julesz (1964) *Binocular depth perception without familiarity cues* Science 145, 356-362
M. Kass (1986) *Computing visual correspondence*, in From Pixels to Predicates: Recent advances in computational and robotic vision, (A.P. Pentland ed.), Ablex Publishing, Norwood NJ, 78-92
A. E. Kayaalp, R. C. Jain (1987) *Using SEM Stereo to extract semiconductor wafer pattern topography* Proc. SPIE 775 18-26
J. H. Lee, J. C. Russ (1989) *Topographic measurement of microelectronic devices with automatic SEM stereo pair matching* J. Comput. Assist. Microscopy 1, 79-90
D. Marr (1982) *Vision*, W. H. Freeman, San Francisco
D. Marr, T. Poggio (1976) *Cooperative computation of stereo disparity*, Science, 194, 283-287
D. Marr, T. Poggio (1979) *A Theory of Human Stereo Vision* Proc. Roy. Soc. Lond. B204 1979 301-328
J. E. W. Mayhew, J. P. Frisby (1981) *Psychophysical and Computational Studies towards a Theory of Human Stereopsis*, in Computer Vision (J. M. Brady, ed.) North Holland, Amsterdam, p. 349-395
G. Medioni, R. Nevatia (1985) *Segment-Based Stereo Matching* Computer Vision Graphics and Image Proc. 31 2-18
H. Mehlo (1983) *Der Einsatz von Grauerwertbildspeichern bei der quantitativen Bildanalyse von Werkstoffgefügen*, Dissertation, Universität Stuttgart
H. P. Moravec (1977) *Towards automatic visual obstacle avoidance* Proc. 5th IJCAI, August 1977, p. 584
R. Nevatia (1982) Machine Perception Prentice Hall, Englewood Cliffs NJ
Y. Nishimoto, S-I. Imaoka, H. Yasukuni (1988) *A PC-Based Interactive Image Processing System for 3-D Analyses of SEM Stereo Images* Proc. IAPR Workshop on CV, Tokyo, 383-387
L. Quam, M. J. Hannah (1974) *Stanford automated photogrammetry research* AIM-254, Stanford AI Lab
S. G. Roberts, T. F. Page (1981) *Microcomputer-based system for stereogrammetric analysis*, J. Microscopy 124, 77
J. C. Russ, T. M. Hare (1986) *Serial sections and stereoscopy - complementary approaches to three-dimensional reconstruction*, Proc. EMSA, San Francisco Press
J. C. Russ, T. M. Hare, R. P. Christensen, T. K. Hare, J. C. Russ (1986) *SEM low magnification stereoscopic technique for mapping surface contours*, J. Microscopy 144, 329
J. C. Russ, H. Palmour III, T. M. Hare (1989) *Direct 3-D pore location measurement in alumina*, J. Microscopy 155, RP1
F. F. Sabins, Jr. (1986) Remote Sensing: Principles and Interpretation second edition, Freeman, New York
K. A. Thompson (1987) *Surface characterization by use of automated stereo analysis and fractals* Microbeam Analysis 1987 (R. H. Geiss, ed.) San Francisco Press,115-116
D. F. Watson G. M. Philip (1984) *Systematic Triangulation* CVGIP 26 217-223

Chapter 12

Serial Sections

In the previous chapter on stereo pair imaging, there was some discussion of measurements of object locations in a transparent medium. Usually, this is restricted to disconnected points (particles or other small objects), or to simple linear or planar features such as fibers or membranes. This is necessary because the complexity of matching points in the two images increases very rapidly when more intricate arrangements of points are encountered, and especially when precedence (one surface lying in front of another and obscuring it) makes it impossible to locate the corresponding point in the second image unequivocally, so that fusion is possible. The final chapter will discuss some of the ways that the human visual system deals with this problem, by grouping algorithms and continuity of lines and surfaces. These are still very difficult to model in a computer image analysis system.

Consequently, stereo images find only limited use for studying the complex three-dimensional arrangement of objects and surfaces that are routinely encountered in many sciences. Applications range from the very small (the distribution of elements at a scale of nanometers in diffused junctions in semiconductor devices, as may be studied with a secondary ion mass spectrometer (SIMS), to the microscopic (arrangements of organelles within a cell, as viewed with the electron or light microscope), to full size (three-dimensional reconstruction of the arrangement of artefacts and layers in an archaeological site) to large (geological studies of sedimentary rocks encountered in exploratory oil drilling). The basic techniques for all of these diverse applications are surprisingly similar.

Obtaining serial section images

The method for obtaining information about the three dimensional structure to be examined is to acquire a series of two-dimensional images (Figure 12-1). Preferably, these will be parallel and equally spaced, and aligned with each other spatially and rotationally. When these desirable attributes cannot be met for one reason or another, greater computational efforts are needed to overcome the problems. The technique is called a "serial section" method because on of the most obvious ways to obtain the desired series of images is to cut the structure into a series of slices, each of which can be imaged separately. There are, of course, many ways to do this. One, involving tomography to obtain nondestructive image sections by computation from various projected images, will be discussed separately in the next chapter.

Mechanical sectioning of structures is most commonly encountered in dealing with light or electron microscope images of biological specimens. This is because thin sections must be cut anyway for normal imaging, and collecting a series of the sections for sequential imaging is a simple additional step. With mechanical sectioning, the usual method is to cut the slices with a microtome (for the light microscope, the slices may be

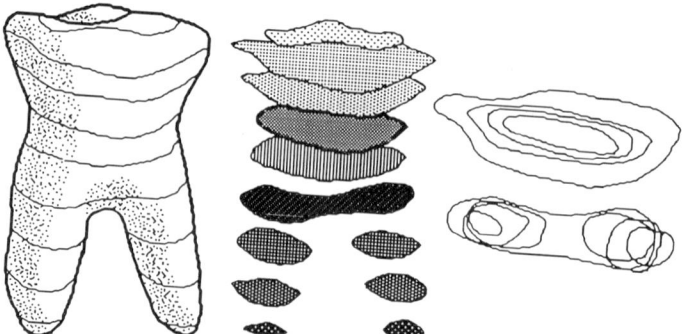

Figure 12-1: Schematic diagram of serial sectioning of an object. The individual section planes (middle) may be viewed as a series of outlines.

several micrometers thick, whereas for the electron microscope an ultra-microtome cuts slices a few tens of nanometers thick). The cutting operation is presumed to introduce minimal distortion or disturbance to the structure, but in fact this can be a major source of error in subsequent reconstruction and measurement. Additional distortions may be produced by chemical fixation and drying procedures applied to organic materials.

This is something of an art, and depends on the speed of the cut, the hardness of the materials, the thickness of the section, the angle of the cutting edge, and so forth. Microtomy can also be applied successfully to polymeric materials and even some of the softer metals. The sections are of finite thickness, so the resulting images are transmission images of the projected internal structure within each slice. In many cases, the slices that are collected for examination are not contiguous but separated by some number of discarded intervening slices. It is most convenient to deal with slices that are uniform in their thickness and spacing, but this is not always the case.

With microtomed sections, a common problem is that the slices are collected in a way that does not assure their consistent alignment. This would make it impossible to interpret the features seen in each slice as part of the same 3-D object (Figure 12-2). Successive slices may be rotated and are generally shifted with respect to each other. This means that

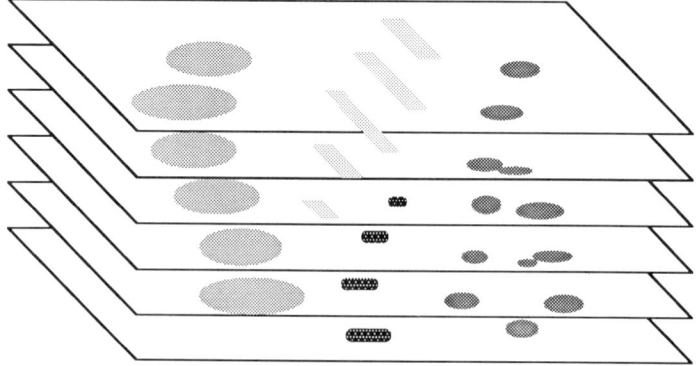

Figure 12-2: Presentation of serial section images so that continuity of features from slice to slice can be seen.

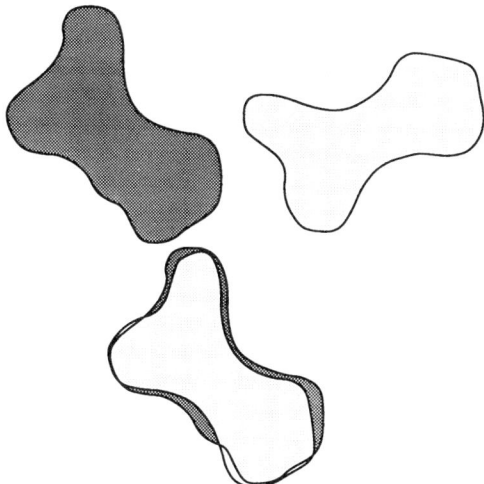

Figure 12-3: Aligning features in sequential slices may require translation and rotation.

the images must be aligned after they have been acquired (Figure 12-3). Sometimes this is done using photographs as an intermediate step, with the user selecting identifiable structures as reference locations in successive images and physically aligning the photographic prints. In other cases, external reference marks are used such as a fine hole drilled with a laser through the specimen (or near it, through the mounting medium which is generally a plastic or resin for biological materials). These holes can be recognized and used for image rectification.

It is also common to find that microtomed slices are distorted by mechanical deformation that causes foreshortening in the direction of the cut, and that the amount of this foreshortening varies (usually in the range of 5 to 15%) as a sequence of slices is cut from a block (usually with the later slices showing more squeezing than the earlier ones). In the process of alignment, either by aligning features in sequential images or by using fiduciary reference marks it is possible to correct this by rubber-sheeting the image, as was discussed previously. An even better method is to record a surface microscopic image of the block before each slice is cut, and use this information to adjust the image from the slice (which usually is a transmission image and has higher resolution) afterwards. However, all of these methods are time consuming and in most cases the corrections are unfortunately not made if the interpretation of the three-dimensional image is intended to be qualitative rather than quantitative.

Acquisition of the images directly from the microscope may produce unaligned images which must later be aligned using rubber-sheeting methods. When it is possible, physical alignment of the slices using the microscope stage or intermediate photographic prints is usually a worthwhile step, and also allows the operator to view the sequential slice images in a way that can eliminate some gross errors of alignment.

For some solid samples, mechanical slicing is impractical, and sequential images must be obtained from surfaces revealed by polishing to remove layers of material (Figure 12-4). Most metal and ceramic samples fall into this category, because of their hardness. So does the normal process of uncovering an archaeological site, which is done by removing layer after layer of material from the surface while recording the location of all interesting objects.

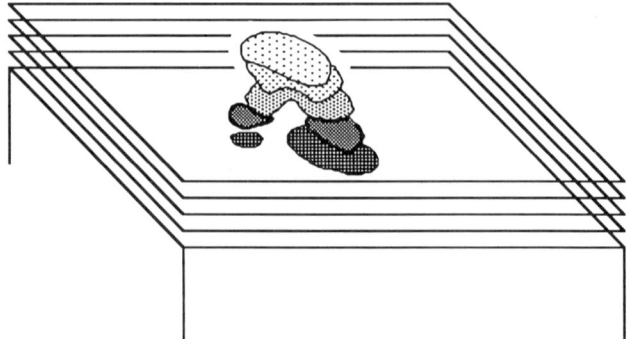

Figure 12-4: Sequential polishing or erosion.

Unlike the cutting of slices, sequential polishing has the disadvantage that the removal is destructive. Consequently, it is essential to record all pertinent image or other data before one layer is removed to reveal the next one. Again, it is desirable to perform the polishing in such a way that images are equally spaced and surfaces are parallel. Of course, in this method the images are of a surface rather than projected through a finite thickness of material. This sometimes simplifies the subsequent reconstruction, as projected images through sections become more difficult to interpret as their thickness increases. Sequential polishing also usually permits the use of external reference marks that facilitate the alignment of the sequential images.

Closely related to mechanical removal of layers by polishing or excavation is in-situ erosion of a surface. The distinction is that the sample is generally imaged or examined as material is removed, or at least does not need to be removed from the microscope for the erosion process to be performed. A very common tool for accomplishing this is ion beam erosion. Moderate energy ion beams are used to remove surface layers from specimens while they are in the vacuum chamber of a scanning electron microscope (SEM) or similar instrument such as an Auger electron microscope. This reveals a deeper layer which can then be immediately imaged. The process can be repeated as necessary to reveal the internal structure of the specimen.

The rate of removal of material with the ion beam is generally calibrated independently, for instance using the SIMS which will be described below, when absolute depth information is required. However, in many cases the specimen itself provides adequate markers for depth (for instance, with semiconductor devices the depth of various structures is considered known) and only relative position of intermediate layers is required for subsequent image interpretation and measurement.

The major problem with ion beam erosion is that it is not uniform. Rates of removal vary significantly for the different materials that may comprise real, complex structures. They even vary for a given material as a function of crystallographic orientation. Any kind of internal structure that is to be studied in its three-dimensional configuration generally is composed of different materials, and these will usually erode at different rates. The result is that the sequential images obtained are not planar and not necessarily equally spaced. Very little work has been done on good methods to correct for this distortion in performing three-dimensional reconstructions from such serial section images. This makes the resulting data qualitatively useful but limits the ability to obtain quantitative dimensions from it.

The Secondary Ion Mass Spectrometer (SIMS) or ion microscope combines ion beam erosion of the sample surface with the process of imaging. The ions which are eroded from the surface are directly imaged, usually after selection in a mass spectrometer to choose a particular element or isotope of interest (or to rapidly sequence through several such elements). Because the rate of image acquisition is fast enough to represent each few atomic layers within the sample, and switching between elements is also very fast, these instruments can produce enormous collections of serial section images (Figure 12-5). We will see below how these can be recalled for presentation in various ways, and for some measurements of elemental distribution or dimension. However, little quantitative morphological or structural measurement use has been made of them to date (as compared to compositional information).

Optical sectioning

Physical destruction of the sample, either by erosion of the surface or even cutting slices for separate examination, is not always required. In addition to the tomographic methods discussed in the next part of this chapter, it is often possible to perform a nondestructive serial sectioning method by selectively focussing the microscope at varying depths in the sample. The light microscope, for example, has a comparatively shallow depth of field at high magnification (about 1 micrometer at 1000x is typical). If the specimen matrix is transparent, then it is possible to focus at different depths in the sample below the surface to acquire sequential images. These can then be used for serial section reconstruction. This is called optical sectioning.

Unfortunately, the light microscope has only limited ability to perform optical sectioning for two reasons. First, at high magnification where the depth of field is small, the working distance (distance from the objective lens to the plane of focus) is also small. This limits the depth to which the method can be carried. Even more important, the light passing from the plane of focus deep in the sample to the lens must pass through the intervening sample matrix, which causes light scattering and generally degrades the

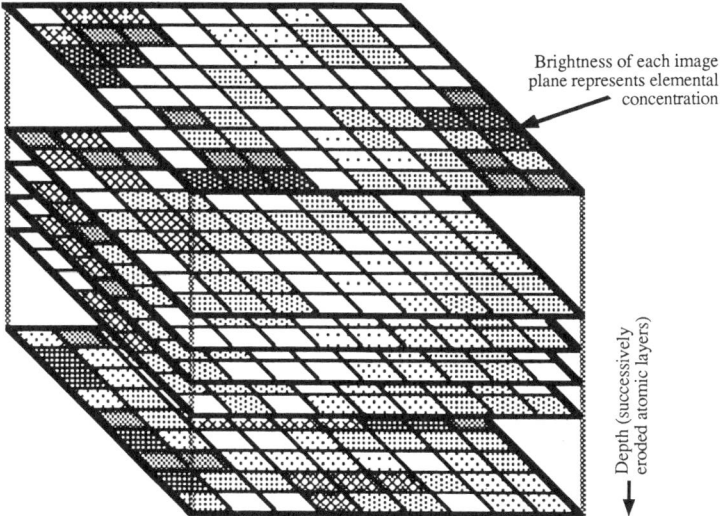

Figure 12-5: Secondary Ion Mass Spectrometer (SIMS) imaging produces a series of closely spaced elemental intensity images.

quality and resolution of the image. This latter problem can be overcome by using a scanning method. The confocal scanning light microscope uses an intense collimated light source (which may come from a laser) on the sample. With this illumination, light is focussed at a particular point in the specimen (and at a particular depth), and only light reflected from that point is measured for storage at the corresponding pixel in the image. Hence light scattered from other points and other depths is not included, and does not contribute to a high overall background illumination level that obscures detail. The image is built up point-by-point either by scanning the specimen or the incident illumination (Minsky, 1957; Sheppard & Choudry, 1977; Wilson & Sheppard, 1984; Sheppard, 1986; Wilson, 1989).

In one of its earliest forms, the collimation of the light source was accomplished by a series of holes in a rotating disk in the microscope optics. Light passes through a hole to illuminate the sample, and is returned through a conjugate hole to be recorded in the image. Multiple holes can be used to speed the collection of the image, provided that the holes are far enough separated to prevent light entering through one hole from being scattered back through another. In this form, the instrument was called the tandem scanning reflected light microscope (Boyde et al., 1987; Brakenhoff et al., 1987; Petran et al., 1985). Subsequent models have used laser illumination with either optical or specimen scanning. Optical scanning can use either mirrors or acousto-optical devices. The tight collimation and focussing of the optics to prevent scattered light from returning to the detector gives rise to the name "confocal" to describe the optics. Mapping of surface relief has also been accomplished with this method, since the maximum reflected signal is obtained when the surface is at the focal distance. Scanning in depth at each point and recording the position of the maximum gives the desired information (Awamura, 1987).

Another very important confocal scanning light microscope technique uses fluorescence. Various fluorescent molecule probes can be used in biological systems to label specific organic molecules, or sites of particular chemical bonds (e.g., S=S bonds). Others are used to indicate physiological activity (e.g., pH). These probes have bonds which absorb light (usually in a vibrational or rotational mode) and then re-emit light of a longer wavelength. It is necessary to select the laser light wavelength to match the absorption band of the fluorescent probe (Waggoner & Ernst, 1989). By using a detector sensitive to the emitted light, the spatial distribution and intensity of the probe can be recorded. Conventional light microscopes are also used for fluorescence microscopy, but the confocal light microscope is far superior in both lateral and depth resolution (Boyde et al., 1989; v.d. Voort & Brakenhoff, 1989; Wallen et al., 1989)).

Regardless of how the microscope designers implemented the confocal principle, the result is an image that largely overcomes the degradation of contrast and resolution with depth, and produces subsurface images to considerable depth in materials that are transparent or even only slightly translucent (e.g., bone, wood, ceramics). The principle area of application has, of course, been to biological tissues. These images are well suited to interpretation and measurement using serial section methods, although they often require processing before measurement because the contrast mechanisms are complex and may not provide uniform demarcation of edges, and because partially out-of-focus features may be present to obscure some detail.

The electron microscopes (SEM and TEM) are generally not of much use for this technique. The SEM can be used with in-situ or sequential erosion, of course, but its images are basically surface images except in a few isolated instances (such as electron beam induced conductivity in semiconductor junctions below the surface), and in those cases the depth of the image information depends on the sample and is not readily varied. The TEM has such a great depth of field that even quite thick samples are viewed as a single projected image, and no optical sectioning is practical (however, stereo viewing of internal structure in thick sections as discussed in Chapter 11, or tomographic reconstruction from a series of projections as discussed in Chapter 13, can be used with the TEM).

However, there is another type of microscope that does lend itself to this application (Briggs, 1985). The acoustic microscope is a relatively new and little used tool. It employs high frequency sound waves, much like sonar, to produce images with resolution comparable to that of a light microscope. The sound waves are generated by piezoelectric transducer and focussed by a spherical cavity that acts as a lens. Coupled into the sample surface by a fluid such as water, the sound waves scatter from any internal structure in a way analogous to the scattering of light. In a sense, the acoustic microscope is a direct analogue of the confocal light microscope. However, whereas light is scattered at surfaces where the index of refraction changes, sound waves are scattered by variations in Young's modulus. This is the constant that relates stress to strain for a material. It varies for different types of material, and also changes abruptly at internal defects, surfaces or discontinuities.

The result is an echo signal returned to the transducer a (very) short time after the initial sound pulse is generated. By varying the frequency of the sound and the configuration of the objective lens cavity, information from a great range of distance within the sample can be obtained. Of course, this information represents only a single point, so the usually method of scanning to build up a complete image must be employed, This is usually done by traversing the specimen under the microscope objective lens, while storing the image in a form directly accessible to the computer. A series of such images can be used for serial section reconstruction. Because of the rather unusual nature of the way the image is formed, it is often necessary to perform some image processing steps first, as discussed in earlier chapters.

The use of seismic sounding in geological exploration is more-or-less similar to acoustic microscopy in its ability to generate a series of depth images based on discontinuities between types of material. Again, the interpretation of the images in terms of structure can be quite complicated but the reconstruction of the arrangement of objects in the three-dimensional structure from the images is basically the same for all of these techniques.

Presentation of 3-D image information

The principal distinction that must be made amongst images for serial section reconstruction is whether the sections are continuous in space, or separated by intervening gaps. In the case of the SIMS if a single element is being monitored, the brightness information (proportional to elemental concentration) is organized into "voxels" (Figure 12-6). These are volume elements in the same sense that pixels are picture elements, that is, voxels are three-dimensional units of measurement. Ideally, they should be small cubes with the same dimension in depth as their lateral spacing, but in practice this is rarely achieved and they represent either much thinner or thicker regions than a cube. This, combined with the irregularity in erosion rates mentioned above which distorts the sequentially acquired images into nonplanar surfaces, makes the quantitative measurement of distances in these images very difficult.

There are some other cases in which true voxel images can be obtained directly, but these are the exception rather than the rule. Tomographic reconstruction, which is

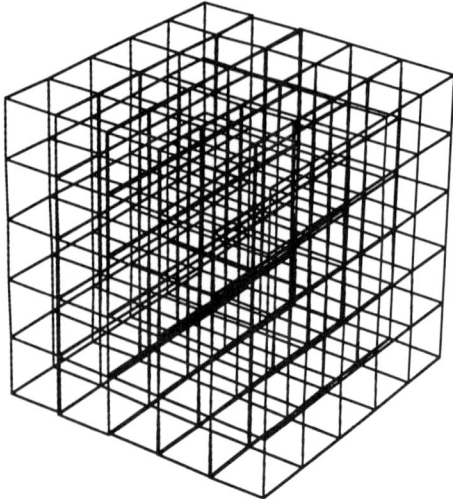

Figure 12-6: Schematic of a 3-D "voxel" array.

discussed in the next chapter, is one method that does so. Usually, serial section images represent planes through the structure that are separated by much more than the lateral spacing of the pixels in each image, and the information from the intervening space between the image planes is lost. This means that serial section reconstruction must interpolate between the planes.

For some purposes, the reconstruction will produce a "voxel" type of representation for display. In other cases, boundaries will be represented directly. The situation is exactly analogous to the distinction that was made in earlier chapters between pixel-based and boundary-based representations for two-dimensional images.

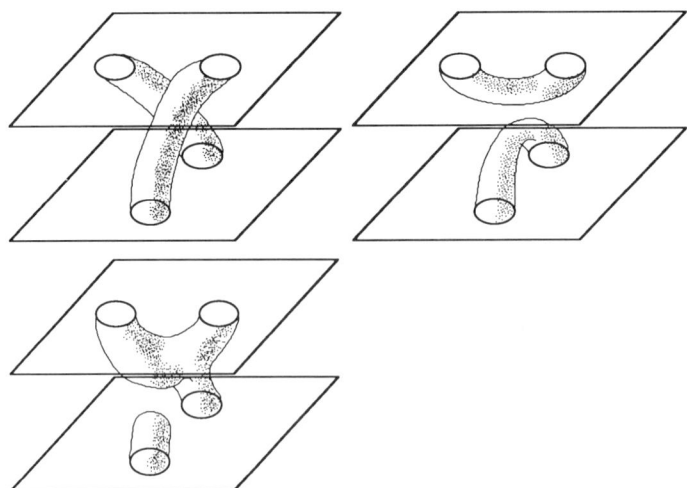

Figure 12-7: Ambiguous serial section images - what happens between the separated planes? Three possible interpretations are shown.

Serial Sections

When serial section images are not continuous, interpolation of the intervening structure must be carried out. Sometimes there is little or no chance for confusion, if structures vary in size or position only gradually through the stack of images. More often, the slices are spaced so that significant changes may occur in the missing space (Figure 12-7). This requires that boundaries or objects in sequential slices be matched together. The logic used in this is usually not the same as that for stereo pair matching, but rather examines the size, color or density, and especially the shape and position of objects in the images. Objects can usually be matched rather quickly in this way, but the matching is not always one-to-one. Indeed, we would hardly be interested in serial sections if the same objects were present in each section image. It is the beginning, ending and branching behavior of solids, surfaces and linear structures that is most interesting in many cases. These events are characterized by the appearance or disappearance of features, or the matching of one feature in one image to more than one in the next.

Sometimes it is possible to interpolate the position of an end or branch point between image planes by considering the shape of the object in the third dimension. If the change in size or position of the objects over several images varies in a smooth way, then extrapolation into the between-slices space can be performed with some confidence. For computer measurement purposes, this is usually performed only on isolated features that have been selected by the operator as being of interest. The full three-dimensional modelling of objects to include ends and branches is discussed below, but its complexity and the amount of computation required makes it little used.

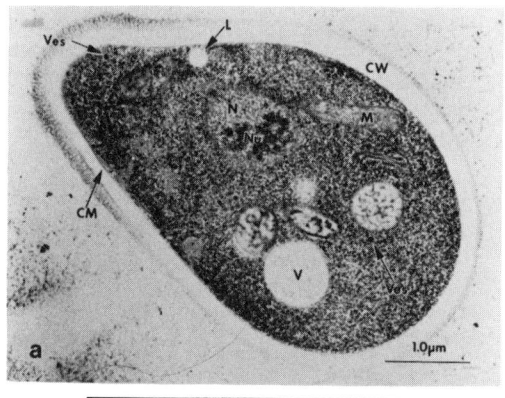

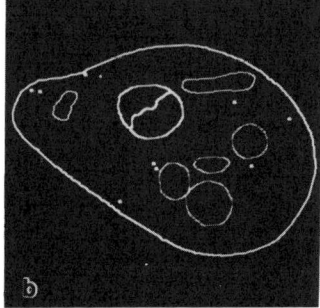

Figure 12-8: Original grey scale image and the user-selected and traced outlines. Note the selection and simplification of feature outlines.

In some cases, there is no obvious unique solution to the matching of objects from one plane to another. In these cases, it is necessary to select from the possibilities using random numbers. The various possibilities can be weighted in proportion to their probability of occurrence if there is some independent knowledge about the general shape of objects, or if there are enough sections through the objects visible in the serial images to allow a judgement about the relative frequency of various possibilities to be made.

The individual serial section images are usually simplified representations of the original full grey scale images (Figure 12-8). In many cases, the selection of features of interest is performed manually and the locations are entered with a drawing tablet. This method is especially used when photographs are used for intermediate storage. The photographic images can be placed on the tablet and outlines marked with the stylus around objects or along boundaries which are to be reconstructed, while ignoring other extraneous information that may be present. The procedure for entry, and the internal storage of the data, is the same as for manual outlining of single images for measurement purposes, and is a boundary-representation. Depending on the way that the serial section information from many slices is later presented, this may be used as a binary representation directly, or converted to a pixel-based representation and implicitly to an array of voxels.

The advantage of the manual entry mode is that it allows great selectivity in the objects, locations and boundaries that are to be used in the reconstruction (Fram et al., 1986). Most images (especially for complex structures such as those from many biological applications) contain a great deal more data than can conveniently be used in reconstructing a three-dimensional structure by serial section methods. Usually only a single class of objects of boundaries are dealt with, and these may represent only a tiny fraction of the information present. Using too many objects or outlines, even with color coding of displays, produces a very visually confusing reconstruction from which little can be learned about the structure.

Manual outlining automatically allows the operator to select only those structures that are interesting and continuous from section to section. It also allows the photographs to be previously compared and aligned, either based on the continuity of features of interest from one section to the next, or based on the alignment of reference or fiducial marks.

However, manual entry of outlines is quite time consuming, and also makes some demands for drawing skills and coordination that some operators would rather avoid. The alternative is to use representations of the individual slice images obtained by the automatic methods described earlier. These may be either boundary- or pixel-representations, with the latter being more common. Thresholding a grey image (either an original, or one appropriately prepared by image processing), is the most common approach. The same techniques for image processing and editing that were described before are still useful. In particular, either features or their edges may be used, with the latter being selected by the use of a gradient operator. Setting of discriminator thresholds by the various manual or automatic methods produces a binary image. In most cases this will include the features of interest along with others that are not intended to be used in the reconstruction. In addition to the various image editing tools that can reshape or remove features from a binary image, it is useful to have the ability to manually select features with a pointing device to eliminate or to select them. This approach combines the best features of manual and automatic entry methods, since it is relatively fast and easy, as compared to manual delineation, while retaining the ability of the operator to select objects of interest.

Of course, in particular cases the use of Boolean logic to combine binary images, or other automatic image editing techniques can be used to isolate the features of interest. In this case, a series of slice images can be quickly assembled.

Aligning slices

As mentioned above, depending upon the method of sample preparation and imaging, the individual section or slice images may or may not be aligned with each other. If they are not, then the next step must be to align them, which in the most general case requires both translation and rotation of the images until continuous features are in registration. It may also require "rubber-sheeting" to correct for variations in magnification, foreshortening, or other problems. This can also be done either by automatic or manual means.

Manual alignment involves one of two common approaches. Both work by aligning one sequential pair of image slices at a time, until the entire stack is finished. One is for the operator to continuously adjust the position and rotation of one slice with respect to its neighbor, while visually watching the display, which shows the two images superimposed (using color, brightness, or other means to distinguish them). When the operator judges them to be best aligned, the position and rotation are recorded with the slice image for future use in constructing the display. This method does not easily accommodate changes in magnification or other image distortions that may be required.

The second method allows the operator to mark several locations in the two images which are to be brought into alignment. A minimum of two points are needed to specify position and rotation, but in general more are used. Either a least-squares minimization of alignment error for all of the points, or a rubber-sheeting operation to exactly align the points and stretch the intervening portions of the image between them can be employed. This method works best when there are registration or fiducial marks on the images to aid in alignment (for instance, laser burn marks through an embedded plastic section before microtome slices are cut). The first method is often preferred by operators who wish to judge the overall alignment of the images by the general match of feature boundaries.

It is possible in principle to automatically carry out either of these kinds of alignment, although in practice the manual methods are more common. The automatic technique can most easily use the presence of discrete registration marks, especially if they are distinguishable from the other features in the images (for instance by color, density, shape, etc.). There may still be some uncertainty as to which registration mark corresponds to which, unless they are individually different in shape, or have a known asymmetric layout on the images. When there is prior planning of registration marks for this method, it is very efficient and easy to apply.

The automatic methods to align general feature boundaries are much more difficult, and require massive computation. However, as a byproduct they produce the complete solid-modelled image that will be discussed below. The first step is to identify features in one image with those in the next. The centroid locations of these features are then used to perform a first-order alignment of rotation and position, using the least-squares minimization of error as mentioned above. In some cases this crude alignment is considered adequate by itself, but there is no fundamental reason to consider this alignment criterion correct.

Refinement of the positions of the images (and more rarely of the relative rotation) is then performed using the outlines of the features. In most cases, the outlines do not exactly match. The area between the pairs of outlines from successive images is taken as

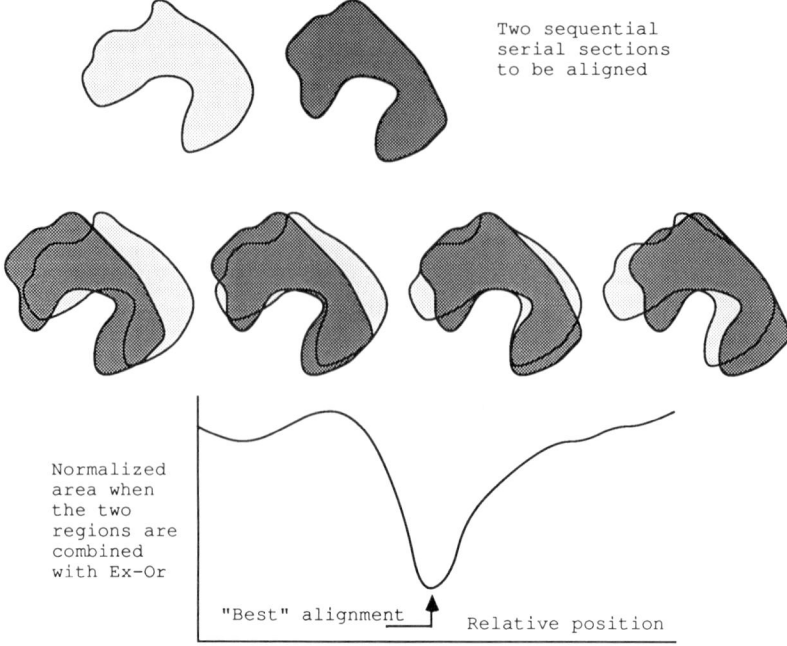

Figure 12-9: Effect of misalignment on "Ex-Or" area between two features.

the parameter to be minimized. This area can be obtained as the Exclusive-OR combination of the two individual images (Figure 12-9). Its minimization by shifting images with respect to each other is approximately equivalent to minimizing the surface area of the solid feature which can be constructed by extending the feature outlines to form a continuous solid. The remaining minimized area can be used in one of the methods described below to estimate the surface area of the solid.

A slight modification of this approach uses not the entire areas of each section, but the outlines of the sections (which represent the three-dimensional surface whose continuity is being used as the alignment criterion). The distances between individual points along the outlines are minimized, usually in a least-squares sense. This requires much more computation.

Further refinement of positional alignment can be accomplished by actually creating the solid model. This is done by selecting points along the feature boundaries in each image slice (when manual entry has been used, these will often be the original boundary-representation points by which the operator entered the feature outlines). These points are then connected together by a series of triangular tiles, as described below. The alignment criterion is to minimize the total area of this surface, an iterative task of considerable magnitude.

All of these automatic methods for feature boundary alignment suffer from the same potential flaw, which is their implicit assumption that the surface area of the solid that can be modelled from the slices should be minimized. There is no physical reason to expect this, and there are simple counter-examples available. For instance, consider serial sections of a uniform diameter cylindrical object which is oriented at an angle to the normal to the section planes. In each section image, the feature will appear as an ellipse.

The automatic alignment procedure will exactly align these ellipses with each other, producing a final solid object that is an elliptical cylinder whose axis is normal to the section planes (Figure 12-10). This cylinder will have a smaller surface area than the actual one, and of course is an incorrect interpretation of the data.

Automatic alignment procedures produce an artificial or implicit symmetry in features, and will suppress torsional offsets as well as the lateral shift shown in the example. Without other features in the image (for instance, intentional discrete registration marks such as holes drilled in the block before the sections are cut), there is no satisfactory way to resolve this problem and avoid the error.

Displays of outline images

Once the array or "stack" of serial section images have been aligned, the operator usually wishes to view a three-dimensional display of the structure that they represent. There are several different ways that these displays can be handled, which vary in terms of the computational effort they require and the ease of interpretation which they afford the viewer. Usually the information at this stage consists of a series of binary images (pixel representations) or feature outlines (boundary representations) for each planar slice, with the X,Y positional offset and rotational angle for each needed for alignment, and the z position of the plane. As noted before, these need not be uniformly spaced, but the information on their separation distances must usually come from some external source as it is not contained within the image itself.

It is worth noting here that we have considered the two-and three-dimensional images without particular regard to how they are stored or encoded. It is clear that a complete voxel representation will require much more storage than storing the feature outlines in boundary-representation form in discrete image planes. Furthermore, the latter format is more directly related to the way in which the data are most commonly entered, when manual outlining is used. However, depending on the mode of data presentation, conversion of the data and considerable computation may be needed to produce a display. We will begin with displays that use the outline information directly.

The simplest display is to present the outlines of the original features for all planes (Wong et al., 1983). Viewed in the original orientation, this is simply a superposition of the aligned images, or at least of the feature outlines ("custers") from the original images (Figure 12-11). Such a presentation is very difficult to interpret because of the overlaps that occur and the inability to distinguish what is in the various levels. Hidden line removal may be needed to clarify the spatial relationships (Figure 12-12).

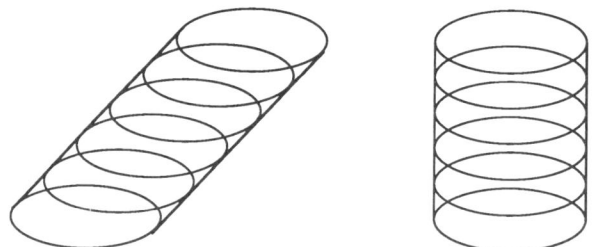

Figure 12-10: Sketch of inclined cylinder misconstrued as normal to section plane.

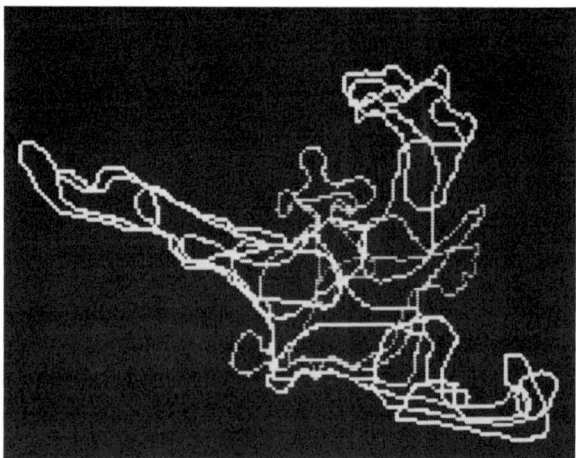

Figure 12-11: Superposition of outlines from several slices through a neuron.

One aid to viewing such a series of levels is to color-code or brightness code the outlines in each level. Sometimes a refinement of this method that is particularly effective is used, in which a different color is assigned to each object (assuming that the feature continuity between planes has been established), with a different shade, saturation or brightness for each plane varying continuously with the z-dimension to indicate relative depth. With this kind of information, the human observer can often visualize the shape, size and inter-relationships between the various objects in the structure, in spite of the fact that this kind of image is foreign to our normal viewing experience. A further refinement of this approach is to allow the user to interactively select which features are to be drawn. The omission of some of the features from the display may make it easier to see those that remain (Johnson & Capowski, 1983, 1985).

Figure 12-13 illustrates this situation with "serial sections" made by cutting one-inch steaks from a side of beef. The grey scale images can be automatically thresholded to obtain images for the various bones and muscles. Then the outlines are color coded for each bone or muscle group, and some are selected for drawing. The serial sections can also be shown as solid planes with hidden lines removed, which also assists in some cases in clarifying 3-D relationships.

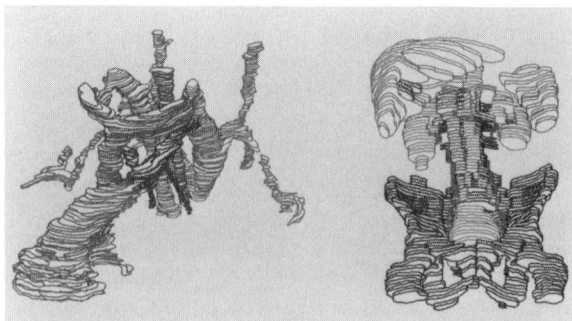

Figure 12-12: Serial section images of selected features in a mouse embryo (TEM images) and human abdomen (CAT scan images) drawn with hidden lines removed (Jandel Corp, Corte Madera, CA).

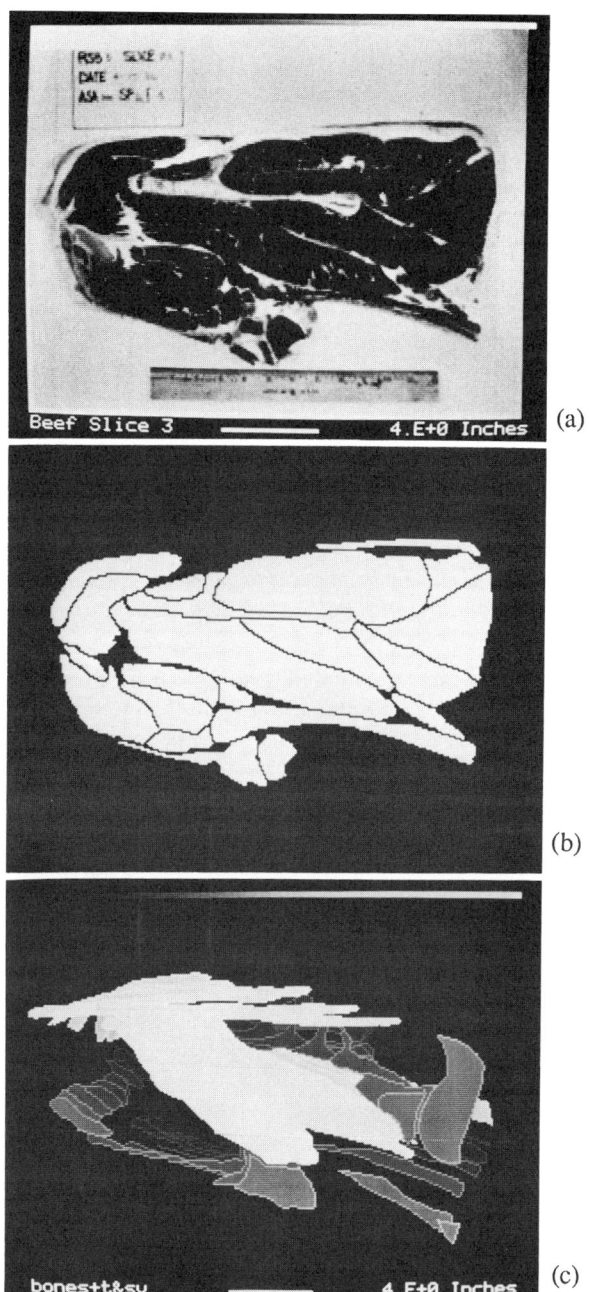

Figure 12-13: Grey scale (a) and binary image (b) of one beef slice, and serial section outlines of selected muscles and bones (c).

It is particularly effective to be able to view the stack of outlines from different directions, instead of just the vertical view along the z axis that corresponds to the original way in which the images were obtained. With the outlines of features encoded by their boundaries, it is a very fast operation to apply a matrix of sine and cosine values to adjust the three-dimensional coordinates of points to any desired set of viewing axes, so that the "wire-frame" outlines can be easily and rapidly rotated, tilted or shifted on the display. This allows the observer to "turn the objects over" in much the same way that we are accustomed to deal with physical objects to examine their shape and structure. It is not entirely clear what choice of manipulations works best for this kind of operation. Control devices such as joysticks, knobs, mice or other pointing devices can be used to vary coordinates, but which coordinates?

Sometimes yaw, tilt and roll with respect to some axes in the stack of images is appropriate, and sometimes it is more intuitive to relate these axes to the viewing screen. Sometimes it works better to specify a viewing location in terms of angles from the center of the stack, and sometimes in terms of the X, Y, Z position of the eye with respect to that center point. Sometimes some other location, such as a particular object, is better selected as the central point. It is no more computational effort to use any of these conventions, and indeed with most computer systems it would be simple to allow control inputs to be assigned to any of these variables, but too much flexibility creates a different problem for the user, namely that the process never develops the kind of intuitive spatial feel that we associate with tasks such as steering a car, or a pilot has for controlling the attitude of an airplane. Great lessons can probably be learned in this regard from the video game manufacturers, who have evaluated an extraordinarily wide variety of control input devices and axes.

When wire frame drawings are rotated or shifted in space, it is also possible to apply a perspective correction so that points farther from the eye are displaced less than ones nearer. For strictly viewing purposes, this seems to be helpful, but for some evaluations of shape it is counter-productive. Some systems allow the amount of perspective distortion to be varied from none (equivalent to viewing the structure from a very great distance through a telephoto lens) to extreme (equivalent to viewing it from a distance of less than its width, through a wide angle lens).

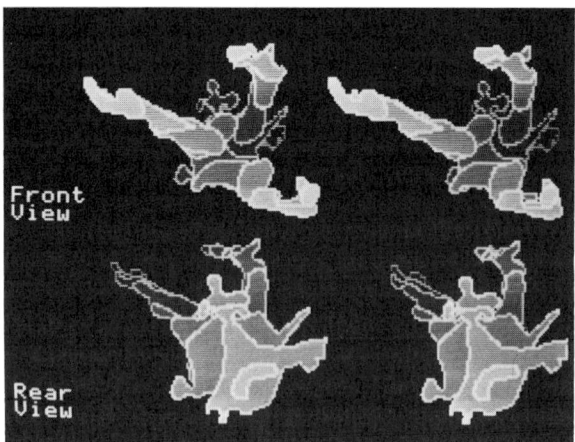

Figure 12-14: Stereo views of section images of neuron (Figure 12-11) using outlines and filled planes.

Another common use of rotated wire frame outline images is to present two such images, with a slight viewing angle between them (Briarty & Jenkins, 1984). With most display screens at a normal viewing angle, about 5-8 degrees works well. If these two images can be presented to the eyes separately, then they can be interpreted as a stereo pair and the depth information extracted by the human visual system as described in the previous chapter (Figure 12-14). This is sometimes accomplished by drawing one view in red and the other in green, overlapped on the screen, and supplying the observer with colored glasses. Other schemes include glasses that are polarized, with two display tubes and a mirror arrangement to bring the images into superposition with appropriate polarizing filters, or simply drawing the two images side by side for the observer to view directly or through a simple stereo viewer.

With all of these methods and some others, quite satisfactory stereo views of the stack can be obtained. A fairly new technique involves special glasses whose lenses can be made transparent or opaque by the computer. Each eye's image can be drawn on the display screen with the corresponding lens transparent, and then the process is repeated separately for the other eye. If done rapidly enough, there is no visual flicker.

If this capability is combined with the ability to rotate the image stack, the interactive ability to study shape and structure becomes rather good. In any case, recording the stereo pair images for publication in which the reader can employ normal stereo viewing techniques is commonly done. This often uses a high-quality plotter rather than the video display, for higher resolution and finer lines that interfere less with each other.

Another novel approach to showing the depth information to the operator's visual system utilizes a single display screen with a mirror inserted in the viewing path. As the lines are drawn on the screen corresponding to a particular z dimension, the mirror is rapidly moved to change the viewing distance from the eye to the screen. This is most easily accomplished by mounting the mirror on a audio speaker cone, whose magnet can be used to vary the position (Fuchs et al., 1982). The result of this synchronous displacement and drawing is to produce an image in the observer's mind that has the proper depth cues for interpretation. An even more specialized type of display that is useful to visualize three-dimensional arrangement of objects is a holographic display, but this requires a very considerable amount of computation to produce (Blackie et al., 1987; Hart & Drinkwater, 1987).

Producing these wire-frame drawings is easiest with boundary-representation images, because the fewest points must have their coordinates transformed to the desired viewing direction. For pixel-based representations, usually only the edge or custer pixels are transformed, although this will sometimes produce broken feature outlines as the viewing direction is changed. In fact, transforming all of the pixels in the original slice images using a transformation matrix will not necessarily produce solid features when they are rotated. If this is important, it is generally necessary to convert each image to chords, convert the end points of the chords to the transformed viewing coordinates, and then connect those points with lines to fill in the space between the end points. This is slower than just drawing the vectors that mark the outlines around features.

The advantage of drawing entire features rather than just their outlines is that in some cases internal structure (varying density or brightness) can also be displayed or that hidden lines and surfaces can be eliminated. It is possible to check all of the points in the wire frame drawing to see which appear, but in most cases it is necessary to calculate where lines will pass behind other lines in slices closer to the view and clip the lines there. Calculating clipping points takes time, so hidden line removal slows down the speed of response of the system when rotating the image stack.

If filled outlines are drawn, it is easy to accomplish hidden line removal by drawing slices from back to front. All lines and features are drawn and filled in, but as planes nearer to the observer are drawn they will overwrite information from ones farther back. Obviously, this only works on a raster display with memory (usually the computer's internal memory or a special memory buffer) where writing to a pixel erases whatever was there before. It will not work with vector displays or with a hardcopy device such as a plotter.

If feature outlines are drawn in whatever color or brightness have been selected, and the internal lines or pixels are drawn in black, it will erase hidden lines leaving a wire frame or outline drawing. Of course, since the drawing and subsequent of erasure of the rear lines takes place each time the stack of images is redrawn, this kind of display is poorly suited to interactive rotation.

If the display is drawn into a normal image buffer, more flexibility is possible. For instance, one view can be created while another is displayed, to avoid the flicker problem. Another approach is to use two image memories, one for points and lines and the other as a "z buffer". As each pixel is drawn in the display image, its z coordinate is encoded in the second image space. This makes it possible to compare the z coordinate of any point before it is drawn to anything that has been previously drawn at that location, and then to skip any pixels whose current contents would have precedence over the new information. This method is especially useful for the surface modelling display to be described below, because the surfaces may intersect. For the simplest case of parallel plane sections, no crossing occurs and the z buffer is not needed since planes can be drawn from back to front.

Surface modelling

It is most desirable in some cases to present the structure information from serial sections in a way most directly familiar to the observer, which means as solid objects with surfaces (Carazo et al., 1987). Outlines in sequential serial section planes can be used as a framework to wrap these models on, much like the construction of a balsa wood and tissue paper model airplane. In fact, the physical construction of models with styrofoam, plexiglass and other materials from serial section photographs has long been a way to present this kind of data, albeit an extremely time consuming one that is being rapidly replaced by computer modelling techniques.

Many small computers have available programs intended for CAD/CAM and similar applications which specifically do this kind of solid surface modelling (Hearn & Baker, 1986). If these are available, the coordinates of points from the boundary-representation outline images in the stack, with identification as to which boundary points in one image are to be connected to points on a feature in the next image, can be passed to these programs to generate the desired solid model. The programs function in a straightforward way that is described below to produce and shade the surfaces. The important and nontrivial first step is to decide on the connections between plane images (Baba et al., 1984).

This is done in exactly the same way as the Voronoi tesselation of surfaces described in Chapter 11, for surface reconstruction from stereo pair elevation data. The only complicating factor is that the points are nonplanar and some triangles can intersect (Russ & Hare, 1986).

As shown in Figure 12-15, points along one outline are connected to the corresponding points along the outline of the same feature in the next plane, forming triangles (the simplest plane figure). These triangles tile the surface space between the

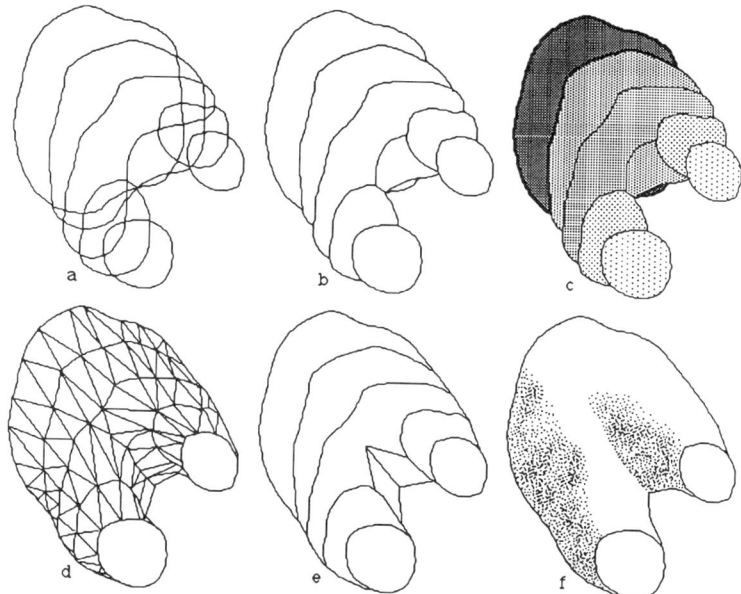

Figure 12-15: Tesselation and shading of an exterior surface: a) original serial section outlines, b) hidden lines removed, c) planes filled and shaded, d) triangles drawn between neighboring planes, e) triangles merged into a smooth surface and hidden surfaces removed, f) surface shaded according to local angle.

outlines, to produce an exterior surface for the solid (Watson & Philip, 1984). The idea of corresponding points is not a simple one. They are not necessarily the nearest points, but rather those points with the nearest value of boundary angle relative to the radius from the centroid of the object.

It is also not entirely straightforward to select the points along the periphery, since there are usually far more digitized points along each outline than are required to produce triangles (this is equivalent to saying that the resolution of the outlines is much finer than the spacing between planes). The most common automatic method is to locate points of maximum local curvature in the outline (as discussed in Chapter 7), and to use enough points to make their average spacing approximately the same as the plane spacing.

When ends of features are encountered (there is no outline in the next plane corresponding to the previous one), either the feature in the last plane in which the object appears is taken as the end of the feature, or a point midway between that plane and the next plane, and usually in line with the centroid of the final object, is used as the termination and all triangles meet there.

When branching occurs, two feature outlines (or more) in one plane must each be connected to the same feature in the preceding plane. When this is done, there are intersections of the triangles between the planes. Wherever these triangular tiles meet, the define the actual saddle surface of the branch. A z buffer is a particularly good way to deal with these intersections, without actually having to calculate the intersections between the planes. This kind of surface tiling was mentioned above in connection with automatic alignment procedures for the outlines. It is computationally very demanding (Christiansen & Sederberg, 1978).

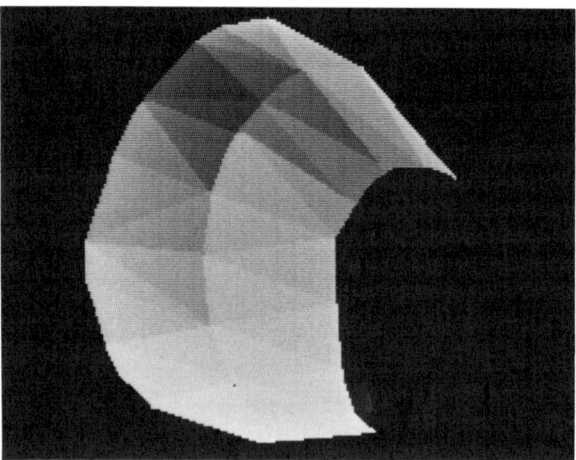

Figure 12-16: Tesselation of the surface between planes, and their shading.

Each triangular tile on the surface has an orientation specified uniquely by its three corner points, which are known. This allows the direct calculation of the brightness of the tile (Figure 12-16). The angle between the surface normal to the tile and the light source can be used to determine the brightness by a Lambertian law, or if required specular brightness reflection can be included in the display. A table look-up method with precalculated values corresponding to various orientation ranges is the fastest way to implement this for any particular light source. It is this kind of shading the converts the

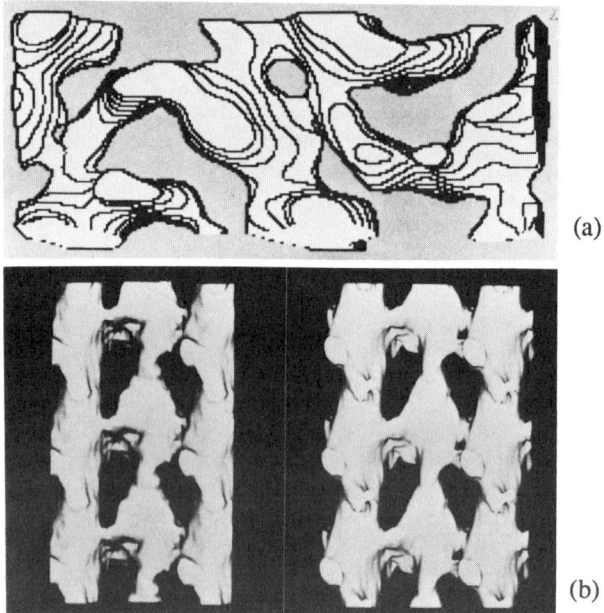

Figure 12-17: TEM reconstruction of molecules using contouring (a) and surface shading (b).

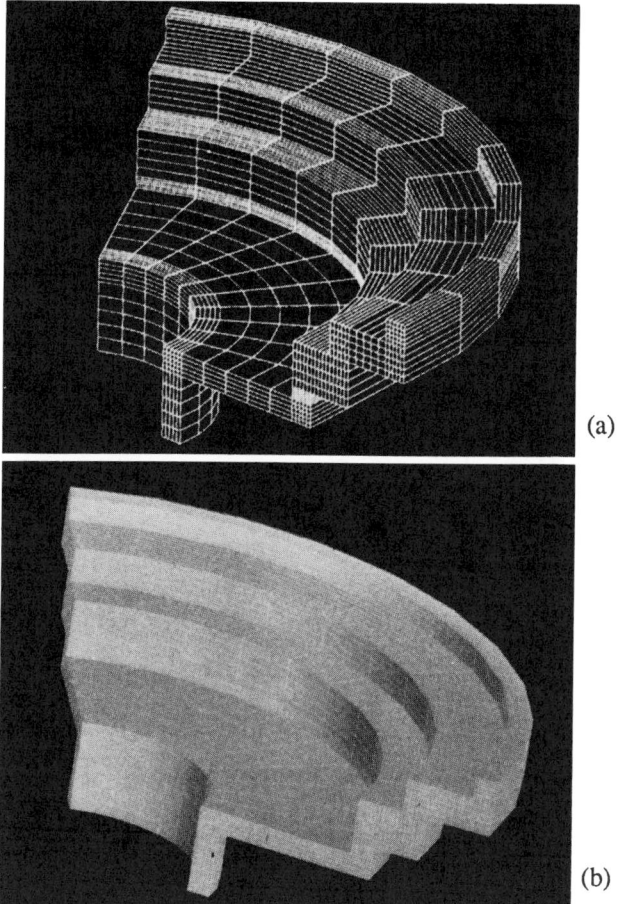

Figure 12-18: Line (a) and facet (b) modelling of a component in CAD/CAM.

serial sections, which are essentially a contour map of the feature surface, to solid objects. Figure 12-17 shows a reconstruction from TEM images using both the contour line and surface shading methods.

Displays of these tiled images are commonly used in CAD/CAM drawing (Figure 12-18). They have a somewhat faceted appearance if the number of tiles is small. Shading algorithms (Gouraud, Phong or other models) that continuously blend the brightness or angles of neighboring facets together to eliminate the facet edges can also be applied, but this kind of surface modelling or "rendering" goes somewhat beyond the intended scope of this text. As an illustration of what is possible in a small computer (Russ & Russ, 1989), Figure 12-19 shows a contour map of a surface from stereo pair synthesis, but which could as easily be the outlines of a stack of serial section images. Interpreting the contour line plot, or its shaded version, is much more difficult that viewing the rendered surface image. Changing the characteristics of the illumination (including its orientation with respect to the surface and viewer) and the surface reflectivity can be done in nearly real time, to assist in viewing.

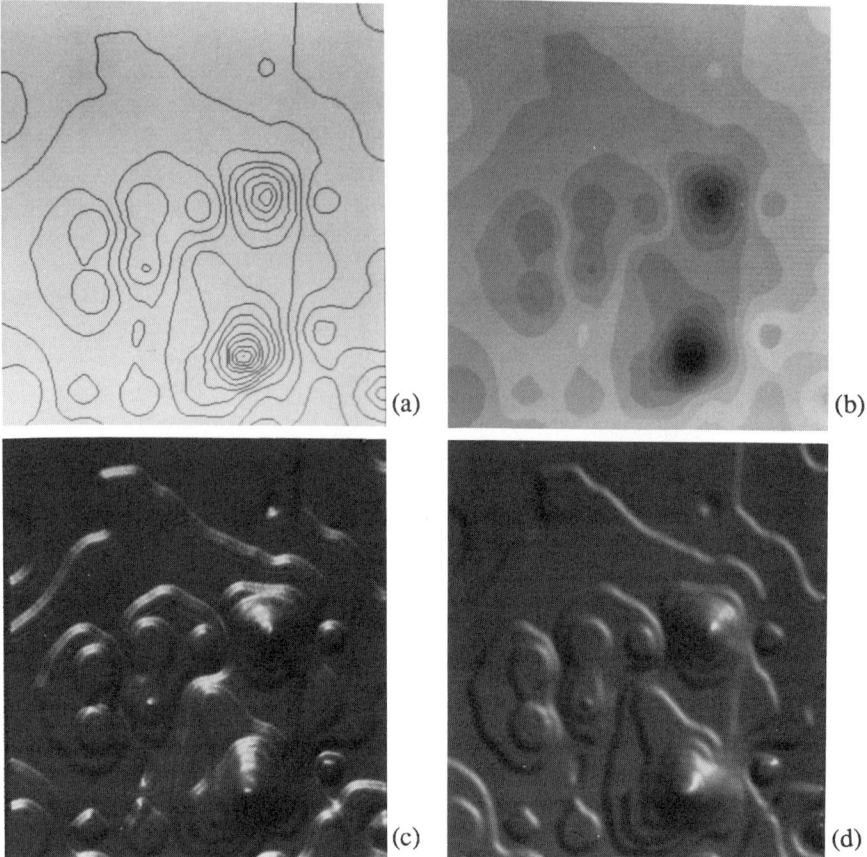

Figure 12-19: Surface rendering of a contour map: a) the original contour map showing lines with 1 μm elevation steps as plotted from stereo pair measurements; b) the same map as (a) with brightness shading to denote elevation; c and d) the same surface rendered with Phong shading using a light source in different positions, and surfaces of different reflectivity.

Computer image modelling is an extensive field in its own right, with its own set of tradeoffs between speed, number of points, horsepower, and display quality. Ray tracing methods can show reflections and transparency that are very difficult to model in serial section displays. Transparency of slices is sometimes used in a gel display, as will be illustrated in the next chapter in connection with tomographic images. When this method is used with serial sections, drawing is done from back to front and the full grey-scale information from each original image is superimposed on images of slices behind it. But the brightness of the current slice is multiplied by a transparency factor (which can vary for different objects) and added to the cumulative brightness of rear slices so that partial images of the rear structures is preserved. These displays are very time consuming to prepare.

Still, even with manageable amounts of computing it is possible to show realistic models of the surfaces of selected features, with color coding and rotation, for even very

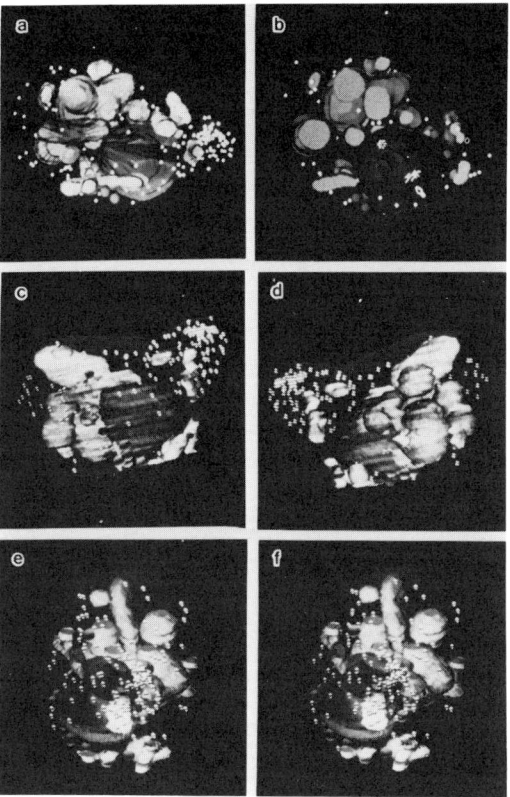

Figure 12-20: Display of selected internal features in a cell showing interpolated and shaded surfaces, viewed in different directions.

complex structures such as the cell in Figure 12-20. And even though the amount of computation is considerable, it is generally much superior in speed, manpower investment, and quality of results to older methods that constructed physical models in wax or plexiglas from serial sections (Figure 12-21).

The interpolation of the surface of the object by this method provides one way to measure the volume of the object(s) in the stack of images. It also permits other measurements such as shape factors which are analogous to those discussed for two-dimensional features in Chapter 7. As for two-dimensional images, many parameters are easier to determine from a pixel (or in this case voxel) based representation as compared to a boundary representation image.

Measurements on surface-modelled objects

Before going on the the voxel-based images, however, it is worth noting that two parameters of great interest, the volume and surface area, can be estimated rather easily from the stack of boundary-representation images. The volume is the integral of area over the z-distance, which can be approximated without bias as the sum of the areas of features in each binary times the spacing between them. This calls for a numerical integration of the areas of the features in each slice, and integration methods such as Simpson's rule are straightforward to apply.

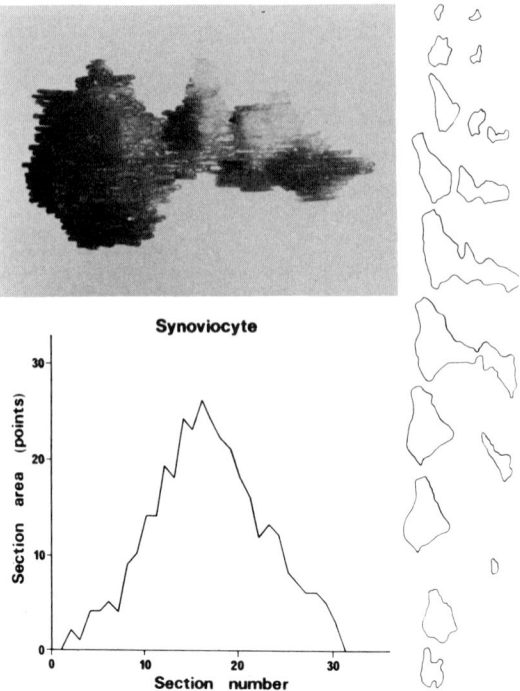

Figure 12-21: A wax serial section model with a plot of the area of each section.

Likewise, the surface area is the integral of the perimeter over the z-direction. If the actual two-dimensional feature perimeters are used, the sum of their product times the plane spacing will underestimate the actual surface area, because it is equivalent to stretching surfaces in straight lines between the plane outlines. This produces a generalized cylinder, whereas most real objects have some irregularity in their surfaces. If it is possible to assume that the object has the same degree of surface roughness in the z-direction as in the lateral directions, then an estimate can be obtained by using the formfactor (defined in Chapter 7) to characterize the irregularity of the outlines, and to define the surface area of the object as that of the cylinder divided by the formfactor. This is a first order correction, just as it is for the surface area of objects discussed in Chapter 7, but it is useful in many cases.

The surface area can also be estimated stereologically by counting the intersections of the scan lines in all planes with each object. This gives the number of intersections per unit length of line, from which the surface area per unit volume of the object is directly calculated as

$$S/V = 4/L_3$$

where L_3 is the mean intercept length of the lines in the feature.

Since the volume is easily obtained, this gives a value for the surface area which is statistically valid provided the object has no preferred orientation with respect to the planes of intersection, which of course is an implicit assumption in all of these calculations. The number of intersections is simply the number of end points in the pixel-based representation of features in the plane images, and can be directly counted from a chord-encoded image (see Chapter 5).

It is not so obvious how to measure the area and volume of features that do not end within the region covered by the planes in the serial section stack. Of course, no absolute value can be assigned to a feature whose total extent is not known. However, a specific surface per unit volume is still meaningful, and can be calculated just a described above.

Since volume and surface area are dimensioned parameters, they can be combined to produce a dimensionless ratio (for instance (volume)$^{1/3}$ / (surface area)$^{1/2}$) which could function as a shape descriptor. However, most of the shape factors and other related parameters are easier to discuss in terms of the voxel based representation of the image.

The drawback of solid modelling of bounding surfaces, or even of displaying individual section outlines with hidden line removal, is that it may hide or obscure important internal detail. Since the display with all of the lines shown is often too confusing to the human viewer to interpret this internal detail, a different approach is often used. This is to insert section planes or cuts through the three-dimensional data to show the intersection of the planar outlines and features with the plane of cut (Figure 12-22). The appearance is much like that produced by physically cutting into the structure,

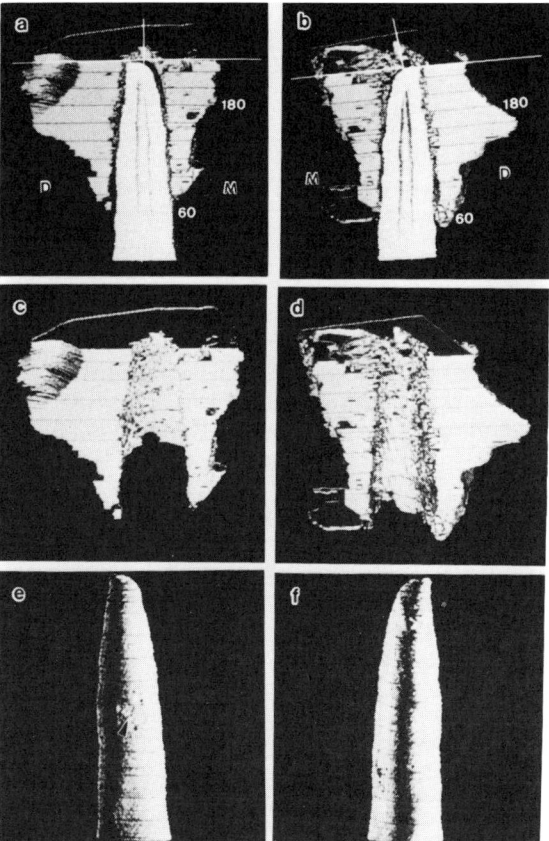

Figure 12-22: Example of cut through stack interpolated by the computer.

and if the plane can be interactively positioned it allows the user to dissect the structure in a powerful way (Baba et al., 1989).

Producing the cut section is generally accomplished for a boundary-coded or outline based image by first operating on each of the individual planes. The intersection of the cutting plane with the image plane produces a line that is added to the image. This requires some interpolation to join it to the original outlines, and some computation to decide where it passes inside features (and should be drawn) and where it lies outside them (and should not be drawn). Then it may also be desirable to remove all portions of the features lying on one side of the line, so that the cut plane will be visible in the three-dimensional reconstruction. In some cases, the display software can simply bypass those lines instead of having to remove them from the data base, but both operations are equivalent in terms of results.

Once this has been done for every plane, the 3-D image display is reconstructed and the result is a view into the internal structural arrangement of the features. From this description it should be apparent that this is not a real-time operation in which the user can freely manipulate the cutting plane, at least not with realistic size data arrays and less than supercomputer facilities. Voxel-based displays make it much easier to interact with the three-dimensional structural data in this way, and are usually preferred when this is important.

Voxel displays

If the serial section image planes are uniformly spaced, and furthermore if the spacing is equal to the lateral pixel resolution within each plane, then combining the images into a voxel ("volume element") representation is particularly natural. When the planar spacing is greater than the latter resolution, either the voxels must be elongated into prisms, or intervening cubic voxels must be interpolated. The later method is more satisfactory since the shape, ends and branches of features can be more continuously modelled. However, it requires significantly more memory, and makes many of the same fundamental assumptions about shape and connectivity.

Voxel-based representation of features cannot be viewed directly unless those voxels which are not part of features are assumed to be transparent. Then a view in any direction can be obtained by projecting the voxels into a plane normal to that viewing direction, for display. In principle such an image could be shaded according to the alignment of surface elements, but because of the number of pixels involved this would be prohibitively time consuming. Projected image views are not usually generated from voxel representation.

Instead, section views are most commonly used. By passing any cutting surface (usually a plane) through the voxel array, an image of the features as they would appear on that plane can be produced by simple interpolation (Figure 12-23). Usually this interpolation selects the voxel that is closest to each pixel's coordinates on the arbitrarily oriented plane, but if desired the same kind of bilinear interpolation used for rubber-sheeting, but extended to three dimensions, can be used.

The user may also select any arbitrary path or plane through the structure to measure density or create an image. This is most conveniently done by using the wire-frame stereo display mode with a pointing device to interactively select points to define the line or plane.

Another method of viewing data is available that produces displays that most users find very readily interpretable. These are sometimes called transparency or gel displays. Each voxel normally has associated with it a value of some measured property, for

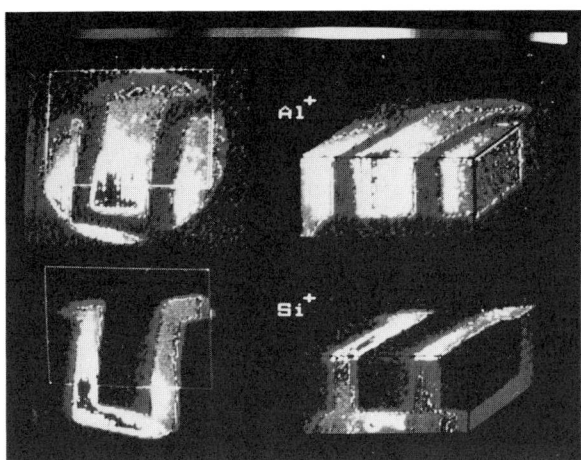

Figure 12-23: SIMS voxel 3-D image showing transverse sections (Bryan 1985).

instance density, color, composition, etc. These are converted to a display transparency coefficient that describes how much of the information coming from voxels behind it in the direction of view will be transmitted.

When the display is then generated, starting from the farthest voxels from the point of view, each voxel's brightness or color is attenuated by the transparency of those in front of it, and mixed with their contribution to brightness or color. The result is an ability to see through some structures to view others, while still retaining a sense of the organization of the various features in three dimensions. Coupled with stereo viewing, this mode of presentation produces extremely life-like impressions of structure. The amount of computation required is within the range of the dedicated computers used for image analysis systems.

This approach has proven especially attractive when used with the confocal light microscope, which has the important advantage of producing a series of focal sections that are in alignment and registration, have a depth of field that is known from the optics of the microscope, and a z position known from the microscope stage (which may be controlled directly by the same computer that acquires and later processes the images). A common and useful equation for depth of field is

$$d = \frac{\lambda \cdot \sqrt{n^2 - (\text{NA})^2}}{(\text{NA})^2}$$

where λ = wavelength (with a typical low-power Ar or He-Ne laser this is between about 360 nm and 682 nm or 0.36 - 0.68 μm), n is the refractive index of the medium (air, water or oil), and NA is the numerical aperture of the lens. For a dry 0.95 NA lens, this indicates a depth of field of 0.24 μm, and with oil immersion lenses having NA values of 1.4 or higher, even smaller depths of field are possible, close to the lateral resolution of conventional light microscope images.

True transmission confocal light microscopy requires a monochromatic light source (a laser) to avoid chromatic aberrations due to the sample itself. The light passes twice through the same points in the sample, with an objective lens and curved mirror below the sample (Figure 12-24), to form the image of each focal section (Figure 12-25). A series of such images can then be combined by using the measured density of each plane

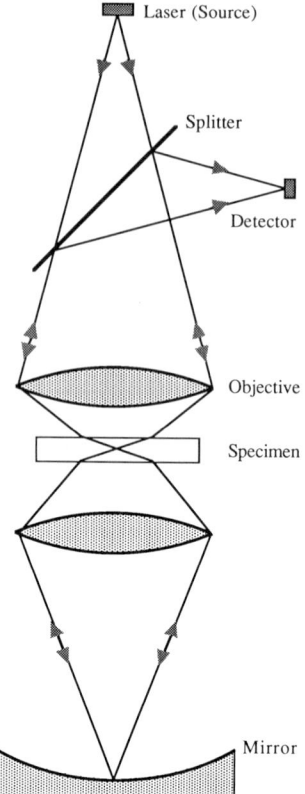

Figure 12-24: Schematic diagram of transmission CSLM optics.

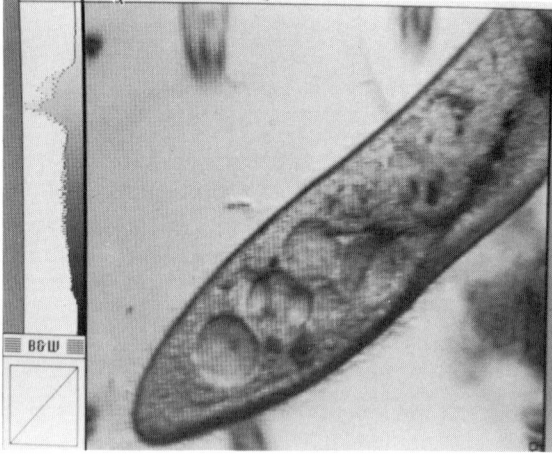

Figure 12-25: One frame from a real-time videotape of swimming paramecium showing internal structure revealed in a single focal section viewed in the transmission CSLM.

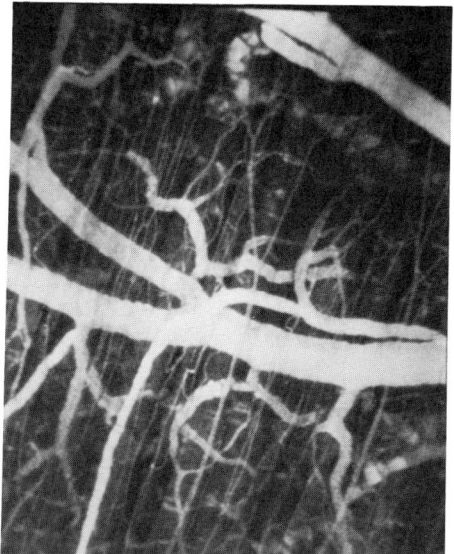 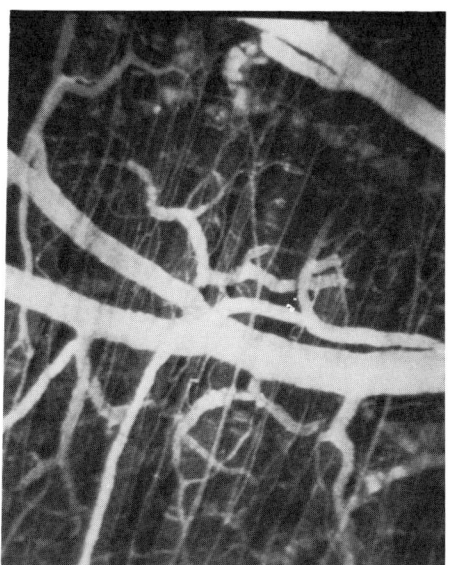

Figure 12-26: Stereo pair views of volumetric reconstruction of canulated hamster dorsal skin flap (specimen courtesy of Dr. K. R. Diller, Dept. of Biomedical Engineering, Univ. of Texas, Austin).

to decrease the brightness of underlying planes, as described. By shifting each plane laterally, the reconstruction can produce views from different directions. These can be used to show the 3-D structure by rotating the entire voxel array, or by creating stereo pair images as shown in Figure 12-26.

Transparency or gel displays are possible with boundary-coded data representations (Nakamae et al., 1985), but require more computation. The bounding surfaces are generated as described above and then given a color and transparency. The projected image is again built from rear to front, using a method much like ray tracing except that no scattering of light need be considered. Interior volumes within the boundaries are not coded, so that only the boundaries have any opacity. The result is an ability to see the bounding surfaces in a true spatial representation.

This approach is less used than voxel transparency displays, and in fact when the bounding surfaces are of interest (rather than the entire features) it is often easier to produce a separate voxel image in which only the boundary pixels are present, and then use this for creating interactive displays. The procedure is a direct three-dimensional analog of obtaining a custer in a two-dimensional pixel-based image. Voxels with all of their neighbors (now considered in the X, Y and Z directions) set are removed, leaving just the outer shell. In fact, there are similar three-dimensional analogs for all of the morphometric operations of erosion, dilation, skeletonization, and so on, that work on voxel images just as were discussed before for pixel images.

Measurements on voxel images

Measurements with voxel images are in most respects directly analogous to those for two-dimensional pixel images, except that they are proportionately more trouble to perform. The volume is simply the sum of points. The length (longest chord) is the distance between the two farthest boundary points, and this can be determined by rotation and saving extreme values, just as in two dimensions.

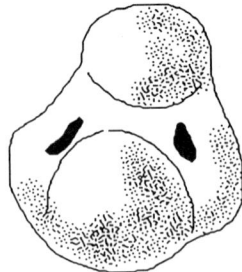

Figure 12-27: The genus of a three-dimensional object is the number of cuts that can be made in the objects skeleton without disconnecting its parts.

Shape, always a complex quantity to deal with, is even more difficult here. Ratios analogous to formfactor, roundness, convexity, etc. can be calculated but there has been very little use of them and so there are few established interpretations or ways to incorporate them in other models (as was done for the two-dimensional case). In three dimensions it is possible to obtain a skeleton using very similar erosion rules to those employed in 2-D, by removing voxels according to their neighbor patterns until only a connected backbone remains (Lobregt et al., 1980; Halford & Preston, 1984). By counting the number of nodes (points with more than 2 neighbors) and ends (points with exactly one neighbor), the genus of the object can be determined (Figures 12-27, 12-28, and 12-29). This describes the topology of the object exactly, which cannot be determined from two-dimensional images.

Two important and related parameters that describe the topology of three-dimensional structures are the number of disconnected parts or objects and their connectivity. The connectivity is the number of redundant connections in the skeleton, or the number of cuts that can be made in the skeleton without increasing the number of parts. An extended pore network may have only a single object with very high connectivity, whereas a distribution of separate particles without holes would have zero connectivity and a large number of objects per unit volume. The number of objects per unit volume was discussed in Chapter 8, and is usually written as N_V. The connectivity per unit volume is C_V, and both have units of length^{-3}.

More detailed study of the topology of a network or skeleton can also be performed. Once it has been formed, the skeleton consists of voxels which have only a few

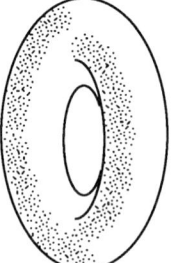

Figure 12-28: The topology of objects is concerned only with their connectivity or genus, not the details of shape. These two shapes are topologically equivalent (genus=1).

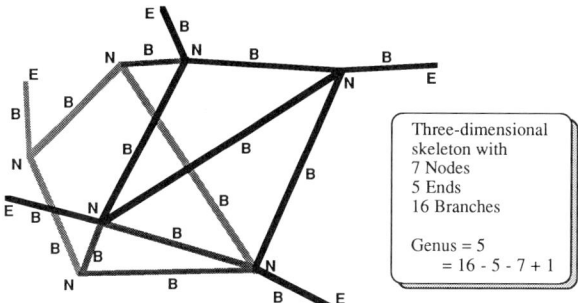

Figure 12-29: The Euler number and its relationship to connectivity and genus.

neighbors. Most have only two. Any voxel with fewer than two neighbors that are part of the feature must be the end of a branch. Ends are important because they correspond to the "points" present in the original two- or three-dimensional feature, and counting them is a first order operation to characterize topological shape which corresponds strongly with human recognition. Likewise, any voxel with more than two neighbors which are part of the feature must be a node, or point where several branches or "links" join. Most of the nodes in typical structures have three links, but it is possible to have as many as 4 in two dimensions, or 6 in three dimensions. We define the degree of the node as the number of links which join there, and measure it by the number of touching neighbor voxels.

Notice that it is not possible to have nodes of degree two, where two links join, in this kind of representation. For some purposes, it would be nice to recognize such locations. For instance, the difference between the letters I and L is the sharp bend in the L which can be considered as a node joining the two straight links. However, determining corners in skeletons or other lines requires defining the change in direction or curvature in a way that can be satisfactorily measured, and which exceeds some arbitrary limit. The measurement is not a local operation, as are all of the other processes used here. Corner-finding methods for chain-coded pixel or voxel lines exist, but require moving along the line with averaging over many points to locate the bend. Strictly speaking, nodes of degree 2 do not exist in a topological sense (the I can be made into an L, or vice versa, without any cutting or joining), and they will be ignored here as well.

We will consider features to be topologically distinct if they contain nodes of different degree, however, even though classic topology would consider them to have the same genus and thus to be equivalent. Consider the shapes shown in Figure 12-30. These are two-dimensional pixel skeletons of the letters O, P, R, and the male and female signs. All are of genus 1 (we can make a single cut without changing the number of separate parts). But as graphs they are all distinct. Indeed, many of the subjects to be discussed in the remainder of this paper are closely related to graph theory and to network analysis as used in operations research (Hillier & Lieberman, 1986). The terminology is not identical; graph theory uses "vertex" for what is here called a node, and "pendant vertex"

Figure 12-30: 2-D skeletons discussed in the text.

for an end point. Links are also known as branches or edges. However, many of the tools developed there have unexpected and powerful uses here. In the terminology of graph theory, the skeleton is a connected, nondirected, simple graph.

The letter O has no vertices and a single link, while the P and R have 1 and 2 nodes of degree two, 2 and 3 links, and 1 and 2 ends, respectively. The female and male symbols differ in their number of ends (3 vs. 2) and links (3 vs. 4). The R and the male symbol both have 2 nodes of degree 3, and both have 2 ends, but the male symbol has one less edge. In other words, based on simple counts of these topological or graph components, these features are all distinct, as indeed a human observer easily recognizes them to be.

Although we consider this skeleton information to be primarily a topological description of shape, and hence only suitable for counting, it is also possible to perform some measurements. The X, Y, Z coordinates of each node and end point can be determined, as can the length of each link. This length can on the one hand be characterized as the Pythagorean distance between the end points. It can also be obtained by summing along the skeleton itself. In classic chain-code for lines composed of 8-connected two-dimensional or 26-connected three-dimensional pixels or voxels, the length would be calculated as shown below, to account for the different distances between each pixel and its touching neighbors

$$\text{Length} = 1 \cdot (\text{number of orthogonal neighbors}) + \sqrt{2} \cdot (\text{number of face diagonal neighbors}) + \sqrt{3} \cdot (\text{number of body diagonal neighbors})$$

However, this simple method does not calculate accurate or unbiased answers for straight lines drawn in 2- or 3-D voxel space using the Bresenham algorithm (which produces the "best" 8- or 26-connected line of voxels closest to the mathematical line. Smeulders (1989) has shown that better results can be obtained with the equations below, which give only a mean error of 2.5 and 3.5% for 2- and 3-D lines, respectively.

$$\text{Length (2D)} = 0.948 \cdot (\text{number of orthogonal neighbors}) + 1.340 \cdot (\text{number of face diagonal neighbors})$$

$$\text{Length (3D)} = 0.877 \cdot (\text{number of orthogonal neighbors}) + 1.342 \cdot (\text{number of face diagonal neighbors}) + 1.647 \cdot (\text{number of body diagonal neighbors})$$

Each link in the skeleton can be characterized by its actual length, the distance between the end points, and the orientation of the line defined by those end points. These values can be used far various statistical descriptions and comparisons as discussed below. If the line was obtained from the medial axis transform and Euclidean distance map (discussed in Chapter 6), then the values of brightness for each voxel in the link also gives the cross-sectional area (or, in 2-D, the width) of the feature. This can be summed along the link in various ways. For permeability studies, the minimum value is often useful as an estimate of the pore throat diameter.

Network analysis

In some cases, the nodes, ends and links can be enumerated and the data about them recorded for statistical and stereological analysis. Different formats for recording the data on the ends, nodes and links offer different advantages in performing subsequent interpretation. An adjacency list is a very compact way to order the information for each node or end. It consists of a list of points, and for each its space coordinates, and a list of

Figure 12-31: Stereo views of a 3-D skeleton, with 18 links, 10 nodes, 6 ends, and genus=3.

the other nodes or ends which are connected to it directly by links. Normally this list contains only the opposite ends of those links which touch the node itself, but it is also possible to construct a more complex table in which the distance to every other node in the structure is listed. This is usually done in terms of the number of links that would have to be traversed to go from one node to the other along a shortest path. This derived information is not immediately evident from the raw data, but can be obtained from the transitivity matrix.

If a complete square matrix is created with its order equal to the number of nodes plus ends enumerated for the skeleton, it can be filled with ones for those combinations of nodes which are connected by a single link, and zeroes elsewhere. This is called a transitivity matrix. In most cases it is very sparse, and well as being symmetrical, and so it may seem to be a poor format for the data. However, the transitivity matrix can be used to easily obtain a great deal of detailed information. Normally, it is first partitioned so that sub-matrices are created for each connected skeleton, which greatly reduces the size of the data structure.

Figure 12-31 shows a skeleton of a 3-D object (a conglomerate of soot particles imaged in a confocal scanning laser microscope). The two views constitute a stereo pair that reveal the full three-dimensional arrangement of the voxels. This skeleton consists of 18 links, 6 ends, and 10 nodes, and has a genus of 3. The organization of these elements is evident in the drawing in Figure 12-32, in which the nodes and ends are numbered. A transitivity matrix for this skeleton is shown in Table 12-1. This is slightly different from the transitivity matrix used in graph theory in that the diagonal contains zeroes. It also differs slightly from the Markov chain used for analysis in operations research in which

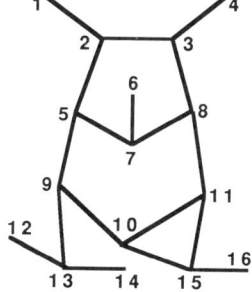

Figure 12-32: The skeleton from Figure 12-31 drawn on a plane, with its nodes numbered.

Table 12-1. The transitivity matrix for the skeleton in Figure 12-31

	1	2	3	4	5	6	7	8	9	10	11	12	13	14	15	16
1	.	1	0	0	0	0	0	0	0	0	0	0	0	0	0	0
2	1	.	1	0	1	0	0	0	0	0	0	0	0	0	0	0
3	0	1	.	1	0	0	0	1	0	0	0	0	0	0	0	0
4	0	0	1	.	0	0	0	0	0	0	0	0	0	0	0	0
5	0	1	0	0	.	0	1	0	1	0	0	0	0	0	0	0
6	0	0	0	0	0	.	1	0	0	0	0	0	0	0	0	0
7	0	0	0	0	1	1	.	1	0	0	0	0	0	0	0	0
8	0	0	1	0	0	0	1	.	0	0	1	0	0	0	0	0
9	0	0	0	0	1	0	0	0	.	1	0	0	1	0	0	0
10	0	0	0	0	0	0	0	0	1	.	1	0	0	0	1	0
11	0	0	0	0	0	0	0	1	0	1	.	0	0	0	1	0
12	0	0	0	0	0	0	0	0	0	0	0	.	1	0	0	0
13	0	0	0	0	0	0	0	0	1	0	0	1	.	1	0	0
14	0	0	0	0	0	0	0	0	0	0	0	0	1	.	0	0
15	0	0	0	0	0	0	0	0	0	1	1	0	0	0	.	1
16	0	0	0	0	0	0	0	0	0	0	0	0	0	0	1	.

fractional values in each row sum to 1. The 1's specifically indicate the pairs of nodes (or ends) that are at the opposite ends of individual links.

When this matrix is multiplied by itself (Table 12-2), the product consists of values for pairs of nodes that are connected by exactly two links. Zeroes are present for all pairs of nodes that are not connected by two links, and numbers greater than 1 indicate multiple paths. In particular, the diagonal of the product matrix gives the degree of each node. Again multiplying the product matrix by the original produces a third matrix in which the numbers indicate paths between pairs of nodes that are exactly three links long (Table 12-3). The appearance of a 2 on the diagonal indicates a loop (which can be traversed in two directions, 10-11-15-10 or 10-15-11-10). If a multiple of 2 appears, it indicates that more than one loop has been found passing through the node with the current length. This procedure can be repeated to find longer paths. However, it is difficult to find information on larger loops than the minimum sizes present.

As defined, the product matrix includes all possible paths including ones that are redundant. With respect to the drawing in Figure 12-32, there is a 3-link path from node 1 to node 2 that goes 1-2-3-2, and this is shown in the degree-3 matrix (along with 1-2-5-2 and 1-2-1-2). One way to eliminate this duplication is to change the matrix multiplication rules to skip indices that have already been used. This prevents counting

Table 12-2. The product of the transitivity matrix in Table 12-1 by itself

	1	2	3	4	5	6	7	8	9	10	11	12	13	14	15	16
1	1	0	1	0	1	0	0	0	0	0	0	0	0	0	0	0
2	0	3	0	1	0	0	1	1	1	0	0	0	0	0	0	0
3	1	0	3	0	1	0	1	0	0	1	0	0	0	0	0	0
4	0	1	0	1	0	0	0	1	0	0	0	0	0	0	0	0
5	1	0	1	0	3	1	0	1	0	1	0	0	1	0	0	0
6	0	0	0	0	1	1	0	1	0	0	0	0	0	0	0	0
7	0	1	1	0	0	0	3	0	1	0	1	0	0	0	0	0
8	0	1	0	1	1	1	0	3	0	1	0	0	0	0	1	0
9	0	1	0	0	0	0	1	0	3	0	1	1	0	1	1	0
10	0	0	0	0	1	0	0	1	0	3	1	0	1	0	1	1
11	0	0	1	0	0	0	1	0	1	1	3	0	0	0	1	1
12	0	0	0	0	0	0	0	0	1	0	0	1	0	1	0	0
13	0	0	0	0	1	0	0	0	1	0	0	0	3	0	0	0
14	0	0	0	0	0	0	0	0	1	0	0	1	0	1	0	0
15	0	0	0	0	0	0	0	1	1	1	1	0	0	0	3	0
16	0	0	0	0	0	0	0	0	0	1	1	0	0	0	0	1

Table 12-3. The degree-three matrix showing paths which are three links long

	1	2	3	4	5	6	7	8	9	10	11	12	13	14	15	16
1	0	3	0	1	0	0	1	1	1	0	0	0	0	0	0	0
2	3	0	5	0	5	1	1	1	0	1	1	0	1	0	0	0
3	0	5	0	3	1	1	1	5	1	1	0	0	0	0	1	0
4	1	0	3	0	1	0	1	0	0	0	1	0	0	0	0	0
5	0	5	1	1	0	0	5	1	5	0	2	1	0	1	1	0
6	0	1	1	0	0	0	3	0	1	0	1	0	0	0	0	0
7	1	1	1	1	5	3	0	5	0	2	0	0	1	0	1	0
8	1	1	5	0	1	0	5	0	2	1	5	0	0	0	1	1
9	1	0	1	0	5	1	0	2	0	5	1	0	5	0	1	1
10	0	1	1	0	0	0	2	1	5	2	5	1	0	1	5	1
11	0	1	0	1	2	1	0	5	1	5	2	0	1	0	5	1
12	0	0	0	0	1	0	0	0	0	1	0	0	3	0	0	0
13	0	1	0	0	0	0	1	0	5	0	1	3	0	3	1	0
14	0	0	0	0	1	0	0	0	0	1	0	0	3	0	0	0
15	0	0	1	0	1	0	1	1	1	5	5	0	1	0	2	3
16	0	0	0	0	0	0	0	1	1	1	1	0	0	0	3	0

paths that go through any node more than once, but it means that the full matrix multiplication must be repeated for each node pair for each number of links, instead of being able to use previous product matrices.

In order to determine the shortest path from one node to another, it is sufficient to keep a separate table of the matrix degree for which a nonzero value first appears in a product matrix. This information is shown in the node/neighbor matrix in Table 12-4, for the same figure. Notice that the diagonal reveals the presence of loops in the structure which connect a node to itself. One loop consists of three links, one of five, and one of six. If there were multiple loops of a given size, they would appear in the product matrices but would not be represented in the path matrix. Each of these different representations of the data are thus useful for extracting different information from the structure. Naturally, blocks of 0's appear in partitioned networks when no path exists from one node or endpoint to another.

The transitivity matrix can also be used for the case in which links are not equivalent, for instance either due to different lengths summed along the skeleton pixels or minimum cross-sectional areas as revealed by the medial axis transform. Instead of ones in the original matrix, fractional values representing the relative "cost" of each connection can be used. These might represent the resistance of the link to the flow of fluid in a pore

Table 12-4. The node/neighbor matrix for the skeleton in Figure 12-31

	1	2	3	4	5	6	7	8	9	10	11	12	13	14	15	16
1	0	1	2	3	2	4	3	3	3	4	4	5	4	5	5	6
2		5	1	2	1	3	2	2	2	3	3	4	3	4	4	5
3			5	1	2	3	2	1	3	3	2	5	4	5	3	4
4				0	3	4	3	2	4	4	3	6	5	6	4	5
5					5	2	1	2	1	2	3	3	2	3	3	4
6						0	1	2	3	4	3	5	4	5	4	5
7							5	1	2	3	2	4	3	4	3	4
8								5	3	2	1	5	4	5	2	3
9									6	1	2	2	1	2	2	3
10										3	1	3	2	3	1	2
11											3	4	3	4	1	2
12												0	1	2	4	5
13													0	1	3	4
14														0	4	5
15															3	1
16																0

Figure 12-33: Two interpenetrating networks obtained by simplifying the skeletons of sintered particles (thin lines) and the surrounding pore network (heavy lines). In general the two networks are different, with a different average coordination number (degree for the node or vertex) for each pore than for each particle, different sizes for loops, and so forth.

structure, electrical resistance in a network, or some other property. This is the classic use of Markov chains in which the matrix values are the individual link flow capacities. Then the sum of successive product matrices gives the integrated resistance between any pair of nodes, summed over all possible paths. Statistical analysis of this data as a function of the physical distance between pairs of pores would be related to overall network performance, permeability of porous materials, etc. The same type of analysis may be useful in studying blood flow in capillary networks, which have been imaged by CSLM in living tissue.

There are many questions that can be asked about 2- and 3-D networks, but it is not always clear how these can be phrased so that numerical answers can be given. The importance of connectivity is well known in many fields. For instance, in materials science, the pore network present during the sintering of ceramics evolves by a reduction in connectivity with time, as links are pinched off. In this type of structure, there are actually two networks present. Figure 12-33 shows a simplified skeleton view of two such interpenetrating networks. The pore and particle networks are dual and interpenetrating, but generally have some quite different properties such as the mean number of branches (the degree) at each node and different link lengths. If the material was pressed before sintering, the orientation of the links in one or both networks may not be isotropic. It is also interesting to determine a frequency distribution for minimal loops in the network as a function of the number of links, and the mean number of links connecting nodes as a function of the metric distance between them. Spatial density (number per unit volume) of specific features, such as nodes of specified degree, can also be obtained.

In geology, the permeability of rock and soil controls the extraction of oil and the flow of ground water. Simple models suggest that this is proportional to the pore throat diameter raised to a power times the number of paths from each pore. We have just seen how this can be estimated from connection tables. These parameters are essentially global, relying on statistical sampling of a portion of a network that extends far beyond the limits of the image space.

Measurement of individual networks that fit completely within the image space can also be performed. Examples of this kind of object-specific measurement include dendritic inclusions in metals, agglomerates of particles, and neurons made visible in brain tissue by staining methods. For instance, it is well known that the number of synapses between neurons changes dramatically in the cortex of young mammalian brains. It is not clear whether statistical information about a reasonable number of such networks, for instance the mean value and standard deviation of link length and orientation, genus, number of nodes, etc. is sufficient to categorize these structures or whether a complete linked graph is needed to reveal important information.

There are some situations in which two-dimensional networks also arise in microscopy. One is the analysis of networks formed by particles in electrorheological fluids. Figure 12-34 shows a light image of a test cell containing small particles whose polar charge causes them to align between two plates when a voltage is applied. The ability of this structure to withstand shear stresses makes fluids suitable as controlled mechanical coupling devices. In the test cell, the thickness of the networks is essentially limited to a single layer of particles. However, as the field strength is increased more particles join the network and the degree of cross-linking between chains increases. This increases the shear stress that can be transmitted, just as a truss structure stiffens bridge construction. By considering the number of paths per unit length of the cell between nodes representing the two charged plates, as a function of the field strength or other variables, network analysis facilitates quantifying this research.

It seems likely that as more full 3-D images are obtained and subjected to analysis, that the tools of graph theory will be increasingly used to determine topological properties of skeletons derived from the images, and these results will shed new light on structural relationships that are hidden in conventional two-dimensional imaging.

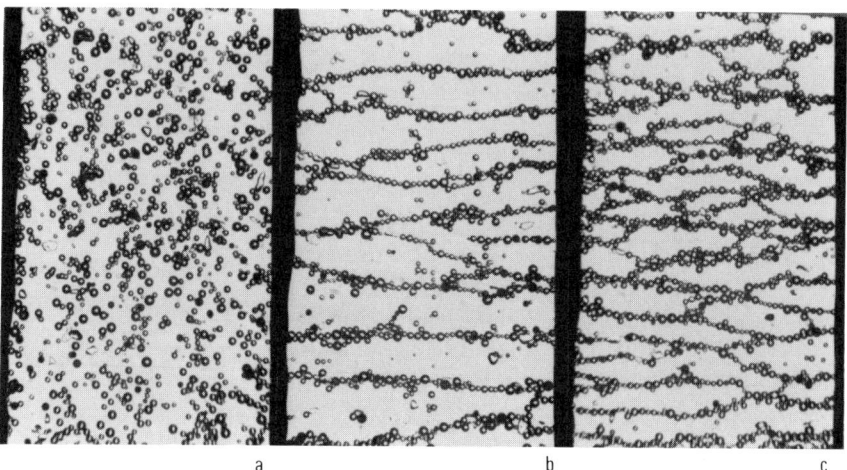

a b c

Figure 12-34: Images of glass spheres (mean diameter 27 µm) in water with no applied field (a) and with two levels of increasing applied DC voltage on the plates at sides of the test cell (b and c). (Courtesy Dr. H. Conrad and Dr. F. Sprecher, Materials Science and Engineering Dept., North Carolina State University, Raleigh NC.)

Connectivity

An important stereological image measurement that is performed with serial section images allows the unbiased estimation of the number N_V objects in a sample (Sterio, 1984). As was explained in Chapter 8, normally the N_V parameter can only be estimated from the information obtained on a single image by making some shape assumptions. The use of a "disector" (Figure 8-9) allows this to be determined directly. By using only two plane section images whose spacing is known, it is possible to count the beginnings of features or interest which occur between the two planes. This number divided by the area represented by the image and the spacing (whose product is the volume between the planes) gives N_V exactly.

It is important in this procedure to use a plane spacing that is sufficiently small that there are no ambiguities between the planes (Davy, 1981). Branches, joins, beginnings and endings must be distinct so that they are not confused. There is also the problem that features that wander out through the edges of the region may be counted erroneously as end points, since the feature is not seen to continue in the next plane. If large areas are used, the magnitude of this effect is small.

The disector can also be used to measure the connectivity of the structure, at least indirectly. Just as we can sweep a line across a two-dimensional images to count tangent points (DeHoff, 1981), to obtain a value for net mean boundary curvature, so we can imagine passing a plane through a three-dimensional structure to count tangencies. There are three types of tangencies that can occur: a) an external tangent point is one in which the plane lies outside the object, and is denoted T^{++} to indicate that the curvature of the boundary of the object is positive (this is the case for a simple convex object); b) an internal tangent point is one in which the plane lies inside the object, and is denoted T^{--} to indicate that the curvature of the boundary of the object is negative (this is the case for points on concave regions of objects); c) a saddle tangent point is one in which the curvature is positive in one direction and negative in the other, so that the tangent plane intersects the surface. It is denoted T^{+-}.

As explained by DeHoff (1987), the tangent counts sample the curvature of the surfaces in proportion to their spherical image or projection. The net volume tangent count is defined as

$$T_{net} = (T^{++}) + (T^{--}) - (T^{+-})$$

For a single object, the net tangent count is equal to the total projected solid angle divided by 2π. For a collection of features, the total projected solid angle is a topological constant that is in turn related to the number of features and their connectivity, giving the result that

$$T_{net} = 2 \cdot (N_V - C_V)$$

The quantity $N_V - C_V$ is the Euler characteristic of the structure (Aigeltinger, 1972). It can be determined with the disector, by counting the three classes of tangencies based on the appearance and disappearance of features and holes in the two parallel sections:

1. A T^{++} event is inferred if a feature appears in one plane that is not a continuation of a feature in the other.
2. A T^{--} event is inferred when an isolated segment of matrix (appearing as a hole in a feature) appears in one plane and not in the other.
3. A T^{+-} event is assigned to each branch or join inferred between the planes, to count the saddle that must be present.

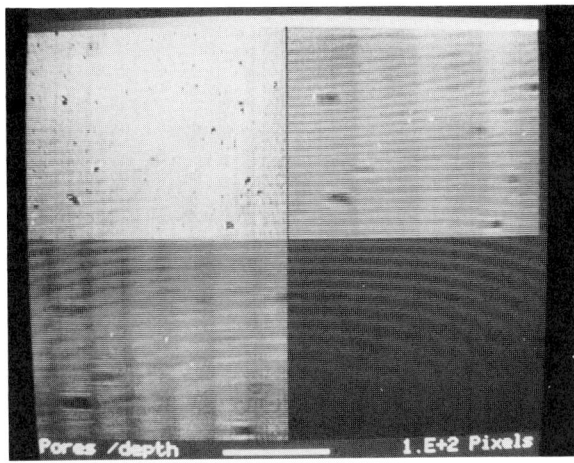

Figure 12-35: Reflected confocal light microscope images focussed at varying depths in a translucent alumina sample, showing pores as bright spots.

From these counts, the value of T_{net} is calculated using the product of the image area and plane separation to obtain the reference volume. The Euler characteristic simplifies to give N_V or C_V in two limiting cases. When the objects are simply connected, the structure consists of N_V objects, and $N_V = T_{net} / 2$. When the structure is a single connected network, the connectivity is just $C_V = -T_{net} / 2$.

The principal global topological property is the Euler or connectivity number, which is formally defined as shown in the equation below in terms of the number of vertices (ends and nodes), edges (links), faces (planar facets formed by the edges), and objects (the number of discrete networks present). For a completely interconnected structure, the connectivity number is equal to $1 - G$, where G is the genus of the bounding surface between the structure and its matrix.

$$\text{Euler Number} = \text{Vertices} - \text{Edges} + \text{Faces} - \text{Objects}$$

The name of this parameter commemorates the tradition that Euler started the study of topology by solving the problem of the seven bridges of Königsberg, a city with two islands in a river and citizenry who tried to find a path that would cross all of the bridges without retracing any steps. The fact that the summation of the degree of all vertices (nodes and ends) in any network must equal twice the number of links, and that therefore the number of odd vertices in any graph must be even, quickly demonstrates for this particular problem that no solution exists.

Of course, in some cases a series of adjacent focal sections can be used to directly determine the important structural information, such as the number of location of features. The confocal scanning light microscope permits recording high resolution, high contrast images in matrices which are not perfectly transparent, because it rejects most of the scattered light along the optical path. Consequently, it is possible to examine materials which are translucent and contain features of interest. Figure 12-35 shows three images at different depths in alumina, in which the bright spots are pores. These result from the sintering history of the sample, and strongly influence its mechanical and optical properties.

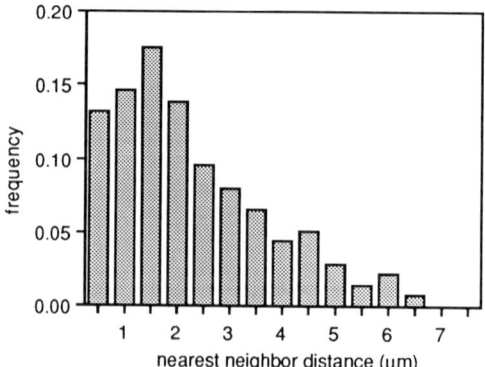

Figure 12-36: Frequency plot for 3-D distance to nearest neighbor, for pores in alumina.

The bright spots result from strong reflection of light at the top of each pore, and can be mapped in three dimensions by scanning the microscope stage in the z direction while digitizing the video signal and recording the x,y position of each spot in the images. The number per unit volume is determined by counting, and in this case the nearest neighbor distance of the pores can be determined and plotted as shown in Figure 12-36 (Russ et al., 1989). Interpretation of the 3-D spatial distribution of objects can be made in detail from such data (Baddeley et al., 1987).

By comparison to the stereoscopy method for locating features discussed in Chapter 11, this technique produces more direct results since the volume can be determined explicitly, but is limited to materials in which images can be obtained, and to the resolution and depth of field of the microscope.

References

E. H. Aigeltinger, K. R. Craig, R. T. DeHoff (1972) *Experimental determination of the topological properties of three dimensional microstructures* J. Microscopy 95 69-81

D. Awamura, T. Ode, M Yonezawa (1987) *Color Laser Microscope* in Imaging Sensors and Displays (C. Freeman, ed.) Proc. SPIE 765, 53-60

N. Baba, M. Naka, Y. Muranaka, S. Nakamura, I. Kino, K. Kanaya (1984) *Computer-aided stereographic representation of an object reconstructed from micrographs of serial thin sections* Micron and Microscopica Acta 15 #4, 221-226

N. Baba, M. Baba, M. Imamura, M. Koga, Y. Ohsumi, M. Osumi, K. Kanaya (1989) *Serial Section Reconstruction Using a Computer Graphics System: Application to Intracellular Structures in Yeast Cells and to the Periodontal Structure of Dogs' Teeth* Journal of Electron Microscopy Technique 11, 16-26

A. J. Baddeley, C. V. Howard, A. Boyde, S. Reid (1987) *Three-dimensional analysis of the spatial distribution of particles using the tandem-scanning reflected light microscope* Acta Stereol. 6 Suppl II, 87-100

R. A. S. Blackie, R. Bagby, L. Wright, J. Drinkwater, S. Hart (1987) *Reconstruction of Three Dimensional Images of Microscopic Objects using Holography* Proc. Royal Microscopical Society 22 #2, 98

A. Boyde, S. J. Jones, T. F. Watson, R. Radcliffe, C. Prescott (1987) *Stereoscopic tandem scanning reflected light microscopy: applications in bone, dental materials and nervous tissue research* Proc. RMS 22/2 91

A. Boyde, S. J. Jones, M. L. Taylor, T. F. Watson, M. MacIntosh, J. Wright (1989) *Fluorescence modes in the tandem scanning microscope*, 2nd Int'l Conference of 3-D Image Processing in Microscopy, Amsterdam (to be published in the Journal of Microscopy)

G. J. Brakenhoff, H. T. M. van der Voort, E. A. van Spronsen, J. Valkenburg, N. Nanninga (1987) *3-D image formation in confocal scanning microscopy and applications in biology* Proc. RMS 22/2 91

L. G. Briarty, P. H. Jenkins (1984) *GRIDSS: an integrated suite of microcomputer programs for three-dimensional graphical reconstruction from serial sections* J Microscopy 134 121-124

A. Briggs (1985) *An introduction to scanning acoustic microscopy* Royal Microscopical Society Handbook #12, Oxford Univ. Press

S. R. Bryan, W. S. Woodward, D. P. Griffis, R. W. Linton (1985) *Secondary Ion Mass Spectrometry / digital imaging for the three dimensional chemical characterization of solid state devices*, J. Vac. Sci. Tech. A3, 2102

J. M. Carazo, J. Jimenez, J. P. Secilla, J. L. Carrascosa (1987) *Computer visualization of biological structure* American Biotechnology Laboratory (Jan 1987) 48-55

H. N. Christiansen, T. W. Sederberg (1978) *Conversion of Complex Contour Line Definitions into Polygonal Element Mosaics* Computer Graphics 12, 187-192

P. Davy (1981) *Interpretation of phases in a material* J. Microscopy 121 (1981) 3-12

R. T. DeHoff (1981) *Stereological meaning of the inflection point count* J. Microscopy 121, 13-19

R. T. DeHoff (1987) *Use of the disector to estimate the Euler characteristic of three-dimensional microstructures* Acta Stereol. 6 Suppl II 133-140

E. K. Fram, S. L. Young, R. E. Albright (1986) *Microcomputer-Based Three-Dimensional Reconstruction and Quantitative Analysis: A General Method*, Acta Stereologica 5#2, 143

H. Fuchs, S. M. Pizer, L. C. Tsai, S. H. Bloomburg, E. R. Heinz (1982) *Adding a true 3-D display to a raster graphics system* IEEE Comput. Graphics Applic. 2 73-78

K. J. Halford, K. Preston (1984) *3-D Skeletonization of Elongated Solids* Computer Vision Graphics and Image Processing 27 #1, 78-91

S. Hart, J. Drinkwater (1987) *A Holographic Display System* Proc. Royal Microscopical Society 22 #2, 93

D. Hearn, M. P. Baker (1986) Computer Graphics, Prentice Hall, Englewood Cliffs NJ

F. S. Hillier, G. J. Lieberman (1986) Introduction to Operations Research, Holden-Day, Oakland, CA, chapters 10 & 15

E. M. Johnson, J. J. Capowski (1983) *A system for the three-dimensional reconstruction of biological structures* Computers in Biomed. Res. 16 79-87

E. M. Johnson, J. J. Capowski (1985) *Principles of reconstruction and three-dimensional display of serial sections using a computer* in The Microcomputer in Cell and Neurobiology Research (R. R. Mize, ed.) Elsevier, New York, 249-263

S. Lobregt, P. W. Verbeek, F. C. A. Groen (1980) *Three-Dimensional Skeletonization: Principle and Algorithm* IEEE Trans. Patt. Anal. Mach. Intell. PAMI-2 #1, 75-77

M. Minsky (1957) *Microscopy Apparatus* U.S. Patent 3,013,467, Dec. 19, 1961

E. Nakamae, K. Harada, K. Kaneda, M. Yasuda, A. Sato (1985) *Reconstruction and semi-transparent stereographic display of an object consisting of multi-surfaces* Trans. Inf. Proc. Soc. of Japan 26, 181-188

M. Petran, M. Hadravsky, J. Benes, R. Kucera, A. Boyde (1985) *The Tandem Scanning Reflected Light Microscope* Proc. Royal Microscopical Society 20 No. 3, 125-139

J. C. Russ, T. M. Hare (1986) *Serial sections and stereoscopy - complementary approaches to three-dimensional reconstruction*, Proc. EMSA, San Francisco Press

J. C. Russ, J. C. Russ (1989) *Image Analysis of Serial Focal Sections from the Confocal Scanning Laser Microscope*, Proc. EMSA, San Francisco Press

J. C. Russ, H. Palmour III, T. M. Hare (1989) *Direct 3-D pore location measurement in alumina*, J. of Microscopy 185, RP1

C. J. R. Sheppard (1986) *Scanned Imagery* J. Physics D Appl. Phys. 19, 2077-2084

C. J. R. Sheppard, A. Choudry (1977) *Image formation in the scanning microscope* Optica Acta 24, 1051-1073

A. W. M. Smeulders, A. L. D. Beckers (1989) *Accurate image measurement methods (applied to 3D length and distance measurements)* Proc. 1st International Conf. on Confocal Microscopy, Academish Medisch Centrum, Amsterdam

D. C. Sterio (1984) *The unbiased estimation of number and sizes of arbitrary particles using a disector* J. Microscopy 134 127-135

H. T. M. van der Voort, G. J. Brakenhoff (1989) *A numerical analysis of the 3-D image formation in a high aperture confocal fluorescence microscope*, 2nd Int'l Conference of 3-D Image Processing in Microscopy, Amsterdam (to be published in the Journal of Microscopy)

A. Waggoner, L. Ernst (1989) *Fluorescent Probes and the Future of 3-D Imaging*, 2nd Int'l Conference of 3-D Image Processing in Microscopy, Amsterdam (to be published in the Journal of Microscopy)

P. Wallen, L. Brodin, K. Carlsson, A. Liljeborg, K. Mossberg, M. Ericsson, S. Grillner, T. Hökfelt, Y. Ohta (1989) *Confocal laser scanning microscopy utilized for 3-D imaging of fluorescent neurons in the central nervous system* European Journal of Cell Biology 48 suppl 25 43-46

D. F. Watson, G. M. Philip (1984) *Systematic Triangulation* Computer Vision Graphics and Image Processing 26, 217-23

T. Wilson (1989) *Three-dimensional imaging in confocal systems* J. Microscopy 153 161-169

T. Wilson, C. J. R. Sheppard (1984) Theory and Practice of Scanning Optical Microscopy Academic Press, London

Y-M Wong, R. P. Thompsons, L. Cobb, T. P. Fitzharris (1983) *Computer Reconstruction of Serial Sections* Computers Biomed Res. 16, 580-586

Chapter 13

Tomography

The previous chapter described the use of serial sections to study the three-dimensional structure of solids. There are things that can be learned from serial sections that are inaccessible from single two dimensional images (for instance, topological information). Also, with appropriate image displays (using the extensive power of computer graphics to shade surfaces, generate stereo pairs, and so forth), it is possible to assist the user in visualizing structure in ways that make good use of the capabilities of human vision. The difficulty inherent in the serial section process is obtaining the images. Usually, this either requires that the specimen be transparent to whatever is used for imaging (light, electrons, sound waves, etc.) and that sequential image planes can be focussed with the microscope optics, or that physical sectioning or erosion of the sample be performed, which is destructive.

Samples that are opaque to light or electrons can usually be penetrated by X-rays, but there are no practical lenses for X-rays that would permit focussing on a single plane at a selected depth, so X-rays cannot be used directly to produce serial section images. The best-known uses of X-rays for studying internal structure are the projected images employed in the medical and dental professions. However, in recent years the most publicized of the methods for reconstructing 2- and 3-D sections has been the "CAT" scanner. Computer Assisted (or Cross-Axial) Tomography is based, as we shall see, on fundamentals derived 70 years ago (Radon, 1917), but only developed to a practical level within the last 20 years. Developments have continued both on the theory, instrumentation and applications.

Reconstruction

The basic principles of tomographic reconstruction are rather simple. It is helpful to imagine the specimen subdivided into a matrix of voxels (volume elements, analogous to 2-D pixels). When X-rays pass through each voxel, they are absorbed according to Beer's Law, $I/I_0 = \exp(-\mu t)$, where μ is a linear attenuation coefficient and t is the dimension of the voxel, as shown in Figure 13-1. The dimension is controlled by the size of the X-ray source and detector, and can range from a few micrometers (and much less if electrons rather than X-rays are used) to centimeters. In typical medical tomography it is about 1 mm.

The linear attenuation coefficient is the product of the density of the material (ρ) and the mass absorption coefficient. The latter is a function of both the concentration of the sample and the energy of the X-rays. This relationship is important and we will return to it later. By measuring the total attenuation of X-rays along many paths through an object, information is obtained about the density within the body. Usually this is illustrated by, and in fact performed within a single transverse section, although information from many

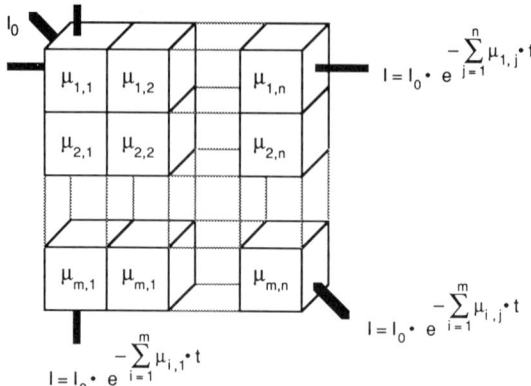

Figure 13-1: Schematic diagram of a single plane of voxels and the total absorption of X-ray beams in different directions.

such parallel sections may be combined to reconstruct 3-D representations of internal structure, and some imaging systems acquire information from several adjacent planes at the same time, using multiple detectors.

From a single projection of density in one direction, it is not possible to locate structures within the object, but with enough such projections the details of the internal organization can be determined.

A simple method for accomplishing this is by "back-projection." The density profile in each direction is projected back along the linear paths. These densities combine (not necessarily linearly) to produce an approximate map of the original structure, as shown in Figure 13-2. The resulting image is significantly blurred because of the overlap of information from many projections, and usually frequency filtering of the projections is used to increase image sharpness. The equivalent of back projection was originally accomplished by mechanical devices that scanned the source and film (or other detector) to blur objects not in a plane of focus, while producing a sharp image of the plane. The method can also be used to reconstruct section images in the computer, but has now largely been superseded by other, more exact mathematical treatments which do not produce the substantial blurred background that is present with this method (Strid, 1986).

One common basis for modern reconstruction methods relies on complete Fourier processing (Herman, 1980; Morgan, 1983). As shown in Figure 13-3, the density profile obtained in each projection can be transformed (typically by a fast Fourier or FFT algorithm, Johnson & Jain, 1981) to produce values (the Fourier coefficients) along a line at the same angle in the frequency domain. Combining the results from many such projections fills the frequency domain representation, which can then be transformed back (by a two-dimensional FFT) to recover the image of the section in the spatial domain. The use of Fourier transforms on 2-D images was described in Chapter 3. A few of the important parameters are the use of adequate precision for the real and imaginary terms to minimize artefacts in the resulting image, the use of optimized programs to achieve speed (a simple row and then column approach is not as fast as some other schemes), and sometimes the use of special hardware to compute the trigonometric function values.

The theoretical basis for this method was shown by Radon in 1917. Unfortunately, the real application of the method suffers from several problems. First, the individual projections contain a finite amount of noise, which affects the frequency domain representation (it adds high frequency information) and produces artefacts in the image, as does the use of a finite FFT. Finally, the use of a finite number of projections leaves gaps in the frequency domain image which must be filled by interpolation, another source of aliasing and artefacts.

Showing that it was possible to implement the Radon transform with real data, and then developing the equipment to obtain the necessary multiple projections, the computer programs to execute the algorithms, and the display technology to show the resulting images, occupied researchers for many years. A. M. Cormack (who developed a mathematically useful reconstruction method at Tufts University in 1963-4) and G. N. Hounsfield (who designed a working instrument at EMI, Ltd. in England in 1972) shared the 1979 Nobel Prize because of the importance of the technology for medical use (Figure 13-4).

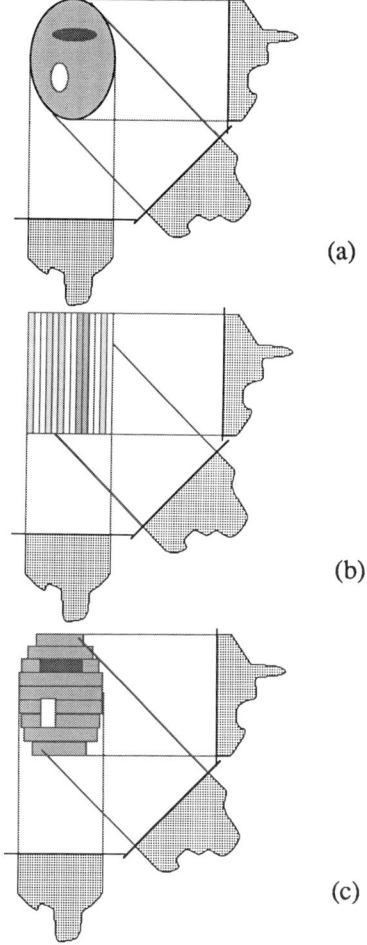

Figure 13-2: Back–projection schematic diagram: a) projections are obtained in many directions through the original object; b) back-projecting from one direction gives some information about the density distribution in the specimen; c) combining many back-projections produces a complete tomogram.

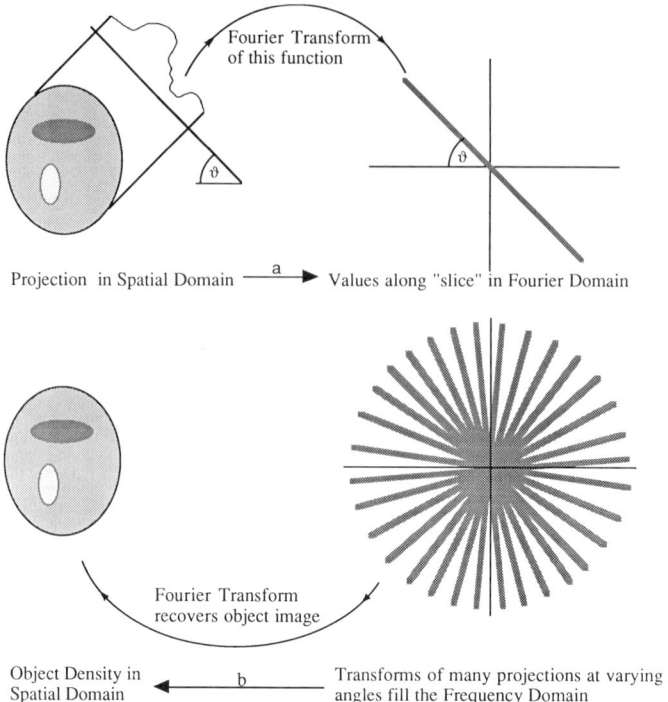

Figure 13-3: Fourier reconstruction: a) each projection is converted to its Fourier transform and plotted along the corresponding direction in frequency space; b) the combined results from many directions are retransformed to construct the tomogram.

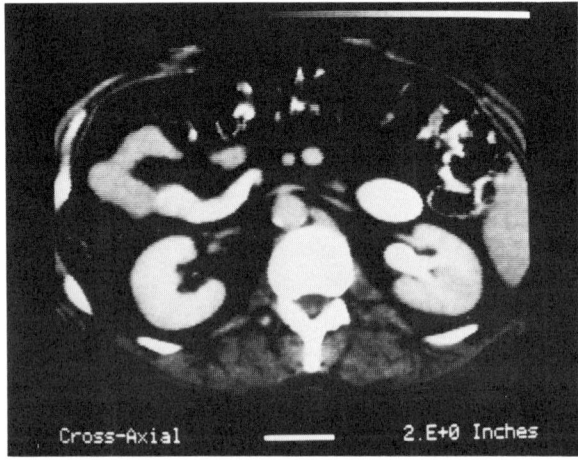

Figure 13-4: Tomogram of lateral section through human torso.

It is also possible to perform a reconstruction using an algebraic technique in which a set of simultaneous equations is solved for the density of each voxel in the section, in terms of the recorded intensity (actually the absorption) along each path from source to detector that has been measured (Gordon, 1974; Censor, 1983; Kak & Slaney, 1988). This involves the inversion of a rather large matrix.

The algebraic reconstruction technique ("ART") sets up a series of equations in which the absorption within voxels along each beam path (generally considered to be proportional to density) are summed. Since the total absorption along that path is known from the measurement, this produces one equation. There are similar equations for each path, and each voxel appears in several such equations. Generally, the contribution of each voxel to each path equation is determined by a weighting function based on the path length through the voxel.

This is equivalent to solving the equation $A^{m \cdot n} x^n = b^m$ where n is the number of voxels, m is the number of projections, and A is the matrix of weights which correspond to how much each voxel contributes to each projection (a matrix which is extremely sparse). The voxel densities are the x values and the projection measurements are the b values. This equation may be either underdetermined or overdetermined if the number of projections does not match the number of voxels; usually the number of projections is far less than the number of voxels.

Kacmarz' method provides simple insight into how this system of equations is solved using ART. An initial value of x is chosen (i.e., each voxel's density is set to some initial value, usually taken as uniform unless prior information is known). Then an iterative procedure is applied in which the next estimate of the x values is calculated as

$$x_{k+1} = x_k + \lambda\, a_i\, (b_i - a_i^T x_k)/\|a_i\|^2$$

This is equivalent to projecting the point onto one of the hyperplanes defined by one of the projection equations. For the simple two-variable case, the process can be visualized as shown in Figure 13-5. The point is projected to one line (given by the equation $a_{1j} x_1 + a_{2j} x_2 = b_j$). The intersection point is then projected in the same way onto the next line, and the process is repeated. In the exactly determined case in this example

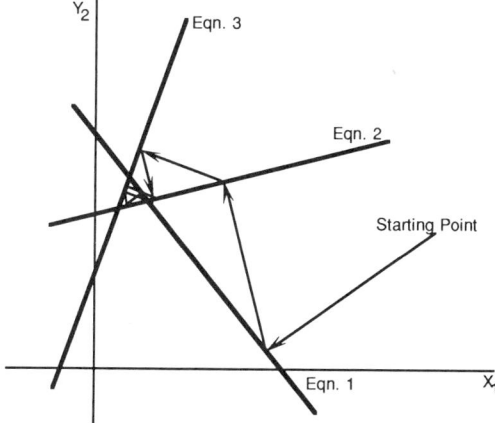

Figure 13-5: Two-dimensional illustration of Kacmarz' method for solving a system of equations.

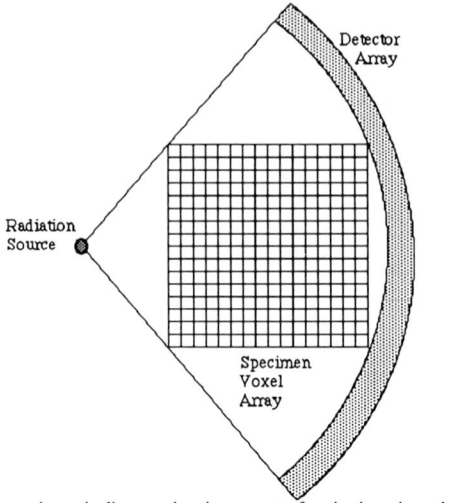

a: schematic diagram showing one set of projections through specimen

```
5  5  5  5  5  5  5  5  5  5  5  5  5  5  5  5
5  5  5  5  5  5  5  5  5  5  5  5  5  5  5  5
5  5  5  5  5  5  5  5  5  5  5  5  5  5  5  5
0  0  0  0  0 20 20 20 20 20 20  0  0  0  0  0
0  0  0  0  0 20 20 20 20 20 20  0  0  0  0  0
0  0  0  0  0 20 20 20 20 20 20  0  0  0  0  0
0  0  0  0  0 20 20 20 20 20 20  0  0  0  0  0
0  0  0  0  0 20 20 20 20 20 20  0  0  0  0  0
0  0  0  0  0 20 20 20 20 20 20  0  0  0  0  0
5  5  5  5  5  5  5  5  5  5  5  5  5  5  5  5
5  5  5  5  5  5  5  5  5  5  5  5  5  5  5  5
5  5  5  5  5  5  5  5  5  5  5  5  5  5  5  5
5  5  5  5  5  5  5  5  5  5  5  5  5  5  5  5
5  5  5  5  5  5  5  5  5  5  5  5  5  5  5  5
5  5  5  5  5  5  5  5  5  5  5  5  5  5  5  5
5  5  5  5  5  5  5  5  5  5  5  5  5  5  5  5
```

b: original voxel densities in a simulated specimen

```
13 19 14  3  0  0  6  8  8  4  0  0  0  9  9  9
 0 12 12  3  2  0  7  7  7  6  0  0  0  3 12 12
 0  1  5  5  3  3 11 10  9  8  1  2  2  1  8 11
 0  0  0  0  3 10 14 14 14 14  6  2  2  1  2 10
 0  0  0  0  4 16 19 16 15 15  8  1  2  1  2  6
 0  2  4  3  5 13 16 17 17 15 10  5  2  3  2  2
 5  3  2  2  3 13 15 15 16 15 14  4  6  3  2  3
 2  2  2  2  3 10 11 12 15 16 16 10  4  5  5  4
 2  2  2  2  4 10 11 12 13 15 16 11  8  7  3  3
 2  0  0  0  2  5  6  8  9 11  8  4  5  6  7
 3  0  0  0  4  4  5  6  7  9 10 11  5  4  3  3
 3  1  0  3  5  4  4  6  7  7  9  9  6  5  7  4
 3  3  1  1  5  4  5  5  7  8  8  9  5  4  4  6
 5  4  3  4  7  6  5  5  5  7  8  9  5  5  5  6
 4  5  5  6  6  6  6  6  5  6  6  7  8  9  5  4  4
 4  5  6  8  6  6  6  5  5  5  8  8  7  5  3  4
```

c: reconstruction after one iteration

```
6 13  9  0  2  4  9  8  8  4  0  0  0  8 10 10
0  7  9  4  3  1  9  6  5  4  0  0  1  2 12 13
0  3  4  2  2  2 13 12  9  8  1  0  0  1  6 11
4  0  0  0  3 11 16 17 16 15  6  0  0  0  0  8
0  0  0  1  5 19 23 16 14 14  9  1  1  0  1  3
0  0  3  4  5 15 17 18 17 14 11  4  1  2  2  1
4  0  2  2  4 13 17 15 16 15 14  3  4  2  1  4
2  1  2  1  3 12 11 13 15 15 16 10  3  3  4  3
1  3  0  1  4 11 11 12 14 14 16 11  8  5  2  1
3  2  1  0  1  2  5  5  6  7  8  7  2  2  5  6
4  1  1  0  5  4  5  6  8  8 10 12  5  4  2  1
5  2  1  3  5  3  4  6  7  7  7 10  6  4  5  4
3  4  2  2  5  4  5  6  7  7  7  9  5  4  3  4
6  3  2  5  8  5  5  4  5  6  7  9  5  4  5  5
5  7  5  6  6  5  6  5  6  5  6  6  9  5  3  3
5  7  8 10  6  5  6  3  4  4  7  7  7  4  3  3
```

d: reconstruction after five iterations

```
3 10  5  1  4  4  9  9  8  5  1  1  2  7  8  8
1  5 10  5  3  0  8  5  4  3  0  1  1  2 10 10
1  2  3  1  1  2 11 12  9  7  0  0  1  3  5  8
1  0  0  0  2 10 18 17 17 15  6  1  0  0  0  5
0  0  0  1  6 21 24 17 15 15 10  1  0  0  0  3
0  0  3  3  6 16 20 17 16 16 12  4  2  2  1  1
2  1  2  0  3 12 19 15 16 17 13  4  3  2  1  3
0  1  1  2  2 13 11 14 15 15 17  9  4  3  4  3
1  3  0  0  3 12 11 13 14 14 17 11  8  5  3  2
2  1  0  0  1  0  3  6  7  8  8  9  2  2  5  6
2  1  2  0  5  3  5  6  8  9 10 12  6  4  2  1
4  2  1  3  5  3  4  7  8  6  7 10  6  4  5  5
2  4  2  2  5  4  5  6  8  7  7  9  5  4  4  4
7  3  1  5  8  5  6  5  6  5  7  9  5  4  5  5
7  8  6  6  7  5  7  4  6  5  5  6 10  5  3  4
7  8 11 12  7  6  6  1  3  4  8  7  7  5  3  3
```

e: reconstruction after fifty iterations

Figure 13.6: Example of Algebraic Reconstruction Method. The simulated specimen contains 16x16 voxels (a total of 256 unknowns). Three projection directions were used (90 degree spacing) with the fan beam geometry shown (a total of 75 equations). (Provided by T. Prettyman, Nuclear Engineering Dept., North Carolina State University.)

there are two lines, and the iteration eventually reaches the answer. The value λ is a "relaxation coefficient" that lies between 0 and 2, and controls the speed of convergence of the answer (λ is set equal to 1 in the figure; when it is very small, the problem becomes equivalent to a conventional least squares solution). So does the order in which the projection equations are applied. These are practical implementation considerations dealt with at length in the literature (Censor, 1983, 1984).

If the problem is overdetermined, the process continues endlessly as the points oscillate within the region bounded by the planes or lines (they do not meet at a single point because of finite errors and noise in the measurements). The use of a stopping criterion is needed to end the process. In the underdetermined case there is only a single line, and the first projection produces a solution, albeit one that depends upon the initial guess. In the normal high-dimensionality case the underdetermined problem still usually produces a resulting image that contain considerable information about the real sample.

There are usually other pieces of information that can be used to constrain the procedure. For instance, the **x** values (voxel densities) must be positive, and are usually known to be less than some maximum value. This limits the area of the plane (in the two-variable example) or of the hyperspace volume in which iteration takes place. If a projection would take the point out of this constrained space, it is usually limited to the boundary. Other constraints such as prior information about the density values or uniformity in specified regions can be incorporated in the solution as well.

As an illustration of the method, consider the small array of voxels in a single plane shown in Figure 13-6. From the given density values (Figure 13.6b), a series of three fan-beam projections were calculated, each using 25 detectors. The resulting number of equations is underdetermined - that is, there are fewer equations (75) than there are unknowns (256). Using the iterative ART solution method, and starting with an assumed uniform density of 10, the calculated image after 1, 5 and 50 iterations is shown in Figure 13-6c-e. Notice that the sharp edges are somewhat distorted and uniform areas show variations (particularly in the corners, which are sampled by fewer equations), but the overall result is quite good.

The major advantage of the ART method arises in situations that create problems for the Fourier method. Missing angles of projection (which may be forced by the geometry of the instrument or specimen), unequal spacing of projections or detectors, and other similar problems can leave portions of frequency space without information. This introduces serious artefacts in the Fourier reconstructed image. The algebraic reconstruction method may also fail to reveal detail in the affected areas, but the problems do not spread to other parts of the image. Also, when the sampling density is low, the algebraic reconstruction methods can still produce an image of somewhat lower resolution (by solving the smaller number of equations for the density of fewer pixels) without introducing artefacts and aliasing.

Instrumentation

The most straightforward calculation of the projection data results when a parallel beam arrangement is used, as shown in Figure 13-7. This requires scanning a source-detector pair across the object being studied, in each of many directions to obtain the necessary multiple projections. The other arrangement, more often used in medical scanners, is a fan beam arrangement in which an array of detectors is used with a single source. This makes the reconstruction mathematics slightly more complicated. If the source and detector array can be rotated around the object (the most common arrangement in commercial medical scanners), a complete set of projections can be obtained for each

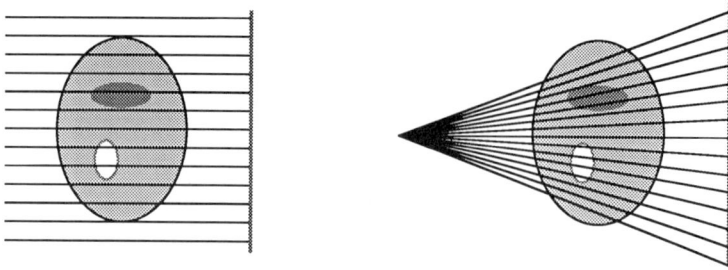

Figure 13-7: Parallel and Fan Beam geometries.

plane section in less than 1 second. Of course, there are countless variations that can be used in terms of whether the object or the equipment moves, whether single or multiple sources are used, and so forth. The various "generations" (currently first through fifth) of medical analyzers are defined by the arrangement of sources and detectors, by what moves, and consequently by how long it takes to acquire an image (Figure 13-8). Most of the industrial tomography is performed with "second generation" technology in which both motion and translation are utilized to obtain the necessary projections.

One rather novel arrangement that has been applied to small specimens that can be placed in the SEM is shown in Figure 13-9 (Sasov, 1985, 1987). The electron beam strikes a target, producing a source of X-rays that pass through the specimen to the EDX detector which measures the absorption of the characteristic line. By scanning the beam, a series of rays are generated which form a complete 1- or 2-D projection of the sample. Rotating the specimen with the SEM stage then provides the necessary multiple projections, from which the internal structure can be imaged with a resolution determined by the size of the X-ray detector and its distance from the sample. The spatial resolution of the information obtained by this setup is typically of the order of 10-20 µm, resulting from the projected size of the detector at the specimen because the source is effectively a point (Figure 13-10)

It is also possible to obtain projected images of X-ray absorption using the X-ray projection microscope, which has a similar spatial resolution. The X-ray projection microscope can provide a direct enlargement of internal structure. Figure 13-11 shows schematically the arrangement used (Cunningham et al. 1986). A microfocus X-ray tube

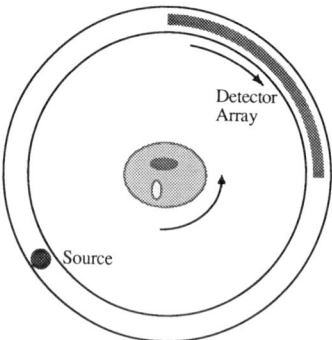

Figure 13-8: Several ways to obtain the necessary projections: 1. move the object or the source and detectors; 2) use single or multiple sources and/or detectors.

Tomography

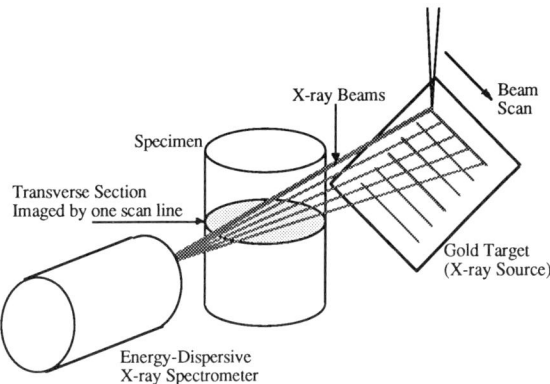

Figure 13-9: Schematic of SEM tomography setup.

acts as a point source, and geometric magnification (typically up to about 100 times) of the image of internal structure in the sample is obtained. Either film recording or a phosphor screen and associated image intensifier and video camera can be used. Modifications can been added to rotate the specimen (to obtain the multiple projections for tomographic reconstruction) and to insert filters in the beam as will be discussed below.

The rather low voltages used (typically about 10 keV.) produce good image contrast based on structure and composition differences (Figure 13-12), and also permit a fine

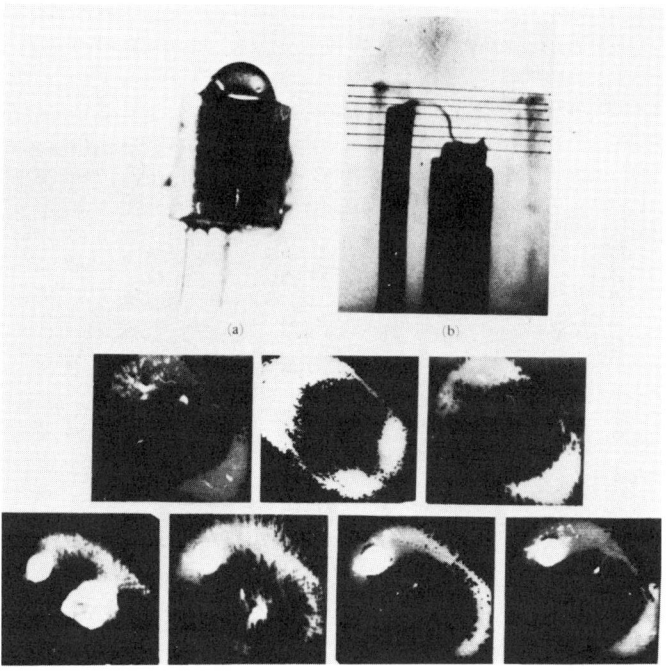

Figure 13-10: Example of micro-tomograms of encapsulated electronic device (Sasov 1985).

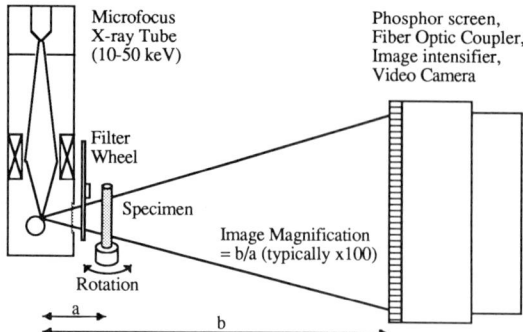

Figure 13-11: Diagram of X-ray projection camera modified for element-specific tomography.

spot (of the order of 1 μm) to be generated in the tube (which defines the ultimate image resolution), but this reduces the specimen thickness that can be penetrated (typically less than 1 mm). Usually this is not a problem, as it is only for moderately thin sections that the overlapped internal structures can be visualized anyway. The use of stereo pairs (produced by recording two images with the specimen tilted or translated) has been used to alleviate this image superposition problem somewhat. Ong (1959) showed that the use of low voltage continuum or monochromatic X-rays generated with a transmission anode microfocus source (based on electron microscope design) could produce simple transmission images with submicrometer resolution, but this was of course before tomographic reconstruction was known. High resolution images suitable for reconstruction with resolution of a few micrometers can also be obtained with high-brightness synchrotron radiation sources (Russ, 1988b; Martz et al., 1989; Dover et al., 1989).

Notice that when images of this sort are obtained in a series of directions through a specimen, tomographic reconstruction of the entire three-dimensional internal structure is possible. This is not simply a series of parallel two-dimensional planes of voxels, but a

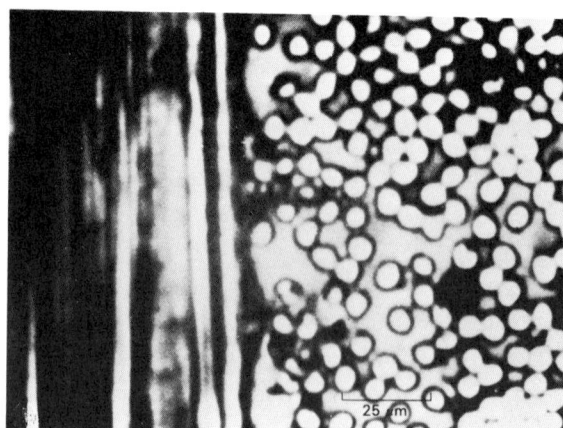

Figure 13-12: X-ray projection microscope image of fibers in a composite (Cunningham et al. 1986).

true three-dimensional array, because the voxel size is the same in both directions (as limited by the image resolution), and the voxels are exactly contiguous in all directions. When the Fourier method is used for reconstruction, both the frequency space and the real space images are three-dimensional and the transform operation is as well.

The contrast mechanism in these X-ray projection images is usually purely absorption of the X-rays. This means that structural differences that produce significant changes in linear absorption coefficient, such as composition or density, are imaged. Rarely do diffraction effects play any role, and then because of the variety of angles at which the X-rays pass through the sample this is only a local effect that will additionally decrease the measured intensity in one direction (and could increase it in some other direction). Consequently, this method has found its greatest application in heterogeneous materials such as composites or materials containing porosity. The intentional use of diffraction to study internal structure in highly perfect crystalline materials, as with a Lang or Berg-Barrett X-ray diffraction setup, is a subject beyond the intended scope of this text, and in any case generally allows the study of structure only in thin specimens or very close to the specimen surface. However, the same technology used to record XRD patterns can also be used to convert X-ray projection microscope images to video signals, which are then easily digitized into the computer and used for tomographic reconstruction.

The techniques become more complicated when the subject being studied is not the human body. For instance, one problem that medical scanners must deal with is collecting data in a short enough time to avoid blur due to the beating of the heart. In some materials applications, much faster motions must be frozen, either by short exposure or stroboscopically (of course, in other cases the object being studied is static and exposures can be accumulated over many days).

Another problem of minor importance in medical applications but very significant for materials objects is the variation of composition as well as density along the beam path. The discussion of absorption earlier in this chapter used μ to indicate the linear X-ray absorption coefficient, which is the product of the density and the mass absorption coefficient. It was noted that the latter value is a function of composition and beam energy. Generally for biological structures, there is only a small variation in the composition of the various voxels in the structure being examined, the energy of the X-rays remains essentially constant, and the imaging that takes place is primarily one of variations in density.

For medical scanners, it is conventional to ignore this subtlety altogether and report the results as "CT numbers". The CT number is a direct measure of X-ray attenuation

$$\text{CT number} = 1000 \cdot \frac{\mu(\text{sample}) - \mu(\text{water})}{\mu(\text{water})}$$

which assigns values of -1000 to air and 0 to water, respectively. Most commercial scanners can handle a range of CT numbers up to about 3-4000.

But when materials specimens, such as rocket motors, artillery shells, drill cores from geological exploration, encapsulated microelectronic devices, and so forth, are examined, there is a much greater range of both CT number and composition that may be present, and furthermore may vary from one voxel to another. This introduces additional complexity. The actual behavior of the matrix mass absorption coefficient is

$$\mu_{\text{total}} = \sum C_i \cdot \mu_i (Z,E)$$

where the summation is over all of the elements present, C is the weight fraction of each element, and the individual mass absorption coefficients are functions of atomic number

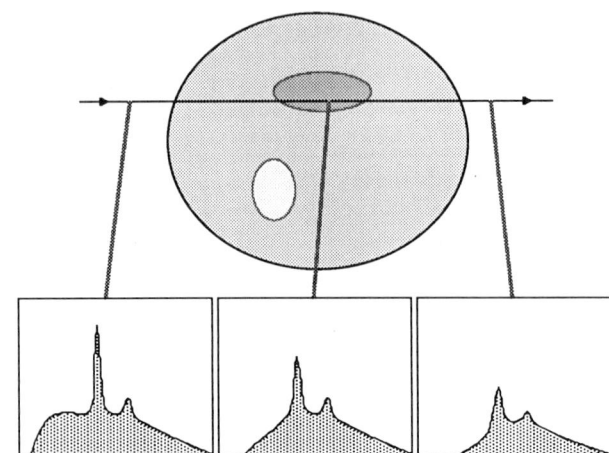

Figure 13-13: Schematic diagram of beam hardening. Absorption of X-rays in the sample changes the energy distribution and hence the linear absorption coefficient.

Z and X-ray energy E. This latter function results because the absorption of the X-rays is caused by the interaction of the photons with bound electrons in the atom. If the photon energy exceeds the binding energy of each particular shell, then it may be absorbed and its energy used to liberate the electron. The probability of this happening increases when the photon energy is close to the binding energy, and of course the binding energies are unique for each element in the periodic table. The result is curves of mass absorption coefficient with energy that can be tabulated or calculated as a function of specimen composition, but show marked variation with the energy of the X-rays.

If the X-ray source generates continuum radiation (i.e., a typical X-ray tube as opposed to isotope sources, etc.), the energy distribution of the radiation changes along the path, which in turn changes the linear absorption coefficient of the X-rays as a function of energy (wavelength) as indicated in Figure 13-13. The total absorption along the path length becomes a very complex function. If the sample is high in absorption coefficient but not highly variable in internal composition or structure, surrounding it with a fluid or fine powder of high absorption material can reduce some of these effects (Sheppard, 1987). More often, the use of different materials or the possible presence of defects within the materials produces a wide variation in composition on a fine structural scale that is to be imaged.

By utilizing monoenergetic X-rays, the absorption coefficient at only a single energy is involved and conventional reconstruction is possible.

When several different elements are present, the spatial distribution of one of them can be mapped by acquiring two images, one using monoenergetic X-rays just above the absorption edge energy of the element, and another with X-rays just below that energy. The latter are absorbed by other elements in the matrix with nearly the same attenuation coefficient as the former, because the absorption coefficient varies gradually with energy except at the absorption edge energies where a different electron shell can be ionized, and so serve to normalize the data. However, there will be a major change in the absorption of the selected element. From the two absorption values, images showing only the selected element can be obtained, and its spatial distribution can be reconstructed. The

Tomography

difference or ratio of the two images, computed on a point-by-point basis, reveals the distribution of just the element of interest (Bigler et al., 1983; Bonse et al., 1986).

This is potentially a very powerful technique, hampered by the practical difficulty of producing high intensity monoenergetic X-ray sources with adjustable energies, unless you have access to a synchrotron. The use of carefully selected filter materials placed in front of an X-ray tube represents one way to accomplish this technique, but with low intensity and correspondingly long imaging times (Russ, 1988b). It is expected that the recording of complete energy spectra for the transmitted beam, referred to above in connection with the beam-hardening problem, will allow this kind of imaging for each element in the specimen (Ryon et al., 1988).

In theory, the recording of complete energy spectra along many beam paths should make it possible to reconstruct the internal composition as well as structural distribution in the solid (sometimes referred to as "tomochemical" analysis). The mathematics to carry out such reconstruction is incompletely developed at this time (Russ, 1988a), but it is clear that in principle there is enough information available. The algebraic reconstruction techniques appear to be more suitable for this type of reconstruction than do the Fourier transform methods.

3-D Imaging

Three-dimensional images are obtained by using a series, or stack of 2-D section images. The display of such an array of information makes great use of modern computer display technology, such as that used in CAD/CAM modelling of surfaces, or the stereo presentation of three-dimensional information. As more information, for instance compositional as well as density data, becomes available, finding ways to display it comprehensibly if not comprehensively will present great challenges. Interactive computer graphics is presently a field of very lively development, and all of the techniques available are relevant to this application in one way or another.

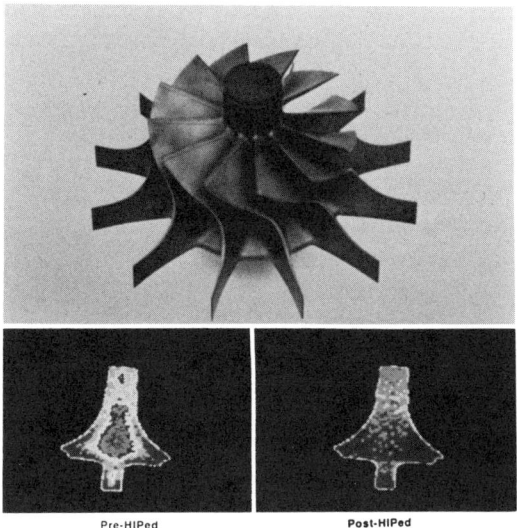

Figure 13-14: Density difference in SiC ceramic rotors before and after hot isostatic pressing (Hunt, 1988).

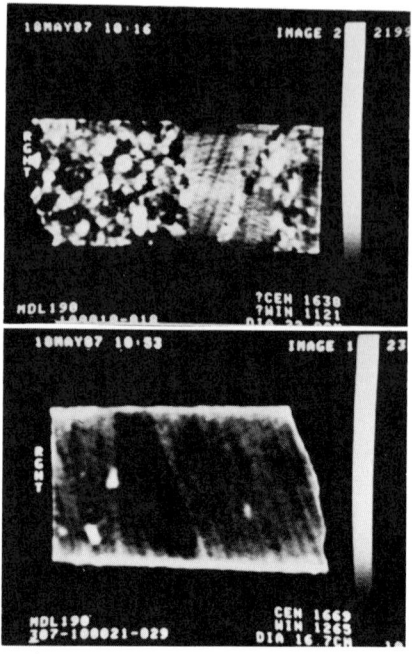

Figure 13-15: CT image of longitudinal section of a drill core showing conglomerated rock and a clay plug (Hunt ,1988).

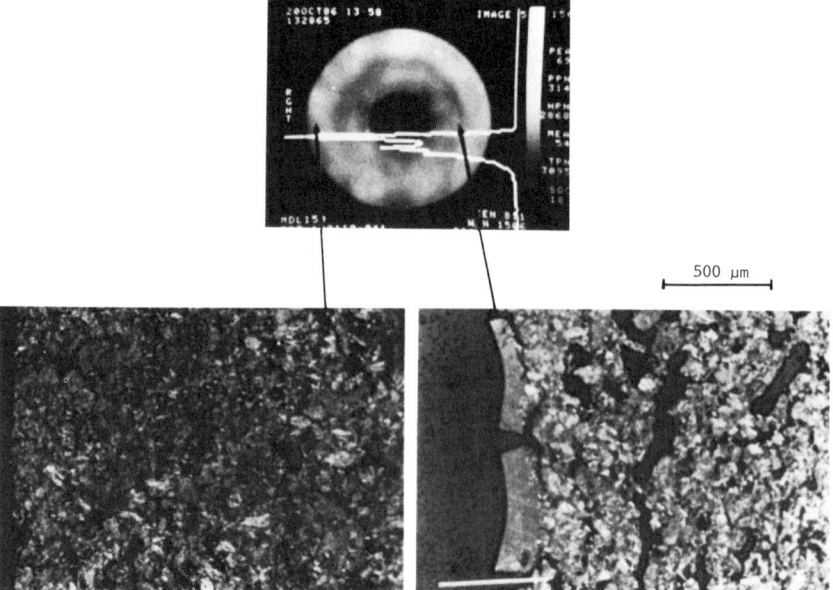

Figure 13-16: Variation in density in a graphite tube, using mercury injection to fill voids, and void in cemented pump shaft (Hunt, 1988).

For the "simple" applications, which just involve imaging internal defects or gross changes in composition or structure, the only image processing operations used are rescaling of brightness and contrast to show the changes or differences of importance (Figure 13-14). In other cases, measurement of the CT number may be correlated with density or composition of the material in an *ad hoc* way, to distinguish regions within the specimen (Figures 13-15). When the natural variation in density is not great enough to show a measurable difference, it is sometimes possible to introduce a penetrant or staining agent to artificially increase the contrast as shown in Figure 13-16 (this method is also used in medical X-ray imaging, of course).

The computed images from tomographic reconstruction may be subjected to normal image analysis methods to enhance particular features (edges, for instance), or to measure certain structures. Revealing internal structure or porosity, texture, etc. is accomplished just as though the images had been obtained by physical sectioning of the sample, subject of course to the limitation of resolution. For instruments based on the technology of medical scanners, this is usually of the order of 1 mm.

Since they are usually obtained as a series of parallel plane images, tomographic reconstructions are of course ideally suited for serial section reconstruction using the methods discussed in the previous chapter (Figure 13-17). It is important to know the thickness and separation of the slices. The voxels are rarely cubic in present systems (the

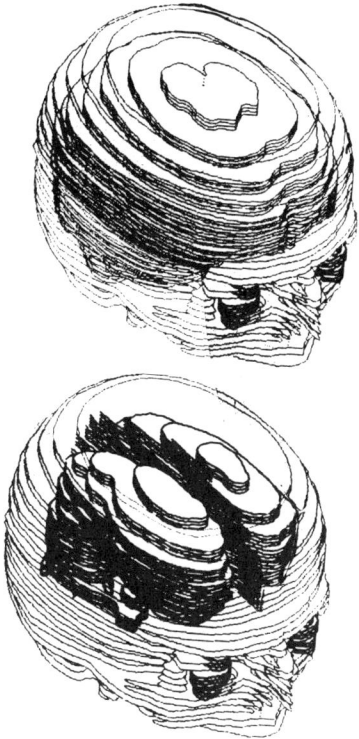

Figure 13-17: Reconstruction of human head with brain and tumor from a series of tomographic slices (Jandel Corp., Corte Madera, CA).

slices are much thicker that the resolution within each slice), and sometimes the slices are partially overlapping.

Tomography performed in the transmission electron microscope is another fairly new offshoot of the basic technique (Crowther et al., 1970; Gordon et al., 1970; Hoppe et al., 1976; Radermacher & Hoppe, 1978; Negeri, 1982; Hegerl et al., 1984; Engel & Massalski, 1984; Cohen et al., 1984; van Heel, 1986; Hegerl, 1989). A complete 2-D projected image in which contrast is based on electron absorption and scattering is obtained for each orientation. However, this is complicated by the fact that the mechanism used to orient the sample gives a conical rather than radial series of projections, and cannot cover all possible angles, so that a large missing segment is present in the Fourier space representation. Techniques to interpolate with minimum artefacts have been developed, and produce complete three-dimensional structural images. The images are two-dimensional, and the reconstruction is three-dimensional. A major consideration is the number of projections required (and the dose of electrons applied to the specimen in the process, especially for biological samples which are easily damaged).

Basically, these reconstructions work by collecting a series of two-dimensional images whose frequency transforms are mapped into planes in a three-dimensional volume. When this volume has been filled by projections (and intervening spaces filled by filtering and interpolation) then a full three-dimensional inverse transform in performed, resulting in a complete three-dimensional image of the original structure. The extension from two to three dimensions for the FFT is straightforward, a direct outgrowth of the step from one to two dimensions. A natural consequence of this method is that the resulting 3-D image has voxels that are exactly cubic, rather than being elongated in one direction. All of the usual display techniques discussed in Chapter 12 are available (Figure 13-18).

This points up the fact that X-rays are not the only radiation useful for tomography. In addition to X-rays, electrons, visible light, ultrasound, neutrons, nuclear magnetic resonance (MRI), electron paramagnetic resonance, and other wave and particle phenomena have all been used. For instance, MRI can map the porosity distribution in green ceramics if they are filled with a marker fluid such as benzene by vacuum

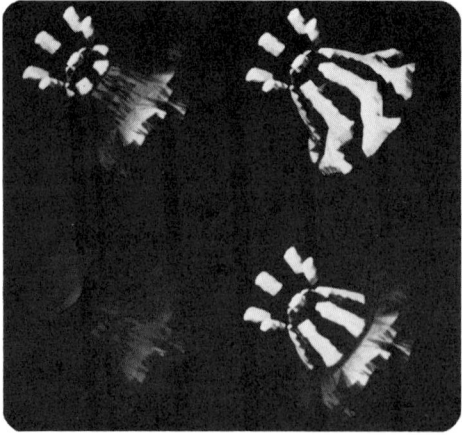

Figure 13-18: TEM tomographic reconstruction of a virus particle, using surface modelling.

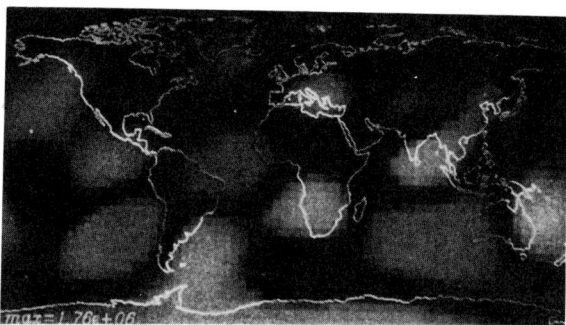

Figure 13-19: Reconstruction of the boundary between the core and mantle, from seismic waves generated by earthquakes.

impregnation, and the recorded image intensity can be interpreted as the local fractional porosity even if the individual submicron pores cannot be resolved. The use of MRI imaging for tomographic reconstruction in medical applications is quite attractive because it avoids the exposure to X-rays, and because it can be used to map the distribution of specific nuclei, such as sodium-23 or phosphorus-31, as well as normal imaging contrast based on hydrogen in water molecules. Even radio waves have been used in astronomical applications of the method, and seismic waves in studies of the internal structure of the earth (Figure 13-19).

The basic techniques for all of these methods are similar, but the sensitivity of the method to various artefacts resulting from finite noise, nonlinear dependence of absorption on density or other structure, and of course the practical implementation in hardware may be quite different.

Traditional absorption tomography measures the attenuation of the primary beam along each path (or, for ultrasound and seismology, the change in time of flight). Emission techniques are also capable of revealing the internal structure (Kak, 1979). One variant is positron emission tomography, in which the decay of radioactive tracers

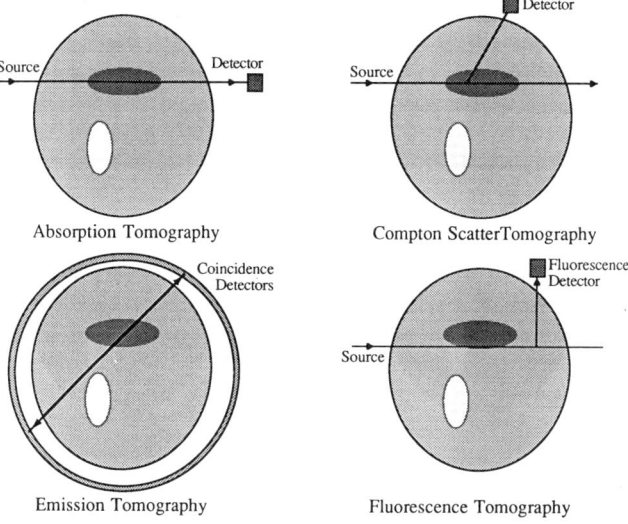

Figure 13-20: Some additional methods for tomography.

produces a positron (β+) which then annihilates with an electron producing a pair of 511 keV X-rays. These travel in opposite directions, and can be detected in coincidence by a pair of detectors which locate the source along a line between them. Many such events allow mapping of the radioactive material in the section.

Another method employs Compton scattering of X-rays from the original beam, with a detector at the side (the position along the beam path is determined from the energy shift of the detected X-ray). In principle this is capable of providing compositional information along the path, to supplement the absorption information. So is using a collimated energy dispersive detector for the fluoresced characteristic X-rays (Nichols et al., 1987). Figure 13-20 illustrates several of these methods.

All of these methods (and many more than can even be listed here) are capable of providing information about the internal structure of objects that can be of great importance in research and industry. The problems of theory, design, and application are much greater for materials than for medical use, because of the greater diversity of objects and composition encountered.

References

E. Bigler et al. (1983) *Quantitative Mapping of Atomic Species by X-ray Absorption Edge Microradiography* Nuclear Instrum. & Methods 208 387-392

U. Bonse et al. (1986) *High Resolution Tomography with Chemical Specificity* Nuclear Instrum. & Methods in Phys. Res A246 644-648

Y. Censor (1983) *Finite Series Expansion Reconstruction Methods* Proc IEEE 71 409-419

Y. Censor (1984) *Row-Action Methods for Huge and Sparse Systems and Their Applications*, SIAM Review 23 #4, 444-466

H. A. Cohen, M. F. Schmid, W. Chiu (1984) *Three-dimensional reconstructions at high resolution* Ultramicroscopy 14, 219-226

R. A. Crowther, D. J. DeRosier, A. Klug (1970) *The reconstruction of a three-dimensional structure from projections and its application to electron microscopy* Proc. Roy. Soc. Lond. A317 319-340

T. G. Cunningham, R. L. Davies, S. C. Graham (1986) *The application of X-ray microscopy in materials science* J. Microscopy 144 pp. 261-275

S. D. Dover, J. C. Elliott, R. Boakes, D. K. Bowen (1989) *Three-dimensional X-ray microscopy with accurate registration of tomographic sections* J. Microscopy 153 187-191

A. Engel, A. Massalski (1984) *3D reconstruction from electron micrographs: Its potential and practical limitations* Ultramicroscopy 13, 71-84

R. Gordon (1974) *A Tutorial on ART (Algebraic Reconstruction Techniques)* IEEE Trans Nucl. Sci. NS-21, 78-93

R. Gordon, R. Bender, G. T. Herman (1970) *Algebraic reconstruction techniques (ART) for three-dimensional electron microscopy and x-ray photography* J. Theor. Biol. 29 471-481

M. van Heel (1986) *Noise limited three-dimensional reconstructions* Optik 73, 83-86

R. Hegerl, W. Hoppe, V. Knauer, D. Typke (1984) *Some aspects of the 3D reconstruction of individual objects* Proc. 8th Eur. Congr. Electron Microsc. 2, 1363-1373

R. Hegerl (1989) *Three-dimensional reconstruction from projections in electron microscopy* European Journal of Cell Biology 48 Supplement 25, 135-138

G. T. Herman (1980) *Image Reconstruction from Projections - The Fundamentals of Computerized Tomography* Academic Press, New York

W. Hoppe, H. J. Schramm, M. Sturm, N. Hunsmann, J. Gassman (1976) *Three-dimensional electron microscopy of individual biological objects Part I: Methods* Z. Naturf. 31a, 645-655

P. K. Hunt, P. Engler, W. D. Friedman (1988) *Industrial applications of X-ray computed tomography* Advances in X-ray Analysis Vol. 31, (C. S. Barrett et al., eds.) Plenum Press, New York NY, 99-105

L. R. Johnson, A. K. Jain (1981) *An Efficient 2-D FFT Algorithm* IEEE Trans Patt. Recog. Mach. Intell. PAMI-3#6, 698

A. C. Kak (1979) *Computed Tomography with X-ray, Emission and Ultrasound Sources* Proc. IEEE 67(9)

A. C. Kak, M. Slaney (1988) Principles of Computerized Tomographic Imaging IEEE publication PC-02071

H. E. Martz, S. G. Azavedo, J. M. Brase, K. E. Waltjen, D. J. Schneberk (1989) *Computed Tomography Systems and their Industrial Applications* International Journal of Radiation Applications and Instrumentation Part A (in press)

C. L. Morgan (1983) *Basic Principles of Computed Tomography* University Park Press, Baltimore

R. Negeri (1982) *3-D reconstruction from electron micrographs* Proc. 10th Intern. Cong. on Electron Microscopy, Hamburg, pp. 545-551

M. C. Nichols, D. R. Boehme, R. W. Ryon, D. Wherry, B. Cross, G. Aden (1987) *Parameters affecting X-ray microfluorescence analysis* Advances in X-ray Analysis Vol. 30, (C. S. Barrett et al., eds.) Plenum Press, New York NY, 45-51

P. S. Ong (1959) Microprojection with X-Rays, Drukkerijen Hoogland en Waltman, Delft Holland (Doctoral thesis, Technische Hogeschool Delft)

M. Radermacher, W. Hoppe (1978) *3-D reconstruction from conically tilted projections* Proc. 9th Int'l. Congr. Electron Microsc. 1, 218-219

J. Radon (1917) *Über die Bestimmung von Funktionen durch ihre Integralwerte längs gewisser Mannigfaltigkeiten* Berlin Sächsische Akad. Wissen. 29 pp. 262-279

J. C. Russ (1988a) *X-ray Imaging of Surface and Internal Structure*, in Advances in X-ray Analysis Vol. 31, (C. S. Barrett et al., eds.) Plenum Press, New York NY, 25-34

J. C. Russ (1988b) *Differential Absorption Three-Dimensional Microtomography* Trans. Amer. Nucl. Soc. 56 Suppl. 3 14

R. W. Ryon, H. E. Martz, J. M. Hernandez, J. J. Haskins, R. A. Day, J. M. Brase (1988) *X-ray Imaging: Status and Trends* Advances in X-ray Analysis Vol. 31, (C. S. Barrett et al., eds.) Plenum Press, New York NY, 35-52

A. Yu. Sasov (1985) *Computerized microtomography in scanning electron microscopy* SEM '85/III, SEM Inc., Chicago, pp. 1109-1120

A. Yu. Sasov (1987) "Microtomography. I: Methods and equipment, II. Examples of applications" J. of Microscopy 147, 169-192

L. M. Sheppard (1987) *Detecting Materials Defects in Real Time* Advanced Materials and Processes (11/87) 53-60

K. G. Strid (1986) *Tomography: Spatial reconstruction from projections* Acta Stereol. 5 no. 2, pp. 103-126

Chapter 14

Lessons from Human Vision

In this concluding chapter, it will be useful to compare some of the image analysis capabilities that have been described to the performance of the human visual system. Several times throughout this text, reference has been made to the methods that appear (based on physiological tests) to be used by humans in dealing with visual information (and indeed other animals - see for example Lettvin et al., 1959; Hubel & Wiesel, 1962). These insights have provided valuable guidelines to methods that are robust, fail gracefully in difficult situations, are usually extremely economical in their computational needs, and above all, work. Examples include stereo pair fusion, feature recognition and identification, correction for shadowing effects, and the location of feature edges even when they are incomplete or obscured in noise.

It is very useful to the hardware and software designer to keep these principles in mind, as they often point the way toward practical implementation in computer-based image analysis systems. But in other cases they do not offer any assistance to the programmer. The human visual system (Figure 14-1) is very different from our present and prospective future image analysis computers. It has an extremely large number of pixels, estimated at about 150,000,000 (compared to 256,000 in the most typical modern computer), and furthermore the density of these pixels is even higher in the fovea, the area near the center of the visual field where we really concentrate on features.

On the other hand, it is possible to process or measure an entire image in a computer-based system, while the human visual system must fixate on a single feature or very small number of closely associated ones at any time. This is why our eyes are constantly jumping from one place to another in the visual field, collecting information on only a few of the features present (which are selected based on a variety of cues and our past experiences).

The brain is massively parallel, with perhaps 25,000 higher level cells performing comparisons between the output from each pixel (and other logic cells connected to it) and various neighboring pixels, all at the same time. It is unlikely that this degree of parallelism has arisen without having a strong connection to the need to extract important information from images, and we must therefore content ourselves to deal with only a small fraction of the data present in computerized images.

Finally, the human visual system is overwhelmingly a comparative system rather than a measurement system. This is clearly true in its estimation of size, position, color, and other objective measures of features. The quality of the results in physiological testing falls off directly with increasing distance (in space or time) or confusion (misalignment, changes in shape, etc.) between objects. Computer-based measurement should in principle be able to outperform humans in this regard, especially in situations where control of the imaging conditions (especially illumination) makes it possible to compare features to internal or remembered standards.

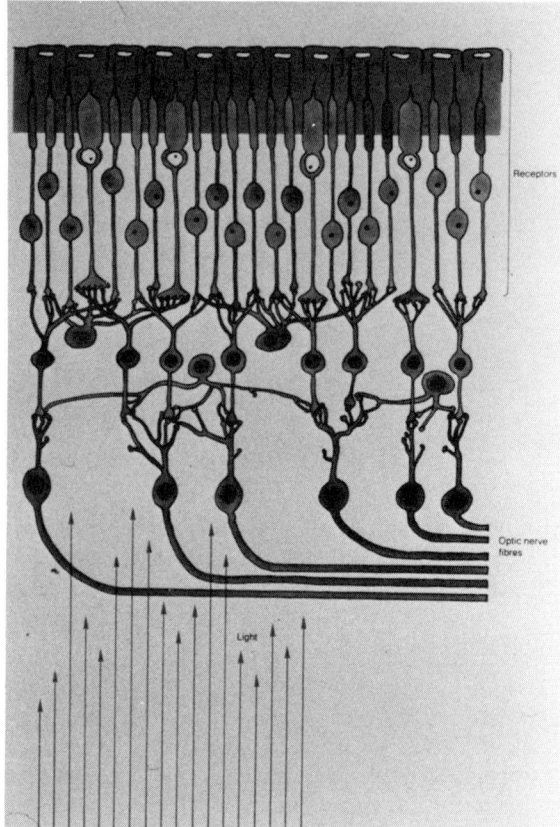

Figure 14-1: The human eye, showing receptors and some of the logical interconnections in the retina.

This is especially true with regard to three-dimensional measurements. Using either stereometry or the other depth cues discussed in Chapters 10-13, we can compute quantitative or metric values for the position of each feature or even each point in an image. There is evidence from physiological tests that human vision does not try to assign such absolute values, but instead uses a simple relative ordering of distance among neighboring points. This does not directly permit comparison of separated points if there are intervening maxima or minima.

Also, the human visual system (and probably that of other mammals and perhaps other animals) distinguished between two types of image interpretation: spatial organization and feature recognition. Low-level information from the retina of the eyes goes to the visual cortex, where further abstraction takes place. This higher-level information then proceeds to both the parietal and temporal lobes. The parietal cortex responds to spatial arrangements, and indeed is closely associated with motor coordination. The temporal cortex is involved with object recognition.

We cannot entirely duplicate the ability of the vision system to recognize features when incomplete or conflicting information is present, or to generalize from a few examples. There is no simple answer to how we recognize a "2" as shown in Figure 14-

Lessons from Human Vision

2. But we can perform a series of objective measurements that, as we saw in Chapter 9, can often accomplish a considerable amount of feature identification.

However, we cannot ignore the lesson of feature recognition. Extracting the important criteria to recognize objects is something that humans do very well. We use all of the details and parameters of features, and then by a process of education decide later which of these are really important. The ability to recognize familiar faces, scenes and objects even after a long period of time, with very low error rates, is perhaps not too surprising in view of the learning and reinforcement that we conceptually understand. But humans also show a surprising ability to generalize from this knowledge and perform similar recognition of things glimpsed only once, under different conditions. It seems that the important details and parameters for recognition that work in the more familiar cases become more-or-less hard-wired into our system so that they are available for use in unique situations as well.

The language of structure

There are, of course, some shortcomings to the human visual system and our ability to interpret images. This is particularly evident when trying to interpret images of 2-dimensional sections through a complex three-dimensional structure. We do not readily comprehend what the underlying structure "looks like" even when we are looking at it. Compare this to our ability to literally "turn over" a complex three-dimensional object in our minds, to visualize its appearance from many different viewpoints even if we have seen it only from one side.

The difference is that we are used to looking at the exterior surfaces of solid objects, and not at internal structure. Whether this particular world view is built into our genes or learned at a young age is unimportant, but we must recognize the bias. In trying to deal with structures, we are forced to rely on mathematical tools to characterize and summarize the information, and the concepts do not translate well into our general experience. Most people have little trouble with the concept of volume fraction (although most are poor at estimating it). Surface area per unit volume is harder to explain, and the mean curvature of surfaces, torsion of lines, and other stereological parameters are really very difficult things to comprehend.

This does not mean that we shouldn't use them. In fact, we have few if any alternative tools to describe structures. What it does mean is that we must be aware that the language of 3-D structure is not a familiar or natural one. We must make special efforts to familiarize ourselves and those with whom we would communicate structural data with this language, and cannot assume a familiar base.

The same limitation is evident in trying to describe the shapes of features, either in two or three dimensions. Whereas size, position and color are all classes of

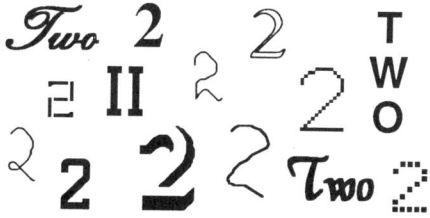

Figure 14-2 : What is a "2"?

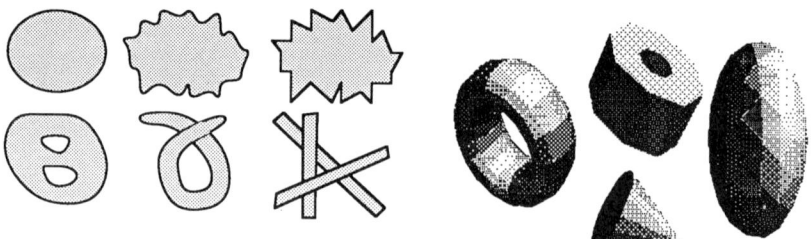

Figure 14-3: A collection of 2- and 3-D shapes. What are the differences? How do we describe them?

measurements with a strong underpinning of common ideas, shape is not. We can agree from our common experience on measures of size such as length, area, and volume, and on measures of position that are either relative (above, behind) or absolute (distance and direction). Description of color and density is complicated somewhat by the ultimate need to resort to wavelengths or coordinates in RGB, HIS or some other explicit quantitative space to define absolute values, but at the qualitative level we all know what orange and violet mean, and understand darker or paler.

This agreement is not present for shape (Figure 14-3), because we have no common terms. Saying that one shape is rounder or more convoluted or is less compact does not communicate a clear and explicit concept. The use of quantitative shape parameters such as formfactor, fractal dimension, or harmonic analysis in which the Fourier coefficients contain a complete description of the shape of a feature (in two or three dimensions), is even further removed from our everyday experience. The bottom line is that communicating an impression of shape from one human observer to another is very difficult.

Further, shape and size interact for human viewers. Comparison of features to determine which is "larger" normally involves a complicated combination of linear dimension, area, aspect ratio and even orientation. For example (Figure 14-4), the state of Georgia is actually larger than Florida in terms of area, but it does not appear so on a map because Florida has a greater maximum chord length. Comparison of Idaho, Oklahoma and Kansas (which are actually rather close in area) is hampered by their different shapes, and the different thickness and orientation of the panhandles.

The use of computer image analysis systems provides partial solutions to these difficulties. It permits quantitative measurements of these parameters, and it allows

Figure 14-4: Map of the United States. Compare the sizes of Georgia and Florida, and of Idaho, Kansas and Oklahoma.

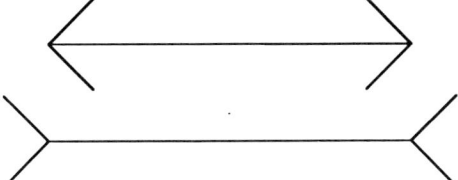

Fig 14-5: Müller-Lyer lines. The horizontal lines actually have the same length.

statistical descriptions and comparisons of populations. Equally important, it can be used to educate the users to learn what the important parameters are, and how to recognize them at a qualitative level. In other words, using computer image analysis makes us better observers.

Illusion

One of the interesting facets of the human visual system that is usually not present in computerized image analysis is the erroneous response to visual cues that we call illusion. There are many different types of illusions, and we will examine a few of them shortly. Beyond the fact that they are intriguing, why should we care about illusion in the context of this book?

One reason to study illusion is to develop better understanding of the physiological basis for human vision, but that is not our purpose here. Another is an interest in using illusion to create tension and dual interpretation in artistic works, or find optimal ways to present data (Tufte, 1983). Again, that is not our goal. Instead, we study illusion because it is closely tied to the algorithms used to process images. These algorithms function at a very low level in the human visual system to extract key information to be passed to higher levels of processing and understanding.

While they go wrong in some situations (creating the illusion), they are extremely efficient in most cases. They produce our ability to fuse stereo pairs, correct for nonuniform shading of surfaces, and extract lines and edges from noise. When these cues are misinterpreted, they often produce very persistent and robust illusions that fool the viewer even when the objective truth is known.

It may seem counterproductive to investigate such illusions. After all, if the computer image analysis systems are not plagued by such errors, why should we seek algorithms that may introduce them? The reason is that these algorithms also allow us to discriminate or measure things that present methods do not handle well or at all. In many cases, particularly for complex images such as the structure in biological material, the computer image analysis system functions as no more than a bookkeeper. Human operators are required to tediously outline features or edit images using their visual processes for interpretation. Eventually it may be possible to adapt or adopt algorithms that will bypass this time-consuming step and allow direct measurement. If the errors due to illusion come along with the algorithms, it may or may not be possible to correct them, but at least we are no worse off than the original results from the human interpretation.

One of the simplest and most common illusions (Figure 14-5) is that two lines of equal length appear unequal if their ends are joined to lines that extend outward or inward. Similarly, curvature or orientation or different thicknesses or densities of the lines can make them appear unequal. It is easy to write a computer program that is not

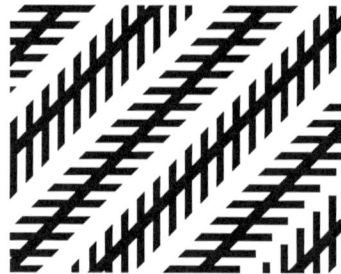

Figure 14-6: Zöllner lines. The oblique lines are actually parallel.

fooled by this illusion, and returns accurate measurements of line length. But why is the human visual system fooled?

There are two slightly different reasons that combine in these cases. For one thing, humans see entire objects, consisting of all of the connected portions of lines. Even when trying to select the principal line for estimation, the line segments connected to the ends of the line act to weight the position of the end points, resulting in the illusion.

Similarly, there is a weighting of orientation, of density, and of angles. We impose Euclidean geometry on the things we see, and prefer to have corners be right angles, or at least multiples of 45 or 60 degrees. In the Figure 14-6, the weighting of orientation by the short vertical and horizontal lines makes the diagonal lines appear to be not parallel. This is accomplished in the visual system by an inhibition mechanism not between adjacent pixels, as mentioned before, but between cells that detect orientation. The vertical bars, for instance, produce an output from cells that detect them which in turn inhibits the output of vertical sensing cells for the diagonal line, making it appear to be rotated to a more horizontal orientation (and *mutatis mutandi* for the diagonal line with horizontal bars).

The same phenomenon is active in the illusion shown in Figure 14-7, in which the shading of circular lines makes them appear to be spiral.

Figure 14-7: A spiral pattern is seen, but the lines are actually circular

Lessons from Human Vision

Figure 14-8: Lines within the two circles do not appear to be horizontal.

How can we make the computer measure these images in the same way? An operator that mimics the response of the human visual system in this case is the Hough transform, which can assign to each point in the image an orientation vector that is the distance-weighted average of all edges present. This tends to extend lines or contract them, depending on the line segments at the end, as in the preceding illusions. It also weights the line orientations in Figures 14-6 and 14-8 so that they appear not to be parallel.

What good is such a foolish transform? Well, another thing that the human visual system does rather well is to supply missing lines (Figure 14-9). We can use the Hough transform to supply missing lines such as grain boundaries, to find linear patterns in noisy data, and to refine jagged boundaries into simplified, smooth object outlines. When images are incomplete or noisy, this may be an important tool. Similar operators are used to find corners, which are also strongly selected by the human visual system.

Inhibition in the human visual system is both lateral (the key to locating edges and corners) and temporal. Temporal inhibition or saturation is easily demonstrated by negative after-images. View a brightly illuminated shape, especially one in color, and then look at a plain grey card. The image will appear in the complementary color. Temporal inhibitions are easily modeled using multiple image planes to contain a time sequence of images, and using weighted subtraction or Boolean logic to combine the images. This makes it possible to follow objects as a function of time, and to interpolate in time just as the human visual system does when watching a movie. It is especially powerful as a motion sensor. It is not surprising, then, that we see moving edges before anything else in an image.

If we knew more about how the eye encodes information on edges and corners, and their relationship to each other, and the visual system's rules by which nearby or similar

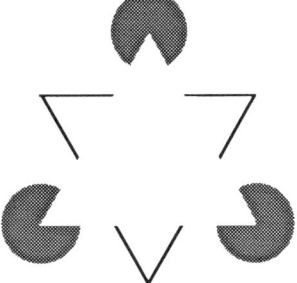

Figure 14-9: Kanizsa's triangle has illusory boundaries and a brightness greater than its surroundings.

Figure 14-10: Grouping of features into objects can be done in more than one way, here based either on which are closer or minimizing sharp angles.

features are connected together to form objects (Figure 14-10), we would advance significantly on the road to recognition of simple objects such as handwritten letters and numerals, which are recognizable to a small child even when written in a variety of styles in ways that intentionally violate all of the conventions which might logically be selected to define the pattern. We will also be able to find better ways to present information to the viewer, especially the sort of multi-dimensional data often encountered in image analysis.

When dealing with three-dimensional images, or more properly with the two-dimensional projection of three-dimensional objects, the human preference for simple shapes and continuous surfaces, mentioned before in the previous chapter, can produce illusions of precedence. In the example shown in Figure 14-11, we "expect" to see a square and a circle, and so assume that the square is in front of the circle. This is closely related to the continuity of surfaces that we interpolate in viewing stereo pairs (Figure 11-15), and the tendency to insert simple geometric shapes and edges into random dot stereograms even where there is no evidence for them.

Connection of lines and dots into simple surfaces is a powerful grouping operation. We are trained by our environment to see surfaces, and this persists (sometimes usefully and sometimes to our detriment) even if no surfaces are present. Figure 11-13 showed a random dot stereogram in which visual simplification imposed a box shape. Figure 14-12 shows an example in which a more-or-less cubic box is perceived due to the pattern of lines, and their spacing generates perspective. Even though we see the local roughness and bends in the "surfaces," we still simplify the already illusionary image into a cube, an even higher level of abstraction.

Temporal illusions of motion are also very common. The familiar "barber pole" with its spiral stripes is a good example. As the pole rotates, we perceive a false motion along the axis of the pole. This results because of local processing considerations in the eye. When an edge moves across a local group of receptors, motion parallel to the edge cannot be detected. Only the component of motion at right angles to the edge produces a response. The signals produced by these local responses are summed and interpreted as

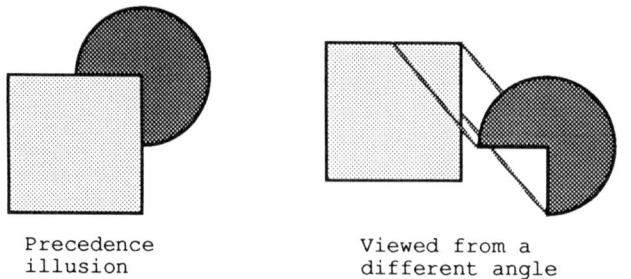

Precedence illusion   Viewed from a different angle

Figure 14-11: Precedence illusion - two objects viewed directly and seen from an angle

Lessons from Human Vision

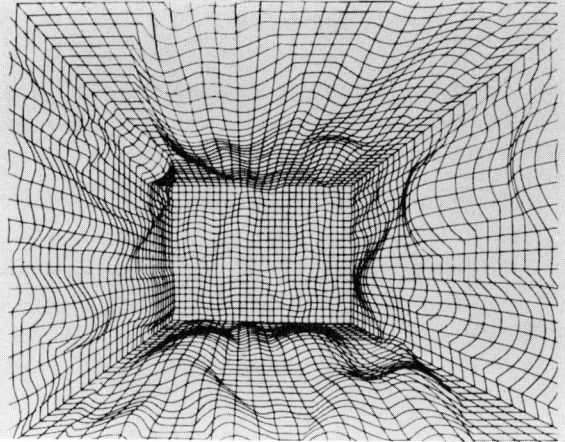

Figure 14-12: A wire mesh cube.

motion of the stripes at right angles to their orientation, producing the illusion of axial motion.

A final example of illusion is brightness aliasing (Figures 14-13 and 14-14). In the Mach bands (Mach, 1906), we perceive a variation in brightness near each boundary that increases the magnitude of the actual step. When such a variation is actually present, as in the Craik-Cornsweet-O'Brien illusion (Craik, 1966; Cornsweet, 1970; O'Brien, 1958), the visual system treats it as an edge and overcompensates, so that the equal brightness areas away from the boundary are judged to be different.

Because of its interest in edges, the visual system preserves or expands brightness changes across discontinuities. But it fills in intervening areas and does not see gradual changes of brightness as indicating edges. This allows seeing objects even in the presence of nonuniform illumination, and together with the ability to normalize for local or overall changes in absolute brightness, allows us to see things under very non-ideal

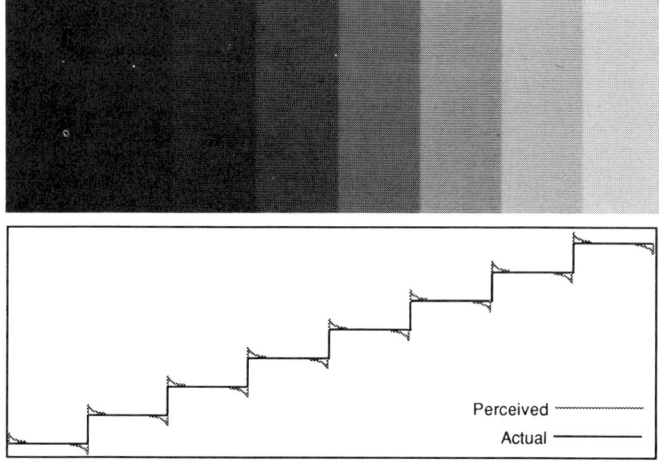

Figure 14-13: Mach bands show the response of the visual system to edge contrast and its effect on uniform regions.

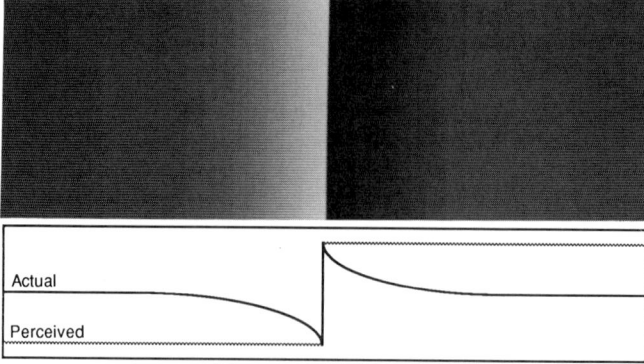

Figure 14-14: The Craik-Cornsweet-O'Brien Illusion, in which a step is perceived.

conditions of lighting. The mechanism is based on lateral inhibition, and can be directly modeled using Laplacian or derivative operators.

There are many other illusions, some famous, which can be explained and modeled using simple computer algorithms, and many more which cannot yet be fully explained. Attempting to understand them may present us with additional image processing, recognition and measurement tools.

Work by computer scientists, neuroanatomists, electronics and device experts, and (not least) image analysts is contributing steady growth to our basic knowledge of human visual processes. They are heavily parallel, hierarchical and probabilistic. They are also marked by extreme economy: very few individual pieces of information are extracted from millions of pixels at the retinal level and communicated to higher levels in the brain, yet these are usually adequate for object identification and scene interpretation. The organization of this information is by feature or object rather than by pixel.

Consideration of these results leads to the selection of a number of useful algorithms for image processing (e.g., spatial domain operators) which can pick out boundaries based on subtle characteristics such as texture (Figure 14-15), join incomplete boundaries, and simplify irregular ones. The ability of humans to extract meaning from extremely complex images (Figure 14-16) and at the same time to be fooled by simple camouflage presents a deep challenge to our understanding of its processes.

While measurement parameters for object size and location are much more accurate with simple computer algorithms than any visual estimate, shape is much more difficult

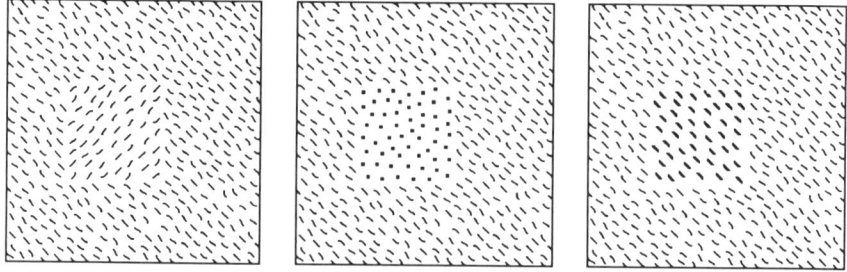

Figure 14-15: Segmentation of the inner square is based upon textural differences.

Figure 14-16: A Dalmatian sniffing leaves.

to quantify. Efforts are still underway to develop algorithms that extract those characteristics of object shape to which the human visual system responds. This is especially true for three dimensional structures, in which some information is either hidden by foreground objects or must be interpolated between successive image planes, and orientation–invariant shape estimators are required.

It is practical now to create expert systems for object recognition and selection. These can be readily trained with small suites of archetypical objects (which by virtue of their human selection may not represent the real world), and will continue to learn from subsequent use. They are quite capable of outperforming the human trainer, and by virtue of yielding results that are subtle and not always obvious to the observer, can be considered to exhibit some degree of artificial intelligence.

More powerful computer architectures are important in dealing with images, which place severe burdens on conventional systems. Massive parallelism is especially suitable for image handling, but it is not easy to take advantage of it with existing programming languages.

Considerable progress in all these areas of computer "seeing" is being made, but few final answers can yet be glimpsed. Developments of software algorithms and hardware architecture can be expected to proceed hand-in-hand into the foreseeable future.

Conclusion

"Seeing" can be initially considered as a visual process, involving light and receptors in the human eye. But our common use of language reveals that we know it means much more than that. We say "I see" to mean "I understand." Seeing, then, involves not simply absorbing photons but extracting some information from the image, and preferably information that describes some useful aspect of the world around us, in a way that helps us to comprehend it.

Computer image analysis is the process of extracting useful information, usually in the form of a small number of measurement parameters, from one or more digitized two-dimensional images. The development of the hardware to acquire these images from raster-scan sources such as video cameras and scanning electron microscopes has flourished in the last few years. The development of the software needed to process, discriminate, measure and interpret the images is still very much in progress.

It will be some time before we can claim to have programmed computer systems to "see" images as humans do. In the meantime, considerable progress has in fact been made in using these automated aids to extract particular types of information from images, and to take advantage of the fact that the computers are much more consistent than most humans. At the same time, the use of computerized image analysis has helped many users of these techniques to become more analytical in their human examination of images, to consider how they view images, the often non-intuitive relationship between two-dimensional images and three-dimensional structure, and how images may be obtained that are more suitable for the automatic methods.

References

T. N. Cornsweet (1970) Visual Perception Academic Press, New York NY

K. J. W. Craik (1966) in The Nature of Psychology (S. Sherwood, ed.) Cambridge Univ. Press, Cambridge U.K.

D. M. Hubel, T. N. Wiesel (1962) *Receptive fields, binocular interaction and functional architecture in the cat's visual cortex* J. Physiol. 160 106-154

J. Y. Lettvin, R. R. Maturana, W. S. McCulloch, W. H. Pitts (1959) *What the Frog's Eye Tells the Frog's Brain* Proc. Inst. Rad. Eng. 47#11, 1940-1951

E. Mach (1906) *Über den Einfluss räumlich und zeitlich variierender Lichtreize auf die Gesichtswarhrnehmung* S.-B. Akad. Wiss. Wien, Math.-Nat. Kl. 115, 633-648

V. O'Brien (1958) *Contour perception, illusion and reality* J. Opt. Soc. Am. 48, 112-119

E. R. Tufte (1983) The Visual Display of Quantitative Information, Graphic Press, Cheshire CT

For further reading

J. A. Anderson, R. Rosenfeld, ed. (1988) Neurocomputing: Foundations of Research MIT Press, Cambridge, MA

D. H. Ballard, C. M. Brown (1982) Computer Vision Prentice Hall, Englewood Cliffs, NJ

V. Braitenberg (1984) Vehicles: Experiments in Synthetic Psychology MIT Press, Cambridge, MA

T. Caelli (1981) Visual Perception: Theory and Practice Pergamon Press, Oxford U.K.

J. P. Frisby (1980) Vision: Illusion, Brain and Mind Oxford Univ. Press, Oxford U.K.

D. H. Hubel (1988) Eye, Brain, and Vision Scientific American Library, W. H. Freeman, New York NY

D. Marr (1982) Vision W. H. Freeman, San Francisco CA

I. Rock (1984) Perception, W. H Freeman, New York NY

Index

Accuracy, 3
Acoustic microscope, 383
Adjacency, 212-217
Algebraic reconstruction, 423-425
Alignment, 378-381, 384-385, 387ff
Analog to digital converter (ADC), 8, 23
Anisotropy, 256-257
Anova test, 243, 276
Area, 111 (*see also* Feature area)
Area fraction, 175
Artificial intelligence, 279, 301-304
Aspect ratio, 201
Atomic number contrast, 101, 318-319
Automatic thresholding, 108-114
Averaging, 25, 42, 44

Back projection, 420-422
Background image, 59
Backscattered electrons, 318-321
Bayesian statistics, 278
Beam hardening, 430
Binary image, 9, 119, 129-172
Boolean logic, 9, 131-135, 171, 388
Boundary area, 225, 227
Boundary representation, 71, 90-92, 97, 106, 393
Branches, 143, 146
Breadth, 186-188
Brightness histogram, 106, 108, 181

Caliper diameter (*see* Feret's diameter)
CCD (*see* Solid state cameras)
Centroid, 210-211
Chain code, 119, 123-125, 180, 184, 205 (*see also* Boundary representation)
Chord encoding (*see* Run length encoding)
Chromaticity, 36
Closing, 138
Cluster analysis, 263-264, 279-282, 303
Clustering, 149
Color coordinates, 19
Color filters, 101
Color images, 17, 19, 56, 117
Confocal light microscope, 343-344, 381-383, 403-404
Connectivity, 122, 130, 406, 408-416
Context line, 274-277, 283-285, 293-298

Contiguity (*see* Adjacency, Connectivity, Neighbor features)
Contour lines, 87, 357, 361
Contour map, 359-363, 398
Convex area, 182-183, 188, 202
Convex perimeter, 183, 188, 201
Correlation, 247, 257-258, 268-269
Counting, 129, 133, 149, 176
Counting methods, 223, 227
Covariance, 151-153
Cross-correlation, 365-367
Cumulative plot, 247-249
Curl, 189
Curvature, 178-180, 192-198, 200, 205, 228, 315-316

Density, 180, 217-218, 419, 425
Depth, 309-310, 343-346, 351
Derived parameters, 182, 194
Descriptive statistics, 242
Difference of Gaussians (DOG), 80, 84
Digitization, 8, 23
Dilation, 136-141, 157
Discriminant function, 274-277, 283-285, 295-298
Discrimination, 9, 57, 99
Disector, 232
Distortion, 22
Distributions, 10, 239, 241, 247-251, 273-278
Division, 58

Edge correction, 31, 176, 216-217, 251-252
Edge following, 87
Edges, 367-370
Electron diffraction patterns, 16, 93
Elevation profile, 357-359, 361
Ellipsoid model, 236-238
Equivalent diameter, 182, 192
Erosion, 135-141, 148, 149, 155
Etching (*see* Erosion)
Euclidean distance map, 148, 154-156
Euler number, 407, 414-415
Expert system, 11, 287-290, 303

Fate tables, 143-144, 156-157
Feature, 71, 99
Feature area, 181

Feature measurements, 10, 107, 181-212, 233-238
Feret's diameter, 183-185, 216, 251
Fiber length, 188
Filling, 130
Filters, 33
Focussing, 67
Formfactor, 191, 201, 203
Fourier transform, 48, 206-210, 421-422
Fractal dimension, 161-168, 218, 331-343
Frequency domain, 48, 55
Fuzzy logic, 290

Gaussian distribution, 242-243
Global measurements, 10, 107, 175-181, 222-231
Gouraud shading, 312, 397
Gradient, 75, 80, 101, 107, 116, 254-255
Grain size, 147-148, 225-226
Grey levels, 27
Grey scale image, 9
Grouping (*see* Neighbor features)

Hardware, 5, 30
Harmonic analysis, 206-210
Hidden line removal, 389, 393-394
High pass filter, 66
Histogram equalization, 34
Holes (within features), 121, 204
Hough transform, 61, 92-95, 181, 197-199, 214, 331, 445
Hue, intensity, saturation (HIS) images, 18, 19, 56, 118

Identification, 11, 202, 208-209, 267ff, 440-441
Illumination, 310-313
Illusion, 443
Image warping (*see* Rubber sheeting)
Intercept length, 228-230, 240
Interesting points, 369
Ion beam erosion, 380
Isometric display, 359

Kernels, 45, 46, 48, 64, 79, 80
Kirsch operator, 79-80, 82, 83, 103
Kolmogorov-Smirnov test, 245-246
Kruskal-Wallis test, 245

Lambert's law, 310-313, 396
Languages, 5, 7, 302
Laplacian of Gaussian (LOG), 80, 84, 87
Laplacian operator, 59, 64-66, 72-74, 81, 102, 448
Learning, 291-299, 304
Length, 124, 178, 184-186
Levelling, 58, 101
Light microscope, 14 (*see also* Confocal light microscope)
Light scattering, 310-316, 334-335
Line-width measurement, 324-330
Location, 210-212

Log-normal distribution, 248-250
Logical combination (*see* Boolean logic)
Look-up table (LUT), 34-36, 38
Low pass filter (*see* Smoothing)

Magnification, 253
Mann-Whitney test (*see* Wilcoxon test)
Manual outlining, 96, 386
Mathematical morphology, 134
Mean diameter, 240
Mean free path, 229-231
Mean intercept length, 178
Mean value, 10, 242-243
Medial axis transform, 148, 194 (*see also* Skeletonization)
Median filter, 46, 74
Metrology (*see* Line-width measurement)
Minkowski height, 192
Moments, 212
Morphological operators, 9, (*see also* Erosion, Dilation, Opening, Closing, Skeletonization)
Mosaic amalgamation, 165, 167-168
Motion flow, 40-41
Multi-spectral images, 16

Neighbor distance, 261-264
Neighbor features, 134, 169-172, 213-215, 446
Neighbor pixels, 45, 134-145, 199-200
Neural nets, 6, 271-272, 304
Nodes, 143, 146
Noise, 45, 56, 129
Nonparametric statistics, 243-247, 276
Number of features, 175-177, 232, 372

Opening, 136
Orientation, 75, 77-78, 211-212
Outline display (*see* Wire frame display)

Parallax, 352-356
Parallel processing, 6 (*see also* Neural nets)
Perimeter, 112, 161-166, 183-184
Periodic structures, 52, 151-153
Pixel, 27, 106, 130-131
Plating (*see* Dilation)
Point operations, 33
Pointing devices, 96, 125-127, 130
Polishing, 377
Populations, 272, 285, 291
Position, 253-257
Precision, 3, 115, 126, 227
Probability, 221, 234
Production rules, 287 (*see also* Expert systems)
Projected image, 11, 221, 419
Pseudo-color, 36

Radius of curvature, 194-195
Range image, 359, 371
Rank filter, 46, 59, 104
Rank tests, 244

Index 453

Recognition, 2, 11, 150, 200, 202, 208-209, 267ff, 440-441
Red, green, blue (RGB) images, 17, 19, 27, 56, 117
Reference area, 129, 133, 149, 175-177, 253
Reflection, 310, 313
Region growing, 87, 89
Regression, 274, 278, 293
Remote sensing, 353-355 (*see also* Satellite images)
Rendering (*see* Surface modelling)
Reproducibility (*see* Precision)
Rescaling, 39, 44
Resolution, 22, 25, 29, 30, 54
Richardson plot, 163-166, 339
Roberts' cross operator, 75-77
Roundness, 201, 203
RS 170 (video) images, 20, 23-24
Rubber sheeting, 17, 38, 42, 61-64, 330
Run length encoding, 119-121

Satellite images, 13, 17, 40, 51, 103
Scanning densitometry, 13, 38
Scanning electron microscope (SEM), 15, 20, 36, 38, 100, 317-330, 335, 354, 383
Secondary electrons, 322-323
Section image, 11, 221 (*see also* serial sections)
Sectioning, 377
Segmentation, 96, 153-160
Serial sections, 377ff, 433
Shading, 57-60, 397-399
Shape, 189-210, 253, 406, 442
Shape-from-shading, 310-313, 320, 324
Sharpening, 51, 64
Size, 239-253
Skeletonization, 141-147, 194-195, 199, 205, 406-408
Skiz, 141, 144
Smoothing, 45-46, 50, 136
Sobel operator, 77-79, 81, 103, 196
Solid state cameras, 22, 58
Spatial domain, 48
Spherical model, 233-235
Standard deviation, 10, 242-243
Statistical analysis, 10, 240ff
Stereo fusion, 41, 351, 356, 363-372
Stereology, 10, 221
Stereoscopy, 11, 316, 329, 351-376, 393

Structured light, 346
Student's t-test, 258-261
Subtraction, 39, 58
Surface area, 190-191, 224, 401
Surface modelling, 394ff
Surfaces, 309ff

Tangent count, 228, 414
Taut string outline (*see* Convex area, Convex perimeter)
Template matching, 150, 267-269, 303
Texture, 102, 218, 314-316, 334-341
Thresholding, 9, 99, 108
Time sequence, 38, 352
Tomography, 12, 419ff
Topography, 11 (*see also* Contour maps, Surface modelling, Surfaces)
Topology, 204, 406-416
Touching features, 95, 153-160, 213
Transfer function, 33
Transmission electron microscope (TEM), 14, 53, 373, 383, 434
Transmission images, 100, 403, 419
Transparent volumes, 372-375

Ultimate eroded points, 154

Varifocal mirror, 393
Vergence, 351-352, 365
Vertical section, 226-227, 240-241
Video (see RS 170)
Vidicon, 21
Visual system, 1, 3, 12, 72, 80, 351, 364, 372, 439ff
Volume, 189-191, 360-363, 399-401, 406
Volume fraction, 222
Volumetric displays, 402-405
Voxel, 383, 402, 433

Watersheds, 153-160
Width, 187-188
Wilcoxon test, 244
Wire frame display, 389-393

X-ray absorption, 419, 429
X-ray imaging, 419, 426-431
X-ray maps, 134, 141